CRC Series in Chromatography

Editors-in-Chief

Gunter Zweig, Ph.D. and Joseph Sherma, Ph.D.

General Data and Principles
Gunter Zweig, Ph.D. and
Joseph Sherma, Ph.D.

Lipids
Helmut K. Mangold, Dr. rer. nat.

Hydrocarbons
Walter L. Zielinski, Jr., Ph.D.

Carbohydrates
Shirley C. Churms, Ph.D.

Inorganics
M. Qureshi, Ph.D.

Drugs
Ram Gupta, Ph.D.

Phenols and Organic Acids
Toshihiko Hanai, Ph.D.

Terpenoids
Carmine J. Coscia, Ph.D.

Amino Acids and Amines
S. Blackburn, Ph.D.

Steroids
Joseph C. Touchstone, Ph.D.

Polymers
Charles G. Smith,
Norman E. Skelly, Ph.D.,
Carl D. Chow, and Richard A. Solomon

Pesticides and Related Organic Chemicals
Joseph Sherma, Ph.D. and
Joanne Follweiler, Ph.D.

Plant Pigments
Hans-Peter Köst, Ph.D.

Nucleic Acids and Related Compounds
Ante M. Krstulovic, Ph.D.

CRC Handbook of Chromatography

Carbohydrates

Volume II

Editor

Shirley C. Churms, Ph.D.
Research Associate
Carbohydrate Research Group
Department of Chemistry
University of Cape Town
Republic of South Africa

Editor-in-Chief

Joseph Sherma, Ph.D.
Charles A. Dana Professor and Head
Department of Chemistry
Lafayette College
Easton, Pennsylvania

CRC Press
Taylor & Francis Group
Boca Raton London New York

CRC Press is an imprint of the
Taylor & Francis Group, an informa business

First published 1991 by CRC Press
Taylor & Francis Group
6000 Broken Sound Parkway NW, Suite 300
Boca Raton, FL 33487-2742

Reissued 2018 by CRC Press

A Library of Congress record exists under LC control number: 80029541

ISBN 13: 978-1-138-50663-3 (hbk)
ISBN 13: 978-1-138-55941-7 (pbk)
ISBN 13: 978-0-203-71273-3 (ebk)

Visit the Taylor & Francis Web site at http://www.taylorandfrancis.com and the CRC Press Web site at http://www.crcpress.com

CRC HANDBOOK OF CHROMATOGRAPHY

SERIES PREFACE

The *CRC Handbook of Chromatography* series began with two volumes that were published in 1972. These volumes, written by Dr. Gunter Zweig and myself, covered all compound types and chromatographic methods. The first volume on carbohydrates by Shirley C. Churms, published in 1982, was the third volume in the series, and the first in the current format in which each book is devoted to a particular class of compounds or chromatographic method. The first volume by Dr. Churms is among the best selling and most highly acclaimed in the entire series, which now numbers 25 volumes overall with the publication of this second book on carbohydrates. Dr. Churms is an internationally recognized expert on the chromatography of carbohydrates and I am gratified that she was willing and able to update her original coverage of the field at this time.

Future volumes in the *Handbook of Chromatography* series are now being planned or written. These include additional volumes on pesticides, polymers, lipids, steroids, and hydrocarbons, and coverage for the first time of toxins, vitamins, and chiral separations. I would appreciate hearing from readers who have suggestions for topics or authors for subsequent volumes in the series. I would also be interested in receiving comments or corrections on the present volume or any others that have already been published.

Joseph Sherma, Ph.D.
Easton, PA
August 1990

PREFACE

Since the compilation of *CRC Handbook of Chromatography: Carbohydrates,* Volume I, which was completed early in 1979, there have been many developments in the field, particularly in liquid chromatography, owing to the rapid proliferation of HPLC methods during the past decade. Gas chromatography, too, has been revolutionized by the increasing use of capillary columns and the advent of chiral phases and some novel methods of derivatization, which have greatly improved resolution of related sugars, including enantiomeric pairs. Derivatization has become important in HPLC, also, owing to the greatly enhanced sensitivity of detection possible by use of chromophoric and, especially, fluorescent derivatives; this applies equally to PC and TLC. In addition to pre- and postcolumn derivatization, other methods of detection have been developed for HPLC: in particular, coupling of HPLC and MS is now a reality and electrochemical methods have become important, especially in the relatively new technique of ion chromatography, the application of which to analysis of mono- and oligosaccharides has produced some remarkable results. Another new technique which shows great promise in the analysis of oligosaccharides and glycoconjugates is supercritical fluid chromatography.

This new handbook, which has been compiled almost entirely from literature published during the period 1979 to 1989, has been structured to take cognizance of these developments. Since HPLC has been the main area of growth during the period under review, the section on liquid chromatography data has been greatly expanded, as have those on detection and derivatization methods in liquid chromatography. The vital importance of mass spectrometry in GC analysis is reflected in the inclusion of relevant mass spectral data for various derivatives used, including some of the newer ones. These new methods of derivatization, and improvements in the older methods, are given in detail in Section III. Although application of supercritical fluid chromatography to carbohydrates is not yet widespread, its potential is great enough to justify the inclusion of a new subsection covering the data available at present, and the recent coupling of SFC to MS is discussed in Section II.

To allow some more space for all this new material and a comprehensive list of important literature references relating to the new methods, it has been decided to omit the separate section on chromatographic materials that was included in Volume I. Developments in this area are so rapid at present that the currency of any such directory will inevitably be of short duration, and therefore to attempt to update the list published in Volume I seemed futile. Instead, relevant details of the packings or plates used are given in the footnotes to tables showing data obtained with these chromatographic materials, and the names of manufacturers, from whom further details can be obtained, are also appended in such footnotes.

Shirley C. Churms, Ph.D.
October 1989

THE EDITOR-IN-CHIEF

Joseph Sherma, Ph.D., received a B.S. in chemistry from Upsala College, East Orange, NJ in 1955 and a Ph.D. in analytical chemistry from Rutgers University, New Brunswick, NJ, in 1958 carrying on his thesis research in ion exchange chromatography under the direction of the late William Rieman III. Dr. Sherma joined the faculty of Lafayette College, Easton, PA in September, 1958, and is presently Charles A. Dana Professor and Head of the Chemistry Department. At Lafayette, he teaches three courses in analytical chemistry.

Dr. Sherma, independently and with others, has written or edited about 350 research papers, chapters, books, and reviews involving chromatography and other analytical methodology. In addition to being Editor-in-Chief of the CRC *Handbook of Chromatography* series, he co-edits the series *Analytical Methods for Pesticides and Plant Growth Regulators,* previously published by Academic Press and now by CRC Press. He is co-editor for residues and trace elements of the *Journal of the Association of Official Analytical Chemists* and a member of the editorial board of the *Journal of Planar Chromatography.* He is consultant on analytical methodology for many companies and federal agencies.

Dr. Sherma has received three awards for superior teaching and scholarship at Lafayette College and the E. Emmet Reid Award for excellence in teaching, presented at the Middle Atlantic Regional meeting of the ACS. He is a member of the ACS, AIC, Phi Lambda Upsilon, Sigma Xi, and AOAC. Dr. Sherma's current research interests are in quantitative TLC, mainly applied to clinical analysis, pesticide residues, lipids, and food additives.

THE EDITOR

Shirley C. Churms (born Macintosh) graduated at the University of Cape Town, South Africa, and then undertook research, at the same University, on cation-exchange processes in aqueous monoethanolamine, for which the degree of Ph.D. was awarded in 1962. She spent her postdoctoral year at the Imperial College of Science and Technology in London, England, as the holder of the Ohio State Fellowship, awarded by the International Federation of University Women; during this period she carried out an extensive survey of the properties of inorganic ion-exchangers and commenced an investigation of cation- and anion-exchange on amphoteric hydrated alumina. In 1964 she returned to the University of Cape Town and during the next 4 years she continued her research on inorganic ion-exchangers, supervised a group of research students working on various aspects of ion-exchange, and lectured in the Department of Chemistry at the University. In 1967 the work of the ion-exchange group was recognized by the South African Chemical Institute in the award of the African Explosives and Chemical Industry medal for one of its publications.

A change in research interests occurred in 1968, when Dr. Churms was appointed Research Associate in the newly formed Carbohydrate Chemistry Research Unit, sponsored by the South African Council for Scientific and Industrial Research, which functioned in the Department of Chemistry (from 1974 until 1988 in the separate Department of Organic Chemistry), under the direction of Professor Alistair M. Stephen. The Unit, which became the Carbohydrate Research Group in 1985, used chromatographic methods, particularly gas chromatography, very extensively in analyses of the products of degradative studies of polysaccharides, and Dr. Churms developed a special interest in the application of gel-permeation (now steric-exclusion) chromatography in these structural studies. Publication of a major review on this topic in *Advances in Carbohydrate Chemistry and Biochemistry* in 1970 resulted in an invitation from Dr. Erich Heftmann to contribute a chapter on chromatography of carbohydrates to the Third Edition of the well-known book on chromatography of which he is the Editor. She has subsequently remained a co-author in the Fourth Edition and, recently, the Fifth Edition of this seminal work. She was Volume Editor, and a major contributor, for Volume I of the present Handbook, published by CRC Press in 1982. To date, Dr. Churms has been the author of seven major reviews and, in addition to the chapters for Heftmann's *Chromatography* already mentioned, co-author (with Professor Stephen) of a further three chapters to be published in books dealing with various aspects of polysaccharide chemistry. She has been author or co-author of over 50 published papers and numerous presentations at national and international conferences, and is currently a member of the Editorial Board for the Symposium Volumes of the *Journal of Chromatography*.

DEDICATION

This volume is dedicated to Professor Alistair M. Stephen, in deep gratitude for many happy and productive years under his inspiring leadership, and to colleagues, past and present, in the Carbohydrate Research Group at the University of Cape Town, whose cooperation and camaraderie have helped to ease the high pressure of carbohydrate chromatography.

Shirley C. Churms
October 1989

ACKNOWLEDGMENTS

The financial support of the University of Cape Town and the Foundation for Research Development of the South African Council for Scientific and Industrial Research is gratefully acknowledged. I thank the University of Cape Town for a grant of study leave during the writing of this book and Professor Alistair M. Stephen for his patience and encouragement, as well as for many helpful discussions of aspects of chromatography. I am indebted also to Mrs. Evelyn Rossmeisl, of the Inter-Library Loans Department in the University of Cape Town Library, for her invaluable assistance in obtaining some of the literature reviewed in this book, and to Dr. Bill Edwards, of the Dionex Corporation, Sunnyvale, California, for kindly sending me some literature on ion chromatography from sources not otherwise accessible to me.

Special thanks are due to Mrs. Patsy Alexander and Mrs. Jean Goode for their excellent processing of the manuscript. The cooperation of the editorial staff of CRC Press is also gratefully acknowledged.

LIST OF ABBREVIATIONS

Ac	= acetyl
Ara	= arabinose
Asn	= asparagine
atm	= atmosphere
CI	= chemical ionization
d	= days
Da	= daltons
DEAE	= diethylamino ethyl
DP	= degree of polymerization
EI	= electron impact
Et	= ethoxyl
f	= furanose
FAB	= fast atom bombardment
fmol	= femtomole
FI	= flame ionization detector
Fru	= fructose
Fuc	= fucose
Gal	= galactose
GalA	= galacturonic acid
GalN	= 2-amino-2-deoxygalactose (galactosamine)
GalNAc	= 2-acetamido-2-deoxygalactose (*N*-acetylgalactosamine)
GalNAc-ol	= 2-acetamido-2-deoxygalactitol (*N*-acetylgalactosaminitol)
GC	= gas chromatography
GLC	= gas liquid chromatography
Glc	= glucose
GlcA	= glucuronic acid
GlcN	= 2-amino-2-deoxyglucose (glucosamine)
GlcNAc	= 2-acetamido-2-deoxyglucose (*N*-acetylglucosamine)
GlcNAc-ol	= 2-acetamido-2-deoxyglucitol (*N*-acetylglucosaminitol)
Glc-ol	= glucitol
Gly	= glycine
h	= hour(s)
HPLC	= high-performance liquid chromatography
I.D.	= inner diameter
k'	= capacity factor = $(t_r - t_o) / t_o$
K_{av}	= SEC partition coefficient = $(V_e - V_o) / (V_t - V_o)$
K_d	= alternative SEC partition coefficient = $(V_e - V_o) / V_i$
kPa	= kilopascal
LC	= liquid chromatography
M	= molecular weight (molar mass); in Table EL 1 = mobility
$\overline{M}_n$	= number-average molecular weight
$\overline{M}_w$	= weight-average molecular weight
Man	= mannose
mCi	= millicurie
Me	= methyl
meq	= milliequivalent
min	= minute(s)
m*M*	= millimolar
mmol	= millimole

MPa	=	megapascal (S.I. unit of pressure)
ms	=	millisecond
MS	=	mass spectrometry
m/z	=	mass/charge ratio
n.a.	=	not available
NeuAc	=	*N*-acetylneuraminic acid (sialic acid)
ng	=	nanogram
nm	=	nanometer
nmol	=	nanomole
p	=	pyranose
PC	=	paper chromatography
pg	=	picogram
pmol	=	picomole
ppb	=	parts per billion
ppm	=	parts per million
r	=	relative retention time
R_F	=	migration distance relative to solvent front
R_{Glc}	=	migration distance relative to glucose
Rha	=	rhamnose
RI	=	refractive index detector
s	=	second
SCOT	=	support-coated open tubular
SEC	=	steric-exclusion chromatography
Ser	=	serine
SFC	=	supercritical fluid chromatography
SS	=	stainless steel
Thr	=	theonine
TLC	=	thin-layer chromatography
TMS	=	trimethylsilyl
t_o	=	retention time of solvent front
t_r	=	retention time
Tyr	=	tyrosine
U	=	units (of enzyme)
UV	=	ultraviolet absorbance
v	=	volume
V_e	=	elution volume
V_i	=	internal solvent volume in SEC
V_o	=	void volume
V_t	=	total volume in SEC
w	=	weight
WCOT	=	wall-coated open tubular
Xyl	=	xylose

Other abbreviations are defined in the test.

TABLE OF CONTENTS

Volume II

Section I

Chromatographic Data

Section I.I.

GAS CHROMATOGRAPHY TABLES

As in Volume I, these tables are grouped according to the type of volatile derivative used in gas chromatography (GC), with the derivatives arranged in alphabetical order for easy reference. In the present volume the emphasis is mainly on capillary GC, although in many cases retention data for chromatography on equivalent packed columns are included for comparison. The new tables do not necessarily supersede those in Volume I; many should be regarded rather as supplementary to the earlier compilations, as indicated in footnotes to the present tables. A few of the derivatives included in Volume I have not been used extensively in recent years and are, therefore, omitted here, while several new derivatives are included for the first time in this volume. Of special interest are those permitting resolution of enantiomers, which has also been achieved by the use of recently developed chiral phases. Tables devoted to this important aspect will be found at the end of Section I.I.

TABLE GC 1
Peracetylated Alditols, Aminodeoxyalditols, and Inositols: GLC on Packed Columns

Packing	P1	P1	P2	P3	P4
Temperature (°C)	T1	T2	T3	195	195
Gas; flow rate (ml/min)	N_2; 40	He; 20	N_2; 20	N_2; 45	N_2; 40
Column					
Length, cm	158	100	183	200	250
Diameter (I.D.), cm	0.4	0.2	0.18	0.3	0.4
Form	Packed	Packed	Packed	Packed	Packed
Material	Glass	Glass	Glass	Glass	SS
Detector	FI	FI	FI	FI	FI
Reference	1	2	3	4	5
Parent compound	r[a]	r[a]	r[b]	r[c]	r[a]
Rhamnose	—	—	—	0.18	—
Fucose	0.32	—	—	0.20	—
Ribose	—	—	—	0.27	—
Arabinose	0.46	0.52	—	0.30	—
Xylose	0.59	0.64	—	0.39	—
Mannose	0.77	0.79	—	0.78	0.69
Galactose	0.83	0.84	—	0.93	0.79
Glucose	0.91	0.91	—	1.00	0.87
2-Amino-2-deoxyglucose	0.57[d]	—	—	0.46[e]	—
2-Amino-2-deoxymannose	—	—	—	0.48[e]	—
2-Amino-2-deoxygalactose	0.73[d]	—	—	0.54[e]	—
2-Acetamido-2-deoxyglucitol	—	—	1.44[f]	—	—
2-Acetamido-2-deoxygalactitol	—	—	1.55[f]	—	—
2-Acetamido-2-deoxymannitol	—	—	1.55[f]	—	—
myo-Inositol	1.00	1.00	—	—	1.00
chiro-Inositol	—	—	—	—	0.72
neo-Inositol	—	—	—	—	0.81
muco-Inositol	—	—	—	—	1.17
scyllo-Inositol	—	—	—	—	1.53

[a] t_r relative to *myo*-inositol hexaacetate (39.2 min,[1] 8.3 min,[2] 25 min[5]).
[b] t_r relative to peracetylated D-*gluco* heptitol (10.5 min).
[c] t_r relative to glucitol hexaacetate (53.94 min).
[d] As 2,5-anhydrohexitol acetates (2,5-anhydromannitol and -talitol from 2-amino-2-deoxyglucose and -galactose) produced by deamination.
[e] *N*-methylated (HCHO/$NaBH_3CN$) between reduction and acetylation.
[f] For simultaneous analysis of acetamidodeoxyhexoses as *O*-methyloxime acetates see Table GC 7, Reference 3.

Packing
- P1 = 3% SP-2340 on Supelcoport® (100—120 mesh).
- P2 = 3% diethylene glycol adipate, stabilized, on Chromosorb® W HP (100—120 mesh).
- P3 = 2% EGSS-X on Chromosorb® W AW DMCS (60—80 mesh).
- P4 = 0.5% QF-1 + 0.5% LAC-2R-446 on Chromosorb® W HP (80—100 mesh).

Temperature
- T1 = 150 → 220°C at 2°C/min.
- T2 = 190 → 260°C at 5°C/min.
- T3 = 210 → 240°C at 2°C/min.

REFERENCES

1. **Anastassiades, T., Puzic, R., and Puzic, O.,** Modification of the simultaneous determination of alditol acetates of neutral and amino sugars by gas-liquid chromatography, *J. Chromatogr.*, 225, 309, 1981.
2. **Lehrfeld, J.,** Gas chromatographic analysis of mixtures containing aldonic acids, alditols, and glucose, *Anal. Chem.*, 56, 1803, 1984.
3. **Mawhinney, T.,** Simultaneous determination of *N*-acetylglucosamine, *N*-acetylgalactosamine, *N*-acetylglucosaminitol and *N*-acetylgalactosaminitol by gas-liquid chromatography, *J. Chromatogr.*, 351, 91, 1986.
4. **Kiho, T., Ukai, S., and Hara, C.,** Simultaneous determination of the alditol acetate derivatives of amino and neutral sugars by gas-liquid chromatography, *J. Chromatogr.*, 369, 415, 1986.
5. **Irving, G. C. J.,** Gas chromatography of inositols as their hexakis-*O*-acetyl derivatives, *J. Chromatogr.*, 205, 460, 1981.

TABLE GC 2
Peracetylated Alditols and Aminodeoxyalditols: GLC on Capillary Columns

Phase	P1	P2	P3	P3	P4	P4	P5	P5	P6	P7	P8	P9
Temperature (°C)	T1	205	T2	T3	T4	T5	T6	T6	T7	275	T8	T9
Gas; flow rate (ml/min)	He; 5	He; 1[a]	H_2; 1[a]	H_2; n.a.	He; 1.5[a]	H_2; 1.3[a]	H_2; 9	H_2; 9.5	He; 10	He; 0.6[a]	He; 0.6[a]	N_2; 2.5[a]
Column												
Length, m	35	25	12	24	25	6	28.5	28.5	24	30	30	25
Diameter (I.D.), mm	0.3	0.25	0.2	0.3	0.25	0.2	0.5	0.5	0.75	0.25	0.25	0.25
Form	Coated	WCOT	WCOT	Coated	Coated	Bonded phase	SCOT	SCOT	Coated	Coated	Bonded phase	WCOT
Material	Glass	Fused silica	Fused silica	Fused silica	Glass	Fused silica	Glass	Glass	Glass	Glass	Fused silica	Glass
Detector	FI	FI	FI	FI	FI	FI	FI	FI	FI	FI	FI	FI
Reference	1	2	3	4	5	6	7	8	9	10	11	12
Parent compound	r[b]	r[b]	r[b]	r[c]	r[c]	r[c]	r[c]	r[c]	r[c]	r[c]	r[b]	r[d]
Glycerol	—	—	—	—	0.12	—	0.08	—	—	—	—	—
Erythritol	0.44	—	—	—	0.27	0.39	0.26	0.28	—	—	0.29[f]	—
2-Deoxy-D-*erythro*-pentose	0.52	0.25	0.23	—	0.37	0.51	0.37	0.39	—	—	—	—
Rhamnose	0.58	0.30	0.31	—	0.38	0.52	0.39	0.41	—	0.55	0.40[f]	—
Fucose	0.59	0.32	0.34	—	0.39	0.54	0.41	0.43	—	0.57	0.44[f]	—
6-Deoxyglucose	—	0.33	0.36	—	0.48	—	—	—	—	—	—	—
Ribose	0.61	0.36	0.38	—	0.49	0.63	0.50	0.53	—	0.63	0.46[f]	—
Arabinose	0.62	0.40	0.42	0.63[e]	0.52	0.65	0.53	0.56	0.35[f]	0.65	0.47[f]	—
1,4-Anhydromannitol[g]	—	—	—	—	0.59	—	—	—	—	—	—	—
1,5-Anhydromannitol[g]	—	—	—	—	0.61	—	—	—	—	—	—	—
Xylose	0.65	0.44	0.48	0.68[e]	0.63	0.72	0.63	0.66	0.48[f]	—	0.53[f]	—
2-Deoxy-D-*arabino*-hexose	0.75	—	—	—	0.64	0.74	0.66	0.68	—	—	—	—
2-Deoxy-D-*lyxo*-hexose	—	0.58	0.62	—	0.65	—	—	—	—	—	—	—
Allose	—	0.74	0.78	—	0.70	0.78	0.72	0.75	—	—	0.77[f]	—
4-*O*-Methylglucose[h]	0.92	0.82	0.84	—	0.80	—	—	—	—	—	—	—
Altrose	—	0.86	0.89	—	0.75	—	—	—	—	—	—	—

Mannose	0.95	0.89	0.92	0.98[e]	0.74	0.83	0.78	0.79	0.77[f]	0.86	0.88[f]	—
Talose	—	—	—	—	—	—	0.79	—	—	—	—	—
Galactose	0.98	1.02	1.01	1.02[e]	0.79	0.87	0.83	0.84	0.87[f]	0.90	0.97[f]	—
Glucose	1.00	1.00	1.00	—	0.86	0.92	0.90	0.90	0.90[f]	0.93	1.00[f]	—
Idose	—	1.08	1.08	—	0.99	—	1.03	—	—	—	—	—
myo-Inositol	—	—	—	1.00[e]	1.00	1.00	1.00	1.00	1.00[f]	1.00	—	—
2,5-Anhydromannose[i]	—	—	—	—	—	—	—	0.63	—	—	—	—
2,5-Anhydrotalose[i]	—	—	—	—	—	—	—	0.76	—	—	—	—
2-Amino-2-deoxyglucose	—	—	1.74	—	—	—	—	—	—	—	—	1.00
2-Amino-2-deoxygalactose	—	—	1.92	—	—	—	—	—	—	—	—	1.22
2-Amino-2-deoxymannose	—	—	1.96	—	—	—	—	—	—	—	—	1.36
2-Amino-2-deoxyallose	—	—	—	—	—	—	—	—	—	—	—	0.83
2-Amino-2-deoxytalose	—	—	—	—	—	—	—	—	—	—	—	1.03
2-Amino-2-deoxyaltrose	—	—	—	—	—	—	—	—	—	—	—	1.05
2-Amino-2-deoxyidose	—	—	—	—	—	—	—	—	—	—	—	1.13
2-Amino-2-deoxyglucose	—	—	—	—	—	—	—	—	—	—	—	1.31
2-Acetamido-2-deoxyglucose	—	—	—	1.45[e]	—	—	—	—	—	—	—	—
2-Acetamido-2-deoxygalactose	—	—	—	1.54[e]	—	—	—	—	—	—	—	—
2-Acetamido-2-deoxymannose	—	—	—	1.56[e]	—	—	—	—	—	—	—	—

[a] Split ratios 1:100,[2,3,10] 1:20,[5] 1:4,[6] 1:50,[11] 1:5.[12]

[b] t_r relative to glucitol hexaacetate (65 min,[1] 25.8 min,[2] 15.2 min,[3] 11.3 min[11]).

[c] t_r relative to *myo*-inositol hexaacetate (18.6 min,[4] 36.8 min,[5] 21.5 min,[6] 18.1 min,[7] 16.7 min,[8] 7.8 min,[9] 7.0 min[10]).

[d] t_r relative to acetylated 2-amino-2-deoxyglucitol (71.5 min.).

[e] For simultaneous analysis of neutral and amino sugars as *O*-methyloxime acetates see Table GC 7, Reference 7.

[f] For simultaneous analysis of alduronic acids as *N*-alkylaldonamide acetates see Table GC 6, References 3 and 4.

[g] Produced on reductive cleavage of polysaccharides.

[h] Produced on carboxyl reduction of 4-*O*-methylglucuronic acid.

[i] Produced on nitrous acid deamination of 2-amino-2-deoxyglucose and -galactose.

TABLE GC 2 (continued)
Peracetylated Alditols and Aminodeoxyalditols: GLC on Capillary Columns

Phase	P1 = 9:1 mixture of *N*-propionyl-L-valine-*tert*-butylamide polysiloxane + Witconal® LA23 (surfactant).
	P2 = FFAP.
	P3 = Carbowax 20M.
	P4 = OV-275.
	P5 = Silar 10C.
	P6 = SP-2330, 1.0-μm film.
	P7 = SP-2380 (stabilized phase, partially cross-linked), 0.20-μm film.
	P8 = DB-1701 (J and W Scientific, Folsom, California).
	P9 = Poly A-103.
Temperature	T1 = 80 → 200°C at 4°C/min; held at 200°C.
	T2 = 185 → 200°C at 1°C/min; held at 200°C for 2 min; 200 → 220°C at 20°C/min.
	T3 = 80 → 180°C at 20°C/min; held at 180°C for 1 min; 180 → 210°C at 2°C/min; held at 210°C for 1 min; 210 → 230°C at 10°C/min; held at 230°C.
	T4 = 165 → 215°C at 20°C/min; held at 215°C.
	T5 = 170°C for 4 min; 170 → 230°C at 4°C/min.
	T6 = 190°C for 4 min; 190 → 230°C at 4°C/min; held at 230°C for 8 min.
	T7 = 200°C for 2 min; 200 → 235°C at 3°C/min.
	T8 = 220 → 270°C at 1°C/min.
	T9 = 175°C for 50 min; 175 → 210°C at 0.5°C/min.

REFERENCES

1. **Green, C., Doctor, V. M., Holzer, C., and Oró, J.,** Separation of neutral and amino sugars by capillary gas chromatography, *J. Chromatogr.*, 207, 268, 1981.
2. **Oshima, R., Yoshikawa, A., and Kumanotani, J.,** High-resolution gas chromatographic separation of alditol acetates on fused-silica wall-coated open-tubular columns, *J. Chromatogr.*, 213, 142, 1981.
3. **Oshima, R., Kumanotani, J., and Watanabe, C.,** Fused-silica capillary gas chromatographic separation of alditol acetates of neutral and amino sugars, *J. Chromatogr.*, 250, 90, 1982.
4. **Neeser, J.-R.,** G.l.c. of *O*-methyloxime and alditol acetate derivatives of neutral sugars, hexosamines and sialic acids, *Carbohydr. Res.*, 138, 189, 1985.
5. **Klok, J., Cox, H. C., de Leeuw, J. W., and Schenck, P. A.,** Analysis of synthetic mixtures of partially methylated alditol acetates by capillary gas chromatography, gas chromatography-electron impact mass spectrometry and gas chromatography-chemical ionization mass spectrometry, *J. Chromatogr.*, 253, 55, 1982.
6. **Blakeney, A. B., Harris, P. J., Henry, R. J., Stone, B. A., and Norris, T.,** Gas chromatography of alditol acetates on a high-polarity bonded-phase vitreous-silica column, *J. Chromatogr.*, 249, 180, 1982.

7. **Blakeney, A. B., Harris, P. J., Henry, R. J., and Stone, B. A.,** A simple and rapid preparation of alditol acetates for monosaccharide analysis, *Carbohydr. Res.*, 113, 291, 1983.
8. **Henry, R. J., Blakeney, A. B., Harris, P. J., and Stone, B. A.,** Detection of neutral and amino sugars from glycoproteins and polysaccharides as their alditol acetates, *J. Chromatogr.*, 256, 419, 1983.
9. **Lehrfeld, J.,** Simultaneous gas-liquid chromatographic determination of aldoses and alduronic acids, *J. Chromatogr.*, 408, 245, 1987.
10. **Sidisky, L. M., Stormer, P. L., Nolan, L., Keeler, M. J., and Bartram, R. J.,** High temperature partially cross-linked cyanosilicone capillary column for general purpose gas chromatography, *J. Chromatogr. Sci.*, 26, 320, 1988.
11. **Walters, J. S. and Hedges, J. I.,** Simultaneous determination of uronic acids and aldoses in plankton, plant tissues, and sediment by capillary gas chromatography of *N*-hexylaldonamide and alditol acetates, *Anal. Chem.*, 60, 988, 1988.
12. **Kontrohr, T. and Kocsis, B.,** Separation of eight epimeric 2-aminoaldohexoses as aminohexitol acetates by gas chromatography and analysis of bacterial heteroglycans containing rare 2-aminoaldohexoses and 2-aminouronic acids, *J. Chromatogr.*, 291, 119, 1984.

TABLE GC 3
Partially Methylated Alditol Acetates: GLC on Capillary Columns

Phase	P1	P1	P2	P3	P4	P5	P6	P7	P8	P9	P10	P11	P12
Temperature	T1	T2	T3	T3	T4	T5	T6	T7	T8	T9	T10	T11	T12
Gas; flow rate (ml/min)	He; 1.5[a]	He; 0.8	He; 0.6	He; 0.6	n.a.[a]	H_2; n.a.	H_2; n.a.	H_2; n.a.	He; 0.8	He; n.a.[a]	He; n.a.	He; n.a.	He; n.a.
Column													
Length, m	25	25	30	30	30	22	20	85	15	30	20	50	50
Diameter (I.D.), mm	0.25	0.22	0.25	0.25	0.28	0.25	0.25	0.25	0.24	0.25	0.30	0.33	0.33
Form	Coated	Bonded phase	Coated	Coated	SCOT	WCOT	WCOT	WCOT	WCOT	WCOT	WCOT	WCOT	Bonded phase
Material	Glass	Fused silica	Glass	Glass	Glass	Glass	Glass	Glass	Fused silica	Fused silica	Glass	Fused silica	Fused silica
Detector	FI	MS	MS	MS	FI	FI	FI	FI	MS	MS	FI	FI	FI
Reference	1	2	3	3	4	5	5	5	6	7	8	9	9
Parent sugar[b]/alditol	r^c	r^c	r^c	r^c	r^c	r^c	r^c	r^c	r^c	r^c	ρ^d	ρ^e	ρ^e
Rhamnose/rhamnitol	1.52	1.25	—	—	—	—	—	—	1.23	1.42	1.89	1.48	1.55
1-*O*-methyl	0.94	—	—	—	—	—	—	—	—	—	—	—	—
2-*O*-methyl	1.30	1.14	—	—	—	—	—	—	1.07	1.24	1.40	1.26	1.14
3-*O*-methyl	1.50	1.22	—	—	—	—	—	—	1.14	1.39	1.55	1.45	1.30

TABLE GC 3 (continued)
Partially Methylated Alditol Acetates: GLC on Capillary Columns

Parent sugar[b]/alditol	r[c]	r[c]	r[c]	r[c]	r[c]	r[c]	r[c]	r[c]	r[c]	r[c]	ρ[d]	ρ[e]	ρ[e]
4-*O*-methyl	1.41	1.18	—	—	—	—	—	—	1.12	1.32	1.47	1.36	1.24
5-*O*-methyl	1.14	—	—	—	—	—	—	—	—	—	—	—	—
1,2-di-*O*-methyl	0.70	—	—	—	—	—	—	—	—	—	—	—	—
1,3-di-*O*-methyl	0.76	—	—	—	—	—	—	—	—	—	—	—	—
1,4-di-*O*-methyl	0.86	—	—	—	—	—	—	—	—	—	—	—	—
1,5-di-*O*-methyl	0.70	—	—	—	—	—	—	—	—	—	—	—	—
2,3-di-*O*-methyl	1.02	0.96	—	—	—	—	—	—	0.91	0.98	0.94	1.00	0.78
2,4-di-*O*-methyl	0.99	0.96	—	—	—	—	—	—	0.93	—	1.00	0.97	0.83
3,4-di-*O*-methyl	0.98	0.94	—	—	—	—	—	—	0.91	0.94	0.89	0.93	—
2,3,4-tri-*O*-methyl	0.60	0.67	—	—	0.54	—	—	—	0.69	0.58	0.54	0.69	0.39
Fucose/fucitol	1.59	1.30	—	—	—	—	1.23	1.11	1.25	1.50	2.08	1.58	1.53
1-*O*-methyl	0.98	—	—	—	—	—	—	—	—	—	—	—	—
2-*O*-methyl	1.38	1.17	—	—	—	—	—	—	1.08	1.30	1.46	1.32	1.15
3-*O*-methyl	1.59	1.26	—	—	—	—	—	—	1.15	1.47	—	1.52	1.32
4-*O*-methyl	1.56	1.25	—	—	—	—	—	—	1.15	1.45	—	—	—
5-*O*-methyl	1.29	—	—	—	—	—	—	—	—	—	—	—	—
1,2-di-*O*-methyl	0.71	—	—	—	—	—	—	—	—	—	—	—	—
1,3-di-*O*-methyl	0.77	—	—	—	—	—	—	—	—	—	—	—	—
1,4-di-*O*-methyl	0.88	—	—	—	—	—	—	—	—	—	—	—	—
1,5-di-*O*-methyl	0.78	—	—	—	—	—	—	—	—	—	—	—	—
2,3-di-*O*-methyl	1.17	1.04	—	—	—	—	—	—	0.93	1.10	1.03	1.11	0.83
2,4-di-*O*-methyl	1.08	1.01	—	—	—	—	—	—	0.96	1.05	1.08	1.07	0.90
3,4-di-*O*-methyl	1.21	1.06	—	—	—	—	—	—	0.97	1.15	—	1.15	0.90
2,3,4-tri-*O*-methyl	0.76	0.78	—	—	0.69	0.82	0.74	0.87	0.75	0.73	0.65	0.79	0.50
Ribose/ribitol	2.00	—	—	—	—	—	—	—	—	—	—	—	—
1-*O*-methyl	1.32	—	—	—	—	—	—	—	—	—	—	—	—
2-*O*-methyl	1.63	—	—	—	—	—	—	—	—	—	—	—	—
3-*O*-methyl	1.59	—	—	—	—	—	—	—	—	—	—	—	—
4-*O*-methyl	1.63	—	—	—	—	—	—	—	—	—	—	—	—
5-*O*-methyl	1.32	—	—	—	—	—	—	—	—	—	—	—	—
1,2-di-*O*-methyl	0.91	—	—	—	—	—	—	—	—	—	—	—	—
1,3-di-*O*-methyl	0.85	—	—	—	—	—	—	—	—	—	—	—	—
1,4-di-*O*-methyl	1.01	—	—	—	—	—	—	—	—	—	—	—	—
1,5-di-*O*-methyl	0.85	—	—	—	—	—	—	—	—	—	—	—	—

2,3-di-*O*-methyl	1.08	—	—	—	—	—	—	—	—	—	—	—	—
2,4-di-*O*-methyl	1.04	—	—	—	—	—	—	—	—	—	—	—	—
2,5-di-*O*-methyl	1.01	—	—	—	—	—	—	—	—	—	—	—	—
3,4-di-*O*-methyl	1.08	—	—	—	—	—	—	—	—	—	—	—	—
3,5-di-*O*-methyl	0.85	—	—	—	—	—	—	—	—	—	—	—	—
4,5-di-*O*-methyl[b]	0.91	—	—	—	—	—	—	—	—	—	—	—	—
1,2,3-tri-*O*-methyl	0.50	—	—	—	—	—	—	—	—	—	—	—	—
1,2,4-tri-*O*-methyl	0.54	—	—	—	—	—	—	—	—	—	—	—	—
1,2,5-tri-*O*-methyl	0.55	—	—	—	—	—	—	—	—	—	—	—	—
1,3,4-tri-*O*-methyl	0.56	—	—	—	—	—	—	—	—	—	—	—	—
1,3,5-tri-*O*-methyl	0.46	—	—	—	—	—	—	—	—	—	—	—	—
1,4,5-tri-*O*-methyl[b]	0.55	—	—	—	—	—	—	—	—	—	—	—	—
2,3,4-tri-*O*-methyl	0.55	—	—	—	—	—	—	—	—	—	—	—	—
2,3,5-tri-*O*-methyl	0.56	—	—	—	—	—	—	—	—	—	—	—	—
2,4,5-tri-*O*-methyl[b]	0.54	—	—	—	—	—	—	—	—	—	—	—	—
3,4,5-tri-*O*-methyl[b]	0.50	—	—	—	—	—	—	—	—	—	—	—	—
Arabinose/arabinitol	2.09	1.49	—	—	—	—	—	—	1.26	1.84	2.62	2.16	1.62
1-*O*-methyl	1.36	—	—	—	—	—	—	—	—	—	—	—	—
2-*O*-methyl	1.75	1.33	—	—	2.23	—	—	—	1.09	1.57	1.82	1.69	1.16
3-*O*-methyl	1.82	1.37	—	—	—	—	—	—	1.12	1.62	1.94	1.79	1.26
4-*O*-methyl	1.80	1.36	—	—	—	—	—	—	1.10	1.60	—	—	—
5-*O*-methyl	1.40	1.17	—	—	—	—	—	—	0.97	1.75	—	1.34	0.91
1,2-di-*O*-methyl	0.97	—	—	—	—	—	—	—	—	—	—	—	—
1,3-di-*O*-methyl	0.94	—	—	—	—	—	—	—	—	—	—	—	—
1,4-di-*O*-methyl	1.08	—	—	—	—	—	—	—	—	—	—	—	—
1,5-di-*O*-methyl	0.86	—	—	—	—	—	—	—	—	—	—	—	—
2,3-di-*O*-methyl	1.26	1.09	—	—	1.32	—	—	—	0.88	1.17	1.12	1.19	0.74
2,4-di-*O*-methyl	1.27	1.10	—	—	—	—	—	—	0.90	1.18	1.20	1.21	0.77
2,5-di-*O*-methyl	1.10	1.01	—	—	1.09	—	—	—	0.81	1.04	0.98	1.07	0.57
3,4-di-*O*-methyl	1.30	1.11	—	—	—	—	—	—	0.90	1.20	1.17	1.44	—
3,5-di-*O*-methyl	0.98	0.94	—	—	0.94	—	—	—	0.80	0.94	0.94	0.98	0.57
4,5-di-*O*-methyl[b]	1.02	—	—	—	—	—	—	—	—	—	—	—	—
1,2,3-tri-*O*-methyl	0.59	—	—	—	—	—	—	—	—	—	—	—	—
1,2,4-tri-*O*-methyl	0.64	—	—	—	—	—	—	—	—	—	—	—	—
1,2,5-tri-*O*-methyl	0.59	—	—	—	—	—	—	—	—	—	—	—	—
1,3,4-tri-*O*-methyl	0.64	—	—	—	—	—	—	—	—	—	—	—	—
1,3,5-tri-*O*-methyl	0.49	—	—	—	—	—	—	—	—	—	—	—	—
1,4,5-tri-*O*-methyl[b]	0.58	—	—	—	—	—	—	—	—	—	—	—	—

TABLE GC 3 (continued)
Partially Methylated Alditol Acetates: GLC on Capillary Columns

Parent sugar[b]/alditol	r[c]	r[c]	r[c]	r[c]	r[c]	r[c]	r[c]	r[c]	r[c]	r[c]	ρ[d]	ρ[e]	ρ[e]
2,3,4-tri-*O*-methyl	0.78	0.81	—	—	—	—	—	—	0.68	0.75	0.67	0.81	0.36
2,3,5-tri-*O*-methyl	0.65	0.70	—	—	0.59	—	—	—	0.60	0.63	0.56	0.73	0.23
2,4,5-tri-*O*-methyl[b]	0.66	—	—	—	—	—	—	—	—	—	—	—	—
3,4,5-tri-*O*-methyl[b]	0.61	—	—	—	—	—	—	—	—	—	—	—	—
Xylose/xylitol	2.55	1.65	—	—	—	—	—	—	1.31	2.14	3.02	2.87	1.74
1-*O*-methyl	1.70	—	—	—	—	—	—	—	—	—	—	—	—
2-*O*-methyl	1.98	1.44	—	—	2.70	—	—	—	1.11	1.73	1.99	1.97	1.22
3-*O*-methyl	1.96	1.43	—	—	2.65	—	—	—	1.12	1.72	1.99	1.95	1.24
4-*O*-methyl	1.98	1.44	—	—	—	—	—	—	1.11	1.73	1.99	1.97	1.22
5-*O*-methyl	1.70	—	—	—	—	—	—	—	—	—	—	—	—
1,2-di-*O*-methyl	1.13	—	—	—	—	—	—	—	—	—	—	—	—
1,3-di-*O*-methyl	1.10	—	—	—	—	—	—	—	—	—	—	—	—
1,4-di-*O*-methyl	1.25	—	—	—	—	—	—	—	—	—	—	—	—
1,5-di-*O*-methyl	1.05	—	—	—	—	—	—	—	—	—	—	—	—
2,3-di-*O*-methyl	1.39	1.18	—	—	1.48	—	—	—	0.89	1.27	1.21	1.32	0.77
2,4-di-*O*-methyl	1.25	1.12	—	—	—	—	—	—	0.87	1.17	1.15	1.20	0.72
2,5-di-*O*-methyl	1.25	—	—	—	—	—	—	—	—	—	—	—	—
3,4-di-*O*-methyl	1.39	1.18	—	—	1.48	—	—	—	0.89	1.27	1.21	1.32	0.71
3,5-di-*O*-methyl	1.10	—	—	—	—	—	—	—	—	—	—	—	—
4,5-di-*O*-methyl[b]	1.13	—	—	—	—	—	—	—	—	—	—	—	—
1,2,3-tri-*O*-methyl	0.65	—	—	—	—	—	—	—	—	—	—	—	—
1,2,4-tri-*O*-methyl	0.65	—	—	—	—	—	—	—	—	—	—	—	—
1,2,5-tri-*O*-methyl	0.68	—	—	—	—	—	—	—	—	—	—	—	—
1,3,4-tri-*O*-methyl	0.71	—	—	—	—	—	—	—	—	—	—	—	—
1,3,5-tri-*O*-methyl	0.59	—	—	—	—	—	—	—	—	—	—	—	—
1,4,5-tri-*O*-methyl[b]	0.68	—	—	—	—	—	—	—	—	—	—	—	—
2,3,4-tri-*O*-methyl	0.80	0.86	—	—	0.74	—	—	—	0.66	0.78	0.69	0.84	0.37
2,3,5-tri-*O*-methyl	0.71	—	—	—	—	—	—	—	—	—	—	—	—
2,4,5-tri-*O*-methyl[b]	0.65	—	—	—	—	—	—	—	—	—	—	—	—
3,4,5-tri-*O*-methyl[b]	0.65	—	—	—	—	—	—	—	—	—	—	—	—
Mannose/mannitol	2.98	1.87	—	—	—	—	2.07	1.37	1.91	2.40	6.63	4.20	4.48
1-*O*-methyl	2.21	—	—	—	—	—	—	—	—	—	—	—	—
2-*O*-methyl	2.64	1.73	—	—	—	1.66	1.91	1.30	1.73	2.23	4.79	3.28	3.46
3-*O*-methyl	2.97	1.84	—	—	—	—	—	—	1.83	2.38	5.49	4.12	3.94

4-*O*-methyl	2.97	1.84	—	—	—	—	—	—	1.83	2.38	5.49	4.12	3.94
5-*O*-methyl	2.64	—	—	—	—	—	—	—	—	—	—	—	—
6-*O*-methyl	2.21	1.56	—	—	—	—	—	—	1.60	1.96	3.53	2.41	2.83
1,2-di-*O*-methyl	1.71	—	—	—	—	—	—	—	—	—	—	—	—
1,3-di-*O*-methyl	1.90	—	—	—	—	—	—	—	—	—	—	—	—
1,4-di-*O*-methyl	2.19	—	—	—	—	—	—	—	—	—	—	—	—
1,5-di-*O*-methyl	1.93	—	—	—	—	—	—	—	—	—	—	—	—
1,6-di-*O*-methyl	1.49	—	—	—	—	—	—	—	—	—	—	—	—
2,3-di-*O*-methyl	2.25	1.56	3.15	2.48	—	—	—	—	1.56	1.99	3.22	2.47	2.67
2,4-di-*O*-methyl	2.35	1.62	3.53	2.81	4.41	1.57	1.82	1.26	1.63	2.05	3.69	2.70	2.97
2,5-di-*O*-methyl	2.28	—	—	—	—	—	—	—	—	—	—	—	—
2,6-di-*O*-methyl	1.93	1.44	2.56	2.21	—	1.43	1.53	1.18	1.43	1.76	2.60	1.99	2.16
3,4-di-*O*-methyl	2.43	1.62	3.45	2.48	4.56	1.59	1.80	1.25	1.60	2.09	3.28	2.77	—
3,5-di-*O*-methyl	2.35	—	—	—	—	—	—	—	—	—	3.69	2.70	2.97
3,6-di-*O*-methyl	2.19	1.52	2.92	2.36	—	1.51	1.66	1.21	1.50	1.93	2.91	2.35	2.43
4,5-di-*O*-methyl[b]	2.25	—	—	—	—	—	—	—	—	—	—	—	—
4,6-di-*O*-methyl	1.90	1.42	2.47	2.21	—	1.41	1.53	1.19	1.46	1.75	2.60	1.96	2.26
5,6-di-*O*-methyl	1.71	—	—	—	—	—	—	—	—	—	—	—	—
1,2,3-tri-*O*-methyl	1.27	—	—	—	—	—	—	—	—	—	—	—	—
1,2,4-tri-*O*-methyl	1.44	—	—	—	—	—	—	—	—	—	—	—	—
1,2,5-tri-*O*-methyl	1.43	—	—	—	—	—	—	—	—	—	—	—	—
1,2,6-tri-*O*-methyl	1.14	—	—	—	—	—	—	—	—	—	—	—	—
1,3,4-tri-*O*-methyl	1.50	—	—	—	—	—	—	—	—	—	—	—	—
1,3,5-tri-*O*-methyl	1.47	—	—	—	—	—	—	—	—	—	—	—	—
1,3,6-tri-*O*-methyl	1.26	—	—	—	—	—	—	—	—	—	—	—	—
1,4,5-tri-*O*-methyl[b]	1.59	—	—	—	—	—	—	—	—	—	—	—	—
1,4,6-tri-*O*-methyl	1.26	—	—	—	—	—	—	—	—	—	—	—	—
1,5,6-tri-*O*-methyl	1.14	—	—	—	—	—	—	—	—	—	—	—	—
2,3,4-tri-*O*-methyl	1.68	1.31	1.85	1.61	2.16	1.32	1.53	1.14	1.33	1.54	1.81	1.63	1.88
2,3,5-tri-*O*-methyl	1.76	—	—	—	—	—	—	—	—	—	—	—	—
2,3,6-tri-*O*-methyl	1.59	1.27	1.75	1.55	2.01	1.30	1.34	1.11	1.26	1.49	1.76	1.54	1.62
2,4,5-tri-*O*-methyl[b]	1.76	—	—	—	—	—	—	—	—	—	—	—	—
2,4,6-tri-*O*-methyl	1.47	1.23	1.68	1.63	—	1.24	1.33	1.12	1.28	1.39	1.80	1.45	1.68
2,5,6-tri-*O*-methyl	1.43	—	—	—	—	—	—	—	—	—	—	—	—
3,4,5-tri-*O*-methyl[b]	1.68	—	—	—	—	—	—	—	—	—	—	—	—
3,4,6-tri-*O*-methyl	1.50	1.22	1.61	1.44	1.81	1.26	1.31	1.10	1.24	1.41	1.62	1.44	1.59
3,5,6-tri-*O*-methyl	1.44	—	—	—	—	—	—	—	—	—	—	—	—
4,5,6-tri-*O*-methyl[b]	1.27	—	—	—	—	—	—	—	—	—	—	—	—

TABLE GC 3 (continued)
Partially Methylated Alditol Acetates: GLC on Capillary Columns

Parent sugar[b]/alditol	r[c]	r[c]	r[c]	r[c]	r[c]	r[c]	r[c]	r[c]	r[c]	r[c]	ρ[d]	ρ[e]	ρ[e]
1,3,4,5-tetra-*O*-methyl[b]	0.99	—	—	—	—	—	—	—	—	—	—	—	—
1,3,4,6-tetra-*O*-methyl	0.80	—	—	—	—	—	—	—	—	—	—	—	—
2,3,4,6-tetra-*O*-methyl	0.99	0.95	0.98	1.00	0.98	0.99	1.00	1.00	1.01	0.98	1.00	0.98	1.00
2,3,5,6-tetra-*O*-methyl	1.01	—	—	—	—	—	—	—	—	—	—	—	—
Galactose/galactitol	3.19	1.93	—	—	—	—	2.13	1.38	1.97	2.50	7.42	4.72	4.78
1-*O*-methyl	2.34	—	—	—	—	—	—	—	—	—	—	—	—
2-*O*-methyl	2.80	1.78	—	—	—	—	—	—	1.78	2.33	5.18	3.69	3.65
3-*O*-methyl	3.18	1.90	—	—	—	—	—	—	1.87	2.48	6.18	4.64	4.17
4-*O*-methyl	3.18	1.90	—	—	—	—	—	—	1.87	2.48	6.18	4.64	4.17
5-*O*-methyl	2.80	—	—	—	—	—	—	—	—	—	—	—	—
6-*O*-methyl	2.34	1.62	—	—	—	—	—	—	1.63	2.12	4.10	2.65	2.92
1,2-di-*O*-methyl	1.82	—	—	—	—	—	—	—	—	—	—	—	—
1,3-di-*O*-methyl	2.04	—	—	—	—	—	—	—	—	—	—	—	—
1,4-di-*O*-methyl	2.23	—	—	—	—	—	—	—	—	—	—	—	—
1,5-di-*O*-methyl	2.04	—	—	—	—	—	—	—	—	—	—	—	—
1,6-di-*O*-methyl	1.61	—	—	—	—	—	—	—	—	—	—	—	—
2,3-di-*O*-methyl	2.52	1.66	3.70	2.70	—	—	—	—	1.61	2.17	3.62	2.97	2.84
2,4-di-*O*-methyl	2.54	1.68	3.97	3.16	5.28	1.63	1.88	1.28	1.68	2.19	—	3.13	—
2,5-di-*O*-methyl	2.48	—	—	—	—	—	—	—	—	—	—	—	—
2,6-di-*O*-methyl	2.04	1.48	2.62	2.32	3.19	—	—	—	1.45	1.85	2.74	2.14	2.20
3,4-di-*O*-methyl	2.68	1.72	4.27	3.02	—	—	—	—	1.67	2.25	—	—	—
3,5-di-*O*-methyl	2.54	—	—	—	—	—	—	—	—	—	—	—	—
3,6-di-*O*-methyl	2.23	1.55	3.01	2.50	—	1.52	1.64	1.21	1.51	1.98	3.04	2.43	2.41
4,5-di-*O*-methyl[b]	2.52	—	—	—	—	—	—	—	—	—	—	—	—
4,6-di-*O*-methyl	2.04	1.49	2.67	2.37	—	1.45	1.59	1.20	1.50	1.85	—	2.14	—
5,6-di-*O*-methyl	1.82	—	—	—	—	—	—	—	—	—	—	—	—
1,2,3-tri-*O*-methyl	1.48	—	—	—	—	—	—	—	—	—	—	—	—
1,2,4-tri-*O*-methyl	1.54	—	—	—	—	—	—	—	—	—	—	—	—
1,2,5-tri-*O*-methyl	1.60	—	—	—	—	—	—	—	—	—	—	—	—
1,2,6-tri-*O*-methyl	1.20	—	—	—	—	—	—	—	—	—	—	—	—

1,3,4-tri-*O*-methyl	1.69	—	—	—	—	—	—	—	—	—	—	—	—
1,3,5-tri-*O*-methyl	1.56	—	—	—	—	—	—	—	—	—	—	—	—
1,3,6-tri-*O*-methyl	1.28	—	—	—	—	—	—	—	—	—	—	—	—
1,4,5-tri-*O*-methyl[b]	1.73	—	—	—	—	—	—	—	—	—	—	—	—
1,4,6-tri-*O*-methyl	1.28	—	—	—	—	—	—	—	—	—	—	—	—
1,5,6-tri-*O*-methyl	1.20	—	—	—	—	—	—	—	—	—	—	—	—
2,3,4-tri-*O*-methyl	1.98	1.44	2.33	1.93	—	1.44	1.59	1.18	1.42	1.78	2.38	2.00	2.13
2,3,5-tri-*O*-methyl	1.93	—	2.27	1.94	—	—	—	—	—	—	—	—	—
2,3,6-tri-*O*-methyl	1.73	1.33	1.84	1.58	—	—	—	—	1.26	1.59	1.84	1.69	1.62
2,4,5-tri-*O*-methyl[b]	1.93	—	—	—	—	—	—	—	—	—	—	—	—
2,4,6-tri-*O*-methyl	1.56	1.28	1.74	1.72	1.97	1.28	1.37	1.13	1.30	1.48	1.91	1.54	1.74
2,5,6-tri-*O*-methyl	1.60	—	1.67	1.57	—	—	—	—	—	—	—	—	—
3,4,5-tri-*O*-methyl[b]	1.98	—	—	—	—	—	—	—	—	—	—	—	—
3,4,6-tri-*O*-methyl	1.69	1.32	1.87	1.68	—	1.34	1.40	1.13	1.30	1.57	—	—	—
3,5,6-tri-*O*-methyl	1.54	—	1.64	1.65	—	—	—	—	—	—	—	—	—
4,5,6-tri-*O*-methyl[b]	1.48	—	—	—	—	—	—	—	—	—	—	—	—
1,3,4,5-tetra-*O*-methyl[b]	1.14	—	—	—	—	—	—	—	—	—	—	—	—
1,3,4,6-tetra-*O*-methyl	0.92	—	—	—	—	—	—	—	—	—	—	—	—
2,3,4,6-tetra-*O*-methyl	1.14	1.06	1.11	1.11	1.19	1.09	1.09	1.02	1.06	1.12	1.14	1.10	1.12
2,3,5,6-tetra-*O*-methyl	1.10	1.02	1.05	1.04	—	—	—	—	1.02	—	—	—	—
Glucose/glucitol	3.48	1.97	—	—	—	—	2.18	1.38	1.94	2.60	6.43	5.43	4.62
1-*O*-methyl	2.51	—	—	—	—	—	—	—	—	—	—	—	—
2-*O*-methyl	2.86	1.78	—	—	—	—	—	—	1.76	2.37	5.02	3.81	3.54
3-*O*-methyl	3.03	1.83	—	—	—	—	—	—	1.82	2.44	5.40	4.26	3.80
4-*O*-methyl	3.26	1.90	—	—	—	—	—	—	1.85	2.44	—	—	4.00
5-*O*-methyl	2.91	—	—	—	—	—	—	—	—	—	—	—	—
6-*O*-methyl	2.54	1.66	—	—	4.99	—	—	—	1.62	—	—	—	2.90
1,2-di-*O*-methyl	1.90	—	—	—	—	—	—	—	—	—	—	—	—
1,3-di-*O*-methyl	2.01	—	—	—	—	—	—	—	—	—	—	—	—
1,4-di-*O*-methyl	2.38	—	—	—	—	—	—	—	—	—	—	—	—
1,5-di-*O*-methyl	2.10	—	—	—	—	—	—	—	—	—	—	—	—
1,6-di-*O*-methyl	1.76	—	—	—	—	—	—	—	—	—	—	—	—
2,3-di-*O*-methyl	2.45	1.64	3.56	2.60	4.75	—	—	—	1.58	2.15	3.42	2.85	2.72
2,4-di-*O*-methyl	2.33	1.60	3.34	2.63	4.25	—	—	—	1.60	—	3.42	2.65	2.80

TABLE GC 3 (continued)
Partially Methylated Alditol Acetates: GLC on Capillary Columns

Parent sugar[b]/alditol	r^c	r^c	r^c	r^c	r^c	r^c	r^c	r^c	r^c	r^c	ρ^d	ρ^e	ρ^e
2,5-di-*O*-methyl	2.43	—	—	—	—	—	—	—	—	—	—	—	—
2,6-di-*O*-methyl	2.11	1.51	2.81	2.27	3.39	—	—	—	1.45	1.90	2.71	2.24	2.22
3,4-di-*O*-methyl	2.46	1.64	3.44	2.52	—	—	—	—	1.60	—	—	—	2.81
3,5-di-*O*-methyl	2.50	—	—	—	—	—	—	—	—	—	—	—	—
3,6-di-*O*-methyl	2.22	1.55	2.97	2.38	3.82	—	—	—	1.50	1.97	2.89	2.42	2.37
4,5-di-*O*-methyl[b]	2.42	—	—	—	—	—	—	—	—	—	—	—	—
4,6-di-*O*-methyl	2.15	1.52	2.84	2.29	—	—	—	—	1.49	1.93	—	2.30	2.36
5,6-di-*O*-methyl	1.94	—	—	—	—	—	—	—	—	—	—	—	—
1,2,3-tri-*O*-methyl	1.38	—	—	—	—	—	—	—	—	—	—	—	—
1,2,5-tri-*O*-methyl	1.57	—	—	—	—	—	—	—	—	—	—	—	—
1,2,6-tri-*O*-methyl	1.26	—	—	—	—	—	—	—	—	—	—	—	—
1,3,4-tri-*O*-methyl	1.49	—	—	—	—	—	—	—	—	—	—	—	—
1,3,5-tri-*O*-methyl	1.59	—	—	—	—	—	—	—	—	—	—	—	—
1,3,6-tri-*O*-methyl	1.34	—	—	—	—	—	—	—	—	—	—	—	—
1,4,5-tri-*O*-methyl[b]	1.70	—	—	—	—	—	—	—	—	—	—	—	—
1,4,6-tri-*O*-methyl	1.46	—	—	—	—	—	—	—	—	—	—	—	—
1,5,6-tri-*O*-methyl	1.26	—	—	—	—	—	—	—	—	—	—	—	—
2,3,4-tri-*O*-methyl	1.67	1.34	1.83	1.61	2.19	—	—	—	1.33	—	1.93	1.67	1.86
2,3,5-tri-*O*-methyl	1.72	—	—	—	—	—	—	—	—	—	—	—	—
2,3,6-tri-*O*-methyl	1.77	1.36	1.98	1.61	2.38	—	—	—	1.28	1.64	1.89	1.76	1.67
2,4,6-tri-*O*-methyl	1.45	1.25	1.61	1.53	1.73	1.23	1.28	1.11	1.24	1.41	1.68	1.43	1.60
2,5,6-tri-*O*-methyl	1.56	—	—	—	—	—	—	—	—	—	—	—	—
3,4,5-tri-*O*-methyl[b]	1.73	—	—	—	—	—	—	—	—	—	—	—	—
3,4,6-tri-*O*-methyl	1.51	1.26	1.62	1.46	—	1.25	1.29	1.10	1.23	1.44	—	1.48	—
3,5,6-tri-*O*-methyl	1.54	—	—	—	—	—	—	—	—	—	—	—	—
4,5,6-tri-*O*-methyl[b]	1.38	—	—	—	—	—	—	—	—	—	—	—	—
1,3,4,5-tetra-*O*-methyl[b]	0.99	—	—	—	—	—	—	—	—	—	—	—	—
1,3,4,6-tetra-*O*-methyl	0.81	—	—	—	—	—	—	—	—	—	—	—	—
2,3,4,6-tetra-*O*-methyl	1.00	1.00	1.00	1.00	1.00	1.00	1.00	1.00	1.00	1.00	1.00	1.00	1.00
2-Acetamido-2-deoxyglucose	—	—	—	—	—	—	2.99	1.53	—	—	—	—	—
3-*O*-methyl	—	—	—	—	—	—	2.86	1.52	—	—	—	—	—

6-*O*-methyl	—	—	—	—	—	—	2.75	1.50	—	—	—	—	—
3,4-di-*O*-methyl	—	—	—	—	—	—	2.76	1.47	—	—	—	—	—
3,6-di-*O*-methyl	—	—	—	—	—	2.00	2.47	1.42	—	—	—	—	—
4,6-di-*O*-methyl	—	—	—	—	—	—	2.65	1.47	—	—	—	—	—
3,4,6-tri-*O*-methyl	—	—	—	—	—	1.78	2.18	1.34	—	—	—	—	—
1,3,5,6-tetra-*O*-methyl	—	—	—	—	—	1.45	1.66	1.19	—	—	—	—	—
2-Acetamido-2-deoxygalactose	—	—	—	—	—	—	3.09	1.56	—	—	—	—	—
3,4,6-tri-*O*-methyl	—	—	—	—	—	1.90	2.36	1.38	—	—	—	—	—

[a] Split ratios 1:20,[1] 1:60,[4] 1:50.[7]
[b] In Reference 1, derivatives were prepared by Haworth methylation of alditols; some obviously have no counterpart among methylated sugars.
[c] t_r relative to 1,5-di-*O*-acetyl-2,3,4,6-tetra-*O*-methyl-D-glucitol (ca. 9 min,[1] 18 min[4]).
[d] ρ = retention coefficient, based on retention time relative to two widely separated standards, 1,4-di-*O*-acetyl-2,3,4,6-tetra-*O*-methyl-D-glucitol (ρ = 1.00) and peracetylated 2-*O*-methyl-*chiro*-inositol (quebrachitol; ρ = 4.30); ρ values calculated by computer program.
[e] ρ values calculated from retention times relative to 1,4-di-*O*-acetyl-2,3,4,6-tetra-*O*-methylallitol (ρ = 0.80 on P11, 0.83 on P12) and peracetylated 2-*O*-methyl-*chiro*-inositol (ρ = 4.00 on P1, 3.00 on P12).

Phase	P1 = OV-275.	P2 = OV-225.
	P3 = OV-17.	P4 = Silar 10C.
	P5 = Silar 9CP.	P6 = Dexsil 410.
	P7 = OV-101.	P8 = SP-2100, 0.25-μm film.
	P9 = SP-2330, 0.20-μm film.	P10 = SP-1000, 0.22-μm film.
	P11 = CP-Sil88, 0.22-μm film.	P12 = OV-1, 0.5-μm film.
Temperature	T1 = 165 → 215°C at 2°C/min; held at 215°C.	
	T2 = 150 → 250°C at 4°C/min; held at 250°C for 10 min.	
	T3 = injection temperature 50°C; 50 → 182°C at 40°C/min; held at 182°C.	
	T4 = 150 → 190°C at 1°C/min.	
	T5 = 100 → 230°C at 1°C/min.	
	T6 = 130 → 240°C at 2°C/min.	
	T7 = 100 → 240°C at 3°C/min.	
	T8 = 120 → 200°C at 2°C/min.	
	T9 = 160 → 210°C at 2°C/min; 210 → 240°C at 5°C/min.	
	T10 = injection temperature 60°C; operating temperatures 206°C.	
	T11 = injection temperature 60°C; operating temperature 210°C.	
	T12 = injection temperature 60°C; operating temperature 195°C.	

TABLE GC 3 (continued)
Partially Methylated Alditol Acetates: GLC on Capillary Columns

REFERENCES

1. **Klok, J., Cox, H. C., de Leeuw, J. W., and Schenck, P. A.,** Analysis of synthetic mixture of partially methylated alditol acetates by capillary gas chromatography, gas chromatography-electron impact mass spectrometry, and gas chromatography-chemical ionization mass spectrometry, *J. Chromatogr.*, 253, 55, 1982.
2. **Bacic, A., Harris, P. J., Hak, E. W., and Clarke, A. E.,** Capillary gas chromatography of partially methylated alditol acetates on a high-polarity bonded phase vitreous-silica column, *J. Chromatogr.*, 315, 373, 1984.
3. **Barreto-Bergter, E., Hogge, L., and Gorin, P. A. J.,** Gas-liquid chromatography of partially methylated alditol acetates on capillary columns of OV-17 and OV-225, *Carbohydr. Res.*, 97, 147, 1981.
4. **Shibuya, N.,** Gas-liquid chromatographic analysis of partially methylated alditol acetates on a glass capillary column, *J. Chromatogr.*, 208, 96, 1981.
5. **Geyer, R., Geyer, H., Kühnhardt, S., Mink, W., and Stirm, S.,** Capillary gas chromatography of methylhexitol acetates obtained upon methylation of *N*-glycosidically linked glycoprotein oligosaccharides, *Anal. Biochem.*, 121, 263, 1982.
6. **Harris, P. J., Bacic, A., and Clarke, A. E.,** Capillary gas chromatography of partially methylated alditol acetates on a SP-2100 wall-coated open-tubular column, *J. Chromatogr.*, 350, 304, 1985.
7. **Shea, E. M. and Carpita, N. C.,** Separation of partially methylated alditol acetates on SP-2330 and HP-1 vitreous silica capillary columns, *J. Chromatogr.*, 445, 424, 1988.
8. **Lomax, J. A. and Conchie, J.,** Separation of methylated alditol acetates by glass capillary gas chromatography and their identification by computer, *J. Chromatogr.*, 236, 385, 1982.
9. **Lomax, J. A., Gordon, A. H., and Chesson, A.,** A multiple-column approach to the methylation analysis of plant cell-walls, *Carbohydr. Res.*, 138, 177, 1985.

TABLE GC 4
Acetylated Aldononitriles from Neutral and Aminodeoxy Sugars

Packing	P1	P2	P3	P4	P5	P6	P7
Temperature	T1	T2	T3	T4	T5	T6	T7
Gas; flow rate (ml/min)	N_2; 32	N_2; 22	N_2; 24	N_2; 30	He; 0.5[a]	N_2; 2[a]	N_2; 2[a]
Column							
Length, cm	123	123	182	200	5000	2500	2700
Diameter (I.D.), cm	0.2	0.2	0.32	0.2	0.02	0.05	0.03
Form	Packed	Packed	Packed	Packed	Capillary	Capillary	Capillary
Material	Glass	Glass	Ni	Glass	Fused silica	Glass	Glass
Detector	FI	FI	FI	FI	FI	FI	FI
Reference	1	1	2	3	4	5	5
Parent sugar	r[b]	r[b]	r[b]	r[b]	r[b]	r[b]	r[b]
D-Erythrose	0.26	0.27	—	—	—	—	—
2,6-Dideoxy-D-*ribo*-hexose	0.45	0.45	—	—	—	—	—
2-Deoxy-D-*erythro*-pentose	0.58	0.49	—	—	0.53	—	—
L-Rhamnose	0.50	0.59	—	0.37	0.61	—	—
L-Fucose	0.61	0.64	0.54[c]	0.46	0.63	—	—
D-Ribose	0.61	0.62	—	0.47	0.60	—	—
D-Lyxose	0.65	0.64	—	0.50	—	—	—
D-Arabinose	0.68	0.64	0.60[c]	0.54	0.61	0.62	0.77
D-Xylose	0.74	0.68	0.68[c]	0.62	0.62	0.63	0.80
2-Deoxy-D-*arabino*-hexose	0.91	0.86	—	—	0.77	—	—
2-Deoxy-D-*lyxo*-hexose	0.96	0.86	—	—	—	—	—
D-Allose	0.90	0.93	—	—	—	—	—
D-Mannose	0.93	0.97	0.97[c]	0.88	0.93	0.96	0.98
D-Talose	0.93	0.97	—	—	—	—	—
D-Glucose	1.00	1.00	1.00[c]	1.00	1.00	1.00	1.00
D-Galactosoe	1.04	1.03	1.06[c]	1.05	1.05	1.06	1.06
L-Idose	1.07	1.07	—	—	—	—	—
D-*glycero*-D-*gluco*-heptose	1.28	1.30	—	—	—	—	—
2,5-Anhydromannose[d]	—	—	—	0.58	—	—	—
2,5-Anhydrotalose[d]	—	—	—	0.75	—	—	—
2-Amino-2-deoxy-D-glucose	—	—	—	—	1.20	—	—
2-Amino-2-deoxy-D-mannose	—	—	—	—	1.30	—	—
2-Amino-2-deoxy-D-galactose	—	—	—	—	1.34	—	—
2-Acetamido-2-deoxy-D-glucose	1.63	1.22	—	—	—	—	—
2-Acetamido-2-deoxy-D-galactose	—	1.34	—	—	—	—	—

[a] Split ratios 1:35,[4] 1:20.[5]

[b] t_r relative to peracetylated gluconoitrile (20.9 min on P1,[1] 15.1 min on P2,[1] 6.7 min,[2] 25.7 min,[3] 14.2 min,[4] 10.4 min on P6,[5] 10.0 min on P7[5]).

[c] For simultaneous analysis of amino sugars as *O*-methyloxime acetates see Table GC 7, Reference 2.

[d] From nitrous acid deamination of 2-amino-2-deoxyglucose and -galactose.

TABLE GC 4 (continued)
Acetylated Aldononitriles from Neutral and Aminodeoxy Sugars

Packing	P1 = 3% neopentyl glycol succinate on Chromosorb® W (60 — 80 mesh).
	P2 = 2% OV-17 on Chromosorb® W HP (80 — 100 mesh).
	P3 = 1% diethylene glycol adipate on Chromosorb® W HP (100 — 120 mesh).
	P4 = 3% OV-225 on Gas-Chrom® Q (120 — 120 mesh).
	P5 = OV-1, 0.11 μm film.
	P6 = OV-17
	P7 = OV-73
Temperature	T1 = 140 → 250°C at 3°C/min.
	T2 = 130 → 300°C at 5°C/min.
	T3 = 170°C for 2 min; 170 → 240°C at 10°C/min.
	T4 = 170 → 180°C at 1°C/min; 180 → 200°C at 2°C/min; held at 200°C for 10 min.
	T5 = 175°C for 4 min; 175 → 260°C at 4°C/min; held at 260°C for 5 min.
	T6 = 120°C for 3 min; 120 → 210°C at 12°C/min.

REFERENCES

1. **Seymour, F. R., Chen, E. C. M., and Bishop, S. H.,** Identification of aldoses by use of their peracetylated aldononitrile derivatives: a g.l.c.-m.s. approach, *Carbohydr. Res.*, 73, 19, 1979.
2. **Mawhinney, T. P., Feather, M. S., Barbero, G. J., and Martinez, J. R.,** The rapid quantitative determination of neutral sugars (as aldononitrile acetates) and amino sugars (as *O*-methyloxime acetates) in glycoproteins by gas-liquid chromatography, *Anal. Biochem.*, 101, 112, 1980.
3. **Turner, S. H. and Cherniak, R.,** Total characterization of polysaccharides by gas-liquid chromatography, *Carbohydr. Res.*, 95, 137, 1981.
4. **Guerrant, G. O. and Moss, C. W.,** Determination of monosaccharides as aldononitrile, *O*-methyloxime, alditol, and cyclitol acetate derivatives by gas chromatography, *Anal. Chem.*, 56, 633, 1984.
5. **Li, H.-P.** Capillary gas chromatography of neutral sugars as their aldononitrile acetates from the hydrolysate of corn bran residues, *J. Chromatogr.*, 410, 484, 1987.

TABLE GC 5
Acetylated Aldononitriles of Methylated Sugars

Packing	P1	P2	P3	P4	P4	P5	P6	P7	P8	P9
Temperature (°C)	165	165	180	T1	T2	T3	T4	150	T5	T6
Gas; flow rate (ml/min)	N_2; 8	N_2; 6	n.a.	n.a.	n.a.	n.a.	n.a.	He; n.a.[a]	He; n.a.[a]	He; n.a.[a]
Column										
Length, cm	300	300	180	180	180	200	200	2500	2500	3000
Diameter (I.D.), cm	0.32	0.32	0.2	0.2	0.2	0.4	0.4	0.02	0.02	0.025
Form	Packed	Packed	Packed	Packed	Packed	Packed	Packed	Capillary	Capillary	Capillary
Material	SS	SS	Glass	Glass	Glass	Glass	Glass	Fused silica	Fused silica	Fused silica
Detector	FI	FI	FI	FI	FI	FI	FI	FI	FI	FI
Reference	1	1	2	2	2	3	3	4	4	4
Parent sugar	**r[b]**	**r[b]**	**r[c]**	**r[c]**	**r[c]**	**r[d]**	**r[d]**	**r[e]**	**r[e]**	**r[e]**
L-Rhamnose	4.73	3.56	—	—	—	—	—	—	—	—
2-*O*-methyl	3.03	2.56	—	—	—	—	—	—	—	—
3-*O*-methyl	3.94	3.28	—	—	—	—	—	—	—	—
4-*O*-methyl	3.85	3.10	—	—	—	—	—	—	—	—
2,3-di-*O*-methyl	1.82	1.64	—	—	—	—	—	—	—	—
2,4-di-*O*-methyl	1.94	1.79	—	—	—	—	—	—	—	—
3,4-di-*O*-methyl	2.79	2.38	—	—	—	—	—	—	—	—
2,3,4-tri-*O*-methyl	1.00	1.00	—	—	—	—	—	—	—	—
D-Galactose	—	—	8.18	2.22	2.70	—	—	—	—	—
2-*O*-methyl	—	—	5.09	1.97	2.33	—	—	—	—	—
3-*O*-methyl	—	—	6.36	2.06	2.48	—	—	—	—	—
4-*O*-methyl	—	—	7.00	2.06	2.48	—	—	—	—	—
6-*O*-methyl	—	—	3.45	1.84	2.11	—	—	—	—	—
2,3-di-*O*-methyl	—	—	3.45	1.84	2.11	—	—	—	—	—
2,4-di-*O*-methyl	—	—	3.45	1.87	1.92	—	—	—	—	—
2,6-di-*O*-methyl	—	—	2.27	1.64	1.78	—	—	—	—	—
3,4-di-*O*-methyl	—	—	3.45	—	—	—	—	—	—	—
3,6-di-*O*-methyl	—	—	3.45	1.84	2.11	—	—	—	—	—
4,6-di-*O*-methyl	—	—	2.45	1.64	1.85	—	—	—	—	—
2,3,4-tri-*O*-methyl	—	—	1.73	—	—	—	—	—	—	—

TABLE GC 5 (continued)
Acetylated Aldononitriles of Methylated Sugars

Parent sugar	r[b]	r[b]	r[c]	r[c]	r[c]	r[d]	r[d]	r[e]	r[e]	r[e]
2,3,6-tri-*O*-methyl	—	—	1.45	1.45	1.48	—	—	—	—	—
2,4,6-tri-*O*-methyl	—	—	1.27	1.42	1.37	—	—	—	—	—
3,4,6-tri-*O*-methyl	—	—	1.82	1.52	1.63	—	—	—	—	—
2,3,4,6-tetra-*O*-methyl	—	—	1.00	1.00	1.00	—	—	—	—	—
D-Glucose	—	—	—	—	—	2.85	—	—	—	—
2-*O*-methyl	—	—	—	—	—	2.26	1.89	—	—	—
3-*O*-methyl	—	—	—	—	—	2.69	2.37	—	—	—
4-*O*-methyl	—	—	—	—	—	2.44	—	—	—	—
6-*O*-methyl	—	—	—	—	—	2.36	—	—	—	—
2,3-di-*O*-methyl	—	—	—	—	—	2.14	1.75	2.49	3.31	1.38
2,4-di-*O*-methyl	—	—	—	—	—	1.92	1.62	2.27	2.83	1.36
2,6-di-*O*-methyl	—	—	—	—	—	1.74	1.51	—	—	—
3,4-di-*O*-methyl	—	—	—	—	—	1.74	1.47	2.77	3.19	1.42
3,6-di-*O*-methyl	—	—	—	—	—	2.05	1.72	—	—	—
4,6-di-*O*-methyl	—	—	—	—	—	1.92	1.62	—	—	—
2,3,4-tri-*O*-methyl	—	—	—	—	—	1.53	1.34	1.64	2.26	1.23
2,3,6-tri-*O*-methyl	—	—	—	—	—	1.59	1.38	1.53	2.38	1.19
2,4,6-tri-*O*-methyl	—	—	—	—	—	1.27	1.22	1.36	1.71	1.15
3,4,6-tri-*O*-methyl	—	—	—	—	—	1.43	1.24	1.59	1.96	—
2,3,4,6-tetra-*O*-methyl	—	—	—	—	—	1.00	1.00	1.00	1.00	1.00

Note: This table supplements Table GC 6 in *CRC Handbook of Chromatography: Carbohydrates,* Volume I.

[a] Split ratio 1:100.

[b] t_r relative to 5-*O*-acetyl-2,3,4-tri-*O*-methyl-L-rhamnononitrile (9.35 min on P1, 16.8 min on P2).

[c] t_r relative to 5-*O*-acetyl-2,3,4,6-tetra-*O*-methyl-D-galactononitrile; all t_r values compared with those of two widely separated standards (5-*O*-acetyl-2,3,4,6-tetra-*O*-methyl-D-galactononitrile and penta-*O*-acetyl-D-galactononitrile) to lessen aging effect, i.e., decrease in t_r with aging of column.

[d] t_r relative to 5-*O*-acetyl-2,3,4,6-tetra-*O*-methyl-D-gluconoitrile (14 min on P5, 18.5 min on P6); all t_r values compared with that of *myo*-inositol hexaacetate (48 min).

[e] t_r relative to 5-*O*-acetyl-2,3,4,6-tetra-*O*-methyl-D-glucononitrile (4.9 min on P7, 6.4 min on P8, 16.3 min on P9); no internal standard.

Packing	P1 = 3% QF-1 on Gas-Chrom® Q (100 — 120 mesh).
	P2 = 1% XE-60 on Gas-Chrom® Q (100 — 120 mesh).
	P3 = 3% ECNSS-M on Gas-Chrom® Q (100 — 120 mesh).
	P4 = 3% OV-225 on Gas-Chrom® Q (100 — 120 mesh).
	P5 = 5% OV-225 on Chromosorb® W AW DMCS (100 — 120 mesh).
	P6 = 3% SP-2340 on Supelcoport® (100 — 120 mesh).
	P7 = SP-2100.
	P8 = Carbowax 20M.
	P9 = SE-54.
Temperature	T1 = 150 → 245°C at 8°C/min; held at 245°C.
	T2 = 185 → 245°C at 4°C/min; held at 245°C.
	T3 = 160 → 210°C at 1°C/min.
	T4 = 150 → 225°C at 2°C/min.
	T5 = 135°C for 5 min; 135 → 165°C at 4°C/min.
	T6 = 120°C for 5 min; 120 → 250°C at 4°C/min.

REFERENCES

1. **Janeček, F., Toman, R., Karácsonyi, S., and Anderle, D.,** Gas-liquid chromatographic separation of methyl ethers of L-rhamnose as their methyl glycosides, trifluoroacetylated L-rhamnitols, and acetylated L-rhamnononitriles, *J. Chromatogr.*., 173, 408, 1979.
2. **Stortz, C. A., Matulewicz, M. C., and Cerezo, A. S.,** Separation and identification of *O*-acetyl-*O*-methyl galactononitriles by gas-liquid chromatography and mass spectrometry, *Carbohydr. Res.*, 111, 31, 1982.
3. **Tanner, G. R. and Morrison, I. M.,** Gas chromatography-mass spectrometry of partially methylated glycoses as their aldononitrile peracetates, *J. Chromatogr.*, 299, 252, 1984.
4. **Slodki, M. E., England, R. E., Plattner, R. D., and Dick, W. E., Jr.,** Methylation analyses of NRRL dextrans by capillary gas-liquid chromatography, *Carbohydr. Res.*, 156, 199, 1986.

TABLE GC 6
Acetylated *N*-Alkylaldonamides Derived from Aldonic and Alduronic Acids[a]

Packing		P1	P1	P2	P3
Temperature		T1	T2	T3	T4
Gas; flow rate (ml/min)		He; 20	He; 20	He; 10	He; 12[b]
Column					
Length, cm		100	100	2400	3000
Diameter (I.D.), cm		0.2	0.2	0.075	0.025
Form		Packed	Packed	Capillary	Bonded phase
Material		Glass	Glass	Glass	Fused silica
Detector		FI	FI	FI	FI
Reference		1	2	3	4
***N*-Alkylaldonamide**		r[c]	r[d]	r[c]	r[e]
D-Ribonamide,	*N*-(1-propyl)	1.25	1.70	1.42	—
	N-(1-butyl)	1.31	—	—	—
	N-(1-pentyl)	1.37	—	—	—
	N-(1-hexyl)	1.44	—	—	—
D-Xylonamide,	*N*-(1-propyl)	1.47	2.17	—	—
	N-(1-butyl)	1.52	—	—	—
	N-(1-pentyl)	1.57	—	—	—
	N-(1-hexyl)	1.64	—	—	—
L-Mannonamide,	*N*-(1-propyl)	1.68	2.75	2.06[f]	—
	N-(1-butyl)	1.73	2.87	—	—
	N-(1-pentyl)	1.81	3.07	—	—
	N-(1-hexyl)	1.85	—	—	3.00[f]
D-Gluconamide,	*N*-(1-propyl)	1.77	3.01	2.24[f]	—
	N-(1-butyl)	1.81	3.15	—	—
	N-(1-pentyl)	1.88	3.40	—	—
	N-(1-hexyl)	1.92	—	—	3.10[f]
D-Galactonamide,	*N*-(1-propyl)	1.87	3.44	2.46[f]	—
	N-(1-butyl)	1.93	3.63	—	—
	N-(1-pentyl)	2.00	3.90	—	—
	N-(1-hexyl)	2.01	—	—	3.30[f]
L-Gulonamide,	*N*-(1-propyl)	—	—	2.35[f]	—
	N-(1-hexyl)	—	—	—	3.17[f]
L-Idonamide,	*N*-((1-hexyl)	—	—	—	3.19[f]

[a] Simultaneous analysis of neutral sugars (as alditol acetates) and sugar acids is possible by this method; see data for sugars in Tables GC 1 (Reference 2) and GC 2 (References 9 and 11).
[b] Split ratio 1:50.
[c] t_r relative to *myo*-inositol hexaacetate (88.3 min,[1] 7.8 min[3]).
[d] t_r relative to mannitol hexaacetate (3.1 min).
[e] t_r relative to glucitol hexaacetate (11.3 min).
[f] *N*-alkyl-D-mannonamide, -L-gulonamide, -D-gluconamide, -L-galactonamide, and -L-idonamide produced from D-mannuronic acid, D-glucuronic acid, L-guluronic acid, D-galacturonic acid, and L-iduronic acid, respectively, by $NaBH_4$ reduction of uronates prior to lactonization and treatment with alkylamine.

Packing	P1 = 3% SP-2340 on Supelcoport® (100 — 120 mesh).
	P2 = SP-2330, 1.0-μm film.
	P3 = DB-1701 (J and W Scientific).
Temperature	T1 = 190 → 260°C at 5°C/min.
	T2 = 220°C for 2 min, 220 → 250°C at 32°C/min.
	T3 = 200°C for 2 min; 200 → 235°C at 3°C/min; held at 235°C.
	T4 = 220 → 270°C at 1°C/min.

TABLE GC 6 (continued)
Acetylated *N*-Alkylaldonamides Derived from Aldonic and Alduronic Acids[a]

REFERENCES

1. **Lehrfield, J.,** Gas chromatographic analysis of mixtures containing aldonic acids, alditols, and glucose, *Anal. Chem.*, 56, 1803, 1984.
2. **Lehrfield, J.,** G. l. c. determination of aldonic acids as acetylated aldonamides, *Carbohydr. Res.*, 135, 179, 1985.
3. **Lehrfield, J.,** Simultaneous gas-liquid chromatographic determination of aldoses and alduronic acids, *J. Chromatogr.*, 408, 245, 1987.
4. **Walters, J. S. and Hedges, J. I.,** Simultaneous determination of uronic acids and aldoses in plankton, plant tissues, and sediment by capillary gas chromatography of *N*-hexyaldonamide and alditol acetates, *Anal. Chem.*, 60, 988, 1988.

TABLE GC 7
Acetylated Oxime Derivatives of Neutral and Amino Sugars

Packing	P1	P2	P3	P4	P5	P6	P7	P7	P8	P9
Temperature	T1	T2	T3	T4	T5	T6	T7	T7	T8	T9
Gas; flow rate (ml/min)	N_2; 22	N_2; 24	N_2; 20	N_2; 0.55[a]	He; 0.4	He; 0.5[a]	H_2; n.a.	H_2; n.a.	N_2; 35	He; 3
Column										
Length, cm	123	183	183	1000	1100	5000	2500	2400	200	1500
Diameter (I.D.), cm	0.2	0.18	0.18	0.021	0.021	0.02	0.03	0.03	0.2	0.032
Form	Packed	Packed	Packed	Capillary	Capillary	Capillary	Capillary	Capillary	Packed	WCOT
Material	Glass	Ni	Glass	Fused silica	Fused silica	Fused silica	Fused silica	Fused silica	Glass	Fused silica
Detector	FI	FI	FI	FI	FI	FI	FI	FI	FI	FI
Reference	1	2	3	4	4	5	6	7	8	8
Parent sugar	**O**[b]	**MO**[b]	**MO**	**MO**	**MO**	**MO**	**MO**	**MO**	**PFBO**	**PFBO**
	r[c]	r[c]	r[d]	r[c]	r[c]	r[f]	r[f]	r[f]	r[g]	r[g]
L-Rhamnose	—	—	—	—	—	0.59; 0.62[h]	0.45; 0.48[h]	0.45;0.47[h]	—	—
L-Fucose	—	—	—	—	—	0.58; 0.63[h]	0.46; 0.51[h]	0.46; 0.50[h]	0.50	0.65; 0.73[h]
2-Deoxy-D-*erythro*-pentose	—	—	—	—	—	0.46	—	—	—	—
D-Ribose	—	—	—	—	—	0.60; 0.61[h]	—	—	—	—
D-Arabinose	—	—	—	—	—	0.59; 0.63[h]	0.52; 0.55[h]	0.50; 0.54[h]	0.67	0.78; 0.86[h]
D-Xylose	—	—	—	—	—	0.65	0.57; 0.59[h]	0.56; 0.59[h]	0.83	0.92; 0.95[h]
2-Deoxy-D-*arabino*-hexose	—	—	—	—	—	0.76	—	—	—	—
D-Mannose	—	—	—	—	—	0.90; 0.94[h]	0.81; 0.86[h]	0.81; 0.86[h]	1.22	1.05; 1.12[h]
D-Glucose	—	—	—	—	—	0.92; 0.95[h]	0.88; 0.89[h]	0.89; 0.90[h]	1.50	1.16; 1.24[h]
D-Galactose	—	—	—	—	—	0.89; 0.96[h]	0.84; 0.92[h]	0.83; 0.92[h]	1.43	1.05; 1.22[h]
D-*erythro*-Pentulose	1.07; 1.10[h]	—	—	—	—	—	—	—	—	—
D-Fructose	1.31	—	—	—	—	0.97;[h] 0.98	—	—	—	—
1,3,4-tri-*O*-methyl	0.96; 0.97[h]	—	—	—	—	—	—	—	—	—
3,4,6-tri-*O*-methyl	0.98	—	—	—	—	—	—	—	—	—
1,3,4,6-tetra-*O*-methyl	0.66;[h] 1.70	—	—	—	—	—	—	—	—	—
D-Tagatose	1.31; 1.35[h]	—	—	—	—	—	—	—	—	—
L-Sorbose	1.35	—	—	—	—	—	—	—	—	—
D-*gluco*-Heptulose	—	—	—	—	—	1.21; 1.23[h]	—	—	—	—
D-*manno*-Heptulose	1.52; 1.56[h]	—	—	—	—	1.24; 1.26[h]	—	—	—	—
2-Amino-2-deoxy-D-glucose	—	1.57[i]	—	0.77	0.75	1.12; 1.15[h]	—	—	—	—

2-Amino-2-deoxy-D-galactose	—	1.72[i]	—	0.80	0.78	1.14; 1.16[h]	—	—	—	—
2-Amino-2-deoxy-D-mannose	—	1.86[i]	—	0.80	0.78	1.14; 1.17[h]	—	—	—	—
2-Acetamido-2-deoxy-D-glucose	—	—	1.14[j]	—	—	—	1.29;[h] 1.30	1.33;[h] 1.34	—	—
2-Acetamido-2-deoxy-D-galactose	—	—	1.24[j]	—	—	—	1.36	1.40	—	—
2-Acetamido-2-deoxy-D-mannose	—	—	1.29[j]	—	—	—	1.36; 1.40[h]	1.41; 1.45[h]	—	—
2-Deoxy-2-glycolyl-amino-D-mannose[k]	—	—	—	—	—	—	—	1.88; 1.98[h]	—	—
N-Acetylneuraminic acid	—	—	—	—	—	1.70	—	—	—	—
Muramic acid	—	—	—	0.94	0.89	1.25	1.64; 1.69[h]	—	—	—
N-Acetylmuramic acid	—	—	—	—	—	1.29; 1.30[h]	—	—	—	—
3-Deoxy-D-*manno*-2-octulosonic acid[l]	—	—	—	—	—	1.35;[h] 1.36	—	—	—	—

[a] Split ratios 1:50,[4] 1:35.[5]
[b] O = oxime, MO = *O*-methyloxime, PFBO = *O*-pentafluorobenzyloxime.
[c] t_r relative to peracetylated glucononitrile (7.2 min,[1] 6.7 min[2]).
[d] t_r relative to peracetylated D-*gluco*-heptitol (10.5 min).
[e] t_r relative to *N*-methylglucamine *O*-methyloxime (15.57 min on P4, 9.96 min on P5).
[f] t_r relative to *myo*-inositol hexaacetate (20.7 min,[5] 21.3 min,[6] 18.6 min[7]).
[g] t_r relative to *meso*-inositol hexaacetate (18 min on P8, 18.5 min on P9).
[h] Major peak.
[i] For simultaneous analysis of neutral sugars as aldononitrile acetates see Table GC 4, Reference 2.
[j] For simultaneous analysis of acetamidodeoxyhexitol acetates see Table GC 1, Reference 3.
[k] Product of cleavage of *N*-glycolylneuraminic acid by neuraminic acid aldolase.
[l] KDO.

Packing
P1 = 2% OV-17 on Chromosorb® W HP (80 — 100 mesh).
P2 = 1% diethylene glycol adipate on Chromosorb® W HP (100 — 120 mesh).
P3 = 3% diethylene glycol adipate on Chromosorb® W HP (100 — 120 mesh).
P4 = OV-101.
P5 = SE-54.
P6 = OV-1, 0.11-μm film.
P7 = Carbowax 20M.

TABLE GC 7 (continued)
Acetylated Oxime Derivatives of Neutral and Amino Sugars

	P8 =	3% SP-2340 on Supelcoport® (100 — 120 mesh).
	P9 =	CP-Sil 88, 0.2-μm film.
Temperature	T1 =	130 → 300°C at 20°C/min.
	T2 =	170°Cc for 2 min; 170 → 240°C at 10°C/min.
	T3 =	210 → 240°C at 2°C/min.
	T4 =	190 → 230°C at 4°C/min.
	T5 =	160 → 200°C at 4°C/min.
	T6 =	175°C for 4 min; 175 → 260°C at 4°C/min; held at 260°C for 5 min.
	T7 =	80 → 180°C at 20°C/min; held at 180°C for 1 min; 180 → 210°C at 2°C/min; held at 210°C for 1 min; 210 → 230°C at 10°C/min; held at 230°C.
	T8 =	240 → 260°C at 4°C/min; held at 260°C.
	T9 =	180°C for 4 min; 180 → 240°C at 5°C/min; held at 240°C for 10 min.

REFERENCES

1. **Seymour, F. R., Chen, E. C. M., and Stouffer, J. E.,** Identification of ketoses by use of their peracetylated oxime derivatives: a g.l.c.-m.s. approach, *Carbohydr. Res.*, 83, 201, 1980.
2. **Mawhinney, T. P., Feather, M. S., Barbero, G. J., and Martinez, J. R.,** The rapid, quantitative determination of neutral sugars (as aldononitrile acetates) and amine sugars (as *O*-methyloxime acetates) in glycoproteins by gas-liquid chromatography, *Anal. Biochem.*, 101, 112, 1980.
3. **Mawhinney, T. P.,** Simultaneous determination of *N*-acetylglucosamine, *N*-acetylgalactosamine, *N*-acetylglucosaminitol and *N*-acetylgalactosaminitol by gas-liquid chromatography, *J. Chromatogr.*, 351, 91, 1986.
4. **Hicks, R. E. and Newell, S. Y.,** An improved gas chromatographic method for measuring glucosamine and muramic acid concentrations, *Anal. Biochem.*, 128, 438, 1983.
5. **Guerrant, G. O. and Moss, C. W.,** Determination of monosaccharides as aldononitrile, *O*-methyloxime, alditol and cyclitol acetate derivatives by gas chromatography, *Anal. Chem.*, 56, 633, 1984.
6. **Neeser, J.-R. and Schweizer, T. F.,** A quantitative determination by capillary gas-liquid chromatography of neutral and amino sugars (as *O*-methyloxime acetates), and a study on hydrolytic conditions for glycoproteins and polysaccharides in order to increase sugar recoveries, *Anal. Biochem.*, 142, 58, 1984.
7. **Neeser, J.-R.,** G.l.c. of *O*-methyloxime and alditol acetate derivatives of neutral sugars, hexosamines, and sialic acid: "one-pot" quantitative determination of the carbohydrate constituents of glycoproteins and a study of the selectivity of alkaline borohydride reductions, *Carbohydr. Res.*, 138, 189, 1985.
8. **Biondi, P. A., Manca, F., Negri, A., Secchi, C., and Montana, M.,** Gas chromatographic analysis of neutral monosaccharides as their *O*-pentafluorobenzyloxime acetates, *J. Chromatogr.*, 411, 275, 1987.

TABLE GC 8
Deoxy(methoxyamino)alditol Derivatives

Phase	P1	P2	P3	P3	P4	P4
Temperature (°C)	T1	T2	T2	T3	T4	260
Gas; pressure (kPa)	H_2; 90[a]	H_2; 50[a]	H_2; 50[a]	H_2; 50[a]	H_2; 70[a]	H_2; 70[a]
Column						
Length, m	25	25	50	25	25	25
Diameter (I.D.), mm	0.18	0.25	0.5	0.25	0.20	0.20
Form	Capillary	Capillary	Capillary	Capillary	Capillary	Capillary
Material	Glass	Glass	Glass	Glass	Glass	Glass
Detector	FI	FI	FI	NPD[b]	MS	FI,NPD[b]
Reference	1	1	2	2	3	3
	TMS[c]	**Ac**[c]	**Me**[c]	**Me**	**Me**	**Me**
Parent sugar	**r**[d]	**r**[d]	**r**[d]	**r**[d]	**r**[e]	**r**[e]
Arabinose	0.43	0.65	0.63	0.55	—	—
Xylose	0.41	0.69	0.65	0.56	—	—
Ribose	0.44	0.71	0.53	0.36	—	—
Rhamnose	0.55	0.63	0.62	0.53	—	—
Fucose	0.57	0.66	0.71	0.65	—	—
Altrose	—	—	0.75	—	—	—
Allose	—	—	0.78	—	—	—
Idose	—	—	0.95	—	—	—
Mannose	0.98	0.98	0.97	0.96	—	—
Glucose	1.00	1.00	1.00	1.00	—	—
Gulose	—	—	1.06	—	—	—
Talose	—	—	1.09	—	—	—
Galactose	1.02	1.03	1.13	1.17	—	—
Cellobiose	—	—	—	—	0.94	0.95
Maltose	—	—	—	—	1.00	1.00
Lactose	—	—	—	—	1.04	1.05
Gentiobiose	—	—	—	—	1.58	1.52
Melibiose	—	—	—	—	1.62	1.54

[a] Split ratios 1:40,[1] 1:50,[2] splitless.[3]
[b] Nitrogen-phosphorus selective detector.
[c] TMS and Ac denote *O*-trimethylsilylated and *O*-acetylated deoxy(methoxyamino)alditols, respectively; Me denotes permethylated deoxy(methylmethoxyamino)alditols, or glycosides from disaccharides.
[d] t_r relative to 1-deoxy-1-methoxyaminoglucitol derivative (28.8 min on P1,[1] 28.2 min on P2,[1] 27.3 min on P3 with T2,[2] 37.0 min with T3.[2])
[e] t_r relative to permethylated deoxy(methylmethoxyamino)alditol glycoside derived from maltose (9.8 min for P4 with T4, 9.2 min at 260°C).

Phase
- P1 = SE-30.
- P2 = OV-101.
- P3 = Carbowax 20M.
- P4 = Superox 0.1.

Temperature
- T1 = 160°C for 10 min; 160 → 170°C at 0.5°C/min.
- T2 = 170°C for 10 min; 170 → 210°C at 2°C/min.
- T3 = 120°C for 5 min; 120 → 160°C at 1°C/min.
- T4 = 240°C for 5 min; 240 → 260°C at 1°C/min.

REFERENCES

1. **Das Neves, H. J. C., Riscado, A. M. V., and Frank, H.,** Derivatives for the analysis of monosaccharides by capillary g.l.c.: trimethylsilylated deoxy(methoxyamino)alditols, *Carbohydr. Res.*, 152, 1, 1986.
2. **Das Neves, H. J. C., Riscado, A. M. V., and Frank, H.,** Single derivatives for GC-MS assay of reducing sugars and selective detection by the nitrogen-phosphorus detector (NPD), *J. High Resolut. Chromatogr. Chromatogr. Commun.*, 9, 662, 1986.
3. **Das Neves, H. J. C. and Riscado, A. M. V.,** Capillary gas chromatography of reducing disaccharides with nitrogen-selective detection and selected-ion monitoring of permethylated deoxy(methylmethoxyamino)alditol glycosides, *J. Chromatogr.*, 367, 135, 1986.

TABLE GC 9
O-Isopropylidene Acetals of Aldoses, Ketoses, and Naturally Occurring *O*-Methylaldoses

Packing		P1	P1	P2	P3
Temperature		T1	T2	T2	T2
Gas; flow rate (ml/min)		N_2; 20	N_2; 20	N_2; 20	N_2; 20
Column					
Length, cm		183	183	183	183
Diameter (I.D.), cm		0.15	0.15	0.15	0.15
Form		Packed	Packed	Packed	Packed
Material		Glass	Glass	Glass	Glass
Detector		FI	FI	FI	FI
Reference		1	2	2	2
Parent sugar	**Main acetal**	**r**[a]	**r**[a]	**r**[a]	**r**[a]
L-Arabinose	1,2:3,4(β-L-Ara*p*)	—	—	—	0.45
D-Arabinose, 2-*O*-methyl	3,4(D-Ara*p*)	—	0.71	0.73	0.72
D-Xylose	1,2:3,5(α-D-Xyl*f*)	—	—	—	0.57
2-*O*-methyl	3,5(D-Xyl*f*)	—	0.72	0.68	0.70
3-*O*-methyl	1,2(α-D-Xyl*f*)	—	0.74	0.74	0.74
L-Rhamnose	2,3(L-Rha*p*)	—	—	—	0.85
L-Fucose	1,2:3,4(α-L-Fuc*p*	—	—	—	0.40
2-*O*-methyl	3,4(L-Fuc*p*)	—	0.69	0.65	0.68
D-Galactose	1,2:3,4(α-D-Gal*p*)	—	—	—	0.91
3-*O*-methyl	1,2:5,6(α-D-Gal*f*)	—	0.82	0.68	0.79
6-*O*-methyl	1,2:3,4(α-D-Gal*p*)	—	0.72	0.55	0.68
D-Glucose	1,2:5,6(α-D-Glc*f*)	—	—	—	0.94
3-*O*-methyl	1,2:5,6(α-D-Glc*f*)	—	0.69	0.52	0.65
D-Mannose	2,3:5,6(D-Man*f*)	1.00	—	—	1.00
L-*erythro*-Pentulose	1,2:3,4(β-L-Ribul*f*)	0.23	—	—	—
D-*threo*-Pentulose	2,3(β-D-Xylul*f*)	0.85	—	—	—
D-Psicose	1,2:3,4(β-D-Psi*f*)	0.65	—	—	—
D-Tagatose	1,2:3,4(α-D-Tag*f*)	0.75	—	—	—
D-Fructose	1,2:4,5(β-D-Fru*p*)	0.80	—	—	—
	2,3:4,5(β-D-Fru*p*)	0.87	—	—	—
L-Sorbose	2,3:4,6(α-L-Sorb*f*)	0.93	—	—	—

Note: This table supplements Table GC 9 in *CRC Handbook of Chromatography: Carbohydrates*, Volume I.

[a] t_r relative to 2,3:5,6-di-*O*-isopropylidene-D-Man*f* (41 min,[1] 22 min on P3[2]).

Packing
P1 = 3% OV-225 on Supelcoport® (100 — 120 mesh).
P2 = 3% ECNSS-M on Gas-Chrom® Q (100 — 120 mesh).
P3 = P1 + P2 (7:4).

Temperature
T1 = 80 → 160°C at 2°C/min.
T2 = 90 → 190°C at 4°C/min.

REFERENCES

1. **Morgenlie, S.,** Gas chromatography-mass spectrometry of hexuloses and pentuloses as their *O*-isopropylidene derivatives: analysis of product mixtures from triose aldol-condensations, *Carbohydr. Res.*, 80, 215, 1980.
2. **Aamlid, K. H. and Morgenlie, S.,** Analysis of mixtures of some mono-*O*-methylaldoses with the common aldoses by g.l.c.-m.s. after isopropylidenation, *Carbohydr. Res.*, 124, 1, 1983.

TABLE GC 10
Methyl Glycosides, Methylated and Acetylated

Packing	P1	P2	P3	P3	P4
Temperature (°C)	T1	T1	T2	170	170
Gas; flow rate (ml/min)	Ar; 60	Ar; 60	N_2; 25	N_2; 20	N_2; 30
Column					
Length, cm	200	200	200	180	300
Diameter (I.D.), cm	0.4	0.4	0.32	0.3	0.3
Form	Packed	Packed	Packed	Packed	Packed
Material	Glass	Glass	SS	Glass	Glass
Detector	FI	FI	FI	FI	FI
Reference	1	1	2	3	4
	Ac[a]	**Ac**		**Ac**	
Methyl glycoside	**ρ**[b]	**ρ**[b]	**r**[c]	**r**[d]	**r**[e]
α-L-Arabinopyranoside	7.17	6.65	—	—	—
2-*O*-methyl	5.12	5.00	—	—	—
3-*O*-methyl	5.64	5.60	—	—	—
4-*O*-methyl	5.86	5.39	—	—	—
2,3-di-*O*-methyl	3.02	3.54	—	—	—
2,4-di-*O*-methyl	3.30	3.42	—	—	—
3,4-di-*O*-methyl	4.53	4.55	—	—	—
2,3,4-tri-*O*-methyl	2.16	2.28	—	—	—
β-L-Arabinopyranoside	7.02	6.18	—	—	—
2-*O*-methyl	6.00	5.60	—	—	—
3-*O*-methyl	4.87	5.00	—	—	—
2,3-di-*O*-methyl	3.20	3.54	—	—	—
2,4-di-*O*-methyl	4.36	4.23	—	—	—
2,3,4-tri-*O*-methyl	2.12	2.28	—	—	—
α-D-Xylopyranoside	7.22	6.06	—	—	—
2-*O*-methyl	6.18	5.50	—	—	—
3-*O*-methyl	5.54	5.00	—	—	—
4-*O*-methyl	5.86	5.05	—	—	—
2,3-di-*O*-methyl	3.70	3.65	—	—	—
2,4-di-*O*-methyl	4.98	4.50	—	—	—
3,4-di-*O*-methyl	2.99	2.67	—	—	—
2,3,4-tri-*O*-methyl	1.31	1.39	—	—	—
β-D-Xylopyranoside	7.30	6.50	—	—	—
2-*O*-methyl	5.30	4.85	—	—	—
3-*O*-methyl	6.12	5.58	—	—	—
4-*O*-methyl	5.66	5.29	—	—	—
2,3-di-*O*-methyl	3.20	3.03	—	—	—
2,4-di-*O*-methyl	4.05	3.86	—	—	—
2,3,4-tri-*O*-methyl	1.00	1.00	—	—	—
α-D-Lyxopyranoside	7.17	6.23	—	—	—
2-*O*-methyl	6.08	5.45	—	—	—
3-*O*-methyl	4.98	4.78	—	—	—
4-*O*-methyl	5.76	5.00	—	—	—
2,3-di-*O*-methyl	3.82	3.86	—	—	—
2,4-di-*O*-methyl	3.94	3.86	—	—	—
3,4-di-*O*-methyl	2.68	2.60	—	—	—
2,3,4-tri-*O*-methyl	1.55	1.57	—	—	—
α-D-Rhamnopyranoside	6.62	5.36	—	—	—
2-*O*-methyl	5.70	4.64	—	—	—
3-*O*-methyl	4.53	3.84	—	—	—
4-*O*-methyl	5.45	4.06	—	—	—
2,3-di-*O*-methyl	3.87	3.22	—	—	—
2,4-di-*O*-methyl	3.60	2.93	—	—	—
3,4-di-*O*-methyl	2.25	2.15	—	—	—

TABLE GC 10 (continued)
Methyl Glycosides, Methylated and Acetylated

Methyl glycoside	Ac[a] ρ[b]	Ac ρ[b]	r[c]	Ac r[d]	r[e]
2,3,4-tri-*O*-methyl	1.16	1.08	—	—	—
α-L-Rhamnopyranoside					
2-*O*-methyl	—	—	9.09	—	—
3-*O*-methyl	—	—	10.18	—	—
4-*O*-methyl	—	—	11.09	—	—
2,3-di-*O*-methyl	—	—	4.27	—	—
2,4-di-*O*-methyl	—	—	3.27	—	—
3,4-di-*O*-methyl	—	—	2.36	—	—
2,3,4-tri-*O*-methyl	—	—	1.00	—	—
α-L-Fucopyranoside	6.82	5.61	—	—	—
2-*O*-methyl	5.86	5.04	—	—	—
3-*O*-methyl	4.65	4.31	—	—	—
4-*O*-methyl	6.02	4.85	—	—	—
2,3-di-*O*-methyl	3.00	2.93	—	—	—
2,4-di-*O*-methyl	4.40	3.86	—	—	—
3,4-di-*O*-methyl	3.87	3.37	—	—	—
2,3,4-tri-*O*-methyl	1.85	1.78	—	—	—
β-L-Fucopyranoside	7.05	6.25	—	—	—
2-*O*-methyl	5.91	4.58	—	—	—
3-*O*-methyl	5.05	4.38	—	—	—
2,3-di-*O*-methyl	2.72	2.81	—	—	—
2,4-di-*O*-methyl	3.89	3.37	—	—	—
2,3,4-tri-*O*-methyl	1.74	1.63	—	—	—
6-Deoxy-α-D-glucopyranoside	6.70	5.43	—	—	—
2-*O*-methyl	5.79	4.85	—	—	—
3-*O*-methyl	5.50	4.51	—	—	—
4-*O*-methyl	5.70	4.35	—	—	—
2,3-di-*O*-methyl	3.57	3.09	—	—	—
2,4-di-*O*-methyl	4.65	3.86	—	—	—
3,4-di-*O*-methyl	2.82	2.60	—	—	—
2,3,4-di-*O*-methyl	—	1.00	—	—	—
6-Deoxy-β-D-glucopyranoside	6.90	5.80	—	—	—
2-*O*-methyl	4.95	4.11	—	—	—
3-*O*-methyl	6.42	4.38	—	—	—
4-*O*-methyl	5.60	4.58	—	—	—
2,3-di-*O*-methyl	3.22	2.70	—	—	—
2,4-di-*O*-methyl	3.80	3.14	—	—	—
2,3,4-tri-*O*-methyl	—	1.30	—	—	—
α-D-Mannopyranoside	9.82	9.18	—	—	—
2-*O*-methyl	8.90	8.66	—	1.81	—
3-*O*-methyl	8.10	7.97	—	1.37	—
4-*O*-methyl	9.29	8.43	—	1.65	—
6-*O*-methyl	8.10	7.40	—	1.00	—
2,3-di-*O*-methyl	7.39	7.40	—	—	11.49
2,4-di-*O*-methyl	7.54	7.40	—	—	7.27
2,6-di-*O*-methyl	7.24	6.89	—	—	9.29
3,4-di-*O*-methyl	6.24	6.05	—	—	5.78
3,6-di-*O*-methyl	6.17	6.05	—	—	8.38
4,6-di-*O*-methyl	7.01	6.28	—	—	10.61
2,3,4-tri-*O*-methyl	5.07	5.36	—	—	2.31
2,3,6-tri-*O*-methyl	5.48	5.49	—	—	3.67
2,4,6-tri-*O*-methyl	5.36	5.16	—	—	2.71
3,4,6-tri-*O*-methyl	3.87	3.62	—	—	2.31
2,3,4,6-tetra-*O*-methyl	2.83	2.90	—	—	1.00
α-D-Galactopyranoside	9.88	9.28	—	—	—
2-*O*-methyl	9.03	8.83	—	—	—

TABLE GC 10 (continued)
Methyl Glycosides, Methylated and Acetylated

Methyl glycoside	Ac[a] ρ[b]	Ac ρ[b]	r[c]	Ac r[d]	r[e]
3-*O*-methyl	8.05	8.15	—	—	—
4-*O*-methyl	9.66	8.88	—	—	—
6-*O*-methyl	8.22	7.65	—	—	—
2,3-di-*O*-methyl	6.39	6.89	—	—	—
2,4-di-*O*-methyl	8.22	7.97	—	—	—
2,6-di-*O*-methyl	7.48	7.22	—	—	—
3,4-di-*O*-methyl	7.75	7.47	—	—	—
3,6-di-*O*-methyl	6.17	6.28	—	—	—
2,3,4-tri-*O*-methyl	5.42	5.94	—	—	—
2,3,6-tri-*O*-methyl	4.52	4.99	—	—	—
2,4,6-tri-*O*-methyl	5.95	5.94	—	—	—
3,4,6-tri-*O*-methyl	5.26	5.24	—	—	—
2,3,4,6-tetra-*O*-methyl	3.00	3.30	—	—	—
β-D-Galactopyranoside	9.94	10.05	—	—	—
2-*O*-methyl	8.28	8.34	—	—	—
3-*O*-methyl	8.92	8.97	—	—	—
4-*O*-methyl	9.36	8.97	—	—	—
6-*O*-methyl	8.45	8.33	—	—	—
2,3-di-*O*-methyl	6.02	6.60	—	—	—
2,4-di-*O*-methyl	7.75	7.68	—	—	—
2,6-di-*O*-methyl	6.65	6.71	—	—	—
3,6-di-*O*-methyl	7.20	7.42	—	—	—
4,6-di-*O*-methyl	7.86	7.42	—	—	—
2,3,6-tri-*O*-methyl	4.24	4.71	—	—	—
2,4,6-tri-*O*-methyl	5.48	5.49	—	—	—
2,3,4,6-tetra-*O*-methyl	3.00	3.25	—	—	—
α-D-Glucopyranoside	9.99	9.53	—	—	—
2-*O*-methyl	9.14	8.97	—	—	—
3-*O*-methyl	9.03	8.72	—	—	—
4-*O*-methyl	9.66	8.88	—	—	—
6-*O*-methyl	8.40	7.70	—	—	—
2,3-di-*O*-methyl	7.20	7.40	—	—	—
2,4-di-*O*-methyl	8.69	8.43	—	—	—
2,6-di-*O*-methyl	7.44	7.22	—	—	—
3,4-di-*O*-methyl	7.14	6.84	—	—	—
3,6-di-*O*-methyl	7.14	6.89	—	—	—
2,3,4-tri-*O*-methyl	5.07	5.17	—	—	—
2,3,6-tri-*O*-methyl	5.36	5.36	—	—	—
2,4,6-tri-*O*-methyl	6.57	6.18	—	—	—
3,4,6-tri-*O*-methyl	4.75	4.14	—	—	—
2,3,4,6-tetra-*O*-methyl	2.83	2.90	—	—	—
β-D-Glucopyranoside	10.02	9.98	—	—	—
2-*O*-methyl	8.32	8.34	—	—	—
3-*O*-methyl	9.66	9.21	—	—	—
4-*O*-methyl	9.94	9.53	—	—	—
6-*O*-methyl	8.32	8.05	—	—	—
2,3-di-*O*-methyl	6.85	6.84	—	—	—
2,4-di-*O*-methyl	7.81	7.55	—	—	—
2,6-di-*O*-methyl	6.65	6.35	—	—	—
3,6-di-*O*-methyl	7.84	7.45	—	—	—
4,6-di-*O*-methyl	7.16	6.84	—	—	—
2,3,4-tri-*O*-methyl	4.52	4.60	—	—	—
2,3,6-tri-*O*-methyl	4.85	4.71	—	—	—
2,4,6-tri-*O*-methyl	5.66	5.36	—	—	—
2,3,4,6-tetra-*O*-methyl	2.25	2.23	—	—	—

Note: This table supplements Table GC 11 in *CRC Handbook of Chromatography: Carbohydrates*, Volume I.

TABLE GC 10 (continued)
Methyl Glycosides, Methylated and Acetylated

[a] Ac = acetylated.
[b] Retention coefficient based on retention times relative to methyl 2,3,4-tri-*O*-methyl-β-D-xylopyranoside (ρ = 1.00) and peracetylated galactononitrile (ρ = 11.00).
[c] t_r relative to methyl 2,3,4-tri-*O*-methyl-α-L-rhamnopyranoside (1.48 min).
[d] t_r relative to methyl 6-*O*-methyl-2,3,4-tri-*O*-acetyl-α-D-mannopyranoside
[e] t_r relative to methyl 2,3,4,6-tetra-*O*-methyl-α-D-mannopyranoside.

Packing	P1 = 3% QF1 on Chromosorb® W (100 — 120 mesh).
	P2 = 3% neopentyl glycol succinate on Chromosorb® W (100 — 120 mesh).
	P3 = 3% ECNSS-M on Chromaton® N AW DMCS (80 — 100 mesh)[2] or Chromosorb® W (60 — 80 mesh).[3]
	P4 = 3% Carbowax 6000 on Chromosorb® W AW DMCS (60 — 80 mesh).
Temperature	T1 = 110 → 230°C at 5°C/min.
	T2 = 110°C for 8 min; 110 → 150°C at 2°C/min.

REFERENCES

1. **Elkin, Y. N.,** Gas chromatographic separation of the methyl ether methyglycopyranoside series hexose, 6-deoxyhexose, and pentose acetates, *J. Chromatogr.*, 180, 163, 1979.
2. **Janeček, F., Toman, R., Karácsonyi, S. and Anderle, D.,** Gas-liquid chromatographic separation of methyl ethers of L-rhamnose as their methyl glycosides, trifluoroacetylated L-rhamnitols and acetylated L-rhamnononitriles,*J. Chromatogr.*, 173, 408, 1979.
3. **Fournet, B., Leroy, Y., Montreuil, J., and Mayer, H.,** Analytical and preparative gas-liquid chromatography of methyl α-D-mannoside monomethyl ethers, *J. Chromatogr.*, 92, 185, 1974.
4. **Fournet, B. and Montreuil, J.,** Analytical and preparative gas-liquid chromatography of methyl α-D-mannoside di-, tri- and tetramethyl ethers, *J. Chromatogr.*, 75, 29, 1973.

TABLE GC 11
Permethylated Oligosaccharide-Alditols and *N*-Trifluoroacetyl Derivatives

Packing	P1	P2	P3
Temperature (°C)	T1	T2	265
Gas; flow rate (ml/min)	He; 40	He; 0.8[a]	n.a.
Column			
Length, cm	100	1000	200
Diameter (I.D.), cm	0.3	0.021	0.2
Form	Packed	WCOT	Packed
Material	Glass	Fused silica	Glass
Detector	FI	FI,ECD[b]	MS
Reference	1	2	3
		NTF[c]	
Parent oligosaccharide	**r[d]**	**r[e]**	**r[d]**
Manα1-3Man	0.31	—	—
GlcNAcβ1-4Man	0.64	—	—
GlcNAcβ11-2Man	0.68	—	—
Galβ1-4GlcNAc	0.71	—	—
Galβ1-4GlcNAcβ1-6GalNAc	—	0.79	—
GlcNAcβ1-2Manα1-3Man	1.18	—	—
GlcNAcβ1-4Manα1-3Man	1.33	—	—
Manα1-3(GlcNAcβ1-4)Man	1.22	—	—
GlcNAcβ1-2(GlcNAcβ1-4)Man	1.56	—	—
GlcNAcβ1-2Manα1-3(GlcNAcβ1-4)Man	3.03[f]	—	—
GlcNAcβ1-4Manα1-3(GlcNAcβ1-4)Man	3.76[f]	—	—

TABLE GC 11 (continued)
Permethylated Oligosaccharide-Alditols and *N*-Trifluoroacetyl Derivatives

Parent oligosaccharide	r[d]	NTF[c] r[e]	r[d]
GalNAcβ1-3Galβ1-4Galβ1-4Glc	—	0.96	—
Galβ1-3GalNAcβ1-4Galβ1-4Glc	—	1.01	—
Galβ1-3GlcNAcβ1-3Galβ1-4Glc	—	1.05	—
Galβ1-4GlcNAcβ1-3Galβ1-4Glc	—	1.07	—
Galβ1-4GlcNAcβ1-6(Galβ1-3)GalNAc	—	1.07	—
Galβ1-4(Fucα1-3)GlcNAcβ1-3Galβ1-4Glc	—	1.27	—
Galβ1-3(Fucα1-4)GlcNAcβ1-3Galβ1-4Glc	—	1.28	—
Fucα1-2Galβ1-3GlcNAcβ1-3Galβ1-4Glc	—	1.38	—
Fucα1-2Galβ1-3(Fucα1-4)GlcNAcβ1-3Galβ1-4Glc	—	1.47	—
GlcNAcβ1-2Manα1-3(GlcNAcβ1-2Manα1-6)Man	—	1.27	—
Galβ1-4GlcNAcβ1-2Manα1-3(Galβ1-4GlcNAcβ1-2Manα1-6)Man	—	2.19[g]	—
NeuAcα2-3Galβ1-4Glc[h]	—	—	2.05
NeuAcα2-6Galβ1-4Glc[h]	—	—	2.65
NeuAcα2-3Galβ1-4GlcNAc[h]	—	—	2.59
NeuAcα2-6Galβ1-4GlcNAc[h]	—	—	2.88

Note: This table supplements Table GC 4 in *CRC Handbook of Chromatography: Carbohydrates*, Volume I.

[a] Split ratio 1:50.
[b] Electron capture detector.
[c] *N*-Trifluoroacetyl derivative, from transamidation of 2-acetamido-2-deoxyhexose residues by trifluoroacetolysis (TFA-TFAA, 1:100).
[d] t_r relative to permethylated maltotritiol.
[e] t_r relative to permethylated isomaltotetraitol (27.5 min).
[f] Initial temperature 200°C.
[g] Temperature increased to 360°C after 54 min.
[h] After de-*N*-acetylation and deamination.

Packing	P1 = 1% Dexsil 300 on Supelcoport® (100 — 120 mesh).
	P2 = OV-101.
	P3 = 2% SE-30 (support not stated).
Temperature	T1 = 150 → 320°C at 4°C/min.
	T2 = 200°C for 2 min; 200 → 350°C at 4°C/min.

REFERENCES

1. **Fournet, B., Dhalluin, J.-M., Strecker, G., Montreuil, J., Bosso, C., and Defaye, J.,** Gas-liquid chromatography and mass spectrometry of oligosaccharides obtained by partial acetolysis of glycans of glycoproteins, *Anal. Biochem.*, 108, 35, 1980.
2. **Nilsson, B. and Zopf, D.,** Gas chromatography and mass spectrometry of hexosamine-containing oligosaccharide alditols as their permethylated *N*-trifluoroacetyl derivatives, *Methods Enzymol*, 86, 46, 1982.
3. **Mononen, I.,** Structural analysis of the neuraminic acid linkages in deaminated trisaccharide-alditols by gas-liquid chromatography-mass spectrometry, *Carbohydr. Res.*, 104, 1, 1982.

TABLE GC 12
Trifluoroacetylated Alditols, Aminodeoxyalditols, and Cyclitols

Packing	P1	P2	P3	P4	P5	P6
Temperature (°C)	120	T1	T2	T3	T4	150
Gas; flow rate (ml/min)	N_2; 50	N_2; 50	N_2; 20	N_2; n.a.	N_2; 1.5[a]	He; 2[a]
Column						
Length, cm	200	150	183	400	5000	2500
Diameter (I.D.), cm	0.3	0.3	0.2	0.3	n.a.	0.02
Form	Packed	Packed	Packed	Packed	WCOT	Bonded phase
Material	Glass	Glass	Glass	Glass	Glass	Fused silica
Detector	FI, NPD[b]	FI	FI	FI	FI	FI
Reference	1	1	2	3	4	5

TABLE GC 12 (continued)
Trifluoroacetylated Alditols, Aminodeoxyalditols, and Cyclitols

Parent compound	r[c]	r[c]	r[c]	r[c]	r[c]	r[c]
Glycerol	—	—	—	0.41	—	—
Erythrose	—	—	0.34	0.66	—	0.40
Threose	—	—	—	0.74	—	0.47
Rhamnose	—	0.54	—	—	—	0.48
Fucose	0.79	0.60	—	—	—	0.55
6-Deoxyglucose	—	—	—	—	—	0.62
Ribose	—	—	0.64	0.86	0.88	0.63
Arabinose[d]	0.74	0.78	0.73	0.92	0.92	0.74
Lyxose[d]	—	—	—	—	—	0.74
Xylose	—	0.85	—	0.96	0.97	0.84
Allose	—	—	—	0.98	—	0.89
Mannose	1.00	1.00	1.00	1.00	1.00	1.00
Altrose[d]	—	—	—	1.03	—	1.10
Talose[d]	—	—	—	—	—	1.10
2-Deoxy-D-*arabino*-hexose	—	—	—	—	—	1.13
Glucose	—	1.13	1.16	1.07	1.09	1.27
Idose	—	—	—	1.08	—	1.32
2-Deoxy-D-*lyxo*-hexose	—	—	—	—	—	1.40
Galactose	1.09	1.18	1.25	1.11	1.14	1.40
2-Amino-2-deoxyglucose	1.85	—	—	—	—	—
2-Amino-2-deoxygalactose	2.09	—	—	—	—	—
2-Amino-2-deoxymannose	2.03	—	—	—	—	—
chiro-Inositol	—	—	0.52	—	—	—
Pinitol	—	—	0.60	—	—	—
Quebrachitol	—	—	0.70	—	—	—
Leucanthemitol	—	—	0.77	—	—	—
Bornesitol	—	—	1.32	—	—	—
Viburnitol	—	—	1.35	—	—	—
myo-Inositol	—	—	1.36	—	—	1.52
scyllo-Inositol	—	—	1.87	—	—	—

Note: This table supplements Table GC 15 in *CRC Handbook of Chromatography: Carbohydrates,* Volume I. Data for GLC of trifluoroacetylated alditols on chiral phases will be found in Table GC 24.

[a] Split ratios 1:15,[4] 1:20.[5]

[b] Nitrogen-phosphorus selective detector used for aminodeoxy sugars.

[c] t_r relative to hexa-*O*-trifluoroacetyl-D-mannitol (3.9 min on P1,[1] 18.1 min on P2,[1] 5.6 min,[2] 14.6 min,[3] 33.2 min,[4] 9.1 min[5]).

[d] Pairs giving identical alditol.

Packing	P1 = 5% OV-101 on Chromosorb® W AW DMCS (60 — 80 mesh).
	P2 = 2% XF-1105 on Gas-Chrom® P (60 — 80 mesh).
	P3 = 3% Dexsil 410 on Chromosorb W HP (80 — 100 mesh).
	P4 = 1% OV-225 on Chromosorb W HP (80 — 100 mesh).
	P5 = OV-225.
	P6 = Shimadzu CBP10 (cyanopropyl bonded phase), 0.25-μm film.
'emperature	T1 = 100 → 160°C at 2°C/min.
	T2 = 100°C for 1.5 min; 100 → 310°C at 3.5°C/min for 3.5 min, 6°C/min for 5 min, 15°C/min for 5 min, 25°C/min for 3.7 min; held at 310°C for 6.3 min.
	T3 = 100°C for 3 min; 100 → 180°C at 5°C/min.
	T4 = 70°C for 2 min; 70 → 180°C at 5°C/min; held at 180°C for 15 min.

REFERENCES

1. **Shinohara, T.,** Use of a flame thermionic detector in the determination of glucosamine and galactosamine in glycoconjugates by gas chromatography, *J. Chromatogr.*, 207, 262, 1981.
2. **Englmaier, P.,** Trifluoroacetylation of carbohydrates for g.l.c. using *N*-methylbis(trifluoroacetamide), *Carbohydr. Res.*, 144, 177, 1985.

3. **Decker, P. and Schweer, H.,** Gas-liquid chromatography on OV-225 of tetroses and aldopentoses as their *O*-methoxime and *O-n*-butoxime pertrifluoroacetyl derivatives and of C_3-C_6 alditol pertrifluoroacetates, *J. Chromatogr.*, 236, 369, 1982.
4. **Decker, P. and Schweer, H.,** G.l.c. of the oxidation products of pentitols and hexitols as trifluoroacetylated *O*-methyloxime and *O*-butyloxime derivatives, *Carbohydr. Res.*, 107, 1, 1982.
5. **Haga, H. and Nakajima, T.,** Analysis of aldoses and alditols by capillary gas chromatography as alditol trifluoroacetates, *Chem. Pharm. Bull.*, 36, 1562, 1988.

TABLE GC 13
Trifluoroacetylated Methyl Glycosides: Capillary GLC of Methanolysates of Lipopolysaccharides and Gangliosides

Phase	P1	P2	P3
Temperature	T1	T2	T3
Gas; flow rate ((ml/min)	He; 1.5	He; 2	n.a.
Column			
Length, m	25	25	25
Diameter (I.D.), mm	0.2	0.22	0.2
Form	Coated	Coated	Bonded phase
Material	Fused silica	Glass	Fused silica
Detector	FI	FI	FI
Reference	1	2,3	4
Parent sugar	r[a]	r[a]	r[a]
3,6-Dideoxy-D-*arabino*-hexose[b]	0.57;[c] 0.67; 0.69	—	—
3,6-Dideoxy-D-*xylo*-hexose[b]	0.65;[c] 0.70; 0.71	—	—
L-Rhamnose	0.68;[c] 0.76	0.67;[c] 0.80	—
L-Fucose	0.70; 0.73;[c] 0.79	0.77; 0.83[c]	—
D-Ribose	0.73; 0.75; 0.76; 0.88[c]	—	—
D-Galactose	0.92; 0.96;[c] 1.01	0.92; 0.97;[c]	0.86;[c] 1.07;[c]
D-Glucose	0.94; 1.00;[c] 1.02	1.00;[c] 1.05	1.00;[c] 1.03
D-Mannose	0.99;[c] 1.05	—	—
D-*glycero*-D-*manno*-Heptose	1.15	1.25	—
L-*glycero*-D-*manno*-Heptose	1.20	1.33	—
2-Amino-2-deoxy-D-glucose	1.29	1.42	—
2-Amino-2-deoxy-D-galactose	1.24; 1.31[c]	1.42	—
3-Deoxy-D-*manno*-2-octulosonic acid[d,e]	1.29; 1.41;[c] 1.45[c]	1.67;[c] 1.70[c]	—
N-Acetylneuraminic acid[e]	1.64	—	1.42

Note: This table supplements Table GC 14 in *CRC Handbook of Chromatography: Carbohydrates*, Volume I. Data for GLC of trifluoroacetylated methyl glycosides on chiral phases will be found in Table GC 24.

[a] t_r relative to major peak for trifluoroacetylated methyl glucoside (9.47 min,[1] 6.0 min,[2,3] 10.3 min[4]).
[b] 3,6-Dideoxy-D-*arabino*-hexose = tyvelose, -D-*xylo*-hexose = abequose.
[c] Major peak.
[d] KDO.
[e] Present in methanolysate as methyl ester methyl glycoside.

Phase	P1 = SE-30.
	P2 = CP-Sil 5.
	P3 = BP-5 (5% phenyl-methyl silicone; S.G.E., Melbourne, Australia).
Temperature	T1 = 90°C for 4 min; 90 → 250°C at 8°C/min.
	T2 = 90°C for 2 min; 90 → 260°C at 9°C/min.
	T3 = 90°C for 2 min; 90 → 290°C at 8°C/min.

REFERENCES

1. **Bryn, K. and Jantzen, E.,** Analysis of lipopolysaccharides by methanolysis, trifluoroacetylation, and gas chromatography on a fused-silica capillary column, *J. Chromatogr.*, 240, 405, 1982.

2. **Brondz, I. and Olsen, I.,** Differentiation between *Actinobacillus actinomycetemcomitans* and *Haemophilus aphrophilus* based on carbohydrates in lipopolysaccharide, *J. Chromatogr.*, 310, 261, 1984.
3. **Brondz, I. and Olsen, I.,** Whole-cell methanolysis as a rapid method for differentiation betweeen *Actinobacillus actinomycetemcomitans* and *Haemophilus aphrophilus,*, *J. Chromatogr.*, 311, 347, 1984.
4. **Laegreid, A., Otnaess, A.-B. K., and Bryn, K.,** Purification of human milk gangliosides by silica gel chromatography and analysis of trifluoroacetate derivatives by gas chromatography, *J. Chromatogr.*, 377, 59, 1986.

TABLE GC 14
Trifluoroacetylated Mono- and Oligosaccharides

Packing	P1	P2
Temperature	T1	T2
Gas; flow rate (ml/min)	N_2; 20	He; 0.6[a]
Column		
Length, cm	183	3000
Diameter (I.D.), cm	0.2	0.026
Form	Packed	Bonded phase
Material	Glass	Fused silica
Detector	FI	FI
Reference	1	2
Parent sugar	**r[b]**	**r[c]**
D-Glucose	0.79; 1.00	—
D-Fructose	0.79	—
Sucrose	2.02	—
Maltose	2.07; 2.12	—
1-Kestose	—	1.04
Melezitose	—	1.13
Fructotriose	2.28	—
Raffinose	2.31	1.00
Fructotetraose	2.44	—
Stachyose	2.47	—
Fructopentaose	2.54	—
Fructohexaose	2.67	—
Glcβ1-3Glcβ1-3Glc	—	1.48; 1.49
Glcβ1-3Glcβ1-4Glc	—	1.35;[d] 1.42
Glcβ1-4Glcβ1-3Glc	—	1.54;[d] 1.57
Glcβ1-4Glcβ1-4Glc	—	1.42;[d] 1.49
Glcα1-4Glcα1-4Glc	—	1.32;[d] 1.34
Glcα1-4Glcα1-6Glc	—	1.37;[d] 1.64
Glcα1-6Glcα1-3Glc	—	1.29;[d] 1.39
Glcα1-6Glcα1-4Glc	—	1.18;[d] 1.21
Glcα1-6Glcα1-6Glc	—	1.06;[d] 1.31
Xylβ1-4Xylβ1-4Xyl	—	1.99;[d] 2.02

Note: This table supplements Table GC 12 in *CRC Handbook of Chromatography: Carbohydrates*, Volume I. Data for GLC of trifluoroacetylated sugars on chiral phases will be found in Table GC 24.

[a] Split ratio 1:100.
[b] t_r relative to trifluoroacetylated β-D-glucose (7.2 min).
[c] t_r relative to trifluoroacetylated raffinose (7.63 min).
[d] Major peak.

TABLE GC 14 (continued)
Trifluoroacetylated Mono- and Oligosaccharides

Packing	P1 = 3% Dexsil 410 on Chromosorb® W HP (80—100 mesh).
	P2 = DB-5 (J and W Scientific), 0.1-μm film.
Temperature	T1 = 100°C for 1.5 min; 100 → 310°C at 3.5°C/min for 3.5 min, 6°C/min for 5 min, 15°C/min for 5 min, 25°C/min for 3.7 min; held at 310°C for 6.3 min.
	T2 = 180°C for 5 min; 180 → 200°C at 5°C/min; held at 200°C.

REFERENCES

1. **Englmaier, P.,** Trifluoroacetylation of carbohydrates for g.l.c. using *N*-methylbis(trifluoroacetamide), *Carbohydr. Res.*, 144, 177, 1985.
2. **Selosse, E. J.-M. and Reilly, P. J.,** Capillary column gas chromatography of triifluoroacetyl trisaccharides, *J. Chromatogr.*, 328, 253, 1985.

TABLE GC 15
Trifluoroacetylated *O*-Alkyloximes of Tri- to Hexoses: Capillary GLC

Phase	OV-225	OV-225	OV-225	OV-225	OV-225	OV-225	OV-225	OV-225
Temperature	T1	T1	T1	T2	T2	T3	T4	T4
Gas; flow rate (ml/min)	He; 1.5[a]	He; 1.5[a]	He; 1.5[a]	N_2; 1.5[a]	N_2; 1.5[a]	He; 1.5[a]	N_2; 2[a]	He; 1.5[a]
Column								
Length, m	50	50	50	50	50	50	50	50
Diameter (I.D.), mm	n.a.	n.a.	n.a.	n.a.	n.a.	n.a.	n.a.	n.a.
Form	WCOT	WCOT	WCOT	WCOT	WCOT	WCOT	WCOT	WCOT
Material	Glass	Glass	Glass	Glass	Glass	Glass	Glass	Glass
Detector	MS	MS	MS	FI	FI	MS	FI	MS
Reference	1,2	1,2	1,2	3,4	3,4	5	6	6
	MO[b]	**MPO[b]**	**BuO[b]**	**MO**	**BuO**	**BuO**	**BuO**	**BuO**
Parent compound	**r[c]**	**r[c]**	**r[c]**	**r[d]**	**r[d]**	**r[e]**	**r[d]**	**r[d]**
DL-Glyceraldehyde	—	—	—	—	—	—	0.56; 0.64	0.56; 0.63
Dihydroxyacetone	—	—	—	—	—	—	0.64	—
D-Erythrose	—	—	—	0.77; 0.80[f]	0.74; 0.81[f]	—	0.73; 0.79[f]	0.73; 0.80[f]
2-*C*-(hydroxymethyl)-DL-glyceraldehyde	—	—	—	—	—	—	0.73	—
D-Threose	—	—	—	0.78; 0.85[f]	0.75; 0.85[f]	—	0.74; 0.84[f]	0.75; 0.85[f]
D-*glycero*-Tetrulose	—	—	—	0.83; 0.86[f]	0.80; 0.85[f]	—	0.78; 0.84[f]	0.79; 0.84[f]
D-Ribose	0.70; 0.77[f]	0.70; 0.81[f]	0.70; 0.79[f]	0.86; 0.92[f]	0.84; 0.91[f]	—	0.82; 0.90[f]	0.83; 0.91[f]
D-Arabinose	0.72; 0.84[f]	0.71; 0.88[f]	0.71; 0.86[f]	0.88; 0.96[f]	0.85; 0.97[f]	—	0.83; 0.96[f]	0.84; 0.98[f]
3-*C*-(hydroxymethyl)-DL-*glycero*-Tetrose	—	—	—	—	—	—	0.88; 0.92	0.89; 0.93
D-Lyxose	0.80; 0.85[f]	0.78; 0.89[f]	0.77; 0.87[f]	0.93; 0.97[f]	0.90; 0.98[f]	—	0.89; 0.98[f]	0.90; 0.99[f]
D-Xylose	0.82; 0.89[f]	0.80; 0.90[f]	0.79; 0.89[f]	0.95; 1.00[f]	0.91; 1.00[f]	—	0.91; 1.00[f]	0.91; 1.00[f]
DL-*threo*-2-Pentulose	—	—	—	0.96;[f] 1.03	0.92;[f] 1.01	0.87;[f] 1.01	0.91;[f] 1.01	0.92;[f] 1.01
DL-*erythro*-2-Pentulose	—	—	—	0.97; 0.98[f]	0.94; 0.96[f]	0.90; 0.93[f]	0.94; 0.95[f]	0.94; 0.97[f]
erythro-3-Pentulose	—	—	—	1.01	0.98	0.96	—	—
D-*threo*-3-Pentulose	—	—	—	1.03	0.99	0.98	—	—
D-Allose	0.81; 0.90[f]	0.79; 0.92[f]	0.77; 0.89[f]	—	—	—	—	0.91; 1.02[f]
D-Altrose	0.85; 0.94[f]	0.83; 0.96[f]	0.81; 0.94[f]	—	—	—	—	0.94; 1.05[f]
D-Mannose	0.88; 0.96[f]	0.88; 0.98[f]	0.85; 0.95[f]	—	—	—	—	0.97; 1.08[f]
D-Gulose	0.91; 1.00[f]	0.89; 0.99[f]	0.86; 1.00[f]	—	—	—	—	0.98; 1.11[f]
D-Talose	0.90; 0.97[f]	0.90; 1.01[f]	0.86; 0.98[f]	—	—	—	—	0.99; 1.09[f]

D-Galactose	0.92; 1.06[f]	0.89; 1.07[f]	0.88; 1.06[f]	—	—	—	—	1.00; 1.17[f]
D-Glucose	0.93; 1.00[f]	0.90; 1.00[f]	0.89; 1.00[f]	—	—	—	—	1.00; 1.11[f]
D-Idose	0.95; 1.06[f]	0.93; 1.06[f]	0.91; 1.06[f]	—	—	—	—	1.03; 1.17[f]
D-Fructose	—	—	—	—	1.01;[f] 1.09	1.00;[f] 1.13	1.03;[f] 1.10	1.02;[f] 1.09
DL-Psicose	—	—	—	—	1.01; 1.04[f]	1.02; 1.07[f]	1.03; 1.05[f]	1.02; 1.05[f]
DL-Sorbose	—	—	—	—	1.06;[f] 1.12	1.09;[f] 1.13	1.07;[f] 1.13	1.06;[f] 1.12
D-*arabino*-3-Hexulose	—	—	—	—	1.09;[f] 1.16	1.14;[f] 1.27	1.09;[f] 1.17	—
DL-*ribo*-3-Hexulose	—	—	—	—	1.11; 1.12	1.17; 1.21	1.12; 1.14	1.11; 1.13
DL-Tagatose	—	—	—	—	1.13; 1.16[f]	1.20; 1.26[f]	1.14; 1.17	1.13; 1.16
DL-*xylo*-3-Hexulose	—	—	—	—	1.13;[f] 1.18	1.21;[f] 1.30	1.14;[f] 1.20	1.13;[f] 1.17
D-*lyxo*-3-Hexulose	—	—	—	—	1.15;[f] 1.17	1.24;[f] 1.28	1.16;[f] 1.18	1.15;[f] 1.17
erythro-2,5-Hexodiulose	—	—	—	—	1.21; 1.26;[f] 1.35	1.36; 1.44;[f] 1.58	—	—
DL-*threo*-2,5-Hexodiulose	—	—	—	—	1.24; 1.30;[f] 1.39	1.41; 1.51;[f] 1.64	—	—

[a] Split ratios 1:10 (References 1, 2, 5, 6, MS detection), 1:15 (References 3, 4), 1:12 (Reference 6, FI detection).
[b] MO = *O*-methyloxime, MPO = *O*-2-methyl-2-propyloxime, BuO = *O*-*n*-butyloxime.
[c] t_r relative to major peak for trifluoroacetylated glucose alkoxime (15.62 min for MO, 15.78 min for MPO, 19.55 min for BuO).[2]
[d] t_r relative to major peak for trifluoroacetylated xylose alkoxime (26.41 min for MO, 31.67 min for BuO,[3] 23.05 and 20.82 min for FI and MS detection[6]).
[e] t_r relative to major peak for trifluoroacetylated *O*-butyloxime of fructose (12.67 min).
[f] Major peak.

Temperature
T1 = 120°C for 2 min; 120 → 180°C at 5°C/min; held at 180°C for 10 min.
T2 = 70°C for 2 min; 70 → 180°C at 5°C/min; held at 180°C for 15 min
T3 = 150°C for 2 min; 150 → 180°C at 5°C/min; held at 180°C for 20 min.
T4 = 100°C for 2 min; 100 → 180°C at 5°C/min; held at 180°C for 15 min.

REFERENCES

1. **Schweer, H.,** Gas chromatography-mass spectrometry of aldoses as *O*-methoxime, *O*-2-methyl-2-propoxime, and *O*-*n*-butoxime pertrifluoroacetyl derivatives on OV-225 with methylpropane as ionization agent. I. Pentoses, *J. Chromatogr.*, 236, 355, 1982.
2. **Schweer, H.,** Gas chromatography-mass spectrometry of aldoses as *O*-methoxime, *O*-2-methyl-2-propoxime, and *O*-*n*-butoxime pertrifluoroacetyl derivatives on OV-225 with methylpropane as ionization agent. II. Hexoses, *J. Chromatogr.*, 236, 361, 1982.
3. **Decker, P. and Schweer, H.,** Gas-liquid chromatography on OV-225 of tetroses and aldopentoses as *O*-methoxime and *O*-*n*-butoxime pertrifluoroacetyl derivatives and of C_3-C_6 alditol pertrifluoroacetates., *J. Chromatogr.*, 236, 369, 1982.
4. **Decker, P. and Schweer, H.,** G.l.c. of the oxidation products of pentitols and hexitols as trifluoroacetylated *O*-methyloxime and *O*-butyloxime derivatives, *Carbohydr. Res.*, 107, 1, 1982.
5. **Schweer, H.,** G.l.c.-m.s. of the oxidation products of pentitols and hexitols as *O*-butyloxime trifluoroacetates, *Carbohydr. Res.*, 111, 1, 1982.
6. **Decker, P., Schweer, H., and Pohlman, R.,** Bioids. X. Identification of formose sugars, presumable prebiotic metabolites, using capillary gas chromatography-mass spectrometry of *n*-butoxime trifluoroacetates on OV-225, *J. Chromatogr.*, 244, 281, 1982.

TABLE GC 16
Trimethylsilylated Alditols, Aminodeoxyalditols, and Cyclitols

Packing	P1	P1	P2	P3	P4	P5
Temperature (°C)	190	T1	T2	T3	T4	T4
Gas; flow rate (ml/min)	n.a.[a]	n.a.	N_2; 15[a]	N_2; 18	N_2; 25	N_2; 25
Column						
Length, cm	2500	3000	2500	183	180	180
Diameter (I.D.), cm	0.023	0.025	0.032	0.18	0.2	0.2
Form	Capillary	Capillary	WCOT	Packed	Packed	Packed
Material	Fused silica	Glass	Fused silica	Glass	SS	SS
Detector	FI	MS	FI	FI	FI	FI
Reference	1	2,3	4	5	6	6
Parent compound	r[b]	r[b]	r[b]	r[c]	r[c]	r[c]
Glycerol	—	0.32	—	—	—	—
Threitol	—	0.54	—	—	—	—
4-deoxy	—	0.34	—	—	—	—
Erythritol	0.23	0.56	—	—	—	—
4-deoxy	—	0.35	—	—	—	—
Xylitol	0.42	0.77	0.59	—	—	—
5-deoxy	—	0.60	—	—	—	—
1,4/2,5-anhydro[d]	—	—	0.32	—	—	—
Arabinitol	0.44	0.78	0.61	—	—	—
5-deoxy	—	0.61	—	—	—	—
1,4-anhydro[d]	—	—	0.29	—	—	—
Ribitol	0.45	0.80	—	—	—	—
2-deoxy	—	0.66	—	—	—	—
Allitol, 6-deoxy	—	0.82	—	—	—	—
Mannitol	0.97	0.98	0.99	—	—	—
6-deoxy	0.52	0.83	—	—	—	—
Gulitol, 6-deoxy	—	0.84	—	—	—	—
Glucitol	1.00	1.00	1.00	—	—	—
1,4-anhydro[d]	—	—	0.64	—	—	—
Galactitol	1.02	—	1.01	—	—	—
1,4/3,6-anhydro[d]	—	—	0.64	—	—	—
6-deoxy	0.55	0.85	0.75	—	—	—
1,4-anhydro-6-deoxy[d]	—	—	0.36	—	—	—
2,5-anhydro-6-deoxy[d]	—	—	0.40	—	—	—
2-Amino-2-deoxyglucitol	—	—	0.96	—	—	—
2-Amino-2-deoxygalactitol	—	—	0.95	—	—	—
2-Acetamido-2-deoxyglucitol	—	—	1.22	1.08[e]	—	—
2-Acetamido-2-deoxymannitol	—	—	—	1.13[e]	—	—
2-Acetamido-2-deoxygalactitol	—	—	1.23	1.13[e]	—	—
1,4-anhydro[d]	—	—	0.95	—	—	—
3,6-anhydro[d]	—	—	0.89	—	—	—
1,4:3,6-dianhydro[d]	—	—	0.50	—	—	—
muco-Inositol	—	—	—	—	0.76	0.45
O-methyl	—	—	—	—	0.59	0.46
chiro-Inositol	—	1.01	—	—	0.82	0.53
Pinitol	—	—	—	—	0.61	0.45
Quebrachitol	—	—	—	—	0.67	0.53
scyllo-Inositol	—	1.05	—	—	0.92	0.87
O-methyl	—	—	—	—	0.82	0.99
myo-Inositol	—	1.12	—	1.00	1.00	1.00
Ononitol	—	—	—	—	0.78	0.83
Sequoyitol	—	—	—	—	0.79	0.84

TABLE GC 16 (continued)
Trimethylsilylated Alditols, Aminodeoxyalditols, and Cyclitols

Parent compound	r[b]	r[b]	r[b]	r[c]	r[c]	r[c]
Bornesitol	—	—	—	—	0.86	1.07
Dambonitol	—	—	—	—	0.74	1.21

Note: This table supplements Table GC 21 in *CRC Handbook of Chromatography: Carbohydrates,* Volume I.

[a] Split ratios 1:150,[1] 1:10.[4]
[b] t_r relative to trimethylsilylated D-glucitol (20.93 min,[2] 40.8 min[3]).
[c] t_r relative to trimethylsilylated *myo*-inositol (13.5 min,[5] 25.5 min on P4[6]).
[d] Formed during methanolysis of oligosaccharide-alditols.
[e] For simultaneous analysis of acetamidodeoxyhexoses as trimethylsilylated *O*-methyloximes see Table GC 22, Reference 3.

Packing P1 = OV-101.
P2 = CP-Sil 5.
P3 = 3% SP-2250 on Supelcoport® (100 — 120 mesh).
P4 = 3% SE-30 on Gas-Chrom® Q (80 — 100 mesh).
P5 = 10% Carbowax 20M on Gas-Chrom Q (80 — 100 mesh).
Temperature T1 = 120 → 260°C at 3°C/min.
T2 = 130 → 220°C at 2°C/min.
T3 = 120 → 220°C at 5°C/min.
T4 = 130 → 190°C at 2°C/min.

REFERENCES

1. **Bradbury, A. G. W., Halliday, D. J., and Medcalf, D. G.,** Separation of monosaccharides as trimethylsilylated alditols on fused-silica capillary columns, *J. Chromatogr.*, 213, 146, 1981.
2. **Niwa, T., Yamamoto, N., Maeda, K., Yamada, K., Ohki, T., and Mori, M.,** Gas chromatographic-mass spectrometric analysis of polyols in urine and serum of uremic patients. Identification of new deoxyalditols and inositol isomers, *J. Chromatogr.*, 277, 25, 1983.
3. **Niwa, T., Yamada, K., Ohki, T., Saito, A., and Mori, M.,** Identification of 6-deoxyallitol and 6-deoxygulitol in human urine. Electron-impact mass spectra of eight isomers of 6-deoxyhexitol, *J. Chromatogr.*, 336, 345, 1984.
4. **Gerwig, G. J., Kamerling, J. P., and Vliegenthart, J. F. G.,** Anhydroalditols in the sugar analysis of methanolysates of alditols and oligosaccharide-alditols, *Carbohydr. Res.*, 129, 149, 1984.
5. **Mawhinney, T. P.,** Simultaneous determination of *N*-acetylglucosamine, *N*-acetylgalactosamine, *N*-acetylglucosaminitol and *N*-acetylgalactosaminitol by gas-liquid chromatography, *J. Chromatogr.*, 351, 167, 1986.
6. **Ford, C. W.** Identification of inositols and their mono-*O*-methyl ethers by gas-liquid chromatography, *J. Chromatogr.*, 333, 167, 1985.

TABLE GC 17
Trimethylsilylated Alditols from Partially Methylated Sugars

Packing	P1	P1	P1
Temperature	T1	T2	T3
Gas; flow rate (ml/min)	He; n.a.	He; n.a.	He; 40
Column			
Length, cm	300	240	300
Diameter (I.D.), cm	0.25	0.25	0.25
Form	Packed[a]	Packed[a]	Packed[a]
Material	Glass	Glass	Glass
Detector	FI	FI	FI
Reference	1	1	2

TABLE GC 17 (continued)
Trimethylsilylated Alditols from Partially Methylated Sugars

Parent sugar	r[b]	r[c]	r[b]
D-Mannose			
2-*O*-methyl	2.47	—	2.44
6-*O*-methyl	2.23	—	2.16
2,4-di-*O*-methyl	1.92	—	1.89
2,6-di-*O*-methyl	2.06	—	2.00
3,4-di-*O*-methyl	1.86	—	1.80
3,6-di-*O*-methyl	1.68	—	1.65
4,6-di-*O*-methyl	1.74	—	1.68
2,3,4-tri-*O*-methyl	1.64	—	1.60
2,3,6-tri-*O*-methyl	1.21	—	1.22
2,4,6-tri-*O*-methyl	1.17	—	1.18
3,4,6-tri-*O*-methyl	1.39	—	1.40
2,3,4,6-tetra-*O*-methyl	1.00	—	1.00
D-Galactose			
2,3,4-tri-*O*-methyl	1.48	—	1.49
2,3,6-tri-*O*-methyl	1.27	—	1.30
2,4,6-tri-*O*-methyl	1.24	—	1.21
3,4,6-tri-*O*-methyl	1.32	—	1.33
2,3,4,6-tetra-*O*-methyl	1.04	—	1.06
2-Acetamido-2-deoxy-D-glucose			
3-*O*-methyl	—	1.22	3.40
3,4-di-*O*-methyl	—	1.12	3.25
3,6-di-*O*-methyl	—	1.06	3.17
3,4,6-tri-*O*-methyl	—	1.00	3.10

Note: This table supplements Table GC 22 in *CRC Handbook of Chromatography: Carbohydrates*, Volume I.

[a] Surface of Chromosorb® support was modified by treatment with polyethylene glycol (mol. wt. 20,000) at 280°C overnight, followed by extraction with methanol, leaving thin film of nonextractable polymer as surface coating.

[b] t_r relative to 1,5-di-*O*-trimethylsilyl-2,3,4,6-tetra-*O*-methyl-D-mannitol (9.2 min).

[c] t_r relative to 2-deoxy-2-*N*-methylacetamido-1,5-di-*O*-trimethylsilyl-3,4,6-tri-*O*-methylglucitol (13.7 min).

Packing	P1 = 0.4% OV-225 on surface-modified Chromosorb® W AW (80 — 100 mesh).
Temperature	T1 = 130°C for 2 min; 130 → 150°C at 1°C/min.
	T2 = 130°C for 2 min; 130 → 205°C at 4°C/min.
	T3 = 130°C for 2 min; 130 → 200°C at 1°C/min.

REFERENCES

1. **Akhrem, A. A., Avvakumov, G. V., and Strel'chyonek, O. A.,** Methylation analysis in glycoprotein chemistry: low-bleeding columns for the gas chromatographic analysis of methylated sugar derivatives, *J. Chromatogr.*, 176, 207, 1979.
2. **Akhrem, A. A., Avvakumov, G. V., Sidorova, I. V., and Strel'chyonok, O. A.,** Methylation analysis in glycoprotein chemistry. General procedure for quantification of the products of solvolysis of permethylated glycopeptides and glycoproteins, *J. Chromatogr.*, 180, 69, 1979.

TABLE GC 18
Trimethylsilylated Aldonic and Aldaric Acids: Capillary GLC of Products of Alkaline Degradation of Carbohydrates

Phase	P1	P2
Temperature	T1	T2
Gas; flow rate (ml/min)	H_2 2[a]	H_2; 2[a]
Column		
Length, m	25	25
Diameter (I.D.), mm	0.32	0.32
Form	Coated	Coated
Material	Fused silica	Fused silica
Detector	FI	FI
Reference	1-4	5
Parent compound	**r[b]**	**r[b]**
Glyceric acid	0.73	0.72
2-*C*-methyl	0.72	0.71
2-Deoxytetronic acid	0.87	—
3-Deoxytetronic acid	0.84	0.83
Erythronic acid	1.00	1.00
Threonic acid	—	1.02
3,4-Dideoxypentonic acid	0.95	—
3-Deoxy-*erythro*-pentonic acid	1.12	—
3-Deoxy-*threo*-pentonic acid	1.14	—
threo-2-Pentulosonic acid	—	1.09
Xylonic acid	—	1.22
Ribonic acid	1.26	—
Lyxonic acid	—	1.24
Arabinonic acid	1.28	—
3,6-Dideoxy-*ribo*-hexonic acid	1.15	—
3,6-Dideoxy-*arabino*-hexonic acid	1.16	—
3,4-Dideoxyhexonic acid	1.25	—
Mannonic acid	1.58	—
Gluconic acid	1.64	—
L-*threo*-2-Hexenono-1,4-lactone[c]	—	1.41
Erythraric acid	1.10	1.05
Threaric acid	1.11	1.07
2,3-Dideoxypentaric acid	1.02	—
2,4-Dideoxypentaric acid	1.03	—
2-Deoxy-*erythro*-pentaric acid	1.15	—
2-Deoxy-*threo*-pentaric acid	1.16	—
3-Deoxy-*erythro*-pentaric acid	1.18	—
3-Deoxy-*threo*-pentaric acid	1.20	—
Arabinaric acid	1.32	—
Xylaric acid	1.34	—
3,4-Dideoxy-*threo*-hexaric acid	1.28	—
3,4-Dideoxy-*erythro*-hexaric acid	1.29	—
3-Deoxy-*xylo*-hexaric acid	1.42	—
3-Deoxy-*lyxo*-hexaric acid	1.44	—
Altraric acid	1.70	—
Galactaric acid	1.73	—

Note: This table supplements Table GC 24 in *CRC Handbook of Chromatography: Carbohydrates*, Volume I.

[a] Split ratio 1:20.
[b] t_r relative to trimethylsilated erythronic acid (6.25 min,[1-4] 6.33 min[5]).
[c] L-Ascorbic acid.

TABLE GC 18 (continued)
Trimethylsilylated Aldonic and Aldaric Acids: Capillary GLC of Products of Alkaline Degradation of Carbohydrates

Packing	P1 = OV-101
	P2 = SE-54.
Temperature	T1 = 100°C for 2 min, 100 → 200°C at 20°C/min; held at 200°C for 5 min.
	T2 = 110°C for 2 min; 100 → 230°C at 20°C/min; held at 230°C for 5 min.

REFERENCES

1. **Hyppänen, T., Sjöström, E., and Vuorinen, T.,** Gas-liquid chromatographic determination of hydroxy carboxylic acids on a fused-silica capillary column, *J. Chromatogr.*, 261, 320, 1983.
2. **Alén, R., Niemelä, K., and Sjöström, E.,** Gas-liquid chromatographic separation of hydroxyl monocarboxylic acids and dicarboxylic acids on a fused-silica capillary column, *J. Chromatogr.*, 301, 273, 1984.
3. **Niemelä, K. and Sjöström, E.,** Non-oxidative and oxidative alkaline degradation of pectic acid, *Carbohydr. Res.*, 144, 87, 1985.
4. **Niemelä, K. and Sjöström, E.,** Non-oxidative and oxidative degradation of D-galacturonic acid with alkali, *Carbohydr Res.*, 144, 93, 1985.
5. **Niemelä, K.,** Oxidative and non-oxidative alkali-catalysed degradation of L-ascorbic acid, *J. Chromatogr.*, 399, 235, 1987.

TABLE GC 19
Trimethylsilylated Diethyl Dithioacetals: GLC of Neutral, Acidic, and Amino Sugars, *O*-Methyl Ethers, and Periodate-Oxidation Products

Packing	P1	P2	P3
Temperature (°C)	225	190	T1
Gas; flow rate (ml/min)	N_2 1-2[a]	N_2; 50	N_2; 50
Column			
Length, cm	5000	100	200
Diameter (I.D.), cm	0.028	0.3	0.3
Form	SCOT	Packed	Packed
Material	Glass	Glass	Glass
Detector	FI	FI	FI
Reference	1-5	4	5
Parent compound	r[b]	r[b]	r[c]
Glycolaldehyde[d]	—	—	0.65
D-Glyceraldehyde[d]	0.24	—	1.00
Glyoxal[d]	—	—	1.24
L-Lactaldehyde[d]	—	—	0.69
2-Hydroxymalonaldehyde[d]	—	—	1.48
D-Threose[d]	0.37	—	—
D-Erythrose	0.38	—	—
D-Ribose	0.67	—	—
D-Xylose	0.67	—	—
2,3-di-*O*-methyl	0.49	—	—
3,4-di-*O*-methyl	0.56	—	—
2,3,4-tri-*O*-methyl	0.43	—	—
L-Arabinose	0.69	—	—
L-Rhamnose	0.85	—	—
2,3,4-tri-*O*-methyl	0.47	—	—

TABLE GC 19 (continued)
Trimethylsilylated Diethyl Dithioacetals: GLC of Neutral, Acidic, and Amino Sugars, *O*-Methyl Ethers, and Periodate-Oxidation Products

Parent compound	r[b]	r[b]	r[c]
L-Fucose	0.90	—	—
2,3,4-tri-*O*-methyl	0.49	—	—
D-Glucose	1.36	—	—
3-*O*-methyl	1.00	—	—
2,3-di-*O*-methyl	0.87	—	—
4,6-di-*O*-methyl	0.80	—	—
2,3,4-tri-*O*-methyl	0.77	—	—
2,3,6-tri-*O*-methyl	0.70	—	—
2,4,6-tri-*O*-methyl	0.70	—	—
3,4,6-tri-*O*-methyl	0.68	—	—
2,3,4,6-Tetra-*O*-methyl	0.60	—	—
D-Mannose	1.39	—	—
2,3-di-*O*-methyl	0.88	—	—
3,4-di-*O*-methyl	0.93	—	—
2,3,4-tri-*O*-methyl	0.79	—	—
2,3,6-tri-*O*-methyl	0.69	—	—
2,4,6-tri-*O*-methyl	0.70	—	—
3,4,6-tri-*O*-methyl	0.67	—	—
2,3,4,6-tetra-*O*-methyl	0.62	—	—
D-Galactose	1.48	—	—
2,4-di-*O*-methyl	0.89	—	—
2,3,4-tri-*O*-methyl	0.75	—	—
2,3,6-tri-*O*-methyl	0.67	—	—
2,4,6-tri-*O*-methyl	0.69	—	—
2,3,4,6-tetra-*O*-methyl	0.59	—	—
D-Glucuronic acid	1.04	—	—
D-Galacturonic acid	1.16	—	—
2,5-Anhydro-D-mannose[e]	0.66	—	—
2,5-Anhydro-D-talose[e]	0.75	—	—
2-Acetamido-2-deoxy-D-glucose	1.69	—	—
2-Acetamido-2-deoxy-D-galactose	1.78	—	—
2-Acetamido-2-deoxy-D-mannose[f]	1.87	2.16	—
6-*O*-acetyl[f]	—	2.55	—
2-Deoxy-2-glycolylamino-D-mannose[f]	—	4.60	—

[a] Split ratio 1:100.
[b] t_r relative to trimethylsilylated diethyl dithioacetal of 3-*O*-methyl-D-glucose (42.5 min,[2] 36.4 min,[3,4] 10.6 min on P2[4]).
[c] t_r relative to trimethylsilylated diethyl dithioacetal of D-glyceraldehyde (17.7 min).
[d] Products of periodate oxidation of glycosides.
[e] Products of nitrous acid deamination of 2-amino-2-deoxy-D-glucose and -galactose.
[f] Products of cleavage by neuraminic acid lyase of *N*-acetyl-, *N,O*-diacetyl-, and *N*-glycolylneuraminic acid.

Packing	P1 = SF-96.
	P2 = 2% OV-1 on Chromosorb W AW DMCS (80 — 100 mesh).
	P3 = 3% OV-1 on Chromosorb W AW DMCS (80 — 100 mesh).
Temperature	T1 = 100 → 250°C at 5°C/min.

REFERENCES

1. **Honda, S., Yamauchi, N., and Kakehi, K.,** Rapid gas chromatographic analysis of aldoses as their diethyl dithioacetal trimethylsilylates, *J. Chromatogr.*, 169, 287, 1979.

2. **Honda, S., Kakehi, K., and Okada, K.,** Convenient method for the gas chromatographic analysis of hexosamines in the presence of neutral monosaccharides and uronic acids, *J. Chromatogr.*, 176, 367, 1979.
3. **Honda S., Nagata, M., and Kakehi, K.,** Rapid gas chromatographic analysis of partially methylated aldoses as trimethylsilylated diethyl dithioacetals, *J. Chromatogr.*, 209, 299, 1981.
4. **Kakehi, K., Maeda, K., Tetramae, M., Honda, S., and Takai, T.,** Analysis of sialic acids by gas chromatography of the mannosamine derivatives releasd by the action of *N*-acetylneuraminate lyase, *J. Chromatogr.*, 272, 1, 1983.
5. **Honda, S., Takeda, K., and Kakehi, K.,** Studies of the structures of the carbohydrate components in plant oligosaccharide glycosides by the dithioacetal method, *Carbohydr. Res.*, 73, 135, 1979.

TABLE GC 20
Trimethylsilylated Methyl Glycosides: GLC of Methanolysates

Packing	P1	P2	P3	P4	P5
Temperature	T1	T2	T2	T3	T4
Gas; flow rate (ml/min)	N_2; 25	n.a.	n.a.	He; n.a.	N_2; 1[a]
Column					
Length, cm	183	200	2500	2500	3000
Diameter (I.D.), cm	0.4	0.2	0.025	0.032	0.025
Form	Packed	Packed	Capillary	WCOT	Bonded phase
Material	Glass	Glass	Glass	Fused silica	Fused silica
Detector	FI	MS	FI	FI	FI
Reference	1	2	2	3	4
Parent sugar	r[b]	r[b]	r[b]	r[b]	r[b]
L-Arabinose	0.62;[c] 0.63; 0.65	—		0.52;[c] 0.53; 0.58	0.46;[c] 0.48
D-Ribose	0.65	—		0.54; 0.57[c]	—
D-Xylose	0.73;[c] 0.75	—		0.66;[c] 0.69	0.61;[c] 0.64
L-Rhamnose	0.66	—		0.57;[c] 0.58	0.50;[c] 0.52
L-Fucose	0.68;[c] 0.70	0.61;[c] 0.64	0.69;[c] 0.72	0.60;[c] 0.63	0.54;[c] 0.57
D-Mannose	0.90;[c] 0.92	0.87	0.90	0.86;[c] 0.91	0.82;[c] 0.87
D-Galactose	0.92; 0.95;[c] 0.98	0.93;[c] 0.96	0.94;[c] 0.97	0.88; 0.92;[c] 0.97	0.83; 0.89;[c] 0.95
D-Glucose	1.00;[c] 1.03	1.00;[c] 1.04	1.00;[c] 1.03	1.00;[c] 1.03	1.00;[c] 1.03
D-Glucuronic acid	—	—	—	1.01; 1.02;[c]	0.78; 1.03; 1.06;[c]
D-Galacturonic acid	—	—	—	0.79;[c] 0.95	0.77;[c] 0.97; 0.98
2-Acetamido-2-deoxy-D-glucose	1.11; 1.14; 1.20[c]	—	—	1.14; 1.20; 1.26[c]	—
2-Acetamido-2-deoxy-D-galactose	1.14; 1.18[c]	—	—	1.13; 1.15; 1.22[c]	—
N-Acetylmuramic acid[d]	1.35	—	—	—	—
N-Acetylneuraminic acid[d]	—	—	—	1.66	—
3-Deoxy-D-*glycero*-D-*galacto*-nonulosonic acid[d,e]	—	1.53	1.42	—	—

Note: This table supplements Table GC 19 in *CRC Handbook of Chromatography: Carbohydrates*, Volume I.

[a] Split ratio 1:5.
[b] t_r relative to major peak for trimethylsilylated methyl glucoside (18.1 min on P3,[2] 9.27 min,[3] 23.41 min[4]).
[c] Major peak.
[d] Present in methanolysate as methyl ester methyl glycosides.
[e] Deaminated *N*-acetylneuraminic acid.

Packing
P1 = 3% SE-30 on Chromosorb® W HP (100 — 120 mesh).
P2 = 2.2% SE-30 on Gas-Chrom® Q (100 — 120 mesh).
P3 = OV-101.

TABLE GC 20 (continued)
Trimethylsilylated Methyl Glycosides: GLC of Methanolysates

	P4 = CP-Sil 5.
	P5 = DB-5 (J and W Scientific).
Temperature	T1 = 80 → 250°C at 2°C/min.
	T2 = 120 → 250°C at 4°C/min.
	T3 = 140°C for 2 min; 140 → 260°C at 8°C/min.
	T4 = 150 → 220°C at 2°C/min.

REFERENCES

1. **Aluyi, H. A. S. and Drucker, D. B.,** Fingerprinting of carbohydrates of *Streptococcus mutans* by combined gas-liquid chromatography-mass spectrometry, *J. Chromatogr.*, 178, 209, 1979.
2. **Mononen, I.,** Quantitative analysis, by gas-liquid chromatography and mass fragmentography, of monosaccharides after methanolysis and deamination, *Carbohydr. Res.*, 88, 39, 1981.
3. **Chaplin, M. F.,** A rapid and sensitive method for the analysis of carbohydrate components in glycoproteins using gas-liquid chromatography, *Anal. Biochem.*, 123, 336, 1982.
4. **Ha, Y. W. and Thomas, R. L.,** Simultaneous determination of neutral sugars and uronic acids in hydrocolloids, *J. Food Sci.*, 53, 574, 1988.

TABLE GC 21
Trimethylsilylated Mono- and Oligosaccharides

Packing	P1	P2	P3	P4	P5	P5	P6	P7
Temperature (°C)	T1	255	220	T2	240	180	170	T3
Gas; flow rate (ml/min)	N_2; 25	Ar; 55	Ar; 55	H_2; na	He; 0.8[a]	N_2; n.a.	N_2; n.a.	N_2; 0.4
Column								
Length, cm	183	270	150	1000	3000	4000	2500	1000
Diameter (I.D.), cm	0.4	0.6	0.6	0.032	0.026	0.018	0.018	0.022
Form	Packed	Packed	Packed	Capillary	Capillary	Capillary	Capillary	WCOT
Material	Glass	Glass	Glass	Glass	Fused silica	Glass	Glass	Fused silica
Detector	FI	FI	FI	FI	FI	FI	FI	FI
Reference	1	2	2	3	4	5,6	6	7
Parent sugar	**r**[b]	**r**[c]	**r**[c]	**r**[c]	**r**[c]	**r**[d]	**r**[q]	**r**[b]
2-Deoxy-D-*erythro*-pentose	0.42; 0.48[f]	—	—	—	—	—	—	—
D-Lyxose	—	—	—	—	—	0.90;[f,g] 0.91;[j] 0.92;[i] 0.93[h]	—	—
D-Arabinose	0.63; 0.72[f]	—	—	0.35; 0.37[f]	—	0.90;[i] 0.91;[f,g] 0.92;[f,h] 0.94[j]	—	0.34(β); 0.37(α)[f]
D-Ribose	0.66;[f] 0.68	—	—	—	—	0.91;[i] 0.92;[j] 0.94;[g] 0.95[f,h]	—	—
D-Xylose	0.71; 0.77[f]	—	—	0.42; 0.48[f]	—	0.91;[i] 0.92;[j] 0.99;[f,g] 1.00[f,h]	—	0.45(α); 0.54(β)[f]
L-Rhamnose	0.64;[f] 0.69	—	—	—	—	—	—	0.35(α);[f] 0.44(β)
L-Fucose	0.65; 0.68; 0.71[f]	—	—	—	—	—	—	—
D-Altrose	—	—	—	—	—	—	0.85;[f,h] 0.86;[g] 0.88;[i] 0.93[j]	—
D-Mannose	0.83;[f] 0.92	—	—	—	—	—	0.87;[f,g] 0.94;[i] 0.95;[h] 0.98;[j]	0.64(α); 0.82(β)
D-Allose	—	—	—	—	—	—	0.88;[g] 0.89;[i] 0.90;[f,h] 0.93[j]	—

D-Gulose	—	—	—	—	—	—	0.88;[f,h] 0.89;[g] 0.92;[j] 0.93[i]	—
D-Idose	—	—	—	—	—	—	0.88;[f,g] 0.89;[j] 0.91;[i] 0.93[h]	—
D-Talose	—	—	—	—	—	—	0.90;[j] 0.91;[f,g] 0.92;[i] 0.96[h]	—
D-Galactose	0.88; 0.90; 0.95[f]	—	—	0.59; 0.63[f]	—	—	0.89[j]; 0.92[g]; 0.95[i]; 0.96[f,h]	0.73(α); 0.82(β)
D-Glucose	0.92; 1.00	—	—	0.61; 0.72[f]	—	—	0.90;[i] 0.91;[j] 0.93;[f,g] 1.00[f,h]	0.78(α); 1.00(β)[f]
D-Fructose	0.95	—	—	—	—	0.98;[i] 0.99;[j] 1.00;[h] 1.03;[g] 1.06;[k]	—	—
L-Sorbose	0.97	—	—	—	—	0.97;[i] 0.98;[j] 1.01;[g] 1.02;[h] 1.05[k]	—	—
D-Tagatose	—	—	—	—	—	0.98;[i] 0.99;[j] 1.00[h] 1.02;[g] 1.03[k]	—	—
D-Psicose	—	—	—	—	—	0.98;[h] 0.99;[i] 1.00;[g] 1.01;[j] 1.03;[k]	—	—
2-Acetamido-2-deoxy-D-galactose	1.04; 1.06[f]	—	—	—	—	—	—	—
Xylobiose	—	—	—	—	0.58(α); 0.93(β)[f]	—	—	—
Lactulose	—	0.90	0.96	—	0.93; 0.96[f]	—	—	—
Lactose	—	1.06; 1.39	0.96; 1.49	1.00; 1.05	1.00(α); 1.44(β)	—	—	—
Sucrose	—	1.00	1.00	1.00	1.00	—	—	—
Maltose	—	1.16; 1.51	1.13; 1.30	1.02; 1.04[f]	1.12(α); 1.27(β)[f]	—	—	—
Cellobiose	—	1.16; 1.51	1.16; 1.67	—	1.14(α); 1.58(β)[f]	—	—	—
Maltulose	—	—	—	—	1.19; 1.21[f]	—	—	—
Turanose	—	1.23	1.26	—	1.24	—	—	—
Nigerose	—	—	—	—	1.24;[f] 1.34	—	—	—
αα-Trehalose	—	1.36	1.32	—	1.33	—	—	—

TABLE GC 21 (continued)
Trimethylsilylated Mono- and Oligosaccharides

Parent sugar	r[b]	r[c]	r[c]	r[c]	r[c]	r[d]	r[d]	r[b]
Kojibiose	—	—	—	—	1.33(β);[f] 1.64(α)	—	—	—
Palatinose	—	1.38	1.35; 1.43	—	1.38;[f] 1.63	—	—	—
Laminaribiose	—	—	—	—	1.51; 1.67[f]	—	—	—
Sophorose	—	—	—	—	1.52(β); 1.76(α)[f]	—	—	—
Melibiose	—	2.01	2.00; 2.20	1.09; 1.11[f]	1.90(α); 2.05(β)[f]	—	—	—
Isomaltose	—	—	—	—	1.98(α); 2.35(β)[f]	—	—	—
Gentiobiose	—	2.21	2.38	—	2.24((α); 2.25(β)[f]	—	—	—
Raffinose	—	—	—	1.31	—	—	—	—
Maltotriose	—	—	—	1.32; 1.34[f]	—	—	—	—
Maltotetraose	—	—	—	1.53	—	—	—	—
Stachyose	—	—	—	1.59	—	—	—	—

Note: This table supplements Tables GC 16 and 17 in *CRC Handbook of Chromatography: Carbohydrates*, Volume I.

[a] Split ratio 1:100.
[b] t_r relative to trimethylsilylated β-D-glucopyranose (18.8 min[7]).
[c] t_r relative to trimethylsilylated sucrose (17 min on P2,[2] 12.6 min on P3,[2] 20 min,[3] 26 min[4]).
[d] Calculated from ratio of retention index to that for trimethylsilylated β-D-xylopyranose.
[e] Calculated from ratio of retention index to that for trimethylsilylated β-D-glucopyranose.
[f] Major peak.
[g] α, pyranose.
[h] β, pyranose.
[i] α, furanose.
[j] β, furanose.
[k] Open chain.

Packing	P1 = 3% SE-30 on Chromosorb® W HP (100 — 120 mesh).
	P2 = 10% OV-17 on Gas-Chrom® Q (80 — 100 mesh).
	P3 = 3% OV-1 on Gas-Chrom® Q (80 — 100 mesh).
	P4 = OV-1 (0.1-to 0.12-μm film).
	P5 = SE-54.
	P6 = OV-225.
	P7 = CP-Sil 5 CB (Chrompack, Middelburg, The Netherlands), 0.12-μm film.
Temperature	T1 = 80 → 250°C at 2°C/min.
	T2 = on-column injection temperature 80°C; held for 1 min; 80 → 140°C at 5°C/min; held for 1 min; 140 → 190°C at 20°C/min; held for 1 min; 190 → 340°C at 8°C/min.
	T3 = 140 → 182°C at 2°C/min; 182 → 290°C at 3.5°C/min.

REFERENCES

1. **Aluyi, H. A. S. and Drucker, D. B.,** Fingerprinting of carbohydrates of *Streptococcus mutans* by combined gas-liquid chromatography-mass spectrometry, *J. Chromatogr.*, 178, 209, 1979.
2. **Laker, M. F.,** Estimation of disaccharides in plasma and urine by gas-liquid chromatography, *J. Carbohydr.*, 163, 9, 1979.
3. **Traitler, H., Del Vedovo, S., and Schweizer, T. F.,** Gas chromatographic separation of sugars by on-column injection on glass capillary columns, *J. High Resolut. Chromatogr. Chromatogr. Commun.*, 7, 558, 1984.
4. **Nikolov, Z. L. and Reilly, P. J.,** Isothermal capillary column gas chromatography of trimethylsilyl disaccharides, *J. Chromatogr.*, 254, 157, 1983.
5. **García-Raso, A., Martínez-Castro, I., Páez, M. I., Sanz, J., García-Raso, J., and Saura-Calixto, F.,** Gas chromatographic behaviour of carbohydrate trimethylsilyl ethers. I. Aldopentoses, *J. Chromatogr.*, 398, 9, 1987.
6. **Páez, M., Martínez-Castro, I. Sanz, J., Olano, A., García-Raso, A., and Saura-Calixto, F.,** Identification of the components of aldoses in a tautomeric equilibrium mixture as their trimethylsilyl ethers by capillary gas chromatography, *Chromatographia*, 23, 43, 1987.
7. **Ford, C. W. and Hartley, R. D.,** Identification of phenols, phenolic acid dimers, and monosaccharides by gas-liquid chromatography on a capillary column, *J. Chromatogr.*, 436, 484, 1988.

TABLE GC 22
Trimethylsilylated Oxime and *O*-Methyloxime Derivatives of Mono- and Oligosaccharides

Packing	P1	P2	P3	P3	P3	P4	P5	P5	P6	P7	P8	P8
Temperature (°C)	T1	T2	T2	T3	T4	T5	180	T6	T6	165	175	T7
Gas; flow rate (ml/min)	He; 30	N_2; 24	N_2; 24	N_2; 18	N_2; 60	He; 2	He; 1	He; 0.8[a]	He; 5	Ar; 0.7	Ar; 0.7	H_2; 0.35[a]
Column												
Length, cm	183	183	183	183	50	3000	5000	6000	4000	3500	5000	2500
Diameter (I.D.), cm	0.32	0.32	0.32	0.32	0.3	0.025	0.02	0.025	0.05	0.03	0.03	0.023
Form	Packed	Packed	Packed	Packed	Packed	Capillary	Capillary	Capillary	Capillary	Capillary	Capillary	Capillary
Material	SS	Ni	Ni	Glass	SS	Glass	Fused silica	Glass	Glass	Glass	Glass	Fused silica
Detector	FI,NPD[b]	FI	FI	FI	FI	MS	MS	FI	FI	FI	FI	FI
Reference	1	2	2	3	4	5	5	6	6	7	7	8
	O[c]	O/MO[c]	O/MO	MO	O	MO	MO	O	O	MO	MO	O
Parent compound	r[d]	r[e]	r[e]	r[f]	r[d]	r[g]	r[g]	r[h]	r[h]	r[g]	r[g]	r[d]
DL-Glyceraldehyde	—	—	—	—	—	—	—	0.76; 0.80[i]	0.75; 0.80[i]	—	—	—
Dihydroxyacetone	—	—	—	—	—	—	—	0.78	0.78	—	—	—
2-C-(hydroxymethyl)-DL-Glyceraldehyde	—	—	—	—	—	—	—	1.01; 1.04[i]	1.01; 1.04[i]	—	—	—
D-*glycero*-Tetrulose	—	—	—	—	—	—	—	1.02; 1.05[i]	1.02; 1.05[i]	—	—	—
D-Erythrose	—	—	—	—	—	—	—	1.05; 1.06[i]	1.05; 1.07[i]	—	—	—
D-Threose	—	—	—	—	—	—	—	1.05; 1.07[i]	1.05; 1.08[i]	—	—	—
2-Deoxy-D-*erythro*-pentose	—	—	—	0.37	—	—	—	—	—	0.34	0.37	—
D-Lyxose	—	—	—	—	—	0.48; 0.50[i]	0.47	—	—	—	—	—
D-Xylose	0.81	—	—	—	—	0.49; 0.51[i]	0.45; 0.46[i]	—	—	—	—	—
D-Arabinose	0.82	—	—	—	0.68	0.51	0.46	—	—	0.43	0.46	—
D-Ribose	0.84	—	—	0.48	—	0.52	0.48	—	—	—	—	—
D-*erythro*-2-Pentulose	—	—	—	—	—	0.53	0.48	—	—	—	—	—
D-*threo*-2-Pentulose	—	—	—	—	—	0.54	0.48	—	—	—	—	—
L-Rhamnose	0.86	—	—	—	0.75	0.56	0.55;[i] 0.59	—	—	0.52;[i] 0.53	0.55;[i] 0.57	—
L-Fucose	—	—	—	0.52	—	0.58;[i] 0.60	0.57;[i] 0.60	—	—	—	—	—
2-Deoxy-D-*arabino*-hexose	—	—	—	0.67	—	0.77; 0.78[i]	0.69	—	—	0.65	0.68	—

D-Allose	—	—	—	—	—	0.91;^{i} 0.95	0.95;^{i} 1.02	—	—	—	—	—
D-Mannose	—	—	—	—	—	0.94;^{i} 0.95	0.97;^{i} 1.02	—	—	—	—	—
D-Talose	—	—	—	—	—	0.97;^{i} 0.99	0.98;^{i} 1.04	—	—	—	—	—
D-Galactose	0.99	—	—	—	—	0.97;^{i} 1.02	0.99;^{i} 1.06	—	—	0.97;^{i} 1.05	0.98;^{i} 1.05	—
D-Glucose	1.00	—	—	0.78	1.00	1.00;^{i} 1.04	1.00;^{i} 1.08	—	—	1.00;^{i} 1.06	1.00;^{i} 1.06	1.00;^{i} 1.06
D-Fructose	—	—	—	0.74	0.90	0.91;^{i} 0.96	0.93;^{i} 0.97	—	—	0.92;^{i} 0.95	0.92;^{i} 0.95	0.91; 0.93
D-Tagatose	—	—	—	—	—	0.86; 0.94^i	0.84; 0.90^i	—	—	—	—	—
L-Sorbose	—	—	—	—	—	0.94; 0.96	0.90; 0.94	—	—	—	—	—
D-*altro*-Heptulose	—	—	—	—	—	1.17	—	—	—	—	—	—
D-*manno*-Heptulose	—	—	—	—	—	1.32	2.15; 2.19^i	—	—	—	—	—
D-*gluco*-Heptulose	—	—	—	—	—	1.34^i 1.39	2.30	—	—	—	—	—
D-Gluconic acid	—	—	—	—	—	1.11	0.88	—	—	—	—	—
D-Glucuronic acid	—	—	—	0.87	—	1.13; 1.20;^{i} 1.23; 1.25	0.84; 0.94; 1.00; 1.11^i	—	—	—	—	—
2-Acetamido-2-deoxy-D-glucose	—	—	—	1.20^j	—	1.45; 1.48^i	1.70; 1.75^i	—	—	—	—	—
2-Acetamido-2-deoxy-D-galactose	—	—	—	1.24^j	—	1.55	1.89	—	—	—	—	—
2-Acetamido-2-deoxy-D-mannose	—	—	—	1.25^j	—	1.50;^{i} 1.52	1.77;^{i} 1.85	—	—	—	—	—
N-Acetylneuraminic acid, methyl ester	—	1.49 (MO); 1.57(O)	1.36 (MO=O)	—	—	—	—	—	—	—	—	—
ethyl ester	—	1.50 (MO); 1.58(O)	1.43(MO); 1.45(O)	—	—	—	—	—	—	—	—	—
N-Glycolylneuraminic acid, methyl ester	—	1.90 (MO); 2.06(O)	1.72(MO); 1.74(O)	—	—	—	—	—	—	—	—	—
ethyl ester	—	1.90 (MO); 2.04(O)	1.80(MO); 1.88(O)	—	—	—	—	—	—	—	—	—
Maltulose	—	—	—	—	—	—	—	—	—	—	—	3.11; 3.14
Nigerose	—	—	—	—	—	—	—	—	—	—	—	3.18;^{i} 3.26
Turanose	—	—	—	—	—	—	—	—	—	—	—	3.19; 3.21
Maltose	—	—	—	—	—	—	—	—	—	—	—	3.20;^{i} 3.24
Cellobiose	—	—	—	—	1.70	—	—	—	—	—	—	—

TABLE GC 22 (continued)
Trimethylsilylated Oxime and *O*-Methyloxime Derivatives of Mono- and Oligosaccharides

Parent compound	O[c]	O/MO[c]	O/MO	MO	O	MO	MO	O	O	MO	MO	O
	r[d]	r[e]	r[e]	r[f]	r[d]	r[g]	r[g]	r[h]	r[h]	r[g]	r[g]	r[d]
Kojibiose	—	—	—	—	—	—	—	—	—	—	—	3.22;[i] 3.24
Palatinose	—	—	—	—	—	—	—	—	—	—	—	3.32; 3.36
Galactobiose[k]	—	—	—	—	1.73	—	—	—	—	—	—	—
Gentiobiose	—	—	—	—	—	—	—	—	—	—	—	3.10; 3.34;[i] 3.42
Melibiose	—	—	—	—	1.81	—	—	—	—	—	—	3.32; 3.44;[i] 3.54
Isomaltose	—	—	—	—	—	—	—	—	—	—	—	3.24; 3.45;[i] 3.54
Maltotriose	—	—	—	—	—	—	—	—	—	—	—	5.35; 6.00;[i] 6.17
Cellotriose	—	—	—	—	2.12	—	—	—	—	—	—	—
Galactotriose[k]	—	—	—	—	2.21	—	—	—	—	—	—	—
Manninotriose[k]	—	—	—	—	2.28	—	—	—	—	—	—	—
Verbascotetraose[k]	—	—	—	—	2.59	—	—	—	—	—	—	—

Note: This table supplements Table GC 20 in *CRC Handbook of Chromatography: Carbohydrates,* Volume I.

[a] Split ratios 1:60,[6] 1:75.[8]
[b] Nitrogen-phosphorus selective detector.
[c] O = oxime, MO = O-methyloxime.
[d] t_r relative to major peak for trimethylsilylated glucose oxime (17.5 min,[1] 6.7 min,[4] 9.6 min[8]).
[e] t_r relative to trimethylsilylated *N*-acetylneuraminic acid methyl ester (5.41 min on P2, 7.00 min on P3).
[f] t_r relative to trimethylsilylated *myo*-inositol (13.5 min).
[g] t_r relative to major peak for trimethylsilylated O-methyloxime of glucose (29 min on P4, 23.7 min on P5,[5], 37.6 min on P8[7]).
[h] t_r relative to trimethylsilylated erythritol (27.3 min on P5, 25.4 min on P6).
[i] Major peak.
[j] For simultaneous analysis of trimethylsilylated acetamidodeoxyalditols see Table GC 16, Reference 5.
[k] Galactobiose and -triose are α-(1 → 6)-linked, manninotriose is Galα1-6Galα1-6Glc, verbascotetraose is Galα-6Galα1-6Galα1-6Glc; related non-reducing oligosaccharides of raffinose family (sucrose, raffinose, stachyose, and verbascose) can be analyzed simultaneously as trimethylsilyl ethers (t_r = 1.61, 2.06, 2.49, 2.78, respectively.)

Packing	P1 = 2% OV-17 on Chromosorb® W HP (800 — 100 mesh).
	P2 = 1.5% SE-52 on Chromosorb® W HP (100 — 120 mesh).
	P3 = 3% SP-2250 on Supelcoport® (100 — 120 mesh,[2,3] 80 — 100 mesh[4]).
	P4 = SP-2250.
	P5 = SP-2100.
	P6 = OV-1.
	P7 = SE-30 containing Silanox 101.
	P8 = OV-101
Temperature	T1 = 120°C for 8 min; 120 → 300°C at 8°C/min.
	T2 = 200°C for 2 min; 200 → 260°C at 6°C/min.
	T3 = 120 → 220°C at 5°C/min.
	T4 = 80 → 380°C at 16°C/min.
	T5 = 75°C for 30 sec; 75 → 150°C at 10°C/min; held at 150°C for 23 min; 150 → 180°C at 20°C/min; held at 180°C.
	T6 = 80 → 275°C at 4°C/min.
	T7 = 180 → 280°C at 3°C/min; held at 280°C for 4 min; 280 → 290°C at 2°C/min; held at 290°C.

REFERENCES

1. **Morita, H. and Montgomery, W. G.,** Gas chromatography of silylated oxime derivatives of peat monosaccharides, *J. Chromatogr.*, 155, 195, 1978.
2. **Mawhinney, T. P., Madson, M. A., Rice, R. H., Feather, M. S., and Barbero, G. J.,** Gas-liquid chromatography and mass-spectral analysis of per-*O*-trimethylsilyl acyclic ketoxime derivatives of neuraminic acid, *Carbohydr. Res.*, 104, 169, 1982.
3. **Mawhinney, T. P.,** Simultaneous determination of *N*-acetylglucosamine, *N*-acetylgalactosamine, *N*-acetylglucosaminitol and *N*-acetylgalactosaminitol by gas-liquid chromatography, *J. Chromatogr.*, 351, 91, 1986.
4. **Molnár-Perl, I., Pintér-Szakács, M., Kövágó, A., and Petróczy, J.,** Gas-liquid chroomatographic determination of the raffinose family of oligosaccharides and their metabolites present in soy beans, *J. Chromatogr.*, 295, 433, 1984.
5. **Pelletier, O. and Cadieux, S.,** Glass capillary or fused-silica gas chromatography-mass spectrometry of several monosaccharides and related sugars: improved resolution, *J. Chromatogr.*, 231, 225, 1982.
6. **Willis, D. E.,** GC analysis of C_2-C_7 carbohydrates as the trimethylsilyl-oxime derivatives on packed and capillary columns, *J. Chromatogr. Sci.*, 21, 132, 1983.
7. **Zegota, H.,** Separation and quantitative determination of fructose as the *O*-methyloxime by gas-liquid chromatography using glass capillary columns, *J. Chromatogr.*, 192, 446, 1980.
8. **Mateo, R., Bosch, F., Pastor, A., and Jimenez, M.,** Capillary column gas chromatographic identification of sugars in honey as trimethylsilyl derivatives, *J. Chromatogr.*, 410, 319, 1987.

TABLE GC 23
Resolution of Enantiomers: Capillary GLC of Chiral Derivatives

Phase	P1	P1	P1	P2	P2	P2	P3	P4	P4	P5
Temperature (°C)	T1	280	280	280	280	250	T2	180	180	200
Gas; flow rate (ml/min)	N_2; 1	H_2; 1.5[a]	H_2; 1.5[a]	H_2; 1.5[a]	H_2; 1.5[a]	H_2; 0.7[a]	He; 0.7[a]	N_2; 2[a]	N_2; 2.5[a]	He; 0.8[a]
Column										
Length, m	25	30	30	30	30	50	25	50	50	50
Diameter (I.D.), mm	0.31	0.25	0.25	0.25	0.25	0.20	0.20	n.a.	n.a.	0.3
Form	Capillary	Capillary	Capillary	Capillary	Capillary	WCOT	WCOT	WCOT	WCOT	SCOT
Material	Glass	Fused silica	Fused silica	Fused silica	Fused silica	Fused silica	Fused silica	Glass	Glass	Glass
Detector	FI	FI	FI	FI	FI	FI	FI	FI	FI	FI
Reference	1	2	2	2	2	3	3	4	5	6
Parent sugar	**TMS(−)-2-BG[b]** r[j]	**(+)-PED,[c] Ac[d]** r[k]	**(+)-PED, TMS[e]** r[k]	**(+)-PED, Ac** r[k]	**(+)-PED, TMS** r[k]	**(−)-MBAA, Ac[f]** r[k]	**(−)-MBAA, TMS** r[k]	**TFA(−)-MenO[g]** r[l]	**TFA(−)-BorO[h]** r[l]	**TMS-MPTC[i]** r[k]
D-Glyceraldehyde	—	—	—	—	—	0.18	0.26	0.30; 0.38	0.30; 0.38[m]	—
L-Glyceraldehyde	—	—	—	—	—	0.19	0.26	0.29; 0.37	0.29; 0.38[m]	—
D-Erythrose	—	—	—	—	—	0.32	0.36	—	—	—
L-Erythrose	—	—	—	—	—	0.31	0.36	—	—	—
D-Arabinose	—	0.58	0.47	0.56	0.46	0.55	0.52	0.45; 0.73	0.47; 0.74[m]	0.67
L-Arabinose	—	0.62	0.50	0.60	0.49	0.54	0.53	0.47; 0.74	0.49; 0.75[m]	0.61
D-Xylose	—	—	—	—	—	—	0.53	0.55; 0.78	0.60; 0.81[m]	0.62
L-Xylose	—	—	—	—	—	—	0.54	0.55; 0.77	0.59; 0.79[m]	0.67
D-Lyxose	—	0.59	0.48	0.57	0.47	—	0.54	0.52; 0.77	0.58; 0.78	0.65
L-Lyxose	—	0.62	0.50	0.61	0.50	—	0.55	0.54; 0.78	0.60; 0.77[m]	0.63
D-Ribose	—	—	—	—	—	0.53	0.55	0.46; 0.62	0.49; 0.62[m]	0.63
L-Ribose	—	—	—	—	—	0.53	0.56	0.43; 0.62	0.46; 0.63	0.70
D-Rhamnose	—	0.57	0.63	0.54	0.62	0.54	0.57	—	—	0.71
L-Rhamnose	—	0.60	0.66	0.57	0.65	0.51	0.58	—	—	0.68
D-Fucose	—	0.62	0.67	0.59	0.67	0.53	0.59	0.38; 0.58	0.39; 0.58	0.73
L-Fucose	—	0.59	0.63	0.57	0.63	0.53	0.61	0.36; 0.57	0.37; 0.58	0.81
6-Deoxy-D-glucose	—	—	—	—	—	—	0.66	—	—	—
L-glucose	—	—	—	—	—	—	0.65	—	—	—

2-Deoxy-D-*erythro*-pentose	—	—	—	—	—	—	0.69	—	—	—
L-*erythro*-pentose	—	—	—	—	—	—	0.69	—	—	—
2,6-Dideoxy-D-*ribo*-hexose	—	—	—	—	—	—	0.67	—	—	—
L-*ribo*-hexose	—	—	—	—	—	—	0.67	—	—	—
D-Fructose	—	—	—	—	—	—	0.78; 0.83	—	—	—
L-Fructose	—	—	—	—	—	—	0.75; 0.79	—	—	—
D-Sorbose	—	—	—	—	—	—	0.94; 0.97	—	—	—
L-Sorbose	—	—	—	—	—	—	0.92; 1.00	—	—	—
D-Tagatose	—	—	—	—	—	—	0.92; 0.94	—	—	—
L-Tagatose	—	—	—	—	—	—	0.90; 0.96	—	—	—
D-Mannose	—	0.95	0.99	0.94	0.98	1.00	0.95	0.62; 0.92	0.69;[m] 0.94	1.01
L-Mannose	—	1.00	0.99	1.00	0.98	0.95	0.93	0.65; 0.93	0.73; 0.91[m]	0.98
D-Glucose	—	1.00	1.00	1.00	1.00	1.00	1.00	0.63; 1.00[m]	0.71; 1.00[m]	1.00
L-Glucose	—	0.92	0.93	0.92	0.92	1.00	0.98	0.64; 1.00[m]	0.70; 0.96[m]	1.07
D-Galactose	—	1.04	0.99	1.03	0.98	0.98	0.96	0.69; 1.13	0.73; 1.16	1.09
L-Galactose	—	0.98	0.94	0.97	0.93	1.00	1.00	0.66; 1.11	0.70; 1.13	1.22
2-Deoxy-D-*lyxo*-hexose	—	—	—	—	—	—	1.02; 1.21	—	—	—
L-*lyxo*-hexose	—	—	—	—	—	—	1.04; 1.23	—	—	—
2-Deoxy-D-*arabino*-hexose	—	—	—	—	—	—	1.20	—	—	—
L-*arabino*-hexose	—	—	—	—	—	—	1.19	—	—	—
D-Allose	—	—	—	—	—	—	1.22	—	—	—
L-Allose	—	—	—	—	—	—	1.17	—	—	—
D-Glucuronic acid	1.08;[m,n] (1.78,[m] 1.83);[o,p] (1.55, 1.57)[o,q]	—	—	—	—	—	—	—	—	—
L-Glucuronic acid	1.11;[m,n] (1.71,[m] 1.83);[o,p] (1.54, 1.59)[o,q]	—	—	—	—	—	—	—	—	—
D-Galacturonic acid	(1.68, 1.74);[o,p] (1.46,[m] 1.63)[o,q]	—	—	—	—	—	—	—	—	—

TABLE GC 23 (continued)
Resolution of Enantiomers: Capillary GLC of Chiral Derivatives

Parent sugar	TMS(−)-2-BG[b] r[j]	(+)-PED,[c] Ac[d] r[k]	(+)-PED, TMS[e] r[k]	(+)-PED, Ac r[k]	(+)-PED, TMS r[k]	(−)-MBAA, Ac[f] r[k]	(−)-MBAA, TMS r[k]	TFA(−)-MenO[g] r[l]	TFA(−)-BorO[h] r[l]	TMS-MPTC[i] r[k]
L-Galacturonic acid	(1.68, 1.79);[o,p] (1.49,[m] 1.59)[o,q]	—	—	—	—	—	—	—	—	—
2-Acetamido-2-deoxy-D-mannose	(1.53,[m] 1.68);[p] (1.79, 1.89)[q]	—	—	—	—	—	0.89	—	—	—
L-mannose	(1.50,[m] 1.66);[p] (1.77, 1.89)[q]	—	—	—	—	—	0.95	—	—	—
2-Acetamido-2-deoxy-D-glucose	(1.90, 1.93[m])[p]	—	—	—	—	—	0.91	—	—	—
L-glucose	(1.89,[m] 1.90)[p]	—	—	—	—	—	0.96	—	—	—
2-Acetamido-2-deoxy-D-galactose	(1.81, 1.86[m])[p] (1.68, 1.76)[q]	—	—	—	—	—	0.98	—	—	—
L-galactose	(1.82,[m] 1.86);[p] (1.68, 1.72)[q]	—	—	—	—	—	1.02	—	—	—

Note: This table supplements Table GC 8 in *CRC Handbook of Chromatography: Carbohydrates,* Volume I.

[a] Split ratios 1:100,[2,3] 1:12,[4] 1:10,[5] 1:75.[6]
[b] Trimethylsilylated (−)-2-butyl glycoside; data for L-enantiomers from (+)-2-butyl D-glycosides [r = r for (−)-2-butyl L-glycosides].
[c] Bis [(+)-1-phenylethyl] dithioacetal.

[d] Acetylated.

[e] Trimethylsilyated.

[f] (L-(−)-α-methylbenzylamino)-deoxyalditol; data for enantiomers not available as such from comparison of r for products with L-methylbenzylamine/$NaBH_3CN$ with r for those from DL-MBA/$NaBH_3CN$.

[g] Trifluoroacetylated (−)-menthyloxime.

[h] Trifluoroacetylated (−)-bornyloxime.

[i] Trimethylsilylated methyl 2-(polyhydroxyalkyl)-thiazolidine-4(*R*)-carboxylate; derivative obtained by reaction of aldose with L-cysteine methyl ester.

[j] t_r relative to trimethylsilylated methyl α-D-galactopyranoside (31 min); cf. data for neutral sugars in Volume I, Table GC 8.

[k] t_r relative to D-glucose derivative (13 min for TMS derivative on P2,[2] 43.26 min for Ac on P2,[3] 16.36 min for TMS on P3,[3] 17.87 min[6]).

[l] t_r relative to major peak for D-glucose oxime (40.30 min,[4] 33.48 min[5].)

[m] Major peak.

[n] Lactone.

[o] Ester.

[p] Pyranose.

[q] Furanose.

Phase	P1 = SE-30.
	P2 = SE-54.
	P3 = Carbowax 20 M.
	P4 = OV-225.
	P5 = OV-17 on Silanox.
Temperature	T1 = 135 → 220°C at 1°C/min.
	T2 = 158°C for aldoses; 155°C for ketoses and 2-deoxyaldoses; 190°C for acetamidodeoxyhexoses.

REFERENCES

1. **Gerwig, G. J., Kamerling, J. P., and Vliegenthart, J. F. G.,** Determination of the absolute configuration of monosaccharides in complex carbohydrates by capillary g.l.c., *Carbohydr. Res.*, 77, 1, 1979.
2. **Little, M. R.,** Separation, by g.l.c., of enantiomeric sugars as diastereoisomeric dithioacetals, *Carbohydr. Res.*, 105, 1, 1982.
3. **Oshima, R., Kumanotani, J., and Watanabe, C.,** Gas-liquid chromatographic resolution of sugar enantiomers as diastereoisomeric methylbenzylaminoalditols, *J. Chromatogr.*, 259, 159, 1983.
4. **Schweer, H.,** Gas chromatographic separation of carbohydrate enantiomers as (−)-menthyloxime pertrifluoroacetates on silicone OV-225, *J. Chromatogr.*, 243, 149, 1982.
5. **Schweer, H.,** Gas chromatographic separation of enantiomeric sugars as diastereomeric trifluoroacetylated (−)-bornyloximes, *J. Chromatogr.*, 259, 164, 1983.
6. **Hara, S., Okabe, H., and Mihashi, K.,** Gas-liquid chromatographic separation of aldose enantiomers as trimethylsilyl ethers of methyl 2-(polyhydroxyalkyl)-thiazolidine-4(*R*)-carboxylates, *Chem. Pharm. Bull.*, 35, 501, 1987.

TABLE GC 24
Resolution of Enantiomers: Capillary GLC on Chiral Phases

Phase	P1	P2	P2	P2	P2	P2	P3	P3	P3	P3	P3	P3
Temperature (°C)	T1	T2	T3	100	120	140	80	90	100	110	115	120
Gas; Pressure (kPa)	He; 103	H_2; 70	H_2; 70	H_2; 70	H_2; 70	H_2; 70	H_2;[a]	H_2;[a]	H_2;[a]	H_2;[a]	H_2;[a]	H_2;[a]
Column												
Length, m	25	40	40	40	40	40	—[b]	—[b]	—[b]	—[b]	—[b]	—[b]
Diameter (I.D.), mm	0.3	0.2	0.2	0.2	0.2	0.2	0.2	0.2	0.2	0.2	0.2	0.2
Form	Capillary	Capillary	Capillary	Capillary	Capillary	Capillary	Capillary	Capillary	Capillary	Capillary	Capillary	Capillary
Material	Glass	Glass	Glass	Glass	Glass	Glass	Glass	Glass	Glass	Glass	Glass	Glass
Detector	FI	FI	FI	FI	FI	FI	FI	FI	FI	FI	FI	FI
Reference	1	2	3	3	3	3	4,5	4,5	4,5	4,5	4,5	4,5
Parent compound	**HFB[c]** **r[e]**	**TFA[d]** **r[f]**	**TFA** **r[f]**	**TFA** **S.F.[g]**	**TFA** **S.F.**	**TFA** **S.F.**	**TFA** **S.F.**	**TFA** **S.F.**	**TFA** **S.F.**	**TFA** **S.F.**	**TFA** **S.F.**	**TFA** **S.F.**
Erythrose	—	—	—	—	—	—	1.033(D)	—	—	—	—	—
Fucose(*p*)	0.79(D);[h] 0.83(L)[h]	—	—	1.035(D)[i]	—	—	—	—	—	—	—	—
Xylose (β-*p*)	—	—	—	1.030(D)	—	—	—	—	—	—	—	—
Arabinose (α-*p*)[j]	0.89(L);[h] 0.93(D)[h]	0.54(L); 0.56(D)	—	1.028(L)	—	—	—	—	—	—	—	—
(α-*f*)	—	0.65(L); 0.66(D)	—	1.017(L)	—	—	—	—	—	—	—	—
(β-*f*)	—	0.80(L); 0.83(D)	—	1.048(L)	—	—	—	—	—	—	—	—
(β-*p*)	—	0.97(D); 1.00(L)	—	1.026(D)	—	—	—	—	—	—	—	—
Ribose (α-*p*)	—	—	—	1.031(D)	—	—	—	—	—	—	1.064(D)	—
(β-*p*)	—	—	—	1.041(L)	—	—	—	—	—	—	1.150(L)	—
(β-*f*)	—	—	—	—	—	—	—	—	—	—	1.020(D)	—
Glucose (α, *p*)	—	—	—	1.036(L)	—	1.071(L)	—	—	1.119(L)	—	1.117(L)	—

(α-f)	—	—	—	—	—	1.044(L)	—	—	—	—	—	—
(β-f)	—	—	—	—	—	1.031(L)	—	—	—	—	—	—
(β-p)	—	—	—	1.060(L)	—	1.140(L)	—	—	1.114(L)	—	1.166(L)	—
Galactose (α-p)	—	—	—	—	—	1.019(D)	—	—	1.070(D)	—	—	1.035(D)
(α-f)	—	—	—	—	—	1.019(D)	—	—	—	—	—	—
(β-f)	—	—	—	—	—	1.045(D)	—	—	—	—	—	1.070(L)
(β-p)	—	—	—	—	—	1.029(D)	—	—	—	—	—	1.059(D)
Mannose (α-p)	1.25(L);[h] 1.31(D)[h]	—	—	—	—	1.036(L)	—	—	1.000	—	—	—
(α-f)	—	—	—	—	—	1.045(L)	—	—	—	—	—	—
(β-p)	—	—	—	—	—	1.247(L)	—	—	1.110(L)	—	—	—
Allose (α-p)	—	—	—	—	—	—	—	—	—	—	—	1.171(D)
(β-p)	—	—	—	—	—	—	—	—	—	—	—	1.064(D)
Talose (α-p)	—	—	—	—	—	—	—	—	—	1.099(D)	—	—
Gulose (α-p)	—	—	—	—	—	—	—	—	—	—	—	1.000
(β-p)	—	—	—	—	—	—	—	—	—	—	—	1.043(D)
Methyl xyloside (α-p)	—	—	—	—	—	—	—	1.020(D)	—	—	—	—
arabinoside (α-f)	—	—	—	1.019(L)	—	—	—	—	—	—	—	—
(α-p)	—	—	—	1.023(L)	—	—	—	1.099(D)	—	—	—	—
(β-p)	—	—	—	1.046(D)	—	—	—	1.093(D)	—	—	—	—
lyxoside (α-p)	—	—	—	1.014(D)	—	—	—	1.019(L)	—	—	—	—
(β-p)	—	—	—	—	—	—	—	1.030(D)	—	—	—	—
riboside (α-p)	—	—	—	1.021(D)	—	—	—	—	1.071(D)	1.075(D)	—	—
(β-p)	—	—	—	—	—	—	—	—	1.043(D)	—	—	—
fucoside (α-f)	—	—	—	—	—	—	—	1.035(D)	—	—	—	—
(α-p)	—	—	—	—	—	—	—	1.057(L)	—	—	—	—
(β-f)	—	—	—	1.014(D)	—	—	—	1.026(D)	—	—	—	—
(β-p)	—	—	—	1.046(D)	—	—	—	1.033(D)	—	—	—	—

TABLE GC 24 (continued)
Resolution of Enantiomers: Capillary GLC on Chiral Phases

Parent compound	HFB[c] r[e]	TFA[d] r[f]	TFA r[f]	TFA S.F.[g]	TFA S.F.	TFA S.F.	TFA S.F.	TFA S.F.	TFA S.F.	TFA S.F.	TFA S.F.	TFA S.F.
mannoside (α-*p*)	—	—	0.92(L); 0.96(D)	—	1.053(L)	—	—	—	1.051(L)	—	—	—
(β-*p*)	—	—	2.18(L); 2.27(D)	—	1.084(L)	—	—	—	—	—	—	—
glucoside (α-*p*)	—	0.97(L); 1.00(D)	0.97(L); 1.00(D)	—	1.032(L)	—	—	—	1.035(L)	1.035(L)	—	—
(β-*p*)	—	1.33(L); 1.37(D)	1.60(L); 1.64(D)	—	1.035(L)	—	—	—	—	—	—	—
galactoside (β-*f*)	—	0.96(D); 0.97(L)	0.88(D,L)	—	1.010(D)	—	—	—	—	—	—	—
(α-*f*)	—	1.01(D); 1.05(L)	—	—	1.044(D)	—	—	—	—	—	—	—
(α-*p*)	—	1.05(D); 1.10(L)	1.03(D); 1.09(L)	—	1.049(D)	—	—	—	1.091(D)	1.070(D)	—	—
(β-*p*)	—	1.60(D); 1.68(L)	2.08(D); 2.20(L)	—	1.089(D)	—	—	—	—	1.106(D)	—	—
idoside (α-*p*)	—	—	—	—	—	—	—	1.040(D)	1.032(D)	—	—	—
(β-*p*)	—	—	—	—	—	—	—	—	1.023(L)	—	—	—
chiro-inositol	1.00(D); 1.06(L)	—	—	—	—	—	—	—	—	—	—	—
Arabinitol	—	—	—	—	—	—	—	—	1.175(D)	1.111(D)	—	—
Glucitol	—	—	—	—	—	—	—	—	1.042(L)	1.033(L)	—	—
Mannitol	—	—	—	—	—	—	—	1.019(D)	1.013(D)	—	—	—
1,5-Anhydro-fucitol[k]	—	—	—	—	—	—	1.035(D)	1.050(D)	—	—	—	—
1,4-Anhydro-ribitol[k]	—	—	—	—	—	—	—	1.064(L)	—	—	—	—
1,5-Anhydro-arabinitol[k]	—	—	—	—	—	—	1.074(D)	1.045(D)	—	—	—	—
1,4-Anhydro-xylitol[k]	—	—	—	—	—	—	1.029(D)	1.030(D)	—	—	—	—
1,5-Anhydro-lyxitol[k]	—	—	—	—	—	—	1.064(D)	—	1.030(D)	—	—	—

1,5-Anhydro-glucitol[k]	—	—	—	—	—	—	—	—	—	—	—	1.037(L)
1,5-Anhydro-galactitol[k]	—	—	—	—	—	—	—	—	—	—	—	1.080(D)
1,5-Anhydro-mannitol[k]	—	—	—	—	—	—	—	—	—	—	—	1.019(D)

[a] 80 kPa[4] or 100 kPa.[5]
[b] 20 m[4] or 40 m.[5]
[c] Heptafluorobutanoic ester.
[d] Trifluoroacetate.
[e] t_r relative to heptafluorobutanoate of D-*chiro*-inositol (11.2 min).
[f] t_r relative to trifluoroacetylated methyl α-D-glucopyranoside (9.4 min,[2] 7.2 min[3]).
[g] Separation factor; the usual symbol, α, is not used here, to avoid confusion with α-anomer.
[h] Single peak for each enantiomer; no resolution of anomers.
[i] Enantiomer eluted first is shown in brackets after separation factor.
[j] For each sugar or glycoside, different forms are listed in order of elution.
[k] Products of reductive cleavage of polysaccharides.

Phase
P1 = Chirasil-Val® (Applied Science Laboratories State College, Pennsylvania); *N-tert*-buty-L-valinamide linked to copolymer of dimethyl- and carboxyalkyl-methylsiloxane.
P2 = XE-60-*S*-valine-*S*-α-phenylethylamide; cyano groups of XE-60 converted to carboxyl groups and coupled to L-valine-*S*-α-phenylethylamide.
P3 = Pentylated α-cyclodextrin [hexakis(2,3,6-tri-*O*-pentyl)-cyclomaltohexaose]; 0.2% solution in CH_2Cl_2 use to coat column after Silanox treatment.

Temperature
T1 = 95°C for 8 min; 95 → 125°C at 3°C/min.
T2 = 100 → 145°C at 3°C/min.
T3 = 120 → 180°C at 3°C/min

REFERENCES

1. **Leavitt, A. L. and Sherman, W. R.,** Direct gas-chromatographic resolution of DL-*myo*-inositol-1-phosphate and other sugar enantiomers as simple derivatives on a chiral capillary column, *Carbohydr. Res.*, 103, 203, 1982.
2. **König, W. A., Benecke, I., and Bretting, H.,** Gas chromatographic separation of carbohydrate enantiomers on a new chiral stationary phase, *Angew. Chem. Int. Ed. Engl.*, 20, 693, 1981.
3. **König, W. A., Benecke, I., and Sievers, S.,** New results in the gas chromatographic separation of enantiomers of hydroxy acids and carbohydrates, *J. Chromatogr.*, 217, 71, 1981.
4. **König, W. A., Lutz, S., Mischnick-Lübecke, P., Brassat, B., and Wenz, G.,** Cyclodextrins as chiral stationary phases in capillary gas chromatography, *J. Chromatogr.*, 447, 193, 1988.
5. **König, W. A., Mischnick-Lübbecke, P., Brassat, B., Lutz, S., and Wenz, G.,** Improved gas chromatographic separation of enantiomeric carbohydrate derivatives using a new chiral stationary phase, *Carbohydr. Res.*, 183, 11, 1988.

Section I.II

SUPERCRITICAL FLUID CHROMATOGRAPHY TABLES

At the time of writing, application of this powerful new technique to carbohydrates is very recent and its potential in this field is not yet fully realized. In this short section the pioneering work of V.N. Reinhold and of T.L. Chester and their co-workers in applying capillary supercritical fluid chromatography (SFC) to the separation of suitably derivatized maltodextrins and glycolipids and the recent success of the former group in using SFC coupled to mass spectroscopy (MS) are presented. For details of the technique, the reader is referred to excellent reviews by Lee and Markides* and by Gere.**

* **Lee, M. L. and Markides, K. E.,** Chromatography with supercritical fluids, *Science*, 235, 1342, 1987.
** **Gere, D. R.,** Supercritical fluid chromatography, *Science*, 222, 253, 1983.

TABLE SFC 1
Permethylated Maltodextrins

Stationary phase	DB-5	DB-5
Mobile phase	CO_2	CO_2
Pressure gradient	p1	p2
Temperature (°C)	120	90
Split ratio	1:5	1:5
Column		
Length, m	10	10
Diameter (I.D.), μm	50	50
Form	Bonded phase	Bonded phase
Material	Fused silica	Fused silica
Detector	FI	MS[a]
Reference	1	2
DP	**r[b]**	**r[b]**
2	1.00	1.00
3	1.16	1.38
4	1.32	1.62
5	1.46	1.81
6	1.58	2.00
7	1.68	2.19
8	1.76	2.38
9	1.84	2.50
10	1.92	2.62
11	2.00	—
12	2.08	—
13	2.16	—
14	2.26	—
15	2.40	—

[a] CI-MS with NH_3 as reagent gas; SFC column coupled to CI chamber through integral pressure restrictor heated to ca 280°C.
[b] t_r relative to permethylated maltose (31.2 min,[1] 24 min[2]).

p1 = density gradient produced by pressure programming: 110 atm (11.14 MPa) for 5 min; 110 → 400 atm (11.14 → 40.52 MPa) at 5 atm (0.51 MPa)/min.
p2 = pressure programming: 100 → 405 atm (10.13 → 41.03 MPa) at 5 atm (0.51 MPa)/min.

REFERENCES

1. **Kuei, J., Her, G.-R., and Reinhold, V. N.,** Supercritical fluid chromatography of glycosphingolipids,, *Anal. Biochem.*, 172, 228, 1988.
2. **Reinhold, V. N., Sheeley, D. M., Kuei, J., and Her, G.-R.,** Analysis of high molecular weight samples on a double-focusing magnetic sector instrument by supercritical fluid chromatography/mass spectrometry, *Anal. Chem.*, 60, 2719, 1988.

TABLE SFC 2
Trimethylsilylated Maltodextrins

Stationary phase	DB-1	DB-5
Mobile phase	CO_2	CO_2
Pressure gradient	p1	p2
Temperature (°C)	89	90
Column		
Length, m	10	10
Diameter (I.D.), μm	50	50
Form	Bonded phase	Bonded phase
Material	Fused silica	Fused silica
Detector	FI	MS[a]
Reference	1	2

DP	r[b]	r[b]
2	1.00; 1.04	1.00; 1.04
3	1.22; 1.27	1.22; 1.28
4	1.42; 1.46	1.42; 1.46
5	1.60; 1.66	1.58; 1.62
6	1.75; 1.79	1.72; 1.76
7	1.92; 1.96	1.87; 1.90
8	2.02; 2.06	1.98; 2.02
9	2.12; 2.16	2.08; 2.12
10	2.25; 2.29	2.19; 2.22
11	2.37; 2.41	2.32; 2.36
12	2.47; 2.50	2.41; 2.44
13	2.56; 2.58	2.49; 2.52
14	2.64; 2.66	2.57; 2.59
15	2.72; 2.74	2.64; 2.66
16	2.79; 2.80	2.76
17	2.89	2.82

[a] CI-MS as described under Table SFC 1.
[b] t_r relative to first peak for trimethylsilylated maltose (24 min,[1] 28.4 min[2]).

p1 = density gradient produced by pressure programming: 115 → 355 atm (11.65 → 35.96 MPa) at 3 atm (0.3 MPa)/min.
P2 = pressure programming: 115 → 400 atm (11.65 → 40.52 MPa) at 3 atm (0.3 MPa)/min.

REFERENCES

1. **Chester, T. L. and Innis, D. P.,** Separation of oligo- and polysaccharides by capillary supercritical fluid chromatography, *J. High Resolut. Chromatogr. Chromatogr. Commun.*, 9, 209, 1986.
2. **Reinhold, V. N., Sheeley, D. M., Kuei, J., and Her, G.-R.,** Analysis of high molecular weight samples on a double focusing magnetic sector instrument by supercritical fluid chromatography/mass spectrometry, *Anal. Chem.*, 60, 2719, 1988.

TABLE SFC 3
Permethylated Glycosphingolipids

Stationary phase	DB-5
Mobile phase	CO_2
Pressure gradient	p1
Temperature (°C)	120
Split ratio	1:5
Column	
Length, m	10
Diameter (I.D.), μm	50
Form	Bonded phase
Material	Fused silica
Detector	FI
Reference	1

Parent compound	**r**[a]
GL[b]	1.70
GL-3	1.67;[c] 1.76;[c] 1.80[d]
GL-4	1.84;[c] 1.85;[c] 1.86;[d] 1.88[c]
Gangliosides	
G_{m1}	1.92; 1.95[e]
G_{D1a},G_{D1b}[f]	2.00; 2.04[e]
G_{T1b}	2.12; 2.16[e]

[a] t_r relative to permethylated maltose (31.2 min).
[b] *N*-Lignoceroyldihydrolactocerebroside; GL-3 is ceramide trihexoside, GL-4 globoside.
[c] Minor peaks due to alkane heterogeneity in fatty acid chains.
[d] Major peak.
[e] Sphingoid heterogeneity; peaks of similar size.
[f] G_{D1a} and G_{D1b} co-eluted.

p1 = density gradient produced by pressure programming: 110 atm (11.14 MPa) for 5 min; 110 → 400 atm (11.14 → 40.52 MPa) at 5 atm (0.51 MPa)/min.

REFERENCE

1. **Kuei, J., Her, G.-R., and Reinhold, V. N.,** Supercritical fluid chromatography of glycosphingolipids, *Anal. Biochem.*, 172, 228, 1988.

Section I.III

LIQUID CHROMATOGRAPHY TABLES

Since the publication of Volume I, liquid chromatography in its various forms has developed more rapidly than any of the other types of chromatography, with the adaptation of most methods to high-performance liquid chromatography (HPLC). This section is, therefore, the major one in the present volume, and the emphasis is on the application of HPLC methods to carbohydrates, except in the few instances, notably in steric-exclusion (formerly called gel-permeation) chromatography, where the classical, low-pressure methods are still of value because of their superior resolution and ready adaptability to preparative chromatography. The main developments covered in this volume are listed below.

1. The production of new silica-based packings with bonded amino and amide phases for normal-phase partition chromatography; these are capable of resolving higher oligosaccharides to degree of polymerization (DP) > 30, cyclic oligosaccharides, and the complex oligosaccharides derived from glycoproteins.
2. The use of reversed-phase packings with water or aqueous solutions as eluents, which is another method permitting resolution of oligosaccharides (to DP 15), cyclic oligosaccharides, and those from glycoproteins, as well as methyl glycosides, methylated sugars, and alditols; under these conditions a form of hydrophobic chromatography is believed to be the predominant mechanism.
3. Reversed-phase chromatography of sugars derivatized so that UV or highly sensitive fluorometric detection is possible, and of partially methylated, partially ethylated oligosaccharide-alditols derived from complex carbohydrates, which can be distinguished by mass spectrometry when HPLC and mass spectrometry (MS) are coupled.
4. Adsorption chromatography on silica gel of sugars derivatized so that sensitive detection is possible, and resolution of enantiomers by HPLC on silica gel.
5. The use of amine modifiers in the mobile phase to convert silica gel packings *in situ* to amine-phase columns with similar resolving power, especially for higher oligosaccharides.
6. The development of new modified-silica packings, such as those with diol, polyol, or cyclodextrin bonded to silica.
7. Chromatography on microparticulate cation-exchange resins, with water or an aqueous solution as eluent and usually at an elevated temperature, unless resolution of anomers is desired; resins are in the H^+, Ca^{2+}, Pb^{2+}, or Ag^+ forms, depending upon the particular application, with mechanisms involving ion exclusion and ligand exchange (ion-moderated partitioning is the generic term used).
8. Use of these HPLC resins with eluents containing both organic and aqueous components, thus permitting the older technique of partition chromatography of sugars and oligosaccharides on ion-exchange resins in aqueous-organic media to be adapted to HPLC.
9. Application of high-performance ion-exchange resins and silica-based ion-exchange phases in rapid, efficient ion-exchange chromatography of acidic and other charged mono- and oligosaccharides and of alditols and sugars as borate complexes.
10. Ion chromatography on pellicular anion-exchange resins, with alkaline eluents at a pH sufficiently high to favor dissociation of the hydroxyl groups in carbohydrates; the use of the pulsed amperometric detector with this method affords a very sensitive analytical technique for carbohydrates, including oligosaccharides (to DP > 40), cyclodextrins, and glycoprotein-derived oligosaccharides, neutral, sialylated, or phosphorylated.

11. In steric-exclusion chromatography, the development of packings such as highly cross-linked agarose gels, various semi-rigid porous polymers, and modified porous silica; this has permitted the use of HPLC equipment in molecular-weight distribution analysis of polysaccharides, which has been facilitated by the application of the so-called "universal calibration".

The tables that follow (LC 1 to 61) illustrate the application of these various modes of liquid chromatography, usually in HPLC systems, to the different classes of carbohydrates. At the end of this section certain other forms of chromatography which do not fit readily into the categories covered by points 1 to 11 are reviewed, and three further tables (LC 62 to 64) are included in these reviews. The miscellaneous topics featured in this subsection are

12. The use of two-dimensional HPLC "mapping" in analysis of complex oligosaccharides from glycoproteins.
13. Other novel methods of LC analysis of carbohydrates, including chromatography of charged molecules on reversed-phase columns by introduction of ion-pairing reagents into the eluent, hydrophobic-interaction chromatography of glycoconjugates, and behavior of glycosaminoglycuronans on packings designed for hydrophobic chromatography; the limited application of hydrodynamic chromatography (field flow fractionation) to polysaccharides is also discussed here.
14. Advances in affinity chromatography: all lectins of importance in carbohydrate chemistry and biochemistry are tabulated together with their binding specificities, not only for sugars as end groups, but also for certain sugar sequences occurring in glycoproteins which have been found to be determinant in lectin-carbohydrate binding. Progress toward the adaptation of affinity chromatography to HPLC, including the use of weakly binding monoclonal antibodies as ligands instead of lectins, is also reviewed.
15. The application of the various modes of chromatography to preparative liquid chromatography of carbohydrates: examples of the use of different chromatographic systems, HPLC, MPLC, and classical, low-pressure methods, are tabulated.

TABLE LC 1
HPLC of Sugars and Polyols on Bonded Amino-Phase Silica-Based Packings

Packing	P1	P2	P3	P3	P4	P4 + P5[a]	P5	P5	P6	P6	P7	P7	P7	P8	P8
Column															
Length, cm	25	25	25	25	30	30[a]	30	30	25	25	25	25	25	25	25
Diameter (I.D.), cm	0.43	0.43	0.46	0.46	0.4	0.39[a]	0.39	0.39	0.46	0.46	0.46	0.46	0.46	0.46	0.46
Material	SS	SS	SS	SS	SS	SS	SS	SS	SS	SS	SS	SS	SS	SS	SS
Solvent	S1	S1	S2	S3	S4	S4	S5	S6	S7	S8	S1	S4	S4	S9	S2
Flow rate (ml/min)	1	1	2	2	0.6	1.7	2.7	2	2	2	1	1	1	1	1
Detection	UV	UV, DI	UV, RI	UV, RI	UV, RI	UV, RI	UV, RI	RI	RI	RI	RI	RI	RI	RI	RI
Reference	1	1	2	2	3	3	4	5	6	6	7[b]	7[b]	8,9	8	9
Compound	**Capacity factor (k′)**														
L-Rhamnose	—	—	1.17	0.94	0.93	0.91	—	—	1.00	0.86	—	—	—	—	—
L-Fucose	1.43	—	—	—	1.40	—	—	—	—	—	—	—	—	—	—
2-Deoxy-D-*arabino*-hexose	—	—	—	—	—	—	—	—	—	—	0.41	0.54	—	—	—
2,6-Dideoxy-D-*ribo*-hexose	—	—	—	—	—	—	—	—	—	—	0.22	0.17	—	—	—
D-Ribose	—	—	—	1.25	1.00	—	—	—	—	—	—	—	—	—	—
D-Xylose	1.46	1.42	1.58	—	1.36	1.20	—	—	—	—	—	—	—	—	—
L-Arabinose	—	—	2.00	—	1.65	1.72	—	—	—	—	—	—	—	—	—
L-Sorbose	—	—	—	2.25	—	—	—	—	—	—	—	—	—	—	—
D-Fructose	1.76	1.70	2.67	2.50	2.02	2.17	4.2	2.0	1.73	1.43	—	—	—	—	—
D-*manno*-Heptulose	—	—	—	—	—	—	7.6	—	—	—	—	—	—	—	—
D-Mannose	1.89	1.80	3.17	—	2.34	2.50	—	—	—	—	—	—	—	—	—
D-Glucose	1.99	1.80	3.67	—	2.64	2.62	6.7	3.0	1.99	2.00	0.62	1.03	—	—	—
D-Galactose	2.10	1.95	4.17	—	2.96	3.04	—	—	—	—	—	—	—	—	—
D-Glucitol	1.98	1.90	—	—	2.78	—	6.1	—	—	—	—	—	—	—	—
D-Mannitol	2.03	1.94	—	—	2.82	—	—	—	—	—	—	—	—	—	—
Galactitol	—	—	—	—	2.89	—	—	—	—	—	—	—	—	—	—
myo-Inositol	—	—	—	—	6.31	6.13	—	—	—	—	1.05	2.28	—	—	—
Xylobiose	—	—	—	—	—	—	—	—	—	—	—	—	1.36	2.32	—

TABLE LC 1 (continued)
HPLC of Sugars and Polyols on Bonded Amino-Phase Silica-Based Packings

Compound	Capacity factor (k')														
Sucrose	2.62	2.27	—	—	4.96	—	18.3	5.3	3.32	2.71	0.83	1.75	2.22	3.49	—
Turanose	—	—	—	—	—	—	—	—	—	—	—	—	2.47	3.80	—
Palatinose	—	—	—	—	—	—	—	—	—	—	—	—	2.50	3.89	—
Maltulose	—	—	—	—	—	—	—	—	—	—	—	—	2.65	4.09	—
Leucrose	—	—	—	—	—	—	—	—	—	—	—	—	2.76	4.20	—
Lactulose	—	—	—	—	—	—	—	—	—	—	—	—	3.03	4.63	—
Laminaribiose	—	—	—	—	—	—	—	—	—	—	—	—	2.41	3.77	—
Nigerose	—	—	—	—	—	—	—	—	—	—	—	—	2.77	4.31	—
Cellobiose	—	—	—	—	6.49	—	—	—	—	—	—	—	2.77	4.36	—
Maltose	2.99	2.53	—	—	6.53	—	—	7.4	3.99	3.28	—	—	2.80	4.38	—
Sophorose	—	—	—	—	—	—	—	—	—	—	—	—	2.92	4.50	—
Kojibiose	—	—	—	—	—	—	—	—	—	—	—	—	3.14	4.85	—
αβ-Trehalose	—	—	—	—	—	—	—	—	—	—	0.98	2.31	2.98	4.61	—
αα-Trehalose	3.11	—	—	—	—	—	—	—	—	—	1.02	2.57	3.22	4.91	—
ββ-Trehalose	—	—	—	—	—	—	—	—	—	—	1.05	2.60	3.42	5.16	—
Lactose	3.38	2.82	—	—	7.54	—	—	—	—	—	—	—	3.31	5.22	—
Isomaltose	—	—	—	—	—	—	—	—	4.49	4.28	—	—	3.51	5.39	—
Gentiobiose	3.47	—	—	—	—	—	—	—	—	—	—	—	3.64	5.65	—
Melibiose	—	—	—	—	—	—	—	—	—	—	—	—	3.94	6.02	—
Xylotriose	—	—	—	—	—	—	—	—	—	—	—	—	2.62	—	2.93
1-Kestose	—	—	—	—	—	—	—	—	—	—	—	—	—	—	5.84
Melezitose	—	—	—	—	—	—	—	—	—	—	—	—	4.51	—	5.87
Raffinose	—	—	—	—	—	—	—	—	—	—	—	—	5.68	—	7.22
Laminaritriose	—	—	—	—	—	—	—	—	—	—	—	—	3.99	—	5.66
3-*O*-Cellobiosyl-D-glucose	—	—	—	—	—	—	—	—	—	—	—	—	—	—	6.61
4-*O*-Laminaribiosyl-D-glucose	—	—	—	—	—	—	—	—	—	—	—	—	—	—	6.67
Maltotriose	—	3.35	—	—	—	—	—	—	7.15	5.71	—	—	5.28	—	7.31
Cellotriose	—	—	—	—	—	—	—	—	—	—	—	—	5.53	—	7.42
3-*O*-Isomaltosyl-D-glucose	—	—	—	—	—	—	—	—	—	—	—	—	6.22	—	8.15

Panose	—	—	—	—	—	—	—	—	—	—	—	—	6.45	—	8.52
Isopanose	—	—	—	—	—	—	—	—	—	—	—	—	6.77	—	8.69
Isomaltotriose	—	—	—	—	—	—	—	—	—	—	—	—	8.23	—	10.22
Stachyose	—	5.55	—	—	—	—	—	—	—	—	—	—	—	—	—

Note: This table supplements Table LC 1 in *CRC Handbook of Chromatography: Carbohydrates,* Volume I.

[a] Double-column system; column P5, of dimensions given here, connected to outlet of P4 (dimensions in preceding heading).
[b] See data on HPLC of these sugars on unmodified silica in Table LC 18 (Reference 2).

Packing
P1 = LiChrosorb®-NH_2 (Merck, Darmstadt, West Germany; 10-μm particles).
P2 = Chromosorb®-NH_2 (Johns, Manville, Denver, CO; 10-μm particles.)
P3 = 5-μm silica with bonded amino groups (Macherey-Nagel, Duren, West Germany No. 711120); now called Nucleosil®-NH_2.
P4 = MicroPak® NH_2-10 (Varian, Sunnyvale, CA; 10-μm particles).
P5 = μ-Bondapakh® Carbohydrate Analysis Column (Waters, Milford, MA; 10-μm particles).
P6 = RSiL®-Carbohydrate Column (Alltech, Deerfield, IL; 10-μm particles).
P7 = Supelcosil® LC-NH_2 (Supelco, Bellefonte, PA; 5-μm particles).
P8 = Zorbax®-NH_2 (DuPont, Wilmington, DE; 7-μm particles).

Solvent
S1 = acetonitrile-water (7:3).
S2 = acetonitrile-water (3:1).
S3 = ethyl acetate-ethanol-water (4:5:1).
S4 = acetonitrile-water (4:1).
S5 = acetonitrile-water (9:1).
S6 = S4 with added ion-pairing reagent, tetrabutylammonium phosphate (PIC A; Waters), 2.5 *M*, to obviate interference by NaCl in sample.
S7 = acetonitrile-water (31:9).
S8 = acetone-ethyl acetate-water (5:3:2).
S9 = acetonitrile-water (77:23).

Detection
D1 = automated tetrazolium blue colorimetric method.

REFERENCES

1. **D'Ambroise, M., Noël, D., and Hanai, T.,** Characterization of bonded-amine packing for liquid chromatography and high-sensitivity determination of carbohydrates, *Carbohydr. Res.*, 79, 1, 1980.
2. **Binder, H.,** Separation of monosaccharides by high-performance liquid chromatography: comparison of ultraviolet and refractive index detection, *J. Chromatogr.*, 189, 414, 1980.
3. **Yang, M. T., Milligan, L. P., and Mathison, G. W.,** Improved sugar separation by high-performance liquid chromatography using porous microparticle carbohydrate columns, *J. Chromatogr.*, 209, 316, 1981.

TABLE LC 1 (continued)
HPLC of Sugars and Polyols on Bonded Amino-Phase Silica-Based Packings

4. **Shaw, P. E. and Wilson, C. W., III,** Separation of sorbitol and *manno*-heptulose from fructose, glucose and sucrose on reversed-phase and amine-modified HPLC columns, *J. Chromatogr. Sci.*, 20, 209, 1982.
5. **Wills, R. B. H., Francke, R. A., and Walker, B. P.,** Analysis of sugars in foods containing sodium chloride by high-performance liquid chromatography, *J. Agric. Food Chem.*, 30, 1242, 1982.
6. **Verzele, M., Simoens, G., and Van Damme, F.,** A critical review of some liquid chromatography systems for the separation of sugars, *Chromatographia*, 23, 292, 1987.
7. **Nikolov, Z. L. and Reilly, P. J.,** Retention of carbohydrates on silica and amine-bonded silica stationary phases: application of the hydration model, *J. Chromatogr.*, 325, 287, 1985.
8. **Nikolov, Z. L., Meagher, M. M., and Reilly, P. J.,** High-performance liquid chromatography of disaccharides on amine-bonded silica columns, *J. Chromatogr.*, 319, 51, 1985.
9. **Nikolov, Z. L., Meagher, M. M., and Reilly, P. J.,** High-performance liquid chromatography of trisaccharides on amine-bonded silica columns, *J. Chromatogr.*, 321, 393, 1985.

TABLE LC 2
HPLC of Glycosides on Amine-Bonded Silica

Packing	P1
Column	
Length, cm	30
Diameter (I.D.), cm	0.39
Material	SS
Solvent	S1
Flow rate (ml/min)	2
Detection	RI, UV
Reference	1

Compound	**r**[a]
D-Glucopyranoside	
methyl α-	1.00
methyl β-	0.97
ethyl β-	0.81
phenyl α-	0.54
phenyl β-	0.51
m-tolyl α-	0.51
m-tolyl β-	0.49
guaiacyl β-	0.55
2-napththyl α-	0.50[b]
2-naphthyl β-	0.47[b]
D-Glucofuranoside	
methyl α-	0.72
methyl β-	0.74
ethyl β-	0.58
phenyl β-	0.46
m-tolyl β-	0.42
guaiacyl β-	0.44
2-naphthyl β-	0.42[b]
D-Galactopyranoside	
methyl α-	1.00
methyl β-	1.09
ethyl α-	0.76
ethyl β-	0.86
phenyl α-	0.53
phenyl β-	0.53
m-tolyl α-	0.49
m-tolyl β-	0.49
guaiacyl α-	0.54
guaiacyl β-	0.59
D-Galactofuranoside	
methyl β-	0.68
ethyl β-	0.59
phenyl β-	0.42
m-tolyl β-	0.40
guaiacyl β-	0.42
D-mannopyranoside	
methyl α-	0.66
D-mannofuranoside	
methyl α-	0.64

[a] t_r relative to methyl α-D-glucopyranoside (7.4 min).
[b] UV detection.

TABLE LC 2 (continued)
HPLC of Glycosides on Amine-Bonded Silica

Packing	P1 = μ-Bondapak® Carbohydrate Analysis Column (Waters, Milford, MA).
Solvent	S1 = acetonitrile-water (9:1).

REFERENCE

1. **Yoshida, K., Kotsubo, K., and Shigematsu, H.,** High-performance liquid chromatography of hexopyranosides and hexofuranosides, *J. Chromatogr.*, 208, 104, 1981.

TABLE LC 3
HPLC of Higher Oligosaccharides on Amine- and Amide-Bonded Silica Packings

Type of oligosaccharide	Malto-				Sophoro-	Laminari-	Gentio-	Isomalto-		Chito-	Xylo-	Fructo
Packing	P1	P2	P3	P4	P2	P2	P2	P2	P5	P5	P1	P1
Column												
Length, cm	25	20	25	25	20	20	20	20	30	30	25	25
Diameter (I.D.), cm	0.43	0.6	0.46	0.46	0.6	0.6	0.6	0.6	0.4	0.4	0.43	0.43
Material	SS	SS	SS	SS	SS	SS	SS	SS	SS	SS	SS	SS
Solvent	S1	S2	S3	S4	S5	S6	S7	S8	S9	S9	S1	S10
Flow rate (ml/min)	1	1	1	1	1	1	1	1	1	1	1	1
Detection	D1	RI	RI	RI	RI	RI	RI	RI	D2	D2	D1	D1
Reference	1	2	3	3	2	2	2	2	4	4	1	5
DP	$k'(t_r)$[a]								r[b]	r[c]	$k'(t_r)$[a]	
1	0.12	—	—	—	—	0.50	—	—	1.00	1.00	—	—
2	0.25	0.43	0.48	0.56	—	0.62	0.38	0.36	1.33	1.17	—	1.33
3	0.50	0.57	0.64	0.92	0.36	0.74	0.57	0.54	1.66	1.33	—	2.00
4	0.75	0.71	0.80	1.28	0.48	0.87	0.76	0.72	2.41	1.66	—	2.60
5	1.20	0.86	0.96	1.66	0.60	1.00	0.95	0.90	3.17	2.00	1.48	3.22
6	1.62	0.98	1.30	2.16	0.71	1.13	1.13	1.08	3.92	(12 min)	1.98	3.85
7	2.00	1.08	1.70	3.00	0.82	1.26	1.30	1.25	4.58		2.60	4.50
8	2.38	1.18	2.08	3.66	0.93	1.40	1.47	1.42	5.16		2.98	5.22
9	2.62	1.28	2.50	4.33	1.04	1.57	1.65	1.60	5.66		3.28	5.78
10	2.92	1.40	2.94	5.00	1.15	1.75	1.83	1.78	6.33		3.53	6.28
11	3.18	1.55	3.44	5.67	1.26	1.93	2.00	1.96	6.67		3.78	6.78
12	3.50	1.71	3.94	6.34	1.38	2.16	2.18	2.14	6.92		4.03	7.12
13	3.70	1.86	4.48	7.00	1.50	2.40	2.36	2.33	7.25		4.30	7.48
14	3.86	2.06	5.16	7.67	1.68	2.70	2.54	2.56	7.50		4.56	7.85
15	4.00	2.28	5.86	8.67	1.86	3.00	2.70	2.80	7.75		4.80	8.20
16	4.16	2.57	6.60	9.67	2.06	3.32	3.00	3.08	8.00		5.02	8.56
17	4.36	2.86	7.57	11.3	2.27	3.64	3.32	3.40	8.28		5.24	8.88
18	4.62	3.14	8.80	13.0	2.48	4.00	3.70	3.80	8.58		5.46	9.20
19	4.80	3.43	10.0	15.0	2.70	(20 min)	4.12	4.32	(51.5 min)			9.50
20	5.00	3.72	11.5	17.0	2.92		4.55	4.92			5.83	9.80

TABLE LC 3 (continued)
HPLC of Higher Oligosaccharides on Amine- and Amide-Bonded Silica Packings

DP	$k'(t_r)$[a]							r[b]	r[c]	$k'(t_r)$[a]	
21	(24 min)	4.00	13.2	19.0	3.22	5.00	5.52			(24 min)	(39 min)
22		4.34	15.2	21.6	3.52	5.64	6.12				
23		4.71	17.6	24.3	3.82	6.28	6.75				
24		5.14	20.0	27.3	4.12	6.98	7.40				
25		5.57	22.8	31.0	4.42	7.76	8.20				
26		6.00	(40.5 min)	(40 min)	4.72	8.56	9.00				
27		6.57			5.10	(34.5 min)	(37.5 min)				
28		7.05			5.50						
29		7.65			6.08						
30		8.26			6.70						
31		(32.5 min)			7.38						
32					8.06						
33					8.74						
34					9.40						
35					10.0						
					(43 min)						

Note: Earlier data on HPLC of malto-, cello-, and chito-oligosaccharides appear in Tables LC 2—4 of *CRC Handbook of Chromatography; Carbohydrates*, Volume I. Further HPLC data for oligosaccharides of Reference 2 will be found in Table LC 12 and TLC data in TLC 5 (present volume).

[a] Retention time of highest oligomer clearly resolved is given in brackets at end of each column.
[b] t_r relative to [^{3}H]glucitol (6 min).
[c] t_r relative to 2-acetamido-2-deoxy-[^{3}H]glucitol (6 min).

Packing
- P1 = Chromosorb®-NH_2 (Johns-Manville; 10-μm particles).
- P2 = ERC®-NH-1171 (Erma Optical Works, Tokyo; 3-μm particles).
- P3 = YMC-Pack® PA-03 (Yamamura Chemical Company, Kyoto); polyamine resin-bonded silica gel.
- P4 = TSK® gel Amide-80 (Toyo Soda, Tokyo); carbamoyl groups bonded to silica gel.
- P5 = MicroPak® AX-5 (Varian; 5-μm particles).

Solvent
- S1 = linear gradient, 70 → 62.5% acetonitrile in water in 30 min.
- S2 = acetonitrile-water (57:43).
- S3 = acetonitrile-water (1:1).
- S4 = acetonitrile-water (53:47).
- S5 = acetonitrile-water (29:21).

	S6 =	acetonitrile-water (3:2).
	S7 =	acetonitrile-water (11:9).
	S8 =	acetonitrile-water (14:11).
	S9 =	linear gradient, acetonitrile initially 65% in water, decreased at 0.5% per minute.
	S10 =	linear gradient, 66 → 57% acetonitrile in water in 40 min.
Detection	D1 =	automated tetrazolium blue colorimetric method.
	D2 =	scintillation counting of samples reduced with sodium borotritiide.

REFERENCES

1. **Noël, D., Hanai, T., and D'Ambroise, M.,** Systematic liquid chromatographic separation of poly-, oligo-, and monosaccharides, *J. Liq. Chromatogr.*, 2, 1325, 1979.
2. **Koizumi, K., Utamura, T., and Okada, Y.,** Analyses of homogeneous D-gluco-oligosaccharides and -polysaccharides (degree of polymerization up to about 35) by high-performance liquid chromatography and thin-layer chromatography, *J. Chromatogr.*, 321, 145, 1985.
3. **Koizumi, K., Utamura, T., Kubota, Y., and Hizukuri, S.,** Two high-performance liquid chromatographic columns for analyses of malto-oligosaccharides, *J. Chromatogr.* 409, 396, 1987.
4. **Mellis, S. J. and Baenziger, J. U.,** Separation of neutral oligosaccharides by high-performance liquid chromatography, *Anal. Biochem.*, 114, 276, 1981.
5. **D'Amboise, M., Noël, D., and Hanai, T.,** Characterization of bonded-amine packing for liquid chromatography and high-sensitivity determination of carbohydrates, *Carbohydr. Res.*, 79, 1, 1980.

TABLE LC 4
HPLC of Cyclic Oligosaccharides on Amine- and Amide-Bonded Silica Packings

Packing	P1	P2	P2	P3	P4	P5	P5	P5	P6	P3
Column										
Length, cm	30	25	25	20	25	25	25	25	16	20
Diameter (I.D.), cm	0.39	0.4	0.4	0.6	0.46	0.46	0.46	0.46	0.46	0.6
Material	SS	SS	SS	SS	SS	SS	SS	SS	SS	SS
Solvent	S1	S2	S3	S3	S4	S3	S2	S5	S6	S7
Flow rate (ml/min)	2	1	1	1	1	1	1	1	1	1
Temperature (°C)	25	ambient	ambient	ambient	ambient	ambient	50	70	ambient	ambient
Detection	RI	RI	RI	RI	RI	RI	RI	RI	RI	RI
Reference	1	2	3	3	3	3	3	3	4	5

TABLE LC 4 (continued)
HPLC of Cyclic Oligosaccharides on Amine- and Amide-Bonded Silica Packings

Compound	k′									
Cyclomaltohexaose[a]	4.16	1.50	1.59	1.03	1.52	2.22	1.97	2.23	1.19	—
6-*O*-α-D-glucosyl-	—	2.10	2.27	1.41	2.14	2.97	2.72	3.16	—	—
6-*O*-α-maltosyl-	—	2.80	2.88	1.71	2.27	3.59	3.42	4.08	—	—
6-*O*-α-maltotriosyl-	—	3.48	3.58	2.05	3.26	4.34	4.27	5.21	—	—
6-*O*-α-maltotetraosyl-	—	—	4.39	2.42	3.94	5.18	5.22	6.56	—	—
6-*O*-α-maltopentaosyl-	—	—	5.34	2.84	4.68	6.21	6.43	8.25	—	—
di-*O*-α-D-glucosyl-	—	—	3.09	1.82	2.86	3.75	3.58	4.23	—	—
di-*O*-α-maltosyl-	—	—	4.60	2.60	4.28	5.31	5.39	6.76	—	—
Cyclomaltoheptaose[a]	5.75	2.00	2.02	1.45	1.89	2.61	2.39	2.81	1.42	—
6-*O*-α-D-glucosyl-	—	2.52	2.86	1.91	2.64	3.50	3.30	3.97	—	—
6-*O*-α-maltosyl-	—	3.20	3.64	2.33	3.29	4.22	4.12	5.06	—	—
6-*O*-α-maltotriosyl-	—	4.28	4.56	2.79	4.03	5.07	5.09	6.44	—	—
6-*O*-α-maltotetraosyl-	—	—	5.57	3.28	4.84	6.04	6.24	8.07	—	—
di-*O*-α-D-glucosyl-	—	3.40	3.94	2.39	3.51	4.43	4.34	5.27	—	—
di-*O*-α-maltosyl-	—	—	6.06	3.42	5.24	6.20	6.52	8.33	—	—
6^A, 6^D-di-*O*-α-maltotriosyl-	—	—	9.53	4.63	7.88	8.94	9.88	13.30	—	—
6-*O*-α-(6′-*O*-α-maltotriosyl)-maltotriosyl-	—	—	9.34	4.60	7.77	9.00	10.09	13.73	—	—
6-*O*-α-(6″-*O*-α-maltotriosyl)-maltotriosyl-	—	—	8.86	4.47	7.51	9.32	10.36	14.10	—	—
Cyclomaltooctaose[a]	7.90	2.28	2.57	1.69	2.38	3.20	3.04	3.60	1.69	—
6-*O*-α-D-glucosyl-	—	3.20	3.62	2.23	3.30	3.67	4.18	5.07	—	—
6-*O*-α-maltosyl-	—	4.28	4.60	2.71	4.14	5.10	5.17	6.46	—	—
6-*O*-α-maltotriosyl-	—	5.40	5.74	3.25	5.07	6.09	6.36	8.15	—	—
6-*O*-α-maltotetraosyl-	—	—	6.99	3.81	6.06	7.22	7.78	10.21	—	—
di-*O*-α-D-glucosyl-	—	—	5.01	2.83	4.40	5.44	5.50	6.87	—	—
di-*O*-α-maltosyl-	—	—	7.69	4.05	6.63	7.55	8.14	10.68	—	—
Cyclosophoraoses[b]										
DP 17	—	—	—	—	—	—	—	—	3.16	1.94
18	—	—	—	—	—	—	—	—	3.76	2.12
19	—	—	—	—	—	—	—	—	4.34	2.19

20	—	—	—	—	—	—	—	—	4.98	2.44
21	—	—	—	—	—	—	—	—	5.72	2.56
22	—	—	—	—	—	—	—	—	6.57	2.88
23	—	—	—	—	—	—	—	—	7.30	3.12
24	—	—	—	—	—	—	—	—	8.11	3.38
25	—	—	—	—	—	—	—	—	8.90	3.62
26	—	—	—	—	—	—	—		10.0	3.94
27	—	—	—	—	—	—	—	—	11.1	4.25
28	—	—	—	—	—	—	—	—	12.2	4.50
29	—	—	—	—	—	—	—	—	13.4	4.88
30	—	—	—	—	—	—	—	—	14.6	5.12
31	—	—	—	—	—	—	—	—	15.8	5.56
32	—	—	—	—	—	—	—	—	17.6	5.94
33	—	—	—	—	—	—	—	—	20.1	6.31
34	—	—	—	—	—	—	—	—	—	6.87
35	—	—	—	—	—	—	—	—	—	7.19
36	—	—	—	—	—	—	—	—	—	7.87
37	—	—	—	—	—	—	—	—	—	8.37
38	—	—	—	—	—	—	—	—	—	8.88
39	—	—	—	—	—	—	—	—	—	9.62
40	—	—	—	—	—	—	—	—	—	10.2

Note: Further HPLC data for these oligosaccharides will be found in Table LC 13, ion chromatography in Table LC 47, and TLC data for cyclodextrins in TLC 6.

[a] Cyclomaltohexaose, -heptaose, and -octaose are generally known, respectively as α-, β, and γ-cyclodextrin.

[b] Produced by *Rhizobium* and *Aqrobacterium* strains.

Packing

P1 = μ-Bondapak® Carbohydrate Analysis Column (Waters; 10-μm particles).
P2 = Hibar LiChrosorb®-NH_2 (Merck).
P3 = ERC®-NH-1171 (Erma Optical Works, Tokyo; 3-μm particles).
P4 = YMC-Pack® PA-03 (Yamamura Chemical Company, Kyoto).
P5 = TSK® gel Amide-80 (Toyo Soda, Tokyo).
P6 = 5-μm silica with bonded aminopropyl groups (0.6 mmol/g); supplied by Phase Separations (Queensferry, U.K.), packed in authors' laboratory.

TABLE LC 4 (continued)
HPLC of Cyclic Oligosaccharides on Amine- and Amide-Bonded Silica Packings

Solvent
S1 = acetonitrile-water (7:3).
S2 = acetonitrile-water (3:2).
S3 = acetonitrile-water (29:21).
S4 = acetonitrile-water (11:9).
S5 = acetonitrile-water (31:19).
S6 = acetonitrile-water (16:9).
S7 = acetonitrile-water (57:43).

REFERENCES

1. **Zsadon, B., Otta, K. H., Tüdös, F., and Szejtli, J.,** Separation of cyclodextrins by high-performance liquid chromatography, *J. Chromatogr.*, 172, 490, 1979.
2. **Koizumi, K., Utamura, T., Kuroyanagi, T., Hizukuri, S., and Abe, J.-I.,** Analysis of branched cyclodextrins by high-performance liquid and thin-layer chromatography, *J. Chromatogr.*, 360, 397, 1986.
3. **Koizumi, K., Utamara, T., Kubota, Y., and Hizukuri, S.,** Two high-performance liquid chromatographic columns for analyses of malto-oligosaccharides, *J. Chromatogr.*, 409, 396, 1987.
4. **Benincasa, M., Cartoni, G. P., Coccioli, F., Rizzo, R., and Zevenhuizen, L. P. T. M.,** High-performance liquid chromatography of cyclic β (1 → 2)-D-glucans (cyclosophoraoses) produced by *Rhizobium meliloti* and *Rhizobium trifolii*, *J. Chromatogr.*, 393, 263, 1987.
5. **Koizumi, K., Okada, Y., Utamura, T., Hisamatsu, M., and Amemura, A.,** Further studies on the separation of cyclic (1 → 2)-β-D-glucans (cyclosophoraoses) produced by *Rhizobium meliloti* IFO 13336, and determination of their degrees of polymerization by high-performance liquid chromatography, *J. Chromatogr.*, 299, 215, 1984.

Data from Reference 3 reprinted with permission of authors and Elsevier Science Publishers.

TABLE LC 5
HPLC of Mono- and Oligosaccharides Derived from Glycoproteins, and of Milk Oligosaccharides, on Amino Phases

Packing	P1	P1	P2	P3	P3
Column					
Length, cm	25	25	25	30	30
Diameter (I.D.), cm	0.4	0.4	0.46	0.4	0.4
Material	SS	SS	SS	SS	SS
Solvent	S1	S2	S3	S4	S5
Flow rate (ml/min)	2	2	1	1.3	1
Detection	UV, D1	UV	UV	UV	UV
Reference	1	1	2	2	3
Compound	**r**[a]	**k'**[b]	**k'**	**k'**[b]	**k'**[b]
Sugars and alditols					
L-Fucose	0.59	—	—	—	—
D-Mannose	0.86	—	—	—	—
D-Glucose	1.00	—	—	—	—
D-Galactose	1.08	—	—	—	—
Galactitol	1.05	—	—	—	—
2-Acetamido-2-deoxy-D-glucose	0.65	—	—	—	—
2-Acetamido-2-deoxy-D-glucitol	0.87	—	—	—	—
2-Acetamido-2-deoxy-D-galactose	0.66	—	—	—	—
2-Acetamido-2-deoxy-D-galactitol	0.79	—	1.7	2.0	—
Oligosaccharides and oligosaccharide-alditols					
Galβ1-3GlcNac	1.53	—	—	—	—
Galβ1-4GlcNAc	1.60	—	—	—	—
Galβ1-6GlcNAc	2.41	—	—	—	—
Galβ1-3GalNAc	2.14	—	—	—	—
Galβ1-3GalNAc-ol	2.01	—	2.3	2.6	—
GlcNAcβ1-2Man	2.27	—	—	—	—
GlcNAcβ1-6Man	2.92	—	—	—	—
GlcNAcβ1-3Gal	2.55	—	—	—	—
GlcNAcβ1-4Gal	2.31	—	—	—	—
GlcNAcβ1-6Gal	3.40	—	—	—	—
GlcNAcβ1-3GalNAc-ol	—	—	2.1	2.3	—
Fucα1-2Galβ1-3GalNAc-ol	—	—	2.4	2.7	—
Fucα1-2Galβ1-4GlcNAc	3.19	—	—	—	—
Galα1-3Galβ1-4GlcNAc	4.12	—	—	—	—
Galβ1-6Galβ1-6GlcNAc	5.43	—	—	—	—
Galβ1-4GlcNAcβ1-4GlcNAc	3.55	—	—	—	—
Galβ1-4GlcNAcβ1-2Man	5.00	—	—	—	—
Galβ1-4GlcNAcβ1-6Man	5.52	—	—	—	—
Galβ1-4GlcNAcβ1-3Gal	5.32	—	—	—	—
Galβ1-4GlcNAcβ1-6Gal	5.85	—	—	—	—
Manα1-3(Manα1-6)Man	5.78	—	—	—	—
GlcNAcβ1-2(GlcNAcβ1-4)Man	4.81	—	—	—	—
GlcNAcβ1-2(GlcNAcβ1-6)Man	5.52	—	—	—	—
GlcNAcβ1-3(GlcNAcβ1-6)Gal	5.64	—	—	—	—
Galβ1-3GlcNAcβ1-3Galβ1-4Glc	6.50	—	—	—	—
Galβ1-3GlcNAcβ1-3Galβ1-4Glc-ol	6.67	—	—	—	—
Galβ1-4GlcNAcβ1-3Galβ1-4Glc	6.52	—	—	—	—
Galβ1-4GlcNAcβ1-3Galβ1-4GlcNAc	5.93	—	—	—	—
Galβ1-4GlcNAcβ1-6(GlcNAcβ1-3)Gal	6.66	—	—	—	—
Galβ1-4GlcNAcβ1-6(Galβ1-3)GalNAc-ol	—	—	3.7	4.6	—
Fucα1-2Galβ1-3GlcNAcβ1-3GalNAc-ol	—		3.1	4.0	—
Fucα1-2Galβ1-3(GlcNAcβ1-6)GalNAc-ol	—	—	2.9	3.6	—
Fucα1-2Galβ1-3(Galβ1-4GlcNAcβ1-6)GalNAc-ol	—	—	4.0	4.8	—
Fucα1-2Galβ1-4GlcNAcβ1-6(Galβ1-3)GalNAc-ol	—	—	4.4	6.2	—

TABLE LC 5 (continued)
HPLC of Mono- and Oligosaccharides Derived from Glycoproteins, and of Milk Oligosaccharides, on Amino Phases

Compound	r[a]	k'[b]	k'	k'[b]	k'[b]
Galβ1-4GlcNAcβ1-2Manα1-3Manβ1-4GlcNAc	—	11.3	—	—	—
Galβ1-4GlcNAcβ1-2(Galβ1-4GlcNAcβ1-4)Man	7.02	—	—	—	—
Galβ1-4GlcNAcβ1-2(Galβ1-4GlcNAcβ1-6)Man	7.28	—	—	—	—
Galβ1-4GlcNAcβ1-3(Galβ1-4GlcNAcβ1-6)Gal	7.40	—	—	—	—
Galβ1-3GlcNAcβ1-3(Galβ1-4GlcNAcβ1-6)Gal	7.40	—	—	—	—
Galβ1-3GlcNAcβ1-3(Galβ1-4GlcNAcβ1-6)Galβ1-4Glc	—	—	—	—	1.70
Galβ1-3GlcNAcβ1-3[Galβ1-4(Fucα1-3)GlcNAcβ1-6]Galβ1-4Glc	—	—	—	—	2.15
Galβ1-3(Fucα1-4)GlcNAcβ1-3[Galβ1-4(Fucα1-3)GlcNAcβ1-6]Galβ1-4Glc	—	—	—	—	3.00
Fucα1-2Galβ1-3(Fucα1-4)GlcNAcβ1-3Galβ1-3GalNAc-ol	—	—	5.1	9.2	—
Fucα1-2Galβ1-3(Fucα1-2Galβ1-4GlcNAcβ1-6)GalNAc-ol	—	—	4.8	8.8	—
Fucα1-2Galβ1-3[Fucα1-2Galβ1-4(Fucα1-3)GlcNAcβ1-6]GalNAc-ol	—	—	5.5	9.3	—
Galβ1-4GlcNAcβ1-2Manα1-3(Galβ1-4GlcNAcβ1-2Manα1-6)Man	8.45	—	—	—	—
Biantennary octasaccharide I[c]	—	17.7	—	—	—
Triantennary decasaccharide II[c]	—	20.0	—	—	—
Triantennary decasaccharide III[c]	—	21.0	—	—	—
Tetraantennary dodecasaccharide IV[c]	—	22.7	—	—	—
		(71 min)		(18 min)	(8 min)

Note: Data on reversed-phase HPLC of reduced oligosaccharides in Reference 2 will be found in Table LC 9.

[a] t_r relative to D-glucose (8.6 min).

[b] t_r of last-eluting oligosaccharide in series is given in brackets at foot of column.

[c] Structures of oligosaccharides I—IV appended.

Packing
- P1 = LiChrosorb®-NH_2 (Merck; 5-μm particles).
- P2 = 605 NH(Alltech; 5-μm particles).
- P3 = MicroPak® AX-5 (Varian; 5-μm particles).

Solvent
- S1 = acetonitrile-15 m*M* aqueous phosphate buffer, pH 5.2; 4:1 for 30 min, then linear gradient with buffer content increasing at 0.5%/min.
- S2 = as S1, but linear gradient with buffer content increasing at 0.3%/min throughout elution.
- S3 = acetonitrile-water (3:2).
- S4 = acetonitrile-1 m*M* aqueous phosphate buffer, pH 5.4 (3:2).
- S5 = acetonitrile-1 m*M* aqueous phosphate buffer, pH 5.4 (1:1).

Detection
- D1 = compounds without UV absorbance detected by phenol-H_2SO_4 colorimetric assay or by scintillation counting after labeling with UDP-[^{14}C]Gal.

REFERENCES

1. **Blanken, W. M., Bergh, M. L.-E., Koppen, P. L., and Van den Eijnden, D. H.,** High-pressure liquid chromatography of neutral oligosaccharides: effects of structural parameters, *Anal. Biochem.*, 145, 322, 1985.
2. **Dua, V. K., Dube, V. E., and Bush, C. A.,** The combination of normal-phase and reverse-phase high-pressure liquid chromatography with NMR for the isolation and characterization of oligosaccharide alditols from ovarian cyst mucins, *Biochim. Biophys. Acta*, 802, 29, 1984.
3. **Dua, V. K., Goso, K., Dube, V. E., and Bush, C. A.,** Characterization of lacto-*N*-hexaose and two fucosylated derivatives from human milk by high-performance liquid chromatography and proton NMR spectroscopy, *J. Chromatogr.*, 328, 259, 1985.

APPENDIX TO TABLE LC 5: STRUCTURES OF NUMBERED COMPOUNDS

I

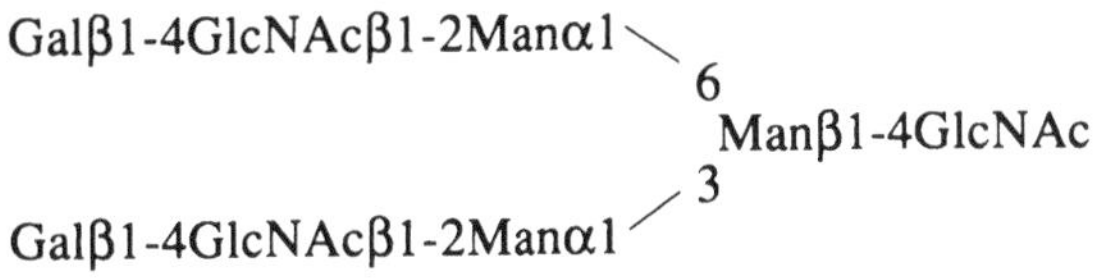

II

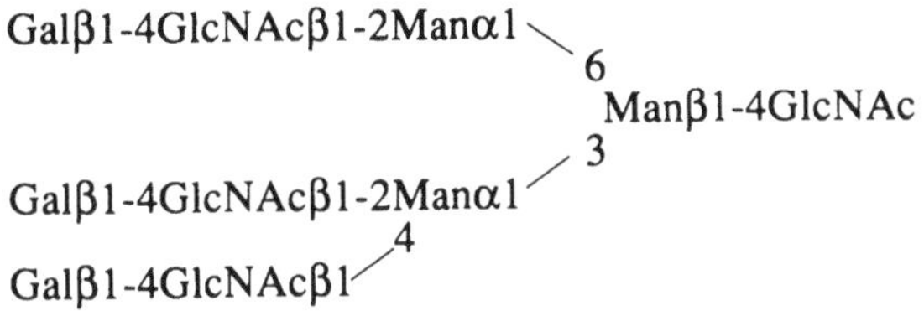

III

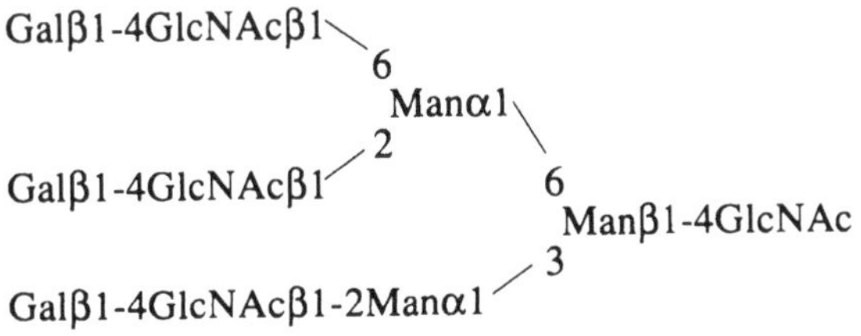

IV

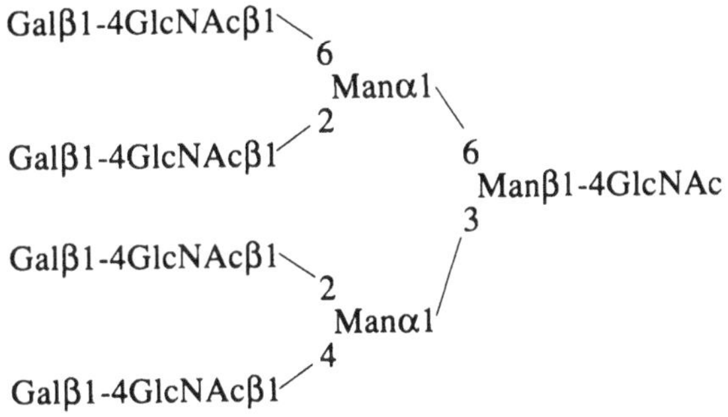

TABLE LC 6
HPLC of Gangliosides on Amino Phases

Packing	P1	P2	P2
Column			
Length, cm	25	20	20
Diameter (I.D.), cm	0.4	0.6	0.6
Material	SS	SS	SS
Solvent	S1	S2	S3
Flow rate (ml/min)	1	2	2
Detection	UV	UV	UV
Reference	1	2	2
Compound	**RRT[a]**	**RRT[a]**	**RRT[b]**
G_{M3}	0.20	—	—
G_{M2}	0.42	—	—
G_{M1}	0.66	0.55	—
Fuc-G_{M1}	0.90	—	—
G_{D1a}	1.00	1.00	—
G_{D1b}	1.18	1.22	—
Fuc-G_{D1b}	1.35	—	—
G_{T1b}	1.50	1.57	—
G_{Q1b}	1.87	2.29	—
G_{Q1c}	—	—	1.00
G_{P1}	—	—	1.56
G_{H1}	—	—	2.34

Note: TLC data for gangliosides will be found in Table LC 7.

[a] RRT = $t_r - t_0$ for compound, relative to $t_r - t_0$ for G_{D1a} (38 min,[1] 13.4 min.[2]

[b] RRT relative to $t_r - t_0$ for G_{Q1c} (12.8 min).

Packing
P1 = LiChrosorb®-NH_2 (Merck; 7-μm particles)
P2 = SSC® NH_2-2201N (5-μm aminopropyl-silica gel; Senshu Science, Tokyo).

Solvent
S1 = (A) acetonitrile-5 m*M* phosphate buffer, pH 5.6 (83:17); (B) acetonitrile-20 m*M* phosphate buffer, pH 5.6 (1:1); elution program: solvent A for 7 min; linear gradient A → A-B (33:17) in 53 min; linear gradient to A-B (9:16) in 20 min.
S2 = (A) acetonitrile-5 m*M* phosphate buffer, pH 5.5 (4:1); (B) acetonitrile-20 m*M* phosphate buffer, pH 5.5 (1:1); elution program: solvent A for 3.5 min; linear gradient A → A-B (9:16) in 26.5 min; isocratic for 10 min.
S3 = (A) acetonitrile-30 m*M* phosphate buffer, pH 5.5 (7:3); (B) acetonitrile-50 m*M* phosphate buffer, pH 5.5 (3:7); elution program: solvent A for 5 min; linear gradient A → A-B (2:3) in 30 min.

REFERENCES

1. **Gazzotti, G., Sonnino, S., and Ghidoni, R.,** Normal-phase high-performance liquid chromatographic separation of non-derivatized ganglioside mixtures, *J. Chromatogr.*, 348, 371, 1985.
2. **Ando, S., Waki, H., and Kon, K.,** High-performance liquid chromatography of underivatized gangliosides, *J. Chromatogr.*, 408, 285, 1987.

TABLE LC 7
HPLC of Methyl Glycosides on Reversed-Phase Packings

Packing		P1	P1	P1	P2	P3
Column						
Length, cm		10	10	10	10	25
Diameter (I.D.), cm		0.8	0.8	0.8	0.46	0.46
Form		Cartridge	Cartridge	Cartridge	Column	Column
Solvent		H_2O	H_2O	H_2O	H_2O	S1
Flow rate (ml/min)		2	2	2.5	0.7	0.5
Detection		RI	RI	RI	RI	RI,D1
Reference		1	2	3	4	5
Sugar	**Methyl glycoside**	**r[a]**	**r[a]**	**k'[b]**	**r[a]**	**k'[b]**
D-Ribose	α-Furanoside	1.23	1.25	—	—	—
	β-Furanoside	1.47	1.42	—	—	—
	α-Pyranoside	2.03	1.93	—	—	—
	β-Pyranoside	2.47	2.32	—	—	—
D-Xylose	α,β-Furanoside	1.17; 1.25	1.05; 1.15	—	—	—
	α-Pyranoside	1.50	1.40	—	1.46	—
	β-Pyranoside	1.77	1.62	—	1.63	—
L-Arabinose	α,β-Furanoside	1.10; 1.23	1.07; 1.21	—	—	—
	α,β-Pyranoside	1.50	1.39	—	—	—
D-Galactose	α,β-Pyranoside	0.83	0.78	—	0.87	—
	α,β-Furanoside	1.03; 1.17	0.96; 1.10	—	1.20	—
D-Glucose	α-Pyranoside	1.00	1.00	—	1.00	—
	β-Pyranoside	1.07	—	—	1.06	—
	α-Furanoside	1.37	1.35	—	—	—
	β-Furanoside	1.47	—	—	—	—
D-Mannose	α-Pyranoside}	1.40	1.31	—	1.33	—
	β-Pyranoside}	1.67; 1.77	1.51	—	1.53	—
	α,β-Furanoside					
D-Fructose	β-Furanoside	0.93	—	—	—	—
	α-Pyranoside	1.13	—	—	—	—
	α-Furanoside	1.27	—	—	—	—
	β-Furanoside	1.83	—	—	—	—
L-Rhamnose	Not assigned	—	3.02; 3.32	—	—	—
L-Fucose	Not assigned	—	1.70; 2.17;	—	2.17; 2.97;	—
			3.15; 3.54		3.27	
D-Glucuronic acid	β-Pyranoside[c]	—	4.54	—	—	3.5
	α-Pyranoside[c]	—	5.22	—	—	4.5
	α,β-Furanoside[c]	—	4.83; 6.25	—	—	—
D-Glucurono-6,3-lactone	β-Furanoside	—	1.06	—	—	1.0
	α-Furanoside	—	1.16	—	—	—
	α,β-Pyranoside	—	2.77[d]	—	—	—
D-Galacturonic acid	α,β-Furanoside[c]	—	3.94	—	—	—
	α,β-Pyranoside[c]	—	4.32; 4.60[d]	—	—	—
D-Mannuronic acid[c]	Not assigned	—	—	7.6;[d] 11.7	—	—
D-Mannurono-6,3-lactone	Not assigned	—	—	0.7; 1.2	—	—
L-Guluronic acid[c]	Not assigned	—	—	6.8; 7.9	—	—
L-Gulurono-6,3-lactone	Not assigned	—	—	3.0; 4.6	—	—
L-Iduronic acid[c]	Not assigned	—	—	—	—	2.5;[d] 2.9
				(13 min)		3.2; 4.8
						(31.7 min)

[a] t_r relative to methyl α-D-glucopyranoside (3.0 min,[1] 3.2 min,[2] 4.2 min[4]).
[b] t_r of latest-eluting peak given in brackets at foot of column.
[c] Methyl ester.
[d] Minor peak.

TABLE LC 7 (continued)
HPLC of Methyl Glycosides on Reversed-Phase Packings

Packing	P1 = Dextropak® (Waters); C_{18}-bonded silica in plastic cartridge operated under radial compression (Waters module RCM 100).
	P2 = CP-Microspher® C_{18} (Chrompack, Middelburg, Netherlands); 3-μm silica with C_{18} bonded phase.
	P3 = ODS-Hypersil® (Shandon, Runcorn, Cheshire, U.K.).
Solvent	S1 = water-methanol (19:1).
Detection	D1 = automated carbole colorimetric method.

REFERENCES

1. **Cheetham, N. W. H. and Sirimanne, P.,** High-performance liquid chromatographic separation of methyl glycosides, *J. Chromatogr.*, 208, 100, 1981.
2. **Cheetham, N. W. H. and Sirimanne, P.,** Methanolysis studies of carbohydrates, using HPLC, *Carbohydr. Res.*, 112, 1, 1983.
3. **Annison, G., Cheetham, N. W. H., and Couperwhite, I.,** Determination of the uronic acid composition of alginates by high-performance liquid chromatography, *J. Chromatogr.*, 264, 137, 1983.
4. **Hjerpe, A., Engfeldt, B., Tsegenidis, T., and Antonopoulos, C. A.,** Separation and determination of neutral monosaccharides using methanolysis and high-performance liquid chromatography, *J. Chromatogr.*, 259, 334, 1983.
5. **Hjerpe, A., Antonopoulos, C. A., Classon, B., Engfeldt, B., and Nurminen, M.,** Uronic acid analysis by high-performance liquid chromatography after methanolysis of glycosaminoglycans, *J. Chromatogr.*, 235, 221, 1982.

TABLE LC 8
HPLC of Partially Methylated Sugars and Polyols on Reversed-Phase Packings

Packing	P1	P2	P2	P3	P3	P3	P3	P3
Column								
Length, cm	30	10	10	25	25	25	25	25
Diameter (I.D.), cm	0.38	0.8	0.8	0.46	0.46	0.46	0.46	0.46
Form	Column	Cartridge	Cartridge	Column	Column	Column	Column	Column
Solvent	S1	S2	S3	H_2O	S4	S5	S6	S7
Flow rate (ml/min)	2.5	1	2	1.8	1.8	1.8	1.8	1.8
Detection	RI	RI, D1	RI, D1	RI	RI	RI	D2	D2
Reference	1	2	2	3	3	3	3	3
Compound	r[a]	RRT[b]	RRT[b]	r[c]	r[d]	r[d]	r[d]	r[d]
D-Xylose								
2,3,4-tri-*O*-methyl	0.85; 1.18	—	—	—	—	—	—	—
D-Mannose								
2,3,6-tri-*O*-methyl	0.85; 1.22(α)	—	—	—	—	—	—	—
2,3,4,6-tetra-*O*-methyl	0.81; 1.30(α)	—	—	—	—	—	—	—
D-Galactose								
2,3-di-*O*-methyl	1.04; 1.44(α)	0.22; 0.30	— 0.88; 1.00	—	—	—	—	—
2,3,6-tri-*O*-methyl	—	0.65; 1.02	0.88; 0.94	—	—	—	—	—
2,4,6-tri-*O*-methyl	—	0.65; 0.92	1.06; 0.94	—	—	—	—	—
2,3,4,6-tetra-*O*-methyl	0.81; 1.30(α)	—	2.00	—	—	—	—	—
D-Glucose								
2-*O*-methyl	0.55	—	—	—	—	—	—	—
3-*O*-methyl	0.55	—	—	—	—	—	—	—
6-*O*-methyl	0.55	—	—	—	—	—	—	—
2,3-di-*O*-methyl	0.85; 0.96(α)	0.40; 0.43	—	—	—	—	—	—
2,4-di-*O*-methyl	—	0.28; 0.32	—	—	—	—	—	—
2,3,4-tri-*O*-methyl	1.30; 1.52(α)	2.32; 3.63	1.50; 2.00	—	—	—	—	—
2,3,6-tri-*O*-methyl	0.81; 1.00(α)	0.87; 1.00	1.00	—	—	—	—	—
2,4,6-tri-*O*-methyl	—	1.16; 1.80	1.06; 1.25	—	—	—	—	—
2,3,4,6-tetra-*O*-methyl	2.44; 2.67(α)	—	4.31; 4.62	—	—	—	—	—

TABLE LC 8 (continued)
HPLC of Partially Methylated Sugars and Polyols on Reversed-Phase Packings

Compound	r[a]	RRT[b]	RRT[b]	r[c]	r[d]	r[d]	r[d]	r[d]
D-Glucitol								
2-*O*-methyl	—	—	—	1.41	0.29	0.62	0.29	0.39
3-*O*-methyl	—	—	—	1.18	—	—	—	—
4-*O*-methyl	—	—	—	1.35	—	—	—	—
6-*O*-methyl	—	—	—	1.59	—	—	—	—
2,3-di-*O*-methyl	—	0.30	—	2.76	0.46	0.76	0.46	0.72
2,6-di-*O*-methyl	—	—	—	—	0.49	—	—	—
4,6-di-*O*-methyl	—	—	—	—	0.72	—	—	—
2,3,4-tri-*O*-methyl	—	0.70	0.94	—	1.00	1.00	1.00	1.00
2,3,6-tri-*O*-methyl	—	0.75	0.94	—	1.13	1.00	1.13	1.05
2,4,6-tri-*O*-methyl	—	1.38	1.12	—	1.68	1.31	—	—
2,3,4,6-tetra-*O*-methyl	—	—	1.69	—	3.54	2.48	2.46	1.94
D-Mannitol								
3-*O*-methyl	—	—	—	1.29	—	—	—	—
4-*O*-methyl	—	—	—	1.29	—	—	—	—
2,3-di-*O*-methyl	—	—	—	—	0.43	—	—	—
2,4-di-*O*-methyl	—	—	—	—	0.80	—	—	—
3,4-di-*O*-methyl	—	—	—	—	0.40	—	—	—
4,6-di-*O*-methyl	—	—	—	—	0.70	—	—	—
2,3,4-tri-*O*-methyl	—	—	—	—	1.01	—	—	—
2,3,6-tri-*O*-methyl	—	—	—	—	1.16	—	—	—
2,4,6-tri-*O*-methyl	—	—	—	—	2.68	1.93	—	—
3,4,6-tri-*O*-methyl	—	—	—	—	1.33	—	—	—
2,3,4,6-tetra-*O*-methyl	—	—	—	—	—	3.00	—	—
Galactitol								
3-*O*-methyl	—	—	—	1.24	—	—	—	—
4-*O*-methyl	—	—	—	1.24	—	—	—	—
6-*O*-methyl	—	—	—	1.41	—	—	—	—
2,3-di-*O*-methyl	—	0.25	—	—	—	—	—	—
2,3,4-tri-*O*-methyl	—	0.72	0.88	—	—	—	—	—
2,3,6-tri-*O*-methyl	—	0.80	0.94	—	—	—	—	—
2,4,6-tri-*O*-methyl	—	0.37	1.00	—	—	—	—	—
2,3,4,6-tetra-*O*-methyl	—	—	2.00	—	—	2.69	—	—

myo-Inositol								
1-*O*-methyl	—	—	—	1.06	—	—	—	—
1,3-di-*O*-methyl	—	—	—	1.65	0.33	—	—	—
1,2,3,5-tetra-*O*-methyl	—	—	—	—	1.22	1.14	—	—
1,3,4,5-tetra-*O*-methyl	—	—	—	—	2.01	1.52	—	—
1,3,4,6-tetra-*O*-methyl	—	—	—	—	1.04	1.07	—	—
1,4,5,6-tetra-*O*-methyl	—	—	—	—	2.90	1.93	—	—
1,2,4,5,6-penta-*O*-methyl	—	—	—	—	—	3.45	—	—

[a] t_r relative to α-anomer of 2,3,6-tri-*O*-methyl-D-glucose (1.35 min).
[b] RRT = t_r-t_o for compound relative to that for α-anomer of 2,3,6-tri-*O*-methyl-D-glucose (6.0 min with S2, 1.6 min with S3).
[c] t_r relative to D-glucitol (1.7 min).
[d] t_r relative to 2,3,4-tri-*O*-methyl-D-glucitol (6.9 min with S4 and S6, 2.9 min with S5, 6.6 min with S7).

Packing
P1 = μ-Bondapak® C_{18} (Waters).
P2 = Dextropak® (Waters).
P3 = Supelcosil® LC-18 (Supelco).

Solvent
S1 = 1% aqueous ammonium acetate, pH 6.9 - ethanol (9:1).
S2 = water-methanol (49:1).
S3 = water-methanol (4:1).
S4 = water-acetonitrile (99:1).
S5 = water-acetonitrile (19:1).
S6 = S4 for 10 min, then S5.
S7 = hyperbolic gradient, water → S5 in 10 min (program no. 5, Waters Model 660 solvent programmer); isocratic with S5 for 5 min.

Detection
D1 = polarimetric detection; $[\alpha]_{350}$ recorded on spectropolarimeter with 10-μl flow-through cell, operating simultaneously with refractometer.
D2 = scintillation counting after labeling by sodium borotritiide reduction.

REFERENCES

1. **Cheetham, N. W. H. and Sirimanne, P.,** Separation of partially methylated sugars by reversed-phase high-performance liquid chromatography, *J. Chromatogr.*, 196, 171, 1980.
2. **Heyraud, A. and Salemis, P.,** Liquid chromatography in the methylation analysis of carbohydrates, and the use of combined refractometric-polarimetric detection, *Carbohydr. Res.*, 107, 123, 1982.
3. **Saadat, S. and Ballou, C. E.,** Separation of *O*-methylhexitols and *O*-methyl-*myo*-inositols by reverse-phase high-performance liquid chromatography, *Carbohydr. Res.*, 119, 248, 1983.

TABLE LC 9
HPLC of Milk Oligosaccharides, Glycopeptides, and Glycoprotein-Derived Reduced Oligosaccharides on Reversed-Phase Packings

Packing	P1	P2	P3	P2	P2
Column					
Length, cm	20	25	25	25	25
Diameter (I.D.), cm	0.8	0.46	0.46	0.46	0.46
Form	Cartridges[a]	Column	Column	Column	Column
Solvent	H_2O	H_2O	H_2O	S1	S1
Flow rate (ml/min)	2	0.5	2	0.5	0.5
Temperature (°C)	Ambient	Ambient	Ambient	25	5
Detection	RI	UV	UV	UV	UV
Reference	1	2	3	4	4
Compound	**k′[b]**	**k′[b]**	**k′[b]**	**k′[b]**	**k′[b]**
Milk oligosaccharides					
Galβ1-3GlcNAcβ1-3Galβ1-4Glc	1.40; 1.91	1.14; 1.48	—	—	—
Galβ1-4GlcNAcβ1-3Galβ1-4Glc	1.23; 1.46	0.94; 1.04	—	—	—
Fucα1-2Galβ1-3GlcNAcβ1-3Galβ1-4Glc	2.51; 3.74	1.80; 2.65	—	—	—
Galβ1-3(Fucα1-4)GlcNAcβ1-3Galβ1-4Glc	0.88; 1.06; 1.23	0.71; 0.86	—	—	—
Galβ1-4(Fucα1-3)GlcNAcβ1-3Galβ1-4Glc	0.88; 1.06; 1.23	0.63; 0.71	—	—	—
Fucα1-2Galβ1-3(Fucα1-4)GlcNAcβ1-3Galβ1-4Glc	0.80; 0.91	0.43; 0.57	—	—	—
Galβ1-3(Fucα1-4)GlcNAcβ1-3Galβ1-4(Fucα1-3)Glc	0.80; 0.91 (16.6min)	— (12.8min)	—	—	—
Reduced oligosaccharides					
Galβ1-3GalNAc-ol	—	—	2.1	—	—
GlcNAcβ1-3GalNAc-ol	—	—	2.5	—	—
Fucα1-2Galβ1-3GalNAc-ol	—	—	9.0	—	—
Galβ1-4GlcNAcβ1-6(Galβ1-3)GalNAc-ol	—	—	2.3	—	—
Fucα1-2Galβ1-3GlcNAcβ1-3GalNAc-ol	—	—	3.5	—	—
Fucα1-2Galβ1-3(GlcNAcβ1-6)GalNAc-ol	—	—	11.1	—	—
Fucα1-2Galβ1-3(Galβ1-4GlcNAcβ1-6)GalNAc-ol	—	—	12.5	—	—

Fucα1-2Galβ1-4GlcNAcβ1-6(Galβ1-3)GalNAc-ol	—	—	3.8	—	—
Fucα1-2Galβ1-3(Fucα1-4)GlcNAcβ1-3Galβ1-3GalNAc-ol	—	—	1.8	—	—
Fucα1-2Galβ1-3(Fucα1-2Galβ1-4GlcNAcβ1-6)GalNAc-ol	—	—	24.6	—	—
Fucα1-2Galβ1-3[Fucα1-2Galβ1-4(Fucα1-3)GlcNAcβ1-6]GalNAc-ol	—	—	19.1 (25.5 min)	—	—
Ovalbumin glycopeptides[c]					
E_3	—	—	—	1.57	2.26
D_3	—	—	—	1.28	1.86
C_3B	—	—	—	1.97 (10.2 min)	2.71 (13.0 min)

Note: Data on normal-phase HPLC of reduced oligosaccharides in Reference 3 will be found in Table LC 5.

[a] Two cartridges, each 10 × 0.8 cm, connected in series.
[b] t_r of last-eluting peak given in brackets at end of each column of data.
[c] Structures of glycopeptides appended.

Packing P1 = Dextropak®; under radial compression (Waters module RCM 100).
P2 = 600 RP (Alltech).
P3 = 605 RP (Alltech; 5-μm particles).
Solvent S1 = aqueous phosphate buffer (1 m*M*, pH 6.0).

REFERENCES

1. **Cheetham, N. W. H. and Dube, V. E.,** Preparation of lacto-*N*-neotetraose from human milk by high-performance liquid chromatography, *J. Chromatogr.*, 262, 426, 1983.
2. **Dua, V. K. and Bush, C. A.,** Identification and fractionation of human milk oligosaccharides by proton-nuclear magnetic resonance spectroscopy and reverse-phase high-performance liquid chromatography, *Anal. Biochem.*, 133, 1, 1983.
3. **Dua, V. K., Dube, V. E., and Bush, C. A.,** The combination of normal-phase and reverse-phase high-pressure liquid chromatography with NMR for the isolation and characterization of oligosaccharide alditols from ovarian cyst mucins, *Biochim. Biophys. Acta*, 802, 29, 1984.
4. **Dua, V. K. and Bush, C. A.,** Resolution of some glycopeptides of hen ovalbumin by reverse-phase high-pressure liquid chromatography, *Anal. Biochem.*, 137, 33, 1984.

Appendix to Table LC 9: Structures of Glycopeptides

E_3

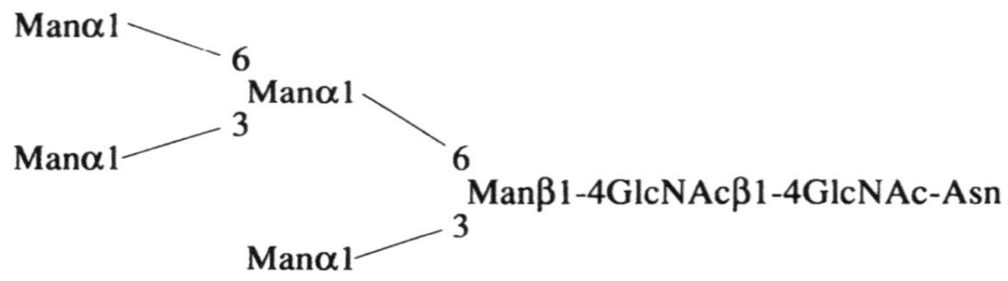

D_3

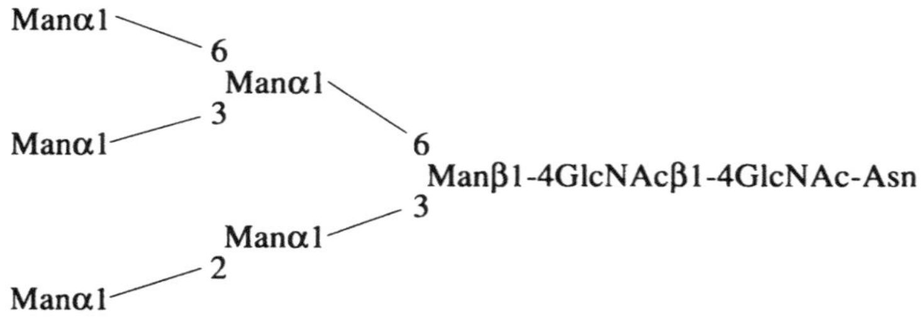

C_3B

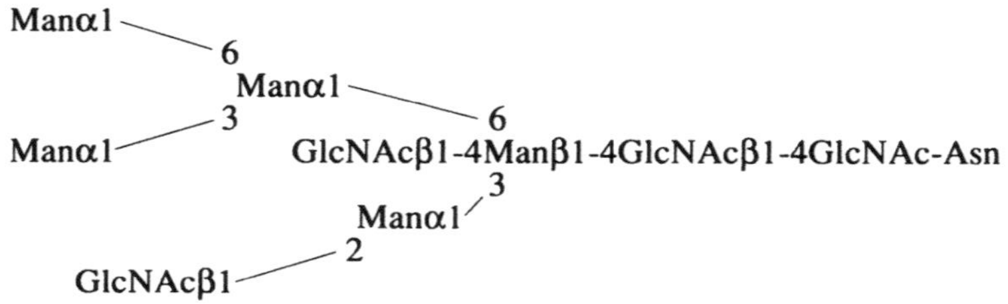

TABLE LC 10
HPLC of Malto-Oligosaccharides and Oligosaccharide-Alditols on Silica-Based Reversed-Phase Packings

Packing	P1	P2	P2	P2	P2	P2	P3	P2
Column								
Length, cm	30	10	10	10	10	10	10	10
Diameter (I.D.), cm	0.39	0.8	0.8	0.8	0.8	0.8	0.8	0.8
Material	SS	Plastic	Plastic	Plastic	Plastic	Plastic	Plastic	Plastic
Form	Column	Cartridge	Cartridge	Cartridge	Cartridge	Cartridge	Cartridge	Cartridge
Solvent	H_2O	H_2O	H_2O	S1	S2	S3	H_2O	S4
Flow rate (ml/min)	0.3	1	2	2	2	2	2	2
Temperature (°C)	15	Ambient	Ambient	Ambient	Ambient	Ambient	Ambient	Ambient
Detection	RI	RI	RI	RI	RI	RI	RI	RI
Reference	1	2	3	3	3	3	3	3
DP	r[a]	r[a]	r[b]	r[b]	r[b]	r[b]	r[b]	r[a]
Oligosaccharides								
1	0.89	0.77	—	—	—	—	—	—
2	1.00	1.00	—	—	—	—	—	1.00
3	1.17	1.26; 1.36	1.00	1.00	1.00	1.00	1.00	1.55; 1.77
4	1.40; 1.48[c]	1.64; 1.90	1.17; 1.30[c]	1.11	1.18	1.24	1.16	2.50; 3.14
5	1.76; 1.84[c]	2.46; 2.72	1.60; 1.80[c]	1.32	1.47	1.64	1.37	4.59; 5.45
6	2.43; 2.55[c]	3.41; 3.74	2.23; 2.40[c]	1.64	1.84	2.17	1.58	—
7	3.05[d]	4.31; 4.54[c]	2.87; 3.03	1.93	2.21	2.67	2.00	—
8	3.54[d]	—	3.37; 3.63	2.25	—	—	2.37	—
9	—	—	4.13; 4.70	2.61	—	—	2.74	—
10	—	—	5.47; 6.40	3.14	—	—	3.26	—
11	—	—	7.53; 8.83	3.86	—	—	4.00	—
12	—	—	10.4;11.8	5.00	—	—	5.21	—
13	—	—	13.9; 15.7	6.71	—	—	6.42	—
14	—	—	—	—	—	—	8.37	—
15	—	—	—	—	—	—	10.2	—
Oligosaccharide-alditols								
1	—	0.77	—	—	—	—	—	—
2	—	0.79	—	—	—	—	—	—
3	—	1.23	—	—	—	—	—	—

TABLE LC 10 (continued)
HPLC of Malto-Oligosaccharides and Oligosaccharide-Alditols on Silica-Based Reversed-Phase Packings

DP	r[a]	r[a]	r[b]	r[b]	r[b]	r[b]	r[b]	r[a]
4	—	1.69	—	—	—	—	—	—
5	—	2.41	—	—	—	—	—	—
6	—	3.31	—	—	—	—	—	—
7	—	4.31	—	—	—	—	—	—
Packing	P4	P4	P4	P5	P6	P6	P7	P7
Column								
Length, cm	15	15	15	15	25	25	25	25
Diameter (I.D.), cm	0.46	0.46	0.46	0.32	0.6	0.6	0.46	0.46
Material	SS	SS	SS	Glass	SS	SS	SS	SS
Form	Column	Column	Column	Column	Column	Column	Column	Column
Solvent	H_2O	S5	S6	H_2O	H_2O	H_2O	H_2O	H_2O
Flow rate (ml/min)	1	1	1	0.23	0.8	0.8	1	1
Temperature (°C)	20	20	20	Ambient	24	60	25	35
Detection	RI	RI	RI	RI	RI	RI	RI	RI
Reference	4	4	4	5	6	6	7	7

DP	r[a]	r[a]	r[a]	r[a]	r[a]	r[a]	r[a]	r[a]
Oligosaccharides								
1	0.89	—	0.98	0.91	0.89	1.00	0.83	0.85
2	1.00	1.00	1.00	1.00	1.00	1.00	1.00	1.00
3	1.17	1.23	1.13	1.14	1.20; 1.24[c]	1.08	1.17; 1.24	1.13; 1.19
4	1.36; 1.42	1.50; 1.61	1.24	1.32	1.39; 1.47	1.17	1.45; 1.59	1.38; 1.50
5	1.65; 1.74	1.96; 2.11	1.43	1.64	1.82; 1.94	1.26	1.90; 2.14	1.73; 1.92
6	2.12; 2.22	2.69; 2.88	1.69[d]	2.00	2.28; 2.40[c]	1.35	2.59; 2.76	2.15; 2.30
7	2.56; 2.90	3.50; 4.17	1.92; 2.03[c]	2.34	2.71[d]	1.44	3.17; 3.38	2.54; 2.69
8	3.31; 3.56	—	2.29; 2.63	2.73	—	1.53	3.76; 4.03	2.92; 3.08
9	—	—	—	—	—	1.80	—	—
10	—	—	—	—	—	1.98	—	—
11	—	—	—	—	—	—	—	—
12	—	—	—	—	—	—	—	—
13	—	—	—	—	—	—	—	—

14	—	—	—	—	—	—	—	—
15	—	—	—	—	—	—	—	—
Oligosaccharide-alditols								
1	—	—	—	—	—	—	—	—
2	—	—	—	—	—	—	—	—
3	—	—	—	—	—	—	—	—
4	—	—	—	—	—	—	—	—
5	—	—	—	—	—	—	—	—
6	—	—	—	—	—	—	—	—
7	—	—	—	—	—	—	—	—

[a] t_r relative to maltose (3.9 min,[2] 2.2 min;[3] 1.91 min in H_2O and $S6^{4-}$, 2.06 min in S5;[4] 4.4 min;[5] 4.3 min at 24°C,[6] 3.7 min at 60°C[6]; 2.9 min at 25°C,[7] 2.6 min at 35°C[7]).
[b] t_r relative to maltotriose (3.0 min in H_2O, 2.8 min in S1, 1.9 min in S2, and with P3, 2.1 min in S3)[3].
[c] Shoulder.
[d] Broad peak due to unresolved anomers.

Packing
P1 = μ-Bondapak®-C_{18} (Waters; 10-μm particles).
P2 = Dextropak® (Waters; 10-μm particles), used under radial compression (Waters module RCM-100).
P3 = Dextropak®, modified by coating with nonionic detergent Triton® X-100 (*p-tert*-octylphenoxypolyethoxyethanol oligomers); aqueous solution of Triton® X-100 (0.1% w/v) pumped through cartridge at 1 ml/min for 300 min before use of cartridge.
P4 = Polygosil® RP-18 (Macherey-Nagel); irregularly shaped particles of average diameter 5 μm.
P5 = Separon® 6 RPS (Laboratory Instrument Works, Prague; 10-μm particles), doped with primary amino groups (0.25 mmol/g).
P6 = Separon® 6 RPS without prior treatment.
P7 = Spherisorb® S 5 ODS2 (Phase Separations, Queensferry, U.K.; 5-μm particles).

Solvent
S1 = water containing tetramethylurea (0.025%).
S2 = 0.1 *M* urea in water.
S3 = 0.1 *M* guanidine hydrochloride in water.
S4 = 1 *M* ammonium sulfate.
S5 = 1 *M* sodium chloride.
S6 = water containing triethylamine (1 m*M*).

REFERENCES

1. **Heyraud, A. and Rinaudo, M.,** Carbohydrate analysis by high pressure liquid chroamtography using water as the eluent, *J. Liq. Chromatogr.*, 3, 721, 1980.
2. **Cheetham, N. W. H, Sirimanne, P., and Day, W. R.,** High-performance liquid chromatographic separation of carbohydrate oligomers, *J. Chromatogr.*, 207, 439, 1981.

TABLE LC 10 (continued)
HPLC of Malto-Oligosaccharides and Oligosaccharide-Alditols on Silica-Based Reversed-Phase Packings

3. **Cheetham, N. W. H. and Teng, G.,** Some applications of reversed-phase high-performance liquid chromatography to oligosaccharide separations, *J. Chromatogr.*, 336, 161, 1984.
4. **Verhaar, L. A. Th., Kuster, B. F. M., and Claessens, H.-A.,** Retention behaviour of carbohydrate oligomers in reversed-phase chromatography, *J. Chromatogr.*, 284, 1, 1984.
5. **Porsch, B.,** High-performance liquid chromatography of oligosaccharides in water on a reversed phase doped with primary amino groups, *J. Chromatogr.*, 320, 408, 1985.
6. **Vrátný, P., Čoupek, J., Vozka, S., and Hostomská, Z.,** Accelerated reversed-phase chromatography of carbohydrate oligomers, *J. Chromatogr.*, 254, 143, 1983.
7. **Rajakylä, E.,** Use of reversed phase chromatography in carbohydrate analysis, *J. Chromatogr.*, 353, 1, 1986.

TABLE LC 11
HPLC of Oligosaccharides of Various Series on Silica-Based Reversed-Phase Packings

Packing	P1	P2	P3	P4	P4	P4	P5	P1	P1	P6	P7	P1	P1	P1	P1
Column															
Length, cm	10	30	15	15	15	15	15	10	10	10	25	10	10	10	10
Diameter (I.D.), cm	0.8	0.39	0.32	0.46	0.46	0.46	0.46	0.8	0.8	0.8	0.4	0.8	0.8	0.8	0.8
Material	Plastic	SS	Glass	SS	SS	SS	SS	Plastic	Plastic	Plastic	SS	Plastic	Plastic	Plastic	Plastic
Form	Cartridge	Column	Column	Column	Column	Column	Column	Cartridge	Cartridge	Cartridge	Column	Cartridge	Cartridge	Cartridge	Cartridge
Solvent	H_2O	H_2O	H_2O	H_2O	S1	S2	S3	H_2O	S4	H_2O	H_2O	S5	H_2O	H_2O	H_2O
Flow rate (ml/min)	1	F1	0.52	1	1	1	1	1	2	2	2	2	2	3	1
Temperature (°C)	Ambient	15	70	20	20	20	20	Ambient	Ambient	Ambient	Ambient	Ambient	Ambient	Ambient	Ambient
Detection	RI	RI	RI	RI	RI	RI	RI	RI	RI	RI	RI	RI	RI	RI	RI
Reference	1	2	3	4	4	4	4	1	5	5	5	5	6	6	1
	Cello-oligosaccharides							Isomalto-							Xylor
Oligosaccharides	r^a	r^a	r^a	r^b	r^b	r^b	r^b	r^c	r^c	r^d	r^c	r^d	r^d	r^e	f
Linear, DP															
1	—	0.88	0.88	0.51	—	—	—	—	—	—	—	—	—	—	—
2	1.00	1.00	1.00	0.63	—	—	0.93	1.00	1.00	—	1.00	—	—	—	1.00
3	2.50; 2.71	1.42; 1.67	1.18	1.00; 1.03	1.00	1.00	1.00	1.48	1.24	1.00	1.20	1.00	1.00	—	1.27
4	7.50; 8.36	2.23; 2.37	1.57	2.36; 2.58	1.60; 1.71	1.44; 1.52	1.20	2.42; 2.52[g]	1.67	1.35	1.46	4.50	1.67	1.00	1.89; 2.03
5	—	4.58; 4.88	2.84	7.08; 7.92	3.21; 3.47	2.52; 2.67	1.78	4.78[h]	2.52	1.95	1.77	—	3.26	1.68	3.70; 4.00
6	—	—	4.80	—	7.33; 7.98	4.98; 5.40	2.53; 2.60[g]	9.30[i]; 9.98[g]	3.95	3.00	2.38	—	—	3.05	—
7	—	—	—	—	—	—	4.07[h]	19.3; 19.9[g]	6.81	4.80[h]	3.31[h]	—	—	5.05	—
8	—	—	—	—	—	—	—	—	—	8.05[h]	4.77[h]	—	—	9.26	—

Branched															
Isomaltotriose,															
3^3-α-D-glucosyl-	—	—	—	—	—	—	—	—	—	—	—	3.80	1.67	0.84	—
3^3-isomaltosyl-	—	—	—	—	—	—	—	—	—	—	—	—	2.86	1.16	—
Isomaltotetraose,															
3^3-α-D-glucosyl-	—	—	—	—	—	—	—	—	—	—	—	—	—	1.42	—
3^3-isomaltosyl-	—	—	—	—	—	—	—	—	—	—	—	—	—	2.00	—
Isomaltopentaose,															
3^3-α-D-glucosyl-	—	—	—	—	—	—	—	—	—	—	—	—	—	2.84	—
3^3-isomaltosyl-	—	—	—	—	—	—	—	—	—	—	—	—	—	3.52	—
Isomaltohexaose,															
3^3-α-D-glucosyl-	—	—	—	—	—	—	—	—	—	—	—	—	—	4.42	—
3^3-isomaltosyl-	—	—	—	—	—	—	—	—	—	—	—	—	—	5.74	—
Isomaltoheptaose,															
3^3-α-D-glucosyl-	—	—	—	—	—	—	—	—	—	—	—	—	—	7.05	—

[a] t_r relative to cellobiose (4.2 min,[1] 6.5 min,[2] 1.7 min[3]).
[b] t_r relative to cellotriose (3.31 min in H_2O, 2.45 min in S1, 2.25 min in S2, 2.0 min on P5 in S3).[4]
[c] t_r relative to isomaltobiose (4.0 min;[1] 2.1 min on P1 in S4,[5] 1.3 min on P7 in H_2O).[5]
[d] t_r relative to isomaltotriose (2.0 min on P1 and P6).[5]
[e] t_r relative to isomaltotetraose (1.9 min).
[f] t_r relative to xylobiose (3.65 min).
[g] Shoulder.
[h] Broad peak due to unresolved anomers.

Packing	P1	=	Dextropak® (Waters; 10-μm particles), used under radial compression (Waters module RCM-100[1,5] or Z-module[6]).
	P2	=	μ-Bondapak® C_{18} (Waters; 10-μm particles).
	P3	=	Separon® 6 RPS (Laboratory Instrument Works, Prague; 5-μm particles).
	P4	=	Polygosil® RP-18 (Macherey-Nagel; irregularly shaped particles, particle diameter 5 μm, pore diameter 60 Å).
	P5	=	Synchropak® R101 (Betron Scientific, Rotterdam; C_{18}-bonded silica, particle diameter 6.5 μm, pore diameter 100Å).
	P6	=	Dextropak®, modified by coating with non-ionic detergent Triton® X-100 (0.1% w/v) pumped through cartridge at 1 ml/min for 260 min before use of cartridge.
	P7	=	LiChrosorb® RP-8 (Merck; C_8-bonded silica, 10-μm particles).
Solvent	S1	=	water-methanol (49:1).
	S2	=	water-methanol (97:3).
	S3	=	water containing 1-pentanol (0.1%).
	S4	=	water containing tetramethylurea (0.025%).
	S5	=	1 *M* ammonium sulfate.
Flow program	F1	=	0.5 ml/min for 10 min, then 1 ml/min.

TABLE LC 11 (continued)
HPLC of Oligosaccharides of Various Series on Silica-Based Reversed-Phase Packings

REFERENCES

1. **Cheetham, N. W. H., Sirimanne, P., and Day, W. R.,** High-performance liquid chromatographic separation of carbohydrate oligomers, *J. Chromatogr.*, 207, 439, 1981.
2. **Heyraud, A. and Rinaudo, M.,** Carbohydrate analysis by high-pressure liquid chromatography using water as the eluent, *J. Liq. Chromatogr.*, 3, 721, 1980.
3. **Vrátný, P., Čoupek, J., Vozka, S., and Hostomská, Z.,** Accelerated reversed-phase chromatography of carbohydrate oligomers, *J. Chromatogr.*, 254, 143, 1983.
4. **Verhaar, L. A. Th., Kuster, B. F. M., and Claessens, H. A.,** Retention behaviour of carbohydrate oligomers in reversed-phase chromatography, *J. Chromatogr.*, 284, 1, 1984.
5. **Cheetham, N. W. H. and Teng, G.,** Some applications of reversed-phase high-performance liquid chromatography to oligosaccharide separations, *J. Chromatogr.*, 336, 161, 1984.
6. **Taylor, C., Cheetham, N. W. H., and Walker, G. S.,** Application of high-performance liquid chromatography to a study of branching in dextrans, *Carbohydr. Res.*, 137, 1, 1985.

TABLE LC 12
HPLC of D-Gluco-Oligosaccharides of Various Series on Polymer-Based Reversed-Phase Packing*

Packing								P1[a]							
Column															
Length, cm								15							
Diameter (I.D.), cm								0.6							
Material								SS							
Form								Column							
Solvent	H_2O	H_2O	S1	H_2O	S1	H_2O	S1	H_2O	S1	H_2O	S1	H_2O	S1	H_2O	S1
Flow rate (ml/min)								1							
Temperature								Ambient							
Detection								RI							
Reference								1							
Series	(1 → 2)-α-D- (Koji-)	(1 → 2)-β-D- (Sophoro-)		(1 → 3)-α-D- (Nigero-)		(1 → 3)-β-D- (Laminari-)		(1 → 4)-α-D- (Malto-)		(1 → 4)-β-D- (Cello-)		(1 → 6)-α-D- (Isomalto-)		(1 → 6)-β-D- (Gentio-)	
DP	$r^b(t_r)^c$														
2	1.00	1.00; 1.06	1.00	1.00; 1.06	1.00	1.00	1.00	1.00	1.00	1.00	1.00	1.00	1.00	1.00; 1.06	1.00
3	1.04	1.23; 1.28	1.17	1.39; 1.64	1.55	1.60; 1.70	1.68	1.10	1.07	1.37	1.41	1.17	1.14	1.24	1.24
4	1.06	1.51	1.44	2.61; 3.03	2.04	4.43; 5.20	4.57	1.20	1.17	2.21; 2.29	2.50	1.41	1.43	1.48; 1.64	1.48
5	1.09	1.71; 1.88	1.69	5.14; 6.94	6.50	(15.6 min)	(12.8 min)	1.33	1.30	4.63; 5.05	5.31	1.76; 1.86	1.78	1.91; 2.15	1.88
6	(2.98 min)	2.11; 2.34	2.14	(25.0 min)	(23.4 min)			1.43; 1.50	1.40	11.5; 12.6	12.8	2.31; 2.48	2.39	2.61; 3.15	2.64
7		2.80; 3.14	2.83					1.60; 1.70	1.57	(48.0 min)	(41.0 min)	3.21; 3.45	3.43	3.64; 4.42	3.67
8		3.66; 4.23	3.78					1.77; 1.90	1.73			4.55; 4.93	5.00	5.12; 6.18	5.15
9		4.80; 5.31	5.05					2.00; 2.13	1.93			6.62; 7.17	7.28	7.39; 8.97	7.42
10		5.88; 6.94	6.61					2.30; 2.43	2.20			9.59; 10.3	10.7	10.7; 13.0	10.9

11	7.77; 8.57	8.55	2.65; 2.80	2.50	14.1; 15.1	15.7	15.4; 18.5	15.9
12	9.82; 11.8	11.6	3.00; 3.17	2.87	(43.8 min)	(44.0 min)	(61.0 min)	(52.5 min)
	(41.4 min)	(42.0 min)						
13			3.40; 3.63	3.27				
14			3.93; 4.27	3.80				
15			4.63; 5.03	4.47				
16			5.53; 6.00	5.33				
17			6.60; 7.07	6.33				
18			7.80; 8.43	7.53				
19			9.23; 10.0	9.00				
20			11.0; 12.0	10.8				
21			13.1; 14.2	12.9				
22			15.6; 17.0	15.5				
23			18.6; 20.2	18.5				
			(60.6 min)	(55.6 min)				

Note: See data for HPLC on amino-bonded silica in Table LC 3.

[a] Conditions listed in center of column headings apply throughout except for solvent, which changes as indicated.
[b] t_r relative to dimer of series or first anomer eluted (β) where anomer resolution occurs.
[c] t_r of highest oligomer eluted is given in brackets at end of each column.

Packing P1 = Asahipak ODP-50 (Asahi Kasei, Tokyo); C_{18}-bonded vinyl alcohol copolymer gel, 5-μm particles.
Solvent S1 = aqueous sodium hydroxide solution, pH 11.

REFERENCE

1. **Koizumi, K. and Utamura, T.,** High-performance liquid chromatography of gluco-oligomers on C_{18}-bonded vinyl alcohol copolymer gel with alkaline eluents, *J. Chromatogr.*, 436, 328, 1988.

TABLE LC 13
HPLC of Cyclic Oligosaccharides on Silica- and Polymer-Based Reversed-Phase Packings

Packing	P1	P1	P1	P1	P1	P1	P2	P3	P4	P5	P5	P5	P1	P6	P7
Column															
Length, cm	25	25	25	25	25	25	15	15	15	15	15	15	25	20	15
Diameter (I.D.), cm	0.4	0.4	0.4	0.4	0.4	0.4	0.46	0.46	0.46	0.6	0.6	0.6	0.4	0.6	0.6
Material	SS	SS	SS	SS	SS	SS	SS	SS	SS	SS	SS	SS	SS	SS	SS

TABLE LC 13 (continued)
HPLC of Cyclic Oligosaccharides on Silica- and Polymer-Based Reversed-Phase Packings

Solvent	S1	S2	S3	S4	S1	S1	S5	S5	S5	S5	S6	S7	S8	S7	S5
Flow rate (ml/min)	1	1	1	1	1	1	1	1	1	1	1	1	0.7	0.7	0.7
Temperature (°C)	28	28	28	28	33	38	25	25	25	25	25	25	Ambient	Ambient	Ambient
Detection	RI	RI	RI	RI	RI	RI	RI	RI	RI	RI	RI	RI	RI	RI	RI
Reference	1	1	1	1	1	1	2	2	2	2	2	2	3	3	3
Compound	k'														
Cyclomaltohexaose[a]	3.54	2.64	2.01	1.85	2.64	1.74	3.89	3.61	3.80	—	2.38	—	—	—	—
6-O-α-D-glucosyl-	2.35	1.72	1.20	1.81	1.72	1.19	3.05	2.83	2.98	—	1.41	—	—	—	—
6-O-α-maltosyl-	2.29	1.57	1.07	0.70	1.64	1.16	3.57	3.30	3.47	—	1.48	—	—	—	—
6-O-α-maltotriosyl-	2.91	1.99	1.36	0.87	2.08	1.42	5.13	4.75	4.98	—	1.77	—	—	—	—
6-O-α-maltotetraosyl-	—	—	—	—	—	—	5.83	5.35	5.60	—	1.96	—	—	—	—
6-O-α-maltopentaosyl-	—	—	—	—	—	—	4.75	4.25	4.48	—	1.47	—	—	—	—
6-O-α-D-glucosyl-	—	—	—	—	—	—	2.14	—	2.21	—	—	—	—	—	—
di-O-α-maltosyl-	—	—	—	—	—	—	3.25	—	3.23	—	—	—	—	—	—
Cyclomaltoheptaose[a]	12.26	9.27	6.75	5.41	8.95	6.01	11.6	11.0	11.8	9.27	—	—	—	—	—
6-O-α-D-glucosyl-	7.17	5.21	3.75	2.52	5.30	3.83	7.85	6.35	7.95	5.00	—	—	—	—	—
6-O-α-maltosyl-	7.01	4.94	3.43	2.29	5.11	3.59	9.22	8.70	9.36	4.87	—	—	—	—	—
6-O-α-maltotriosyl-	8.75	6.05	4.13	2.70	6.28	4.40	12.8	12.3	13.3	6.18	—	—	—	—	—
6-O-α-maltotetraosyl-	—	—	—	—	—	—	14.4	13.8	14.9	6.44	—	—	—	—	—
di-O-α-D-glucosyl-	4.17	2.88	1.97	1.30	3.19	2.33	4.88; 5.13[b]	—	5.07; 5.42[b]	—	—	—	—	—	—
di-O-α-maltosyl-	—	—	—	—	—	—	7.29	—	6.90	—	—	—	—	—	—
Cyclomaltooctaose[a]	3.17	2.37	1.72	1.68	2.36	1.59	2.83	2.42	2.55	—	—	3.07	—	—	—
6-O-α-D-glucosyl-	1.95	1.41	1.00	0.68	1.45	1.05	2.01	1.74	1.83	—	—	1.91	—	—	—
6-O-α-maltosyl-	1.88	1.32	0.91	0.59	1.40	1.00	2.38	2.08	2.17	—	—	2.05	—	—	—
6-O-α-maltotriosyl-	2.30	1.57	1.08	0.68	1.67	1.14	3.21	2.84	2.95	—	—	2.48	—	—	—
6-O-α-maltotetraosyl-	—	—	—	—	—	—	3.62	3.17	3.30	—	—	2.73	—	—	—
di-O-α-D-glucosyl-	—	—	—	—	—	—	1.45	—	1.38	—	—	—	—	—	—
di-O-α-maltosyl-	—	—	—	—	—	—	2.14	—	1.99	—	—	—	—	—	—

Cyclosophoraoses[c]															
DP 17	—	—	—	—	—	—	—	—	—	—	—	—	1.20	0.87	1.00
18	—	—	—	—	—	—	—	—	—	—	—	—	2.60	1.50	1.62
19	—	—	—	—	—	—	—	—	—	—	—	—	3.80	1.87	2.12
20	—	—	—	—	—	—	—	—	—	—	—	—	3.40	1.87	1.87
21	—	—	—	—	—	—	—	—	—	—	—	—	6.01	2.78	3.49
22	—	—	—	—	—	—	—	—	—	—	—	—	4.41	2.37	2.93
23	—	—	—	—	—	—	—	—	—	—	—	—	6.61	3.99	4.62
24	—	—	—	—	—	—	—	—	—	—	—	—	7.41	4.24	4.86
25	—	—	—	—	—	—	—	—	—	—	—	—	7.41	4.62	5.12
26	—	—	—	—	—	—	—	—	—	—	—	—	11.8	6.12	7.74
27	—	—	—	—	—	—	—	—	—	—	—	—	10.6	6.12	7.11
28	—	—	—	—	—	—	—	—	—	—	—	—	14.2	7.73	9.49
29	—	—	—	—	—	—	—	—	—	—	—	—	16.2	9.23	11.2
30	—	—	—	—	—	—	—	—	—	—	—	—	17.0	9.23	11.2
31	—	—	—	—	—	—	—	—	—	—	—	—	25.4	13.2	16.5
32	—	—	—	—	—	—	—	—	—	—	—	—	24.2	13.2	16.0
33	—	—	—	—	—	—	—	—	—	—	—	—	31.4	16.5	20.0
34	—	—	—	—	—	—	—	—	—	—	—	—	38.6	20.6	24.0
35	—	—	—	—	—	—	—	—	—	—	—	—	38.6	20.6	24.0

Note: See data for HPLC of these oligosaccharides on amino phases (Table LC 4).

[a] Cyclomaltohexaose, -heptaose, and -octaose are α-, β-, and γ-cyclodextrins, respectively.
[b] Splitting of peak, due to positional isomers, seen with highly efficient columns.
[c] Produced by *Rhizobium* and *Agrobacterium* strains.

Packing
- P1 = Hibar LiChrosorb® RP-18 (Merck); 5-μm C_{18}-bonded silica.
- P2 = YMC-Pack® AL-302 ODS (Yamamura Chemical Co., Kyoto); 5-μm C_{18}-bonded silica.
- P3 = YMC-Pack® A-302 ODS; as P2, but "endcapped" by treatment with trimethylchlorosilane to reduce proportion of reactive silanol groups.
- P4 = YMC-Pack® AM-302 ODS; P3 treated further to free it completely from unreacted silanol groups.
- P5 = Asahipak® ODP-50 (Asahi Kasei, Tokyo); C_{18}-bonded vinyl alcohol copolymer gel (5-μm particles).
- P6 = ERC®-ODS-1171 (Erma Optical Works, Tokyo); 3-μm C_{18}-bonded silica.
- P7 = YMC-Pack® AL-312 ODS; 5-μm C_{18}-bonded silica.

Solvent
- S1 = water-methanol (23:2).
- S2 = water-methanol (91:9).
- S3 = water-methanol (9:1).

TABLE LC 13 (continued)
HPLC of Cyclic Oligosaccharides on Silica- and Polymer-Based Reversed-Phase Packings

S4 = water-methanol (89:11).
S5 = water-methanol (47:3).
S6 = water-methanol (93:7).
S7 = water-methanol (19:1).
S8 = water-methanol (94.5:5.5).

REFERENCES

1. **Koizumi, K., Utamura, T., Kuroyanagi, T., Hizukuri, S., and Abe, J.-I.,** Analysis of branched cyclodextrins by high-performance liquid and thin-layer chromatography, *J. Chromatogr.*, 360, 397, 1986.
2. **Koizumi, K., Kubota, Y., Okada, Y., Utamura, T., Hizukuri, S., and Abe, J.-I.,** Retention behaviour of cyclodextrins and branched cyclodextrins on reversed-phase columns in high-performance liquid chromatography, *J. Chromatogr.*, 437, 47, 1988.
3. **Koizumi, K., Okada, Y., Utamura, T., Hisamatsu, M., and Amemura, A.,** Further studies on the separation of cyclic (1 → 2)-β-D-glucans (cyclosophoraoses) produced by *Rhizobium meliloti* IFO 13336, and determination of their degrees of polymerization by high-performance liquid chromatography, *J. Chromatogr.*, 299, 215, 1984.

Data from References 1 and 2 reproduced with permission of authors and Elsevier Science Publishers.

TABLE LC 14
Reversed-Phase HPLC of Derivatized Sugars

Packing	P1	P2	P3	P3	P4	P5	P5	P6	P7	P8	P9	P9	P9	P10	P11
Column															
Length, cm	30	25	20	25	25	5	5	15	10	25	25	25	50	25	25
Diameter (I.D.), cm	0.39	0.46	0.46	0.46	0.46	0.46	0.46	0.39	0.8	0.46	0.46	0.46	0.46	0.46	0.46
Material	SS	SS	SS	SS	SS	SS	SS	SS	Plastic	SS	SS	SS	SS	SS	SS
Solvent	S1	S2	S3	S4	S5	S6	S7	S8	S9	S10	S11	S11	S11	S12	S13
Flow rate (ml/min)	1	1	1	1	1	1.5	1.5	1.2	2	1	1.5	F1	0.5	1	1
Temperature (°C)	28	Ambient	Ambient	Ambient	Ambient	40	40	Ambient	Ambient	Ambient	Ambient	Ambient	Ambient	Ambient	Ambient
Technique	T1	T2	T2	T3	T4	T5	T5	T6	T7	T6	T6	T6	T8	T6	T9
Detection	UV, D1	D2	D2	D2	UV	D3	D3	D4	D5	UV	D6	D6	D6	D7	UV
Reference	1	2	2	2	3	4	4	5	6	7	8	8	8	9	10

	Dansylhydrazones[a]					Dabsylhydrazones[b]			DDB[c]	MPGA[d]	Pyridylamino-[e]			NBDG[f]	Benzoyl[g]
Parent sugar	k'	k'	k'	k'	k'	k'	k'	k'	k'	k'	k'	k'	k'	k'	k'
Gentiobiose	—	1.06	3.00	—	—	4.5	3.7	—	—	—	—	—	—	—	—
Lactose	—	1.13	3.49	2.20	—	5.5	4.8	—	—	—	—	—	—	—	—
Maltose	7.8	1.22	3.82	—	—	6.2	5.5	—	—	—	—	—	—	—	—
Cellobiose	—	1.24	3.82	—	—	6.1	5.5	—	—	—	—	—	—	—	—
D-*glycero*-D-*gluco*-Heptose	—	—	—	—	—	—	—	—	—	1.45	—	—	—	—	—
D-Galactose	8.9	1.42	4.32	2.69	2.2	6.6	5.5	6.6	—	1.65	1.0	1.0	1.2	5.7	12.2; 12.5
D-Glucose	9.7	1.46	4.48	—	2.4	7.2	5.5	7.6	—	1.92	1.0	1.1	1.4	6.0	12.2; 12.6
D-Mannose	10.9	1.73	5.32	3.00	3.0	8.1	7.4	8.6	—	2.58	1.2	1.3	1.6	6.6	10.5; 11.8
D-Talose	—	—	—	—	—	—	—	—	—	2.77	—	—	—	—	—
D-Fructose	—	2.03; 2.84	5.99; 10.3	—	—	9.4	8.1; 10.1	—	—	—	—	—	—	—	—
D-Arabinose	12.8	2.04	5.99	3.10	—	8.7	7.4	9.6	—	2.78	—	—	—	—	—
D-Xylose	12.9	2.00	5.99	—	3.6	9.4	8.1	10.2	—	3.14	1.2	1.3	1.8	7.6	9.4
D-Lyxose	13.9	—	—	—	4.2	9.4	8.7; 9.2	—	—	3.09	—	—	—	—	—
D-Ribose	14.2	2.31	7.32	3.60	—	11.0	9.4; 10.9	11.4	—	3.47	1.4	1.5	2.0	8.3	—
2-Deoxy-D-*arabino*-hexose	14.3	2.18	6.49	—	—	11.0	10.7	—	—	—	—	—	—	—	—
2-Deoxy-D-*erythro*-pentose	15.9	2.94	9.15	—	—	13.9	8.1	13.0	—	—	3.2	2.1	4.7	—	—
6-Deoxy-D-glucose	16.9	—	—	—	—	—	—	—	—	5.17	—	—	—	—	—
L-Fucose	16.9	2.54	8.34	—	5.4	11.0	9.6	—	—	4.75	2.2	1.7	2.9	9.3	8.9; 9.7
L-Rhamnose	17.8	2.73	8.65	—	—	12.1	11.5	—	—	5.30	2.4	1.9	3.3	—	—
D-Glyceraldehyde	—	—	—	—	—	—	—	14.0	—	—	—	—	—	—	—
2-Amino-2-deoxy-D-galactose	—	—	—	—	—	—	—	—	—	—	—	—	—	—	5.9; 6.5
2-Amino-2-deoxy-D-glucose	—	—	—	—	—	—	—	—	—	—	—	—	—	—	6.1; 6.8
2-Acetamido-2-deoxy-D-mannose	—	—	—	—	—	—	—	—	—	—	2.8	2.0	3.8	—	—
2-Acetamido-2-D-glucose	—	—	—	—	—	—	—	—	—	—	4.0	2.3	5.4	—	4.3; 4.6
2-Acetamido-2-deoxy-D-galactose	—	—	—	—	—	—	—	—	—	—	4.4	2.5	6.0	—	4.3; 4.5
N-Glycolylneuraminic acid	—	—	—	—	—	—	—	—	2.75; 4.25	—	—	—	—	—	—
N-Acetylneuraminic acid	—	—	—	—	—	—	—	—	3.75; 5.00	—	7.0	—	—	—	—
Methyl-D-glucuronate	19.4	—	—	—	—	—	—	—	—	—	—	—	—	—	—
D-Glucurono-6,3-lactone	20.5	—	—	—	—	—	—	—	—	—	—	—	—	—	—
D-Glucuronic acid[h]	8.5	—	—	—	—	—	—	—	—	—	—	—	—	—	—
1,5-Anhydroidose[i]	—	—	—	—	—	—	—	—	—	—	—	—	—	—	5.8

Note: See data for HPLC of derivatized sugars on silica gel (Table LC 19).

[a] Samples treated with 5-dimethylaminonaphthalene-1-sulfonylhydrazine (dansylhydrazine) + trichloroacetic acid at 65°C for 20 min.

[b] Samples treated with 4'-*N*, *N*-dimethylamino-4-azobenzenesulfonylhydrazine (dabsylhydrazine) + acetic acid at 60°C for 1 h.

[c] Samples treated with 1,2-diamino-4,5-dimethoxybenzene + sulfuric acid, sodium sulfite and β-mercaptoethanol at 60°C for 2.5 h.

[d] Samples converted to *N*-(*p*-methoxyphenyl)glycosylamines by reaction with *p*-anisidine at 60°C for 80 min.

[e] Pyridylamino derivatives prepared by reductive pyridylamination; treatment with 2-aminopyridinium chloride at 100°C for 15 min followed by reaction with sodium cyanoborohydride at 90°C for 8 h.

TABLE LC 14 (continued)
Reversed-Phase HPLC of Derivatized Sugars

[f] Reductive amination (reaction with sodium cyanoborohydride + ammonium sulfate at 100°C for 90 min) followed by treatment of resulting glycamines with 7-fluoro-4-nitrobenz-2-oxa-1,3-diazole (NBD-F) at 60°C for 30 min.

[g] Benzoylation, with benzoic anhydride + 4-dimethylaminopyridine in pyridine, at 37°C for 90 min; preceded by carboxyl reduction of uronic acids (carbodiimide/$NaBH_4$).

[h] For elution of glucuronic acid, 0.01 *M* ammonium sulfate, adjusted to pH 7.0 with aqueous ammonia, replaced water in eluent.

[i] Product of carboxyl reduction of iduronic acid.

Packing	P1	= μ-Bondapak® C_{18} (Waters; 10-μm particles); Sep-Pak C_{18} cartridge (Waters) used in pre-treatment of derivatized sample.
	P2	= Ultrasphere® ODS (Altex, Beckman, San Ramon, CA; 5-μm particles) with Brownlee® (Brownlee Labs, Santa Clara, CA) cartridge (3 × 0.46 cm, Spherisorb ODS, 5-μm) as precolumn.
	P3	= Nucleosil® ODS (Macherey-Nagel; 5-μm particles), with precolumn (4.5 × 0.2 cm) containing same packing.
	P4	= 600 RPB® (Alltech; 10-μm particles).
	P5	= ODS-Hypersil® (Shandon; 3-μm particles).
	P6	= Nova-PAK® C_{18} (Waters; 4-μm particles).
	P7	= Radial-Pak® C_{18} cartridge (Waters; 5-μm particles).
	P8	= LiChrosorb® RP-8 (Merck; 10-μm particles), with Brownlee cartridge as precolumn.
	P9	= P2, without precolumn.
	P10	= TSK-Gel® LS-410A (Toyo Soda, Tokyo).
	P11	= Supelcosil® LC-18 (Supelco), with Brownlee® cartridge (RP-18) as precolumn.
Solvent	S1	= water-acetonitrile (39:11).
	S2	= 0.08 *M* acetic acid-acetonitrile (19:6).
	S3	= 0.08 *M* acetic acid-acetonitrile (79:21).
	S4	= 0.08 *M* acetic acid-acetonitrile; solvent program: % acetonitrile 20 → 30 in 18 min, then 30 → 70 in 2 min; isocratic at 70% acetonitrile for 2 min; 70 → 20% in 2 min.
	S5	= 20% aqueous acetonitrile containing 0.01 *M* formic acid, 0.04 *M* acetic acid, and 1 m*M* triethylamine.
	S6	= 0.08 *M* acetic acid-acetone (3:1).
	S7	= 10 m*M* sodium phosphate buffer, pH 6.5 - acetone (3:1).
	S8	= water-acetonitrile; concave elution gradient, % acetonitrile 22 → 85% in 45 min (program no. 7, Waters Model 660 solvent programmer).
	S9	= water-methanol-acetonitrile (77:15:8).
	S10	= water-acetonitrile (22:3).
	S11	= 0.25 *M* sodium citrate buffer solution, pH 4.0, containing 1.0% acetonitrile.
	S12	= water-acetonitrile (23:2).
	S13	= acetonitrile-water (3:1).

Flow program	F1	= 0.5 ml/min for 18 min, then 1.5 ml/min.
Technique	T1	= Sep-Pak® C_{18} cartridge (Waters), activated by rinsing with acetonitrile (2 ml) then water (1 ml), used to purify derivatized sample before injection; loaded cartridge rinsed with water-acetonitrile (9:1), then dansylhydrazones eluted with water-acetonitrile (3:2).
	T2	= precolumn, conditioned by rinsing with acetonitrile then water, used to purify derivatized sample before injection; sample loaded on precolumn, then precolumn switched into mobile phase stream for 30—90 s; after being switched out, precolumn washed with acetonitrile to remove excess reagent, then conditioned with water prior to loading of next sample.
	T3	= precolumn omitted; unreacted reagent washed from column by rapid increase in acetonitrile concentration at end of elution program (S4).
	T4	= derivatized sample passed through Sep-Pak® C_{18} cartridge before injection; remaining unreacted reagent washed from column at end of elution of dansylhydrazones with S5 by changing solvent to methanol-acetonitrile (4:1).
	T5	= after elution of dabsylhydrazones, column cleaned by passage of acetone-0.08 *M* acetic acid (4:1).
	T6	= no special measures to clean column.
	T7	= column washed with methanol-water (1:1).
	T8	= two columns, each 25 × 0.46 cm, connected in series.
	T9	= derivatized sample passed through Sep-Pak C_{18} cartridge before elution; excess reagent removed by washing cartridge with water-pyridine (9:1), then water; benzoylated derivatives eluted with acetonitrile and solution evaporated; residue redissolved in acetonitrile for injection.
Detection	D1	= fluorescence method; excitation 240 nm, emission 550 nm.
	D2	= fluorescence; excitation 360—380 nm, emission 540 nm.
	D3	= spectrophotometric detection at 485 nm.
	D4	= spectrophotometric detection at 425 nm.
	D5	= fluorescence; excitation 370 nm, emission 453 nm.
	D6	= fluorescence; excitation 320 nm, emission 400 nm.
	D7	= fluorescence; Shimadzu FLD-1 fluorometric detector, cut-off filter EM-5.

REFERENCES

1. **Alpenfels, W. F.,** A rapid and sensitive method for the determination of monosaccharides as their dansyl hydrazones by high-performance liquid chromatography, *Anal. Biochem.*, 114, 153, 1981.
2. **Mopper, K. and Johnson, L.,** Reversed-phase liquid chromatographic analysis of Dns-sugars. Optimization of derivatization and chromatographic procedures and applications to natural samples. *J. Chromatogr.*, 256, 27, 1983.
3. **Eggert, F. M. and Jones, M.,** Measurement of neutral sugars in glycoproteins as dansyl derivatives by automated high-performance liquid chromatography, *J. Chromatogr.*, 333, 123, 1985.
4. **Muramoto, K., Goto, R., and Kamiya, H.,** Analysis of reducing sugars as their chromophoric hydrazones by high-performance liquid chromatography, *Anal. Biochem.*, 162, 435, 1987.
5. **Lin, J.-K. and Wu, S.-S.,** Synthesis of dabsylhydrazine and its use in the chromatographic determination of monosaccharides by thin-layer and high-performance liquid chromatography, *Anal. Chem.*, 59, 1320, 1987.
6. **Hara, S., Yamaguchi, M., Takemori, Y., and Nakamura, M.,** Highly sensitive determination of *N*-acetyl- and *N*-glycolylneuraminic acids in human serum and urine and rat serum by reversed-phase liquid chromatography with fluorescence detection, *J. Chromatogr.*, 377, 111, 1986.

TABLE LC 14 (continued)
Reversed-Phase HPLC of Derivatized Sugars

7. **Batley, M., Redmond, J. W., and Tseng, A.,** Sensitive analysis of aldose sugars by reversed-phase high performance liquid chromatography, *J. Chromatogr.*, 253, 124, 1982.
8. **Takemoto, H., Hase, S., and Ikenaka, T.,** Microquantitative analysis of neutral and amino sugars as fluorescent pyridylamino derivatives by high-performance liquid chromatography, *Anal. Biochem.*, 145, 245, 1985.
9. **Shinomiya, K., Toyoda, H., Akahoshi, A., Ochiai, H., and Imanari, T.,** Fluorometric analysis of neutral sugars as their glycamines by high-performance liquid chromatography and thin-layer chromatography, *J. Chromatogr.*, 387, 481, 1987.
10. **Karamanos, N. K., Hjerpe, A., Tsegenidis, T., Engfeldt, B., and Antonopoulos, C. A.,** Determination of iduronic acid and glucuronic acid in glycosaminoglycans after stoichiometric reduction and depolymerization using high-performance liquid chromatography and ultraviolet detection, *Anal. Biochem.*, 172, 410, 1988.

TABLE LC 15
Reversed-Phase HPLC of Derivatized Glycosides

Packing	P1	P2	P3	P3	P4
Column					
Length, cm	15	15	15	15	10
Diameter (I.D.), cm	0.46	0.21	0.64	0.64	0.8
Material	SS	SS	SS	SS	Plastic
Solvent	S1	S2	S3	S4	S5
Flow rate (ml/min)	1.4	0.6	2	2	1.5
Technique	T1	T2	T3	T3	T3
Detection	UV	UV	UV	UV	UV, RI
Reference	1	2	3	3	4
	Benzylated		**Benzylated**		**Acetyl**
Compound	**r[a]**	**r[a]**	**r[b]**	**r[b]**	**k′**
Methyl glycosides of:					
D-Glucose[c]	0.80; 1.00[d]	0.92; 1.00[d]	—	—	0.55; 0.75(α)
D-Galactose	0.77; 1.02[d]	0.92 1.02[d]	—	—	—
	1.06; 1.09	1.36; 1.44	—	—	—
D-Mannose[c]	0.66; 0.92[d]	0.86; 0.98[d]	—	—	—
L-Fucose	0.50; 0.62[d], 0.74	0.72; 0.79;[d] 0.89	—	—	—
N-Acetylneuraminic acid	0.56	0.79	—	—	—
2-Acetamido-2-deoxy-D-glucose[c]	0.20; 0.30[d]	0.44; 0.52[d]	—	—	—
2-Amino-2-deoxy-D-galactose	—	0.42; 0.50;[d] 0.55; 0.76	—	—	—
2-Acetamido-2-deoxy-D-galactose	0.19; 0.29;[d] 0.34; 0.35	0.41; 0.49;[d] 0.54; 0.75	—	—	—
Other glycosides[c]					
D-glucoside					
Ethyl	—	—	—	—	0.75; 1.13(α)
Propyl	—	—	—	—	1.75; 2.25(α)
Butyl	—	—	—	—	2.63; 3.75(α)
tert-Butyl	—	—	—	—	1.75; 3.06(α)
Phenyl	—	—	—	—	3.13; 4.25(α)
Benzyl	—	—	—	—	1.88; 2.44(α)
p-Chlorophenyl	—	—	—	—	3.25; 4.50(α)
Phenyl-1-thio	—	—	—	—	3.75; 6.25(α)
p-Biphenylyl	—	—	—	—	8.28; 11.81(α)
D-galactoside					
Phenyl	—	—	—	—	1.38; 2.56(α)
2-Phenylethyl	—	—	—	—	2.25(α); 2.32
Phenyl maltoside	—	—	—	—	0.25(α); 0.75
Methyl maltoside	—	—	0.86	—	—
cellobioside	—	—	0.97	—	—
isomaltoside	—	—	1.00	1.00	—
gentiobioside	—	—	1.00	0.92	—
αα-Trehalose	—	—	1.63	—	—
αβ-Trehalose	—	—	1.97	—	—
ββ-Trehalose	—	—	2.09	—	—

Note: See data for HPLC of derivatized glycosides on silica gel (Table LC 21).

[a] t_r relative to major peak for perbenzoylated methyl D-glucopyranoside (38.1 min,[1] 39.9 min[2]).
[b] t_r relative to methyl-2,3,4-tri-*O*-benzyl-6-*O*-(2,3,4,6-tetra-*O*-benzyl-α-D-glucopyranosyl)-α-D-glucopyranoside (8.4 min with S3, 63 min with S4).
[c] Pyranoside.
[d] Major peak.

TABLE LC 15 (continued)
Reversed-Phase HPLC of Derivatized Glycosides

Packing	P1	=	Spherisorb® ODS 11 (Alltech; 3-μm particles).
	P2	=	Supelcosil® LC-18 (Supelco; 5-μm particles).
	P3	=	Zorbax® ODS (DuPont: 7-μm particles).
	P4	=	Radial-Pak® C_{18} (Waters; 10-μm particles), used under radial compression (Waters module RCM-100).
Solvent	S1	=	acetonitrile-water (1:1) for 20 min, then acetonitrile-water (3:2).
	S2	=	acetonitrile-water, linear gradient, 35 → 90% acetonitrile in 65 min; held at 90% acetonitrile for 10 min; re-equilibrated to 35% acetonitrile over 15 min.
	S3	=	methanol-water (19:1).
	S4	=	methanol-water (22:3).
	S5	=	methanol-water (13:7).
Technique	T1	=	derivatized sample applied to Sep-Pak® C_{18} cartridge (Waters), inserted into vacuum manifold; excess reagents and by-products from derivatization reaction removed by washing with water, then benzoylated glycosides eluted with acetonitrile, vacuum being maintained at 100 mm Hg; recovered by evaporation then redissolved in acetonitrile for injection.
	T2	=	derivatized sample purified by solid-phase extraction in a 3-ml Supelclean® LC-18 tube previously conditioned with acetonitrile, then water; loaded tube washed with water to remove reaction by-products, and sample eluted with acetonitrile; after evaporation, residue dissolved in 35% aqueous acetonitrile for injection.
	T3	=	samples previously prepared; no special clean-up procedure described.

REFERENCES

1. **Jentoft, N.**, Analysis of sugars in glycoproteins by high-pressure liquid chromatography, *Anal. Biochem.*, 148, 424, 1985.
2. **Gisch, D. J. and Pearson, J. D.**, Determination of monosaccharides in glycoproteins by reversed-phase high-performance liquid chromatography on 2.1-mm narrow-bore columns, *J. Chromatogr.*, 443, 299, 1988.
3. **Dreux, M., La Fosse, M., Zollo, P. H. A., Pougny, J. R., and Sinaÿ, P.**, Synthesis of glycosides. Direct analysis of intermediate perbenzylated anomers by high-performance liquid chromatography, *J. Chromatogr.*, 204, 207, 1981.
4. **Kulkarni, S. Y., Verma, R., and Pansare, V. S.**, Gas and liquid chromatographic behaviour of some acetylated glucosides, *J. Chromatogr.*, 366, 243, 1986.

TABLE LC 16
Reversed-Phase HPLC of Derivatized Oligosaccharides and Oligosaccharide-Alditols

Packing	P1	P1	P2	P2
Column				
Length, cm	200[a]	200[a]	25	25
Diameter (I.D.), cm	0.32	0.32	0.46	0.46
Material	SS	SS	SS	SS
Solvent	S1	S2	S3	S4
Flow rate (ml/min)	2	1	2	2
Temperature (°C)	65	65	Ambient	Ambient
Technique	T1	T1	T2	T2
Detection	D1	D1	UV	UV
Reference	1	1, 2	3	3
	Acetyl		**Benzoyl**	
Compound	**k′**	**k′**	**k′**	**k′**
αα-Trehalose	1.66	—	7.72	—
Kojibiose	—	—	8.57[b,c]	—
Nigerose	—	—	10.65[b]	—
Maltose	1.64	1.56	—	6.47; 6.71
Maltitol	1.68	—	9.20	7.90
Isomaltose	1.70	—	—	—
Isomaltitol	1.72	—	8.92	—
Gentiobiose	1.58	—	8.55[b,c]	—
Lactose	1.52	—	—	—
Lactitol	1.60	—	—	—
Chitobiose	0.77	—	—	—
Maltotriose	2.41	2.38	—	9.20; 9.42
Maltotriitol	—	—	—	10.33
Isomaltotriitol	—	—	—	10.87
Raffinose	2.47	—	—	—
Maltotetraose	3.24	3.20	—	11.3; 11.5
Maltotetraitol	—	—	—	11.9
Stachyose	3.37	—	—	—
Galβ1-3GlcNAcβ1-3Galβ1-4Glc	2.09	—	—	—
Fucα1-2Galβ1-3GlcNAc-β1-3Galβ1-4Glc	3.16	—	—	—
Maltopentaose	4.18	4.02	—	12.6; 12.7
Maltopentaitol	—	—	—	13.0
Higher malto-saccharides				
DP 6	5.28	5.11	—	13.5(13.8)[b]
7	6.14	5.97	—	14.3(14.6)[b]
8	7.00	6.73	—	14.9(15.3)[b]
9	7.95	7.53	—	16.1[b]
10	8.71	8.23	—	17.0[b]
11	9.48	8.95	—	18.0[b]
12	10.1	9.66	—	19.3[b]
13	10.8	10.4	—	20.8[b]
14	11.5	11.0	—	—
15	12.2	11.6	—	—
16	13.0	12.2	—	—
17	13.8	12.8	—	—
18	14.5	13.4	—	—
19	15.1	14.0	—	—
20	15.6	14.6	—	—
21	16.1	15.1	—	—

TABLE LC 16 (continued)
Reversed-Phase HPLC of Derivatized Oligosaccharides and Oligosaccharide-Alditols

	Acetyl		Benzoyl	
Compound	k'	k'	k'	k'
22	16.6	15.6	—	—
23	17.1	16.1	—	—
24	17.5	16.6	—	—
25	17.9	17.0	—	—
26	18.3	17.4	—	—
27	18.7	17.8	—	—
28	19.2	18.1	—	—
29	19.6	18.4	—	—
30	20.0	18.7	—	—
31	—	19.0	—	—
32	—	19.3	—	—
33	—	19.6	—	—
34	—	19.9	—	—
35	—	20.2	—	—

[a] Two columns, each 100 × 0.32 cm, connected in series.
[b] Oligosaccharide-alditol.
[c] Co-elute with cellobiitol.

Packing	P1 =	Vydac® (Separations Group, Hesperia, CA; pellicular octadecyl support, 30- to 44-μm particles. Bondapak C_{18} Corasil (Waters), 37—50 μm, gave equivalent results.
	P2 =	Ultrasphere® octyl (Altex; C_8-bonded silica).
Solvent	S1 =	water-acetonitrile, elution gradient, 10 → 70% acetonitrile in 80 min (program no 4, Waters model 660 solvent programmer).
	S2 =	as S1, but total gradient time 160 min.
	S3 =	acetonitrile-water (17:3).
	S4 =	acetonitrile-water, linear gradient, 80 → 100% acetonitrile in 15 min, then isocratic with acetonitrile for 10 min.
Technique	T1 =	after completion of gradient elution, column washed with acetonitrile then with water-acetonitrile (9:1).
	T2 =	derivatized sample applied to Sep-Pak® C_{18} cartridge (Waters); loaded cartridge washed with water-pyridine (9:1) then with water; benzoylated oligosaccharides eluted with acetonitrile; eluate evaporated and residue redissolved in acetonitrile for injection.
Detection	D1 =	moving wire detector (Pye Unicam, Model LCM2); UV at 205 nm also used.

REFERENCES

1. **Wells, G. B. and Lester, R. L.,** Rapid separation of acetylated oligosaccharides by reverse-phase high-pressure liquid chromatography, *Anal. Biochem.*, 97, 184, 1979.
2. **Wells, G. B., Kontoyiannidou, V., Turco, S. J., and Lester, R. L.,** Resolution of acetylated oligosaccharides by reverse-phase high-pressure liquid chromatography, *Methods Enzymol.*, 83, 132, 1982.
3. **Daniel, P. F., de Feudis, D. F., Lott, I. T., and McCluer, R. H.,** Quantitative microanalysis of oligosaccharides by high-performance liquid chromatography, *Carbohydr. Res.*, 97, 161, 1981.

TABLE LC 17
Reversed-Phase HPLC of Partially Methylated, Partially Ethylated Reduced Oligosaccharides Derived from Complex Carbohydrates

Packing	P1	P1	P1	P1	P2	P1	P1
Column							
Length, cm	25	25	25	25	25	25	25
Diameter (I.D.), cm	0.46	0.46	0.46	0.46	0.46	0.46	0.46
Material	SS	SS	SS	SS	SS	SS	SS
Solvent	S1	S2	S3	S4	S3	S5	S6
Flow rate (ml/min)	0.5	0.5	0.5	0.5	0.5	0.5	0.5
Technique	T1	T1	T1	T1	T1	T2	T2
Detection	RI	RI	RI	RI	RI	MS	MS
Reference	1, 2	2, 3	3	2	2, 3	3	2, 4
Compound	k′						
Et-6Glc*	7.0	—	—	—	—	—	—
Et-2Gal*	7.6	—	—	—	—	—	—
Xyl-6Glc*	5.2	—	—	—	—	—	—
Et-2Gal-2Xyl*	6.0	—	—	—	—	—	—
Et-4Glc-4Glc*	—	6.6	—	—	—	—	2.0
Et-4Glc-3Gal*	—	—	—	—	—	—	2.5
Et-3Glc-3Glc*	—	—	—	—	—	—	1.9
Et-3Glc-6Glc*	—	—	—	—	—	—	2.1
Et-4(Et-6)Glc-3Glc*	—	—	—	—	—	—	2.7
Et-4(Et-6)Glc-4Glc*	—	—	—	—	—	—	3.0
Et-6Glc-6(Et-4)Glc*	—	—	—	—	—	—	2.9
Et-4(Et-6)Glc-6Gal*	—	—	—	—	4.6	—	—
Et-4GlcA-4Glc*	—	6.8	—	—	4.4	—	—
Et-4(Et-6)Glc-4GlcA*	—	9.0	—	—	—	—	—
Pyr<$^{4}_{6}$Glc-3Glc*	—	—	—	—	—	—	2.3
Et-4GlcA-4GlcA*	—	10.2	—	—	—	—	—
Et-4GlcA-4GlcA-4Glc*	—	—	—	—	—	2.7[a]	—
Et-4Glc-4Glc-4Glc*	—	—	—	—	3.9	2.7	—
Et-4GlcA-4Glc-4Glc*	—	14.2	—	—	—	2.7[a]	—
Et-4GlcA-4Glc-4(Et-6)Glc*	—	—	—	—	—	4.1[a]	—
Et-4Glc-4Glc-3Gal*	—	—	—	—	—	—	3.7
Et-3Glc-3Glc-6Glc*	—	—	—	—	—	—	4.0
Et-3Glc-6Glc-6(Et-4)Glc*	—	—	—	—	—	—	4.4
Et-4(Et-6)Glc-3Glc-3Glc*	—	—	—	—	—	—	4.9
Et-4(Et-6)Glc-4Glc-4Glc*	—	—	—	—	—	—	4.6
Et-4(Et-6)Glc-6Gal-4GlcA*	—	—	7.6	—	—	—	—
Et-6Glc-6(Et-4)Glc-4Glc*	—	—	—	—	—	—	4.5
Pyr<$^{4}_{6}$Glc-3Glc-3Glc*	—	—	—	—	—	—	3.8
Et-4Glc-4Glc-4GlcA*	12.2[a]	—	—	—	—	—	—
Et-4Glc-4(Et-4Glc-6)Glc*	12.8	9.7	—	—	—	—	—
Et-4Glc-4GlcA-4GlcA*	—	18.2	—	—	—	—	—
Et-6Glc-4(Et-6)Glc-4(Et-6)Glc*	16.2	—	—	—	—	—	—
Et-4(Et-6)Glc-4(Et-6)Glc-Glc	15.3	—	—	—	—	—	—
Et-4(Et-2Xyl-6)Glc-4Glc	8.8	—	—	—	—	—	—
Et-6Glc-4(Xyl-6)Glc*	12.8	—	—	—	—	—	—
Xyl-6Glc-4(Et-6)Glc*	13.8	—	—	—	—	—	—
Et-4(Xyl-6)Glc-4(Et-6)Glc-4Glc	11.6	—	—	—	—	—	—
Et-4GlcA-4Glc-4Glc-4Glc*	—	19.7	—	—	—	—	—
Et-4GlcA-4Glc-4(Et-4Glc)6Glc*	19.3[a]	—	—	—	—	—	—
Pyr<$^{4}_{6}$Glc-3Glc-3Glc-6Glc*	—	—	—	—	—	—	5.6
Et-3Glc-6Glc-6(Et-4)Glc-4Glc*	—	—	—	—	—	—	5.8
Et-4Glc-4(Et-4Glc-6)Glc-4GlcA*	18.4[a]	—	—	—	—	—	—
Et-4GlcA-4Glc-4Glc-4GlcA*	17.7[a]	—	—	—	—	—	—

TABLE LC 17 (continued)
Reversed-Phase HPLC of Partially Methylated, Partially Ethylated Reduced Oligosaccharides Derived from Complex Carbohydrates

Compound	k'						
Et-4(Et-6)Glc-6Gal-4GlcA-4Glc*	—	—	8.5	—	—	—	—
Et-6Glc-4(Et-6)Glc-4(Et-6)Glc-4Glc	20.8	—	—	—	—	—	—
Et-6Glc-4(Xyl-6)Glc-4(Et-6)Glc-4Glc	19.0	—	—	—	—	—	—
Et-4(Et-6)Glc-6Gal-4GlcA-4Glc-4Glc*	—	—	11.2	—	—	—	—
Et-4(Et-6)Gal-3(Et-4,Et-6)Glc-4Glc-4Glc-6(Et-4)Glc*	—	—	—	21.3	—	—	—
Xyl-6Glc-4(Xyl-6)Glc-4(Et-6)Glc-4Glc	14.7	—	—	—	—	—	—
Xyl-6Glc-4(Et-6)Glc-4(Et-2Xyl-6)Glc-4Glc	23.8	—	—	—	—	—	—
Et-6Glc-4(Xyl-6)Glc-4(Et-2Xyl-6)Glc-4Glc	28.3	—	—	—	—	—	—

Notes: * denotes reduced end-group of oligosaccharide formed by cleavage interior to residue; labeled with ethoxyl groups at C-1 and C-5. End-groups not thus designated were reducing ends of parent molecules and have methoxyl groups at all positions not involved in linkage. Positions of cleavage exterior to residue are labeled with ethoxyl groups; others not involved in linkages carry methoxyl groups. Pyruvic acid (1-carboxyethylidene) groups (Pyr) surviving hydrolysis are ethyl-esterified. GlcA methyl esters have been reduced with $LiAlD_4$ to 6,6-dideuterioglucosyl residues; C-6 carries an ethoxyl group, except where remethylation has preceded ethylation of the oligosaccharide (footnote a).

[a] Carboxylate-reduced GlcA remethylated; C-6 carries methoxyl group.

Packing P1 = Zorbax® ODS (DuPont), with precolumn (7 × 0.2 cm) of CO:Pell ODS (Whatman Clifton, NJ; pellicular C_{18}-bonded silica).
P2 = Partisil® 5, ODS (Whatman; 5-μm particles).
Solvent S1 = acetonitrile-water (1:1).
S2 = acetonitrile-water (11:9).
S3 = acetonitrile-water (3:2).
S4 = acetonitrile-water (13:7).
S5 = acetonitrile-water, linear gradient, 50 → 70% acetonitrile in 60 min.
S6 = acetonitrile-water, linear gradient, 50 → 65% acetonitrile in 45 min.
Technique T1 = components fractionated by HPLC, but characterized by subsequent GLC-MS of acetylated alditols derived from hydrolysates of isolated fractions.
T2 = HPLC connected directly to mass spectrometer, HPLC solvent serving as CI reactant gas.

REFERENCES

1. **Valent, B. S., Darvill, A. G., McNeil, M., Robertsen, B. K., and Albersheim, P.,** A general and sensitive chemical method for sequencing the glycosyl residues of complex carbohydrates, *Carbohydr. Res.*, 79, 165, 1980.
2. **McNeil, M., Darvill, A. G., Åman, P., Franzén, L.-E., and Albersheim, P.,** Structural analysis of complex carbohydrates using high-performance liquid chromatography, gas chromatography and mass spectrometry, *Methods Enzymol.*, 83, 3, 1982.
3. **Åman, P., Franzén, L.-E., Darvill, J. E., McNeil, M., Darvill, A. G., and Albersheim, P.,** The structure of the acidic polysaccharide secreted by *Rhizobium phaseoli* strain 127 K38, *Carbohydr. Res.*, 103, 77, 1982.
4. **Åman, P., McNeil, M., Franzén, L.-E., Darvill, A. G., and Albersheim, P.,** Structural elucidation, using HPLC-MS and GLC-MS, of the acidic polysaccharide secreted by *Rhizobium meliloti* strain 1021, *Carbohydr. Res.*, 95, 263, 1981.

TABLE LC 18
HPLC of Sugars and Polyol Glycosides on Silica Gel

Packing	P1	P1	P1	P2	P2
Column					
Length, cm	25	25	25	25	25
Diameter (I.D.), cm	0.46	0.46	0.46	0.46	0.46
Material	SS	SS	SS	SS	SS
Solvent	S1	S2	S3	S4	S5
Flow rate (ml/min)	1	1.5	1	1	1
Detection	RI	RI	RI	RI	RI
Reference	1	1	1	2[a]	2[a]
Compound	k'				
2,6-Dideoxy-D-*ribo*-hexose	—	—	—	0.12	0.10
2-Deoxy-D-*arabino*-hexose	—	—	—	0.28	0.46
L-Rhamnose	0.67	—	—	—	—
D-Xylose	0.79	—	—	—	—
L-Arabinose	0.92	—	—	—	—
D-Mannose	1.13	—	—	—	—
D-Glucose	1.34	—	0.48	0.42	0.90
Sucrose	—	1.50	—	0.65	1.56
Maltose	—	1.78	0.67	—	—
Lactose	—	2.67	—	—	—
αβ-Trehalose	—	—	—	0.80	2.26
αα-Trehalose	—	—	—	0.82	2.52
ββ-Trehalose	—	—	—	0.85	2.56
myo-Inositol	—	—	—	0.87	2.02
Nigerosylerythritol[b]	—	—	1.93	—	—
Nigerotriosylerythritol[b]	—	—	2.81	—	—
Maltotriose	—	—	0.89	—	—
Maltotetraose	—	—	1.15	—	—
Maltopentaose	—	—	1.52	—	—
Maltohexaose	—	—	2.07	—	—

Note: This table supplements Table LC 5 in *CRC Handbook of Chromatography: Carbohydrates*, Volume I.

[a] See data on HPLC of these sugars on amine-bonded silica in Table LC 1 (Reference 7).

[b] Products of Smith degradation of glucans linked α-(1→3) and α-(1→4).

Packing P1 = Aquasil® SS-452 N (Sensyu Scientific Co., Tokyo).
P2 = Supelcosil® LC-Si (Supelco; 5 μm particles).
Solvent S1 = ethyl formate-methanol-water (12:3:1).
S2 = ethyl acetate-methanol-water (12:3:1).
S3 = ethyl acetate-methanol-water (7:3:2).
S4 = acetonitrile-water (4:1).
S5 = acetonitrile-water (9:1).

REFERENCES

1. **Iwata, S., Narui, T., Takahashi, K., and Shibata, S.,** Separation of carbohydrates by LC in an aqueous silica gel column, *Carbohydr. Res.*, 133, 157, 1984.
2. **Nikolov, Z. L. and Reilly, P. J.,** Retention of carbohydrates on silica and amine-bonded stationary phases: application of the hydration model, *J. Chromatogr.*, 325, 287, 1985.

TABLE LC 19
HPLC of Derivatized Sugars and Alditols on Silica Gel

Packing	P1	P2	P2	P3	P4	P5	P5	P6	P5
Column									
Length, cm	30	15	15	25	25	25	25	50	25
Diameter (I.D.), cm	0.39	0.8	0.45	0.46	0.46	0.4	0.4	0.21	0.46
Material	SS	SS	SS	SS	SS	SS	SS	SS	SS
Solvent	S1	S2	S3	S4	S5	S6	S7	S8	S9
Flow rate (ml/min)	1	2.5	1.25	1.5	1.5	1.5	1.5	0.5	2
Temperature (°C)	Ambient	Ambient	Ambient	35	Ambient	27	27	Ambient	Ambient
Detection	UV	UV	UV	UV	UV	D1	D1	D2	D2
Reference	1	2	2	3	4,5	6	6	7	8
	Benzoylated			NB[a]	DMPS[b]	DNS-hydra-zones[c]		DNP-hydrazones[d]	
Parent compound	r[e]	k′	k′	k′	k′	k′	k′	k′	k′
Sugars									
D-Glyceraldehyde	—	—	—	—	—	1.02	—	7.00; 7.33[f] 9.83	—
D-Erythrose	—	—	—	—	—	—	—		—
D-Fucose	0.77; 0.88; 1.12[f,g]	—	—	—	2.4; 2.6;[f] 3.3;[f] 3.7	1.33	0.33	—	0.87
L-Rhamnose	0.77;[f] 0.92; 1.48; 2.31[g]	—	—	—	2.8; 3.2; 3.7[f]	1.95	0.66	—	—
D-Xylose	0.86[g]	—	—	—	2.9;[f] 4.1	2.39	—	—	1.52
D-Ribose	0.85; 0.90;[f] 1.03[g]	—	—	—	3.6; 4.0[f]	1.11	—	—	—
L-Arabinose	0.77; 0.93[f] 0.99; 1.10[g]	—	—	—	4.1; 4.5; 5.9	2.17	—	11.3	—
D-Lyxose	0.75;[f] 0.86[g]	—	—	—	4.1;[f] 4.9	—	—	—	—
D-Altrose	—	—	—	—	2.6; 3.7; 5.5	—	—	—	—
D-Glucose	1.20;[f] 1.35[g]	—	—	6.3	2.4; 2.8[f]	4.73	2.10	18.0	5.48
D-Galactose	1.01; 1.11;[f] 1.21; 1.27[g]	—	—	—	3.0;[f] 3.3; 3.8	4.81	2.55	—	4.93
D-Mannose	1.09; 1.17;[f] 1.63; 1.99[g]	—	—	—	3.0;[f] 3.5	4.90	2.76	—	2.95
D-Tagatose	—	—	—	—	1.9; 2.3;[f] 2.8;[f] 3.2	—	—	—	—
D-Fructose	1.01;[f] 1.08; 1.24; 1.31; 1.50[g]	—	—	6.7; 7.1; 7.3; 7.9; 10.4	2.3; 3.1;[f] 3.2; 3.9	—	—	—	—
2-Amino-2-deoxy-D-galactose[h]	1.62; 1.86; 2.28[f]	—	—	—	—	—	—	—	—
2-Amino-2-deoxy-D-glucose[h]	1.99; 2.29; 3.55;[f] 3.63; 3.83	—	—	—	—	—	—	—	—

2-Acetamido-2-deoxy-D-galactose	1.24; 1.68; 2.54[f]	—	—	—	—	—	—	—	—
2-Acetamido-2-deoxy-D-glucose	2.06; 3.04[f]	—	—	—	—	5.43	—	—	—
2-Acetamido-2-deoxy-D-mannose	—	—	—	—	—	5.87	—	—	—
Sucrose	2.21	—	—	—	2.7	—	—	—	—
Maltose	2.95; 3.17[f]	—	—	—	2.5;[f] 3.7	7.19	5.87	—	—
Cellobiose	—	—	—	—	—	7.81	6.65	—	—
Gentiobiose	—	—	—	—	—	—	7.76	—	—
Lactose	3.30	—	—	—	2.8; 4.4[f]	—	8.20	—	—
Alditols									
Erythritol	—	—	—	—	1.7	—	—	—	—
2,6-Dideoxy-D-*ribo*-hexitol	—	2.3	3.5	—	—	—	—	—	—
2-Deoxy-D-*erythro*-pentitol	—	3.0	4.5	—	—	—	—	—	—
Rhamnitol	—	4.2	5.8	—	—	—	—	—	—
Fucitol	—	4.4	6.1	—	—	—	—	—	—
Ribitol	—	4.6	6.4	—	1.5	—	—	—	—
Xylitol	0.98	4.9	6.7	—	1.7	—	—	—	—
D-Arabinitol	—	5.1	6.9	—	2.1	—	—	—	—
D-Glucitol	1.28	8.4	9.6	9.7	1.5	—	—	—	—
3-*O*-methyl-	—	5.2	7.2	—	—	—	—	—	—
4-*O*-methyl-	—	6.3	8.1	—	—	—	—	—	—
Allitol	—	7.2	8.6	—	—	—	—	—	—
Altritol	—	8.0	9.4	—	—	—	—	—	—
D-Mannitol	1.30	9.0	10.1	9.0	2.3	—	—	—	—
Galactitol	1.30	9.4	10.4	—	1.5	—	—	—	—
2-Amino-2-deoxy-D-glucitol	—	—	14.8	—	—	—	—	—	—
2-Amino-2-deoxy-D-mannitol	—	—	15.7	—	—	—	—	—	—
2-Amino-2-deoxy-D-galactitol	—	—	16.2	—	—	—	—	—	—
2-Acetamido-2-deoxy-D-mannitol	—	—	23.2	—	—	—	—	—	—
2-Acetamido-2-deoxy-D-glucitol	—	—	23.6	—	—	—	—	—	—
2-Acetamido-2-deoxy-D-galactitol	—	—	26.8	—	—	—	—	—	—

Note: Data on HPLC of acetylated carbohydrates on silica gel, and earlier data for 4-nitrobenzoates, will be found in *CRC Handbook of Chromatography: Carbohydrates,* Volume I, Tables LC 7—9, data for reversed-phase HPLC of derivatized sugars in Table LC 14, present volume.

[a] 4-Nitrobenzoates.

[b] Dimethylphenylsilyl derivatives, prepared by heating sample, in dry *N,N*-dimethylformamide containing imidazole, at 100°C for 1 h, cooling to 0—4°C, then treating with chlorodimethylphenylsilane for 6 h at ambient temperature.

TABLE LC 19 (continued)
HPLC of Derivatized Sugars and Alditols on Silica Gel

[c] Dansylhydrazones; samples treated with 5-dimethylaminonaphthalene-1-sulfonylhydrazine (dansylhydrazine) + trichloroacetic acid at 50°C for 90 min.
[d] 4-Dinitrophenylhydrazones; samples treated with 2,4-dinitrophenylhydrazine hydrochloride in 1,2-dimethoxyethane at 25°C for 3 h,[7] or with reagent solution + trifluoroacetic acid at 65°C for 90 min.[8]
[e] t_r relative to methyl 2,3,4,6-tetra-*O*-benzoyl-α-D-glucopyranoside (9.8 min).
[f] Major peak.
[g] Sugar heated in pyridine at 60°C for 1 h, to ensure attainment of isomeric equilibrium, before benzoylation.
[h] Hydrochloride.

Packing	P1 =	μ-Porasil® (Waters; 10 μm particles).
	P2 =	Develosil® 60-3 (Nomura Chemcials, Aichi, Japan; 3-μm particles).
	P3 =	Zorbax® SIL (Du Pont; 6-μm particles).
	P4 =	Partisil® 5 (Whatman; 5-μm particles).
	P5 =	LiChrosorb® SI-100 (Merck).
	P6 =	LiChrospher® SI-100 (Merck).
Solvent	S1 =	*n*-hexane-ethyl acetate (5:1).
	S2 =	*n*-hexane-dioxane-dichloromethane (30:4:1).
	S3 =	*n*-hexane-dioxane-dichloromethane, linear gradient, 22:2:1→4:2:1 in 80 min.
	S4 =	*n*-hexane-chloroform-acetonitrile-water (10:3:1.9:0.1).
	S5 =	*n*-hexane-ethyl acetate (99:1 for monosaccharides, 197:3 for disaccharides, 199:1 for alditols).
	S6 =	0.5% water in chloroform-ethanol (23:2) for 6 min, then 0.6% water in chloroform-ethanol (4:1) for 20 min.
	S7 =	0.5% water in chloroform-ethanol (22.3) for 4 min, then 0.6% water in chloroform-ethanol (4:1) for 25 min.
	S8 =	gradient elution, chloroform-methanol (9:1) added continuously to *iso*-octane-chloroform (19:1) for 20 min, then isocratic for 20 min.
	S9 =	0.7% water in chloroform-methanol (23:1:1.9).
Detection	D1 =	fluorescence; excitation 350 nm, emission 500 nm.
	D2 =	spectrophotometric detection at 352 nm.

REFERENCES

1. **White, C. A., Kennedy, J. F., and Golding, B. T.,** Analysis of derivatives of carbohydrates by high-pressure liquid chromatography, *Carbohydr. Res.*, 76, 1, 1979.
2. **Oshima, R. and Kumanotani, J.,** Determination of neutral, amino and *N*-acetyl amino sugars as alditol benzoates by liquid-solid chromatography, *J. Chromatogr.*, 265, 335, 1983.
3. **Petchey, M. and Crabbe, M. J. C.,** Analysis of carbohydrates in lens, erythrocytes, and plasma by high-performance liquid chromatography of nitrobenzoate derivatives, *J. Chromatogr.*, 307, 180, 1984.

4. **White, C. A., Vass, S. W., Kennedy, J. F., and Large, D. G.,** High-pressure liquid chromatography of dimethylphenylsilyl derivatives of some monosaccharides, *Carbohydr. Res.*, 119, 241, 1983.
5. **White, C. A., Vass, S. W., Kennedy, J. F., and Large, D. G.,** Analysis of phenyldimethylsilyl derivatives of monosaccharides and their role in high-performance liquid chromatography of carbohydrates, *J. Chromatogr.*, 264, 99, 1983.
6. **Takeda, M., Maeda, M., and Tsuji, A.,** Fluorescence high-performance liquid chromatography of reducing sugars using Dns-hydrazine as a pre-labelling reagent, *J. Chromatogr.*, 244, 347, 1982.
7. **Honda, S. and Kakehi, K.,** Periodate oxidation analysis of carbohydrates. VII. High-performance liquid chromatographic determination of conjugated aldehydes in products of periodate oxidation of carbohydrates by dual-wavelength detection of their 2,4-dinitrophenylhydrazones, *J. Chromatogr.*, 152, 405, 1978.
8. **Karamanos, N. K., Tsegenidis, T., and Antonopoulos, C. A.,** Analysis of neutral sugars as dinitrophenyl-hydrazones by high performance liquid chromatography, *J. Chromatogr.*, 405, 221, 1987.

TABLE LC 20
HPLC of Aldoses on Silica Gel as Derived Diastereoisomeric 1-(*N*-Acetyl-α-Methylbenzylamino)-1-Deoxyalditol Acetates:[a] Resolution of Enantiomers

Packing	P1
Column	
Length, cm	15
Diameter (I.D.), cm	0.46
Material	SS
Solvent	S1
Flow rate (ml/min)	1.25
Detection	UV
Reference	1

	k'	
Sugar	D	L
Glyceraldehyde[b]	6.95	7.13
Erythrose[b]	9.26	9.04
2-Deoxy-*erythro*-pentose[b]	11.1	11.1
Ribose[b]	11.2	10.2
Arabinose	9.34	11.1
Xylose	12.9	11.6
Lyxose	10.0	10.8
Rhamnose[b]	6.39	7.39
Fucose	8.48	6.74
2-Deoxy-*lyxo*-hexose[b]	12.1	12.1
Glucose	12.9	12.5
Mannose[b]	9.78	11.3
Galactose	13.1	11.1
4-*O*-Methylglucose[b]	16.5	14.5

[a] Derivatives prepared by reductive amination of sugars with chiral L-(−)-methylbenzylamine (MBA) in presence of sodium cyanoborohydride.

[b] Only one enantiomer available; peak for other enantiomer found by comparison of chromatograms for products of reaction with DL-(±)-MBA with those produced by L-(−)-MBA.

Packing P1 = Develosil® 60-3, 3 μm (Nomura Chemicals, Aichi, Japan).
Solvent S1 = *n*-hexane-ethanol (19:1).

REFERENCE

1. **Oshima, R., Yamauchi, Y., and Kumanotani, J.,** Resolution of the enantiomers of aldoses by liquid chromatography of diastereoisomeric 1-(*N*-acetyl-α-methylbenzylamino)-1-deoxyalditol acetates, *Carbohydr. Res.*, 107, 169, 1982.

TABLE LC 21
HPLC of Derivatized Glycosides on Silica Gel

Packing	P1	P2	P2	P1	P1	P3
Column						
Length, cm	30	10	10	30	30	25
Diameter (I.D.), cm	0.39	0.8	0.8	0.39	0.39	0.46
Material	SS	Plastic	Plastic	SS	SS	SS
Solvent	S1	S2	S3	S4	S5	S6
Flow rate (ml/min)	1	2	2	0.3	1.2	1.5
Detection	UV	UV	UV	UV	UV	UV
Reference	1	2	2	3	3	4,5
	Benzoyl		Acetyl			DMPS[a]
Compound	r[b]	r[b]	r[c]	k′	k′	k′
D-Glucoside						
Methyl	1.00(α); 1.41	1.00(α); 1.36; 1.17(α)[d]; 1.28[d]	1.00(α); 1.12; 0.97(α);[d] 0.98[d]	0.71(α); 0.84	—	2.3; 5.6(α)
Ethyl	—	—	0.98(β); 0.89(β)[d]	0.54(α); 0.69	—	—
Propyl	—	—	—	0.41(α); 0.53	—	—
Butyl	—	—	—	0.35(α); 0.46	—	—
tert-Butyl	—	—	—	0.46(α); 0.60	—	—
Hexyl	—	—	—	0.29(α); 0.49	—	—
Cetyl	—	—	—	0.20(α); 0.41	—	—
Phenyl	—	0.62(α); 0.66; 0.64(β)[d]	0.62(α); 0.77; 0.71(β)[d]	—	0.28(α); 0.49	—
Benzyl	—	—	—	—	2.43(α); 2.74	—
m-Tolyl	—	0.61(α); 0.75; 0.64(β)[d]	0.60(α); 0.71; 0.66(β)[d]	—	—	—
Guaiacyl	—	—	1.02(β); 1.03(β)[d]	—	—	—
2-Naphthyl	—	0.61(α); 0.62; 0.54(β)[d]	0.58(α); 0.72; 0.66(β)[d]	—	—	—
p-Chlorophenyl	—	—	—	—	0.31(α); 0.43	—
Phenyl-1-thio	—	—	—	—	0.37(α); 0.58	—
p-Biphenylyl	—	—	—	—	0.54(α); 0.77	—
D-Galactoside						
Methyl	0.94(α); 1.25	1.02(α); 1.18; 1.07(β)[d]	0.92(α); 1.05; 1.14(β)[d]	—	—	3.8; 8.4(α)
Ethyl	—	0.86(α); 1.09; 0.95(β)[d]	0.80(α); 0.94; 1.00(β)[d]	—	—	—

TABLE LC 21 (continued)
HPLC of Derivatized Glycosides on Silica Gel

Compound	Benzoyl		Acetyl			DMPS[a]
	r[b]	r[b]	r[c]	k′	k′	k′
Phenyl	—	0.62(α); 0.78; 0.66(β)[d]	0.62(α); 0.72; 0.78(β)[d]	—	1.20(α); 1.85	—
m-Tolyl	—	0.59(α); 0.68; 0.59(β)[d]	0.55(α); 0.69; 0.75(β)[d]	—	—	—
Guaiacyl	—	1.05(α); 1.18; 1.12(β)[d]	0.82(α); 1.00; 1.18(β)[d]	—	—	—
2-Phenylethyl	—	—	—	—	2.49(α); 2.90	—
D-Mannoside						
Methyl	1.10(α)	1.25(α); 1.14(α)[d]	1.05(α); 1.46; 0.92(α)[d]	—	—	—
D-Xylopyranoside						
Methyl	0.73(α); 0.87	—	—	—	—	3.6; 6.4(α)
Phenyl maltoside	—	—	—	0.73; 0.81(α)	—	—

Note: See data for reversed-phase HPLC of derivatized glycosides (Table LC 15).

[a] Dimethylphenylsilyl derivatives; see footnote b to Table LC 19.
[b] t_r relative to methyl 2,3,4,6-tetra-*O*-benzoyl-α-D-glucopyranoside (9.8 min,[1] 5.6 min[2]).
[c] t_r relative to methyl 2,3,4,6-tetra-*O*-acetyl-α-D-glucopyranoside (6.5 min).
[d] Furanoside.

Packing P1 = μ-Porasil® (Waters; 10-μm particles).
P2 = Radial-Pak® B cartridge (Waters; 10-μm silica).
P3 = Partisil® 5 (Whatman; 5-μm particles).
Solvent S1 = *n*-hexane-ethyl acetate (5:1).
S2 = benzene-ethyl acetate (99:1).
S3 = benzene-ethyl acetate (9:1).
S4 = light petroleum (b.p. 60—80°C)-ethyl acetate (1:1).
S5 = chloroform-carbon tetrachloride (3:2).
S6 = *n*-hexane-ethyl acetate (49:1).

REFERENCES

1. **White, C. A., Kennedy, J. F., and Golding, B. T.,** Analysis of derivatives of carbohydrates by high-pressure liquid chromatography, *Carbohydr. Res.*, 76, 1, 1979.
2. **Yoshida, K., Kotsubo, K., and Shigematsu, H.,** High-performance liquid chromatography of hexopyranosides and hexofuranosides, *J. Chromatogr.*, 208, 104, 1981.
3. **Kulkarni, S. Y., Verma, R., and Pansare, V. S.,** Gas and liquid chromatographic behaviour of some acetylated glucosides, *J. Chromatogr.*, 366, 243, 1986.
4. **White, C. A., Vass, S. W., Kennedy, J. F., and Large, D. G.,** High-pressure liquid chromatography of dimethylphenylsilyl derivatives of some monosaccharides, *Carbohydr. Res.*, 119, 241, 1983.
5. **White, C. A., Vass, S. W., Kennedy, J. F., and Large, D. G.,** Analysis of phenyldimethylsilyl derivatives of monosaccharides and their role in high-performance liquid chromatography of carbohydrates, *J. Chromatogr.*, 264, 99, 1983.

TABLE LC 22
HPLC of Glycolipids on Silica Gel

Packing	P1	P1	P1	P2	P3	P4
Column						
Length, cm	50	50	50	50	25	25
Diameter (I.D.), cm	0.21	0.21	0.21	0.21	0.46	0.5
Material	SS	SS	SS	SS	SS	SS
Solvent	S1	S2	S3	S4	S5	S6
Flow rate (ml/min)	2	2	2	2	0.5	2
Technique	T1	T1	T2	T3	T4	T5
Detection	D1	D2	D2	D2	UV[a]	D3
Reference	1	2	3	4	5	6
	Benzoylated				Ac-NB[b]	Free
Compound	k'	k'	k'	k'	k'	k'
Glycosphingolipids						
Glucosylceramide	3.32	7.00	6.75	—	4.80	—
Lactosylceramide	5.02	9.60	9.00	—	6.50	—
Lactotriaosylceramide	—	—	—	—	7.48	—
Globotriaosylceramide	6.16	11.4	11.2	—	8.26	—
Neolactotetraosylceramide	—	—	—	—	8.93	—
Globotetraosylceramide	8.54	13.6	14.2	—	9.82	—
Monosialogangliosides						
G_{M4}	—	—	—	3.30;[c] 3.38	—	—
G_{M3}	—	—	—	4.75;[d] 4.90	—	—
G_{M2}	—	—	—	5.88	—	—
G_{M1}	—	—	—	6.75	—	—
Cereal glycolipids						
Monogalactosyldiacylglycerol	—	—	—	—	—	3.44
Monogalactosylmonoacylglycerol	—	—	—	—	—	4.55
Digalactosyldiacylglycerol	—	—	—	—	—	7.33
Digalactosylmonoacylglycerol	—	—	—	—	—	8.09

Note: Earlier data on HPLC of glycoplipids on silica gel will be found in *CRC Handbook of Chromatography: Carbohydrates*, Volume I, Table LC 10.

[a] Standard wavelength (254 nm).
[b] *O*-acetyl-*N*-4-nitrobenzoyl derivatives.
[c] Minor peak; lipid moiety heterogeneous.
[d] Shoulder; lipid moiety heterogeneous.

Packing
- P1 = Zipax® (DuPont; pellicular silica gel, 27-μm particles).
- P2 = LiChrospher® SI 4000 (Merck; 10-μm particles).
- P3 = Zorbax® SIL (DuPont; 6-μm particles).
- P4 = Spherisorb®-S (Phase Separations; 3-μm particles).

Solvent
- S1 = *n*-hexane-aqueous ethyl acetate, linear gradient, 2 → 17% aqueous ethyl acetate in 10 min, then isocratic for 2 min; aqueous ethyl acetate prepared by mixing dry with water-saturated ethyl acetate in ratio 5:1.
- S2 = *n*-hexane-dioxane, linear gradient, 1 → 20% dioxane in 13 min, then isocratic for 5 min.
- S3 = *n*-hexane-dioxane, linear gradient, 2.5 → 25% dioxane in 13 min, then isocratic for 5 min.
- S4 = *n*-hexane-dioxane, linear gradient, 7 → 23% dioxane in 18 min.
- S5 = 1% 2-propanol in *n*-hexane-dichloroethane (2:1) for 5 min; linear gradient, 1 → 5% 2-propanol in *n*-hexane-dichloroethane (2:1) for 55 min; isocratic for 10 min.
- S6 = (*A*) *n*-hexane-butan-2-one-acetic acid (35:65:0.4);
  (*B*) *n*-hexane-chloroform-2-propanol-aqueous buffer (42:5:45:3), aqueous buffer consisting of 0.5 m*M* serine adjusted to pH 7.5 with ethylamine;
  (*C*) as *B* but 32:5:50:8; linear gradient 100% *A* → 100% *B* in 8 min, then 100% *B* → 100% *C* in 23 min, to elute phospholipids.

TABLE LC 22 (continued)
HPLC of Glycolipids on Silica Gel

Technique T1 = samples benzoylated by treatment with 10% benzoyl chloride in pyridine at 37°C for 16 h.

T2 = samples benzoylated by treatment with 20% benzoic anhydride and 5% *N,N*-dimethyl-4-aminopyridine (catalyst) in pyridine at 37°C for 4 h; this method prevents *N*-benzoylation.

T3 = samples benzoylated with benzoyl chloride in pyridine at 60°C for 1 h; products in benzene, applied to micro-column of Unisil (silicic acid), column washed repeatedly with benzene (5 ml total), then benzoylated products eluted with 2.5 ml benzene-methanol (9:1); eluate evaporated under nitrogen at room temperature, residue dissolved in carbon tetrachloride for injection.

T4 = samples acetylated, then treated with 4-nitrobenzoyl chloride in pyridine at 60°C for 6 h; after removal of pyridine, residue dissolved in methanol (with sonication) and water (20% total) added; applied to Sep-Pak® C_{18} cartridge (Waters), prewashed with chloroform-methanol (1:1), then methanol-water (4:1), and impurities removed by further washing of cartridge with methanol-water (4:1); products eluted with chloroform-methanol (1:1), eluate evaporated under nitrogen at room temperature, residue dissolved in *n*-hexane-dichloroethane (2:1) for injection.

T5 = lipids, extracted from wheat flour with water-saturated 1-butanol, applied to column without derivatization.

Detection D1 = UV absorbance measured at 280 nm; detection limit 70 pmol.

D2 = UV absorbance measured at 230 nm, with 12-fold increase in sensitivity over D1; to obviate drifting baseline due to UV absorption by dioxane during elution, eluent passed through reference cell before entering column.

D3 = ACS 750/14 mass detector (Applied Chromatography Systems, Macclesfield, Cheshire, U.K.); solvent emerging from column evaporated in stream of compressed air (40 psi = 275 kPa), solute passed through light beam and light-scattering (related to mass of material) measured.

REFERENCES

1. **Ullman, M. D. and McCluer, R. H.,** Quantitative analysis of plasma neutral glycosphingolipids by high-performance liquid chromatography of their perbenzoyl derivatives, *J. Lipid Res.*, 18, 371, 1977.
2. **Ullman, M. D. and McCluer, R. H.,** Quantitative microanalysis of perbenzoylated neutral glycosphingolipids by high-performance liquid chromatography with detection at 230 nm, *J. Lipid Res.*, 19, 910, 1978.
3. **Gross, S. K. and McCluer, R. H.,** High-performance liquid chromatographic analysis of neutral glycosphingolipids as their per-*O*-benzoyl derivatives, *Anal. Biochem.*, 102, 429, 1980.
4. **Bremer, E. G., Gross, S. K., and McCluer, R. H.,** Quantitative analysis of monosialogangliosides by high-performance liquid chromatography of their perbenzoyl derivatives, *J. Lipid Res.*, 20, 1028, 1979.
5. **Suzuki, A., Kundu, S. K., and Marcus, D. M.,** An improved technique for separation of neutral glycosphingolipids by high-performance liquid chromatography, *J. Lipid Res.*, 21, 473, 1980.
6. **Christie, W. W. and Morrison, W. R.,** Separation of complex lipids of cereals by high-performance liquid chromatography with mass detection, *J. Chromatogr.*, 436, 510, 1988.

TABLE LC 23
HPLC of Sugars and Polyols on Amine-Modified Silica Gel

Packing	P1	P1	P2	P2	P2	P2	P1	P1	P3	P4
Column										
Length, cm	24	24	10	10	10	10	25	25	25	25
Diameter (I.D.), cm	0.7	0.7	0.8	0.8	0.8	0.8	0.4	0.4	0.46	0.46
Material	SS	SS	Plastic	Plastic	Plastic	Plastic	SS	SS	SS	SS
Form	Column	Column	Cartridge	Cartridge	Cartridge	Cartridge	Column	Column	Column	Column
Solvent	S1	S2	S3	S3	S4	S5	S2	S6	S7	S8
Flow rate (ml/min)	2	2	2	2	2	2	3	3	2	2
Temperature (°C)	Ambient	Ambient	26	23	23	23	Ambient	Ambient	Ambient	Ambient
Detection	RI	RI	RI	RI	RI	RI	RI, D1	RI, D1	RI	RI
Reference	1	1	2	3	3	3	4	4	5	5
Compound	k′									
D-Glyceraldehyde	—	—	—	—	0.0[a]	—	—	—	—	—
D-Erythrose	—	—	—	—	0.72	—	—	—	—	—
D-Ribose	—	—	—	—	1.08	—	1.28	1.63	—	—
D-Xylose	—	—	2.09	2.00	1.39	—	1.71	2.40	—	—
D-Lyxose	—	—	—	—	1.58	—	—	—	—	—
L-Arabinose	—	—	—	—	1.77	—	2.09	2.88	—	—
2-Deoxy-D-*arabino*-hexose	—	—	—	—	0.0[a]	—	—	—	—	—
2-Deoxy-D-*erythro*-pentose	—	—	—	—	1.59	—	—	—	—	—
6-Deoxy-D-Glucose	—	—	—	—	1.18	—	—	—	—	—
L-Rhamnose	—	—	1.45	—	1.21	—	1.17	1.50	—	—
L-Fucose	—	—	1.88	—	1.33	—	—	—	—	—
D-Talose	—	—	—	—	1.56	—	—	—	—	—
D-Altrose	—	—	—	—	1.87	—	—	—	—	
D-Mannose	—	—	—	—	2.04	—	2.83	4.42	—	—
D-Glucose	2.87	3.09	4.05	3.50	2.22	0.75	3.27	5.32	2.16	2.16
D-Allose	—	—	—	—	2.33	—	—	—	—	—
D-Galactose	—	—	—	3.50	2.62	—	3.62	5.81	—	—
D-*erythro*-Pentulose	—	—	—	—	1.04	—	—	—	—	—
D-*threo*-Pentulose	—	—	—	—	1.08	—	—	—	—	—

D-Tagatose	—	—	—	—	1.71	—	—	—	—	—
D-Fructose	2.21	2.31	2.90	2.75	1.80	0.50	2.33	3.46	1.83	1.67
L-Sorbose	—	—	—	—	1.80	—	2.35	3.55	—	—
D-*manno*-Heptulose	—	—	—	—	2.35	—	—	—	—	—
D-*altro*-Heptulose, 2,7-anhydro-	—	—	2.33	—	1.62	—	—	—	—	—
D-*glycero*-D-*gluco*-Heptose	—	—	—	—	2.55	—	—	—	—	—
Sucrose	3.99	4.82	6.52	5.50	3.39	1.25	5.15	10.1	2.83	2.50
Turanose	—	—	—	—	3.51	—	—	—	—	—
Palatinose	—	—	—	—	3.72	—	—	—	—	—
Lactulose	—	—	—	—	4.20	—	—	—	—	—
Maltose	—	6.58	8.62	9.01	4.35	—	7.04	14.5	3.33	3.00
Cellobiose	—	—	8.40	—	4.92	—	6.93	14.2	—	—
Lactose	6.32	7.84	9.83	9.01	5.09	—	8.37	17.3	—	—
αα-Trehalose	—	—	9.29	—	4.45	—	—	—	—	—
Gentiobiose	—	—	—	—	5.39	—	—	—	—	—
Melibiose	—	—	—	—	5.53	—	9.53	21.0	—	—
Melezitose	—	—	—	12.5	6.11	—	—	—	—	—
Raffinose	8.98	—	17.6	—	7.46	2.75	—	—	—	—
Maltotriose	—	—	—	—	7.80	—	14.9	—	5.00	4.67
Isomaltotriose	—	—	—	—	11.0	—	—	—	—	—
Stachyose	—	—	—	—	15.6	7.25	—	—	—	—
Polyols										
Ethylene glycol	—	—	0.50	0.50	0.47	—	—	—	—	—
Glycerol	—	—	1.02	1.02	0.80	—	—	—	—	—
Erythritol	—	—	1.59	—	1.14	—	—	—	—	—
Threitol	—	—	1.64	—	1.18	—	—	—	—	—
Ribitol	—	—	2.29	—	1.54	—	—	—	—	—
Arabinitol	—	—	2.50	—	1.39	—	—	—	—	—
Xylitol	—	—	—	—	1.64	—	—	—	—	—
D-Glucitol	—	—	3.45	3.50	2.16	—	—	—	—	—
D-Mannitol	—	—	3.62	—	2.15	—	—	—	—	—
Galactitol	—	—	3.59	—	2.24	—	—	—	—	—
Perseitol	—	—	—	—	3.03	—	—	—	—	—
myo-Inositol	—	—	8.26	—	4.46	—	—	—	—	—
Galactinol[b]	—	—	—	—	8.21	—	—	—	—	—

[a] Co-elutes with mobile phase.
[b] α-D-Galactopyranosyl-(1→1)-L-*myo*-inositol.

TABLE LC 23 (continued)
HPLC of Sugars and Polyols on Amine-Modified Silica Gel

Packing	P1	= LiChrosorb® SI60 (Merck; 5-μm particles), with small precolumn of same silica gel, impregnated with HPLC Amine Modifier I (NATEC, Hamburg) by passage of ca. 500 ml acetonitrile-water (4:1) containing 0.1% amine modifier through column, then same solvent with 0.01% amine modifier for 2 h.
	P2	= Radial-Pak® B silica cartridge (Waters), used under radial compression (Waters module RCM-100) and protected by precolumn of AX/Corasil (Waters, pellicular silica); impregnated with tetraethylenepentamine (TEPA) by passage of 50 ml acetonitrile-water (7:3) containing 0.1% TEPA, pH 9.2, through column, then of eluting solvent with 0.02% TEPA, pH 8.9, overnight before use.
	P3	= ROSiL® (Alltech; 5-μm spherical particles), impregnated with TEPA.
	P4	= as P3, but impregnated with piperazine.
Solvent	S1	= acetonitrile-water (7:3) containing 0.01% of HPLC Amine Modifier I.
	S2	= acetonitrile-water (3:1) containing 0.01% of HPLC Amine Modifier I.
	S3	= acetonitrile-water (81:19), pH 8.9, containing 0.02% of TEPA.
	S4	= acetonitrile-water (3:1), pH 8.9, containing 0.02% of TEPA.
	S5	= acetonitrile-water-methanol (3:1:1), pH 8.9, containing 0.02% of TEPA.
	S6	= acetonitrile-water (4:1) containing 0.01% of HPLC Amine Modifier I.
	S7	= acetonitrile-water (13:7) containing 0.01% of TEPA.
	S8	= acetonitrile-water (13:7) containing piperazine (0.44 mg/l) as amine modifier.
Detection	D1	= automated tetrazolium blue colorimetric method, with prior hydrolysis of sucrose by β-fructosidase.

REFERENCES

1. **Aitzetmüller, K.,** Sugar analysis by high-performance liquid chromatography using silica columns, *J. Chromatogr.*, 156, 354, 1978.
2. **Hendrix, D. L., Lee, R. E., Jr., Baust, J. G., and James, H.,** Separation of carbohydrates and polyols by a radially compressed high-performance liquid chromatographic silica column modified with tetraethylenepentamine, *J. Chromatogr.*, 210, 45, 1981.
3. **Baust, J. G., Lee, R. E., Jr., Rojas, R. R., Hendrix, D. L., Friday, D., and James, H.,** Comparative separation of low molecular-weight carbohydrates and polyols by high-performance liquid chromatography: radially compressed amine-modified silica *versus* ion exchange, *J. Chromatogr.*, 261, 65, 1983.
4. **Wight, A. W. and Van Niekerk, P. J.,** A sensitive and selective method for the determination of reducing sugars and sucrose in food and plant material by high performance liquid chromatography, *Food Chem.*, 10, 211, 1983.
5. **Verzele, M., Simoens, G., and Van Damme, F.,** A critical review of some liquid chromatography systems for the separation of sugars, *Chromatographia*, 23, 292, 1987.

TABLE LC 24
HPLC of Higher Oligosaccharides on Amine-Modified Silica Gel

Packing	P1	P2	P3	P4	P4	P4	P4	P3	P4	P4	P5
Column											
Length, cm	10	10	10	10	20	20	20	20	20	20	25
Diameter (I.D.), cm	0.5	0.5	0.5	0.5	0.8	0.8	0.8	0.8	0.8	0.8	0.4
Material	SS	SS	SS	SS	SS	SS	SS	SS	SS	SS	SS
Solvent	S1	S2	S3	S4	S4	S5	S6	S7	S6	S6	S8
Flow rate (ml/min)	1	1	1	1	1.5	2	1.5	2	2	1	1
Temperature (°C)	—[a]	—[a]	—[a]	—[a]	—[a]	—[a]	—[a]	—[a]	—[a]	—[a]	35
Detection	RI	RI	RI	RI	RI	RI	RI	RI	RI	RI	RI
Reference	1	1	1	1	1	1	1	1	1	1	2
	Malto-oligosaccharides										Fructans
DP	**k'**	**k'**	**k'**	**k'**	**k'**	**k'**	**k'**	**k'**	**k'**	**k'**	**k'**
1	0.9	0.9	0.9	1.0	0.82	0.93	0.60	0.55	0.50	0.53	—
2	1.2	1.2	1.2	1.5	1.10	1.33	0.76	0.68	0.66	0.67	0.70
3	1.7	1.7	1.7	2.0	1.41	1.90	0.95	0.81	0.81	0.80	0.85
4	2.3	2.4	2.3	2.8	1.81	2.57	1.15	1.00	1.00	1.00	1.00
5	3.1	3.1	3.0	3.8	2.31	3.47	1.38	1.18	1.17	1.20	1.15
6	4.0	4.0	3.8	5.0	2.85	4.40	1.60	1.36	1.33	1.43	1.30
7	5.0	5.0	4.7	6.3	3.41	5.60	1.85	1.54	1.49	1.67	1.45
8	6.2	6.2	5.8	7.9	4.13	7.07	2.10	1.72	1.65	1.93	1.60
9	—	—	—	—	—	—	—	2.00	1.92	2.26	1.75
10	—	—	—	—	—	—	—	2.29	2.18	2.60	1.90
11	—	—	—	—	—	—	—	2.58	2.46	3.00	2.10
12	—	—	—	—	—	—	—	3.00	2.75	3.40	2.30
13	—	—	—	—	—	—	—	3.40	3.08	3.86	2.60
14	—	—	—	—	—	—	—	3.70	3.50	4.33	2.95
15	—	—	—	—	—	—	—	4.00	3.92	4.86	3.20
16	—	—	—	—	—	—	—	4.48	4.34	5.53	3.60
17	—	—	—	—	—	—	—	4.94	4.76	6.20	4.00
18	—	—	—	—	—	—	—	5.40	5.30	7.00	4.50
19	—	—	—	—	—	—	—	5.87	5.84	7.93	5.02

TABLE LC 24 (continued)
HPLC of Higher Oligosaccharides on Amine-Modified Silica Gel

	Malto-oligosaccharides										Fructans
DP	k'	k'	k'	k'	k'	k'	k'	k'	k'	k'	k'
20	—	—	—	—	—	—	—	6.95	6.64	8.73	5.62
21	—	—	—	—	—	—	—	7.94	7.50	—	6.31
22	—	—	—	—	—	—	—	8.70	8.37	—	7.00
23	—	—	—	—	—	—	—	9.44	9.22	—	7.71
24	—	—	—	—	—	—	—	—	—	—	8.42
25	—	—	—	—	—	—	—	—	—	—	9.38
26	—	—	—	—	—	—	—	—	—	—	10.5
27	—	—	—	—	—	—	—	—	—	—	11.6
28	—	—	—	—	—	—	—	—	—	—	12.8
29	—	—	—	—	—	—	—	—	—	—	14.0
30	—	—	—	—	—	—	—	—	—	—	15.2
								(39 min)[b]	(38.6 min)[b]	(73 min)[b]	(35 min)[b]

[a] Ambient temperature.
[b] t_r for highest member of series resolved.

Packing P1 = irregular silica, 5-μm (H.S. Chromatography Packings, U.K.), with precolumn (5 × 0.5 cm) of Lichroprep® Si 60 (Merck; 15—25 μm), impregnated with tetraethylenepentamine by passage of 300 ml acetonitrile-water (11:9) containing 0.1% of amine.
P2 = as P1, but impregnated with pentaethylenehexamine.
P3 = as P1, but impregnated with polyamine modifier (H.S. Chromatography Packings, U.K.).
P4 = as P1, but impregnated with 1,4-diaminobutane.
P5 = Spherisorb®, 5 μm (Knauer, Bad Honburg, West Germany), impregnated with 1,4-diaminobutane by overnight washing of column with acetonitrile-water (1:1) containing 0.2% of 1,4-diaminobutane and 0.2% of polyethylene glycol 35,000.

Solvent S1 = acetonitrile-water (11:9) containing 0.01% of tetraethylenepentamine.
S2 = as S1, but containing pentaethylenehexamine as modifier.
S3 = as S1, but containing polyamine modifier.
S4 = as S1, but containing 1,4-diaminobutane as modifier.
S5 = acetonitrile-water (3:2) containing 0.01% of 1,4-diaminobutane.
S6 = acetonitrile-water (1:1) containing 0.01% of 1,4-diaminobutane.
S7 = as S6, but containing polyamine modifier.
S8 = acetonitrile-water (1:1) containing 0.02% of 1,4-diaminobutane and 0.2% of polyethylene glycol 35,000.

REFERENCES

1. **White, C. A., Corran, P. H., and Kennedy, J. F.,** Analysis of underivatised D-gluco-oligosaccharides (DP2-20) by high-pressure liquid chromatography, *Carbohydr. Res.*, 87, 165, 1980.
2. **Praznik, W., Beck, R. H. F., and Nitsch, E.,** Determination of fructan oligomers of degree of polymerization 2-30 by high performance liquid chromatography, *J. Chromatogr.*, 303, 417, 1984.

Data from Reference 1 reproduced with permission of authors and Elsevier Science Publishers.

TABLE LC 25
HPLC of Sugars and Alditols on Other Modified Silica Packings: Copper-Loaded, Diol-, and Polyol-Modified Silica Gel

	Cu-loaded			Diol-	Polyol-		
Packing	**P1**	**P1**	**P1**	**P2**	**P3**	**P3**	**P3**
Column							
Length, cm	20	15	15	25	25	25	25
Diameter (I.D.), cm	0.48	0.48	0.48	0.46	0.46	0.46	0.46
Material	SS	SS	SS	SS	SS	SS	SS
Solvent	S1	S2	S3	S4	S5	S6	S7
Flow rate (ml/min)	2	2	2	1.5	1.33	1	0.5
Temperature (°C)	20	30	30	40	—	Ambient	—
Detection	UV	RI	RI	RI	RI	RI	RI
Reference	1	2	2	3	4	4	5
Compound	k'						
L-Rhamnose	1.00	—	—	—	—	1.00	1.33
L-Arabinose	—	—	—	—	—	1.53	—
D-Xylose	1.37	0.69	0.73	—	—	—	—
D-Fructose	2.25	—	—	1.01	—	1.84	2.44
D-Glucose	2.62	1.77	1.96	1.37	—	2.31	3.11
D-Mannose	—	—	—	—	—	1.91	—
D-Galactose	—	—	—	1.33	—	2.16	—
2-Amino-2-deoxy-D-glucose	—	3.07	2.46	—	—	—	—
Sucrose	3.75	2.85	3.09	2.56	—	2.94	4.44
Lactulose	—	—	—	2.69	—	—	—
Maltose	4.78	4.00	4.42	3.07	3.80	4.27	—
Lactose	7.37	5.31	—	3.21	4.30	4.59	6.33
Maltotriose	—	—	—	—	6.20	7.83	—
Raffinose	—	—	—	—	6.80	—	—
D-Glucitol	—	—	—	1.33	—	2.28	—
D-Mannitol	—	—	—	1.40	—	—	—

Packing P1 = Partisil® 5 (Whatman; 5- to 7-μm particles) impregnated with Cu^{2+} by shaking for 10 min with ammoniated $CuSO_4$ solution, (Cu^{2+} 0.1 *M*, NH_3 *M*); copper silicate gel washed with 1 *M* NH_3 solution, then with water, and dried at 100°C for 12 h.
P2 = LiChrosorb® DIOL (Merck; 10-μm particles).
P3 = Polyol® RSiL (Alltech); irregularly shaped silica particles (5 μm) with bonded polyol phase.

Solvent S1 = acetonitrile-water (3:1) containing $CuSO_4$ (0.02 m*M*) and ammonia (1.5 *M*).
S2 = acetonitrile-water (3:1) containing ammonia (0.46 *M*).
S3 = as S2, but with ammonia 0.92 *M*.
S4 = acetonitrile-water (17:3) containing 0.1% of diisopropylethylamine.
S5 = acetonitrile-water (68.6:31.4) containing 0.1% of triethylamine.
S6 = acetonitrile-water (7:3) containing 0.1% of triethylamine.
S7 = acetonitrile-water (3:1) containing 0.075% of triethylamine.

REFERENCES

1. **Leonard, J. L., Guyon, F., and Fabiani, P.,** High-performance liquid chromatography of sugars on copper(II)-modified silica gel, *Chromatographia,* 18, 600, 1984.
2. **Guyon, F., Foucault, A., Caude, M., and Rosset, P.,** Separation of sugars by HPLC on copper silicate gel, *Carbohydr. Res.,* 140, 135, 1985.
3. **Brons, C. and Olieman, C.,** Study of the high-performance liquid chromatographic separation of reducing sugars, applied to the determination of lactose in milk, *J. Chromatogr.,* 259, 79, 1983.
4. **Verzele, M. and Van Damme, F.,** Polyol bonded to silica gel as stationary phase for high-performance liquid chromatography, *J. Chromatogr.,* 362, 23, 1986.
5. **Verzele, M., Simoens, G., and Van Damme, F.,** A critical review of some liquid chromatography systems for the separation of sugars, *Chromatographia,* 23, 292, 1987.

TABLE LC 26
HPLC of Sugars, Polyols, and Cyclodextrins on Cyclodextrin-Bonded Silica Gel

Packing	P1	P1	P2	P2	P2	P2
Column						
Length, cm	25	25	25	25	25	25
Diameter (I.D.), cm	0.46	0.46	0.46	0.46	0.46	0.46
Material	SS	SS	SS	SS	SS	SS
Solvent	S1	S2	S3	S4	S5	S6
Flow rate (ml/min)	1.5	1.5	1.5	1.5	1	2
Detection	RI, D1	RI, D1	RI, D1	RI, D1	D1	RI
Reference	1	1	1	1	1	1
Compound			k′			
D-Glyceraldehyde	—	0.33	0.71	0.53	—	—
D-Erythrose	0.74	0.24	0.48	0.41	—	—
L-Rhamnose	—	0.32	0.73	—	—	—
6-Deoxy-D-glucose	1.04	0.40	0.82	—	—	—
6-Deoxy-D-galactose	—	0.57	0.97	—	—	—
2-Deoxy-D-*erythro*-pentose	0.86	0.34	0.64	—	0.70	—
2-Deoxy-D-*arabino*-hexose	1.15	0.43	0.93	—	—	—
2-Deoxy-D-*lyxo*-hexose	—	0.48	0.99	—	—	—
D-Ribose	1.01	0.37	0.78	0.53	0.90	—
D-Lyxose	1.17	0.43	0.93	0.61	—	—
D-Xylose	1.20	0.46	0.96	0.66	1.10	—
L-Arabinose	1.35	0.57	1.05	0.72	—	—
D-Tagatose	1.40	0.51	1.16	0.73	—	—
D-Fructose	1.50	0.60	1.25	0.81	—	—
L-Sorbose	1.52	0.53	1.29	0.81	1.70	—
D-Talose	1.27	0.41	1.02	0.61	1.30	—
D-Mannose	1.54	0.59	1.34	—	—	—
D-Allose	—	0.65	1.34	—	—	—
D-Glucose	1.74	0.70	1.55	0.96	2.20	—
D-Galactose	1.83	—	1.65	1.06	—	—
D-*gluco*-D-*gulo*-Heptose	—	0.75	1.94	1.04	—	—
Sucrose	2.40	0.82	2.71	1.33	5.41	—
Turanose	2.48	0.92	3.05	1.46	6.11	—
Cellobiose	2.64	1.16	3.30	1.60	—	—
Maltose	2.70	1.01	3.33	1.62	6.41	—
Lactose	2.97	1.26	3.84	1.95	6.81	—
Lactulose	—	—	—	1.74	—	—
Melibiose	3.30	1.43	4.51	2.23	7.61	—
Gentiobiose	—	—	4.12	2.00	—	—
Melezitose	3.68	1.34	5.76	2.38	8.92	—
Maltotriose	—	1.42	6.83	2.67	9.42	—
Raffinose	4.26	1.54	7.19	2.88	—	—
Stachyose	7.34	2.81	17.6	7.23	11.0	—
Cyclomaltohexaose(α-CD)	—	2.56	—	—	11.3	—
Cyclomaltoheptaose(β-CD)	—	3.29	—	—	12.1	—
Cyclomaltooctaose(γ-CD)	—	4.20	—	—	12.7	—
Polyols						
Glycerol	0.94	0.40	0.65	—	—	0.2
Erythritol	1.23	0.50	0.94	—	—	0.4
Threitol	—	—	0.95	—	—	—
Ribitol	1.48	0.57	1.25	—	—	0.7
Arabinitol	1.53	0.62	1.29	—	—	0.8
Xylitol	1.54	0.65	1.33	—	—	—
D-Glucitol	1.86	0.77	1.75	—	2.80	1.4
D-Mannitol	1.88	—	1.76	—	—	—
Galactitol	2.01	0.80	1.80	—	—	1.6
myo-Inositol	2.91	1.67	3.10	—	—	3.0
Maltitol	2.97	0.93	3.76	—	—	6.2

TABLE LC 26 (continued)
HPLC of Sugars, Polyols, and Cyclodextrins on Cyclodextrin-Bonded Silica Gel

Packing	P1 =	Cyclobond® III (Advanced Separation Technologies, Whippany, NJ); α-cyclodextrin bonded to silica gel (5-μm spherical particles).
	P2 =	Cyclobond® I; β-cyclodextrin bonded to same silica gel.
Solvent	S1 =	acetonitrile-water (4:1).
	S2 =	acetone-water (17:3).
	S3 =	acetonitrile-water (17:3).
	S4 =	acetone-water (9:1).
	S5 =	acetonitrile-water (23:2) for 6 min, then linear gradient. 92 → 70% acetonitrile in 30 min; isocratic for 20 min.
	S6 =	acetonitrile-methanol-water (19:1:1).
Detection	D1 =	UV at 195 nm.

REFERENCE

1. **Armstrong, D. W. and Jin, H. L.,** Evaluation of the liquid chromatographic separation of monosaccharides, disaccharides, trisaccharides, tetrasaccharides, deoxysaccharides and sugar alcohols with stable cyclodextrin bonded phase columns, *J. Chromatogr.*, 462, 219, 1989.

TABLE LC 27
HPLC by Ion-Moderated Partitioning: Sugars and Polyols on Cation-Exchange Resins (H^+-Form)

Packing	P1	P2	P2	P2	P2	P2
Column						
Length, cm	60	30	30	30	30	30
Diameter (I.D.), cm	0.76	0.78	0.78	0.78	0.78	0.78
Material	SS	SS	SS	SS	SS	SS
Solvent	S1	S2	S2	S2	S2	S2
Flow rate (ml/min)	0.8	0.6	0.6	0.6	0.5	0.5
Temperature (°C)	Ambient	65	40	70	70	60
Detection	RI, UV[a]	RI	UV[a]	RI	RI	RI
Reference	1	2	2	3	4	4
Compound	r[b]			k'		
Polyols						
Glycerol	—	—	—	—	1.15	1.15
Erythritol	—	—	—	—	1.00	0.97
Ribitol	—	—	—	—	0.81	0.80
Arabinitol	—	—	—	—	0.86	0.84
Xylitol	—	—	—	—	0.89	0.88
D-Glucitol	—	—	—	—	0.74	0.72
D-Mannitol	1.05	—	—	—	—	—
Galactitol	—	—	—	—	0.74	0.72
meso-Inositol	—	—	—	—	0.61	0.60
Sugars						
DL-Glyceraldehyde	—	—	—	0.88	0.88	0.84
D-Xylose	1.03	1.02	—	—	0.65	0.64
L-Arabinose	1.12	1.12	—	—	0.78	0.76
D-Ribose	1.15	—	—	—	0.82	0.82
2-Deoxy-D-*erythro*-pentose	1.22	—	—	—	1.06	1.06

TABLE LC 27 (continued)
HPLC by Ion-Moderated Partitioning: Sugars and Polyols on Cation-Exchange Resins (H^+-Form)

Compound	r^b			k'		
2-Deoxy-D-*arabino*-hexose	—	—	—	—	0.74	0.72
2-Deoxy-D-*lyxo*-hexose	—	—	—	—	0.89	0.88
L-Rhamnose	1.05	1.10	—	—	0.74	0.72
L-Fucose	1.13	1.18	1.25	—	0.89	0.88
2,6-Dideoxy-D-*ribo*-hexose	—	1.30	—	—	—	—
D-Glucose	1.00	1.00	1.00	0.54	0.53	0.52
D-Mannose	1.03	0.99	—	—	0.63	0.62
D-Galactose	1.05	1.03	1.07	—	0.63	0.62
D-*threo*-Pentulose	—	—	—	—	0.74	0.72
L-Sorbose	1.01	0.98	—	—	0.58	0.55
D-Fructose	1.06	1.02	—	0.67	0.65	0.64
2-Acetamido-2-deoxy-D-glucose	1.19	—	1.32	—	—	—
2-Acetamido-2-deoxy-D-galactose	—	—	1.50	—	—	—
N-Acetylneuraminic acid	—	—	0.89	—	—	—
Trehalose	—	0.79	—	—	—	—
Maltose	0.90	0.78	—	—	—	—
Cellobiose	—	0.77	—	0.23	0.22	0.21
Lactose	0.92	0.80	—	—	—	—
Melibiose	—	0.78	—	—	—	—
Maltotriose	—	0.71	—	—	—	—
Raffinose	0.84	0.75	—	—	—	—
Melezitose	—	0.77	—	—	—	—
Stachyose	—	0.68	—	—	—	—

[a] Absorbance measured at 205 nm[1] or 193 nm.[2]
[b] t_r relative to D-glucose (17 min,[1] 10.16 min,[2] 9.33 min at 40°C[2]).

Packing P1 = TSK-Gel® LS-212 (Toyo Soda, Tokyo), 12-μm resin, 21% cross-linked, Na^+-form, converted to H^+-form by washing with 1% orthophoshoric acid (500 ml) then 0.1% orthophosphoric acid (500 ml).
P2 = Aminex® HPX-87H (Bio-Rad, Richmond, CA), 9-μm, 8% cross-linked resin, H^+-form; precolumn (4 × 0.46 cm) of Cation H^+ resin (Bio-Rad).
Solvent S1 = 0.1 *M* orthophosphoric acid
S2 = 5 m*M* sulfuric acid

REFERENCES

1. **Oshima, R., Kurosu, Y., and Kumanotani, J.,** Separation of acidic and neutral saccharides by high-performance gel chromatography on cation-exchange resins in the H^+ form with acidic eluents, *J. Chromatogr.*, 179, 376, 1979.
2. Bio-Rad Catalog L (Bio-Rad Laboratories, Richmond, CA), 87, 1986.
3. **Bonn, G., Pecina, R., Burtscher, E., and Bobleter, O.,** Separation of wood degradation products by high-performance liquid chromatography, *J. Chromatogr.*, 287, 215, 1984.
4. **Pecina, R., Bonn, G., Burtscher, E., and Bobleter, O.,** High-performance liquid chromatographic elution behaviour of alcohols, aldehydes, ketones, organic acids and carbohydrates on a strong cation-exchange stationary phase, *J. Chromatogr.*, 287, 245, 1984.

TABLE LC 28
HPLC of Carbohydrate Acids and Lactones on H⁺-Form Resins

Packing	P1	P2	P3	P3	P3
Column					
Length, cm	60	50	30	30	30
Diameter (I.D.), cm	0.76	0.9	0.78	0.78	0.78
Material	SS	SS	SS	SS	SS
Solvent	S1	S2	S3	S4	S4
Flow rate (ml/min)	0.8	1.5	0.7	0.6	0.6
Temperature (°C)	Ambient	80	30	35	25
Detection	RI, UV[a]	UV[a]	RI, D1	UV[a]	UV[a]
Reference	1	2	3	4	4,5
Compound	**r[b]**	**k'**	**r[b]**	**k'**	**k'**
D-Arabinonic acid	—	1.00	—	—	—
D-Ribonic acid	—	1.05	—	0.45	0.43
D-Gluconic acid	1.01	—	—	0.39	0.37
2-Ketogluconic acid	—	—	—	0.26	0.27
5-Ketogluconic acid	—	—	—	0.33	0.31
D-Galactonic acid	—	—	—	0.36	0.33
L-Mannonic acid	—	—	—	0.42	0.38
D-*glycero*-D-*gulo*-Heptonic acid	—	—	—	0.33	0.31
D-Glucuronic acid	0.96	—	0.89	0.26	0.25
D-Galacturonic acid	1.02	—	1.02	0.36	0.36
D-Mannuronic acid	—	—	—	0.33	0.31
D-Ribono-1,4-lactone	—	—	—	0.85	0.87
D-Glucono-1,5-lactone	—	—	1.02	0.38	0.37
D-Galactono-1,4-lactone	—	—	1.11	0.53	0.53
L-Mannono-1,4-lactone	—	—	1.22	0.73	0.73
L-Gulono-1,4-lactone	—	—	1.22	—	—
D-*glycero*-D-*gulo*-Heptono-1,4-lactone	—	—	—	0.58	0.58
D-Glucurono-6,3-lactone	1.25	—	1.33	0.80	0.83
D-Mannurono-6,3-lactone	—	—	—	0.90	0.92
L-*threo*-Hexenono-1,4-lactone[c]	—	—	1.11	0.63	0.60
D-*erythro*-2-Hexenono-1,4-lactone[d]	—	—	1.16	0.69	0.71

[a] Absorbance measured at 205 nm,[1] 210 nm,[2] 220 nm.[4,5]
[b] t_r relative to D-glucose (17 min,[1] 8 min[3]).
[c] L-Ascorbic acid (vitamin C).
[d] D-Erythorbic acid (isoascorbic acid).

Packing P1 = TSK-Gel® LS-212 (Toyo Soda, Tokyo), 12-μm resin, 21% cross-linked; Na⁺-form converted to H⁺-form (see Table LC 27).
P2 = HC-X8.00 (Hamilton, Reno, NV); particle diameter 10—15 μm, 8% cross-linked cation-exchange resin, H⁺-form.
P3 = Aminex® HPX-87H (Bio-Rad), 9-μm, 8% cross-linked cation-exchange resin, H⁺-form; precolumn (4 × 0.46 cm) of HPX-85-H⁺ or Cation H⁺ resin (Bio-Rad).

Solvent S1 = 0.1 *M* orthophosphoric acid.
S2 = 1.5 m*M* sulfuric acid.
S3 = 0.1 *M* formic acid.
S4 = 4.5 m*M* sulfuric acid.

Detection D1 = amperometric detector, Model LC-3, with glassy carbon electrode flow cell Type LC-16 (Bioanalytical Systems, West Lafayette, IN) used for electroactive L-ascorbic and D-erythorbic acids; other acids and lactones present did not give signal (all were detected by RI).

TABLE LC 28 (continued)
HPLC of Carbohydrate Acids and Lactones on H^+-Form Resins

REFERENCES

1. **Oshima, R., Kurosu, Y., and Kumanotani, J.,** Separation of acidic and neutral saccharides by high-performance gel chromatography on cation-exchange resins in the H^+ form with acidic eluents, *J. Chromatogr.*, 179, 376, 1979.
2. **Rajakylä, E.,** Separation and determination of some organic acids and their sodium salts by high-performance liquid chromatography, *J. Chromatogr.*, 218, 695, 1981.
3. **Grün, M. and Loewus, F. A.,** Determination of ascorbic acid in algae by high-performance liquid chromatography on strong cation-exchange resin with electrochemical detection, *Anal. Biochem.*, 130, 191, 1983.
4. **Hicks, K. B., Lim, P. C., and Haas, M. J.,** Analysis of uronic and aldonic acids, their lactones, and related compounds by high-performance liquid chromatography on cation-exchange resins, *J. Chromatogr.*, 319, 159, 1985.
5. **Hicks, K. B.,** A simple LC method for the direct analysis of sugar acids and their lactones, *Carbohydr. Res.*, 145, 312, 1986.

TABLE LC 29
HPLC of Higher Oligosaccharides on H^+-Form Cation-Exchange Resin

Packing	P1	P1	P1	P1	P1	P1	P1	P1
Column								
Length, cm	30	30	30	30	30	30	30	30
Diameter (I.D.), cm	0.78	0.78	0.78	0.78	1.0	1.0	1.0	1.0
Material	SS	SS	SS	SS	SS	SS	SS	SS
Solvent	S1	S1	S1	S1	S1	S1	S1	S1
Flow rate (ml/min)	0.2	0.4	0.2	0.4	0.5	0.5	0.5	0.5
Temperature (°C)	76	76	88	88	85	85	85	85
Detection	RI	RI	RI	RI	RI	RI	RI	RI
Reference	1	1	1	1	2	2	2	2
	Malto-oligosaccharides					Cello-	Chito-	PGA[a]
DP	**k′**	**k′**	**k′**	**k′**	**k′**	**k′**	**k′**	**k′**
1	1.87	2.22	1.83	1.89	2.16	2.16	2.53	1.92
2	1.56	2.02	1.60	1.60	1.84	1.73	2.31	1.40
3	1.33	1.77	1.33	1.30	1.52	1.39	2.03	1.04
4	1.13	1.50	1.13	1.10	1.26	1.10	1.84	0.80
5	0.93	1.24	0.93	0.91	1.08	0.89	1.65	0.56
6	0.80	1.05	0.80	0.78	0.90	0.68	1.52	0.44
7	0.67	0.91	0.70	0.65	0.76	0.54	1.39	0.32
8	0.55	0.78	0.60	0.58	0.62	0.40	1.27	0.24
9	0.48	0.65	0.50	0.51	0.52	0.32	—	0.16
10	0.40	0.58	0.43	0.44	0.44	—	—	—
11	0.33	0.51	0.36	0.37	0.36	—	—	—
12	0.27	0.45	0.30	0.32	0.28	—	—	—
	(53.8 min)[b]	(30.6 min)[b]	(53.1 min)[b]	(27.5 min)[b]	(37.6 min)[b]	(37.6 min)[b]	(42 min)[b]	(34.8 min)[b]

[a] Oligosaccharides from polygalacturonic acid.
[b] r_r for monomer given in brackets at foot of each column.

Packing P1 = Aminex® HPX-22H (Bio-Rad), 20- to 25-μm, 2% cross-linked cation-exchange resin, H^+-form; precolumn (4 × 0.46 cm) of Cation H^+ (Bio-Rad).
Solvent S1 = 5 m*M* sulfuric acid.

REFERENCES

1. **Derler, H., Hörmeyer, H. F., and Bonn, G.,** High-performance liquid chromatographic analysis of oligosaccharides. I. Separation on an ion-exchange stationary phase of low cross-linking, *J. Chromatogr.*, 440, 281, 1988.
2. **Hicks, K. B. and Hotchkiss, A. T., Jr.,** High-performance liquid chromatography of plant-derived oligosaccharides on a new cation-exchange resin stationary phase: HPX-22H, *J. Chromatogr.*, 441, 382, 1988.

TABLE LC 30
Ion-Moderated Partitioning: HPLC of Sugars, Polyols, and Methyl Glycosides on Ca^{2+}-Form Cation-Exchange Resins

Packing	P1	P2	P2	P2	P3	P3	P3	P3	P3	P3
Column										
Length, cm	25	30	30	30	30	30	30	30	30	30
Diameter (I.D.), cm	0.46	0.65	0.65	0.65	0.78	0.78	0.78	0.78	0.78	0.78
Material	SS	SS	SS	SS	SS	SS	SS	SS	SS	SS
Solvent	S1	S2	S3	H_2O[a]	H_2O[a]	H_2O[a]	S4	H_2O[a]	H_2O[a]	H_2O
Flow rate (ml/min)	0.5	0.5	0.5	0.4	0.6	0.6	1.0	1.0	0.65	0.3
Temperature (°C)	45	90	90	65	85	85	85	85	24	1.5
Detection	RI	RI	RI	UV[b]	RI	RI	RI	RI	RI	RI
Reference	1	2	3	4	5	6	6	7	7	8
Compound	r[c]		k′			r[c]		r[d]		r[e]
Polyols										
Glycerol	—	1.42	2.01	2.22	—	1.54	—	—	—	—
Erythritol	—	—	1.82	2.05	—	—	—	—	—	—
Threitol	—	—	2.02	—	—	—	—	—	—	—
Arabinitol	—	—	1.88	2.36	—	—	—	—	—	—
Ribitol	—	1.29	1.77	—	—	—	1.16	—	—	—
Xylitol	—	1.68	—	2.90	—	—	—	—	—	—
D-Mannitol	1.74	1.43	1.79	2.25	3.40	1.57	—	—	—	—
Galactitol	—	—	—	2.68	—	—	—	—	—	—
D-Glucitol	2.32	1.67	1.97	2.88	4.47	2.07	—	—	—	—
scyllo-Inositol	—	—	—	—	—	—	—	0.85	0.79	—
myo-Inositol	—	—	1.77	1.80	—	—	—	1.00	1.00	—

TABLE LC 30 (continued)
Ion-Moderated Partitioning: HPLC of Sugars, Polyols, and Methyl Glycosides on Ca^{2+}-Form Cation-Exchange Resins

Compound	r^c		k'			r^c		r^d		r^e
chiro-Inositol	—	—	—	—	—	—	—	1.00	1.07	—
muco-Inositol	—	—	—	—	—	—	—	1.17	1.36	—
neo-Inositol	—	—	—	—	—	—	—	1.29	1.58	—
epi-Inositol	—	—	—	—	—	—	—	2.50	5.05	—
allo-Inositol	—	—	—	—	—	—	—	2.62	6.41	—
cis-Inositol	—	—	—	—	—	—	—	11.7	—	—
Sugars										
D-Glyceraldehyde	—	—	2.02	—	—	—	—	—	—	—
D-Erythrose	—	—	2.34	—	—	—	—	—	—	—
D-Xylose	—	1.10	2.32	—	—	1.10	—	—	—	0.90; 1.22(α)
L-Arabinose	—	1.23	2.37	—	—	1.26	—	—	—	1.31(α); 1.78
D-Lyxose	—	1.27	—	—	—	—	—	—	—	—
D-Ribose	—	1.75	2.47	3.31	—	1.97	—	—	—	—
2-Deoxy-D-*erythro*-pentose	—	—	1.75	—	—	—	—	—	—	—
2-Deoxy-D-*arabino*-hexose	—	1.04	1.38	—	—	—	—	—	—	0.91
6-Deoxy-D-glucose	—	—	1.46	—	—	—	—	—	—	0.83; 1.09(α)
L-Rhamnose	—	1.12	1.50	—	—	1.16	—	—	—	—
L-Fucose	—	—	1.75	—	—	1.27	—	—	—	1.09; 1.40(α)
2,6-Dideoxy-D-*ribo*-hexose	—	—	—	—	—	1.28	—	—	—	0.94(α); 1.12
D-Glucose	1.00	1.00	1.36	1.27	1.59	1.00	1.00	—	—	0.83; 1.00(α)
D-Galactose	—	1.11	1.55	1.52	—	1.13	—	—	—	1.00; 1.24(α)
D-Mannose	1.19	1.13	1.58	1.57	1.92	1.15	1.10	—	—	1.03(α); 1.23
D-Altrose	—	—	1.54	—	—	—	—	—	—	—
D-Allose	—	—	1.82	—	—	—	—	—	—	—
D-Idose	—	—	1.73	—	—	—	—	—	—	—
D-Talose	—	—	2.34	—	—	—	—	—	—	—
D-*glycero*-D-*gluco*-Heptose	—	—	1.60	—	—	1.40	—	—	—	—
D-*threo*-Pentulose	—	1.24	1.55	—	—	—	—	—	—	—
D-*erythro*-Pentulose	—	—	2.17	—	—	—	—	—	—	—
L-Sorbose	—	1.10	1.49	1.49	—	1.12	—	—	—	—

D-Fructose	1.29	1.19	1.62	1.77	2.29	1.25	1.20	—	—	—
D-Tagatose	—	—	1.76	—	—	—	—	—	—	—
D-*manno*-Heptulose	—	—	1.41	—	—	—	—	—	—	—
D-*altro*-Heptulose,2,7-anhy-dro-	—	—	1.68	—	—	—	—	—	—	—
Sucrose	0.81	0.80	0.87	—	1.11	0.82	—	—	—	—
Palatinose	—	—	0.89	—	—	—	—	—	—	—
Turanose	—	—	0.88	—	—	—	—	—	—	—
Trehalose	—	—	0.83	—	—	0.83	—	—	—	—
Cellobiose	—	0.78	0.83	—	—	0.81	—	—	—	0.73; 0.76(α)
Maltose	—	0.80	0.88	—	1.13	0.83	0.85	—	—	0.73; 0.83(α)
Gentiobiose	—	—	0.79	—	—	—	—	—	—	—
Melibiose	—	0.82	0.91	—	—	0.85	—	—	—	—
Lactose	—	0.83	0.93	—	1.22	0.86	—	—	—	0.77; 0.86(α)
Lactulose	—	0.90	1.03	—	1.42	—	—	—	—	—
Maltotriose	—	0.71	0.64	—	—	0.75	—	—	—	—
Isomaltotriose	—	—	0.61	—	—	—	—	—	—	—
Melezitose	—	—	0.61	—	—	0.75	0.75	—	—	—
Raffinose	—	0.70	0.61	—	—	0.76	—	—	—	—
Stachyose	—	0.64	0.45	—	—	0.71	—	—	—	—
Methyl glycosides										
Methyl α-D-xylopyranoside	—	—	—	—	—	—	—	—	—	1.23
β-D-xylopyranoside	—	—	—	—	—	—	—	—	—	1.11
α-D-glucopyranoside	—	—	—	—	—	—	—	—	—	0.91
β-D-glucopyranoside	—	—	—	—	—	—	—	—	—	0.84
α-D-galactopyranoside	—	—	—	—	—	—	—	—	—	1.01
β-D-galactopyranoside	—	—	—	—	—	—	—	—	—	0.97
α-D-mannopyranoside	—	—	—	—	—	—	—	—	—	1.35

[a] Distilled, deionized water, filtered through a Millipore HA filter (0.45 μm)[4,5] or Nylon 66 (0.2 μm),[7] or doubly distilled.[6]

[b] Absorbance measured at 190 nm.

[c] t_r relative to D-glucose (3.87 min,[1] 8.47 min,[2] 10.87 min at 0.6 ml/min, 6.67 min at 1.0 ml/min[6]).

[d] t_r relative to *myo*-inositol (8.13 min at 85°C, 1 ml/min, 12.80 min at 24°C, 0.65 ml/min).

[e] t_r relative to peak for α-D-glucose (25.27 min).

TABLE LC 30 (continued)
Ion-Moderated Partitioning: HPLC of Sugars, Polyols, and Methyl Glycosides on Ca^{2+}-Form Cation-Exchange Resins

Packing	P1 =	Aminex® A-5 (Bio-Rad), 13-μm, 8% cross-linked cation-exchange resin, converted to Ca^{2+}-form by washing with 0.2 *M* $CaCl_2$ solution.
	P2 =	Sugar-Pak® I (Waters), microparticulate cation-exchange resin, Ca^{2+}-form.
	P3 =	Aminex® HPX-87C (Bio-Rad), 9-μm, 8% cross-linked cation-exchange resin, Ca^{2+}-form; used with precolumn (4 × 0.46 ml of Carbo-C® (Bio-Rad).
Solvent	S1 =	water containing triethylamine (1 m*M*).
	S2 =	water containing calcium disodium EDTA (50 mg/l).
	S3 =	water containing calcium acetate (0.1 m*M*).
	S4 =	0.01 *M* calcium sulfate, pH 5.5.

REFERENCES

1. **Verhaar, L. A. Th. and Kuster, B. F. M.,** Improved column efficiency in chromatographic analysis of sugars on cation-exchange resins by use of water-triethylamine eluents, *J. Chromatogr.*, 210, 279, 1981.
2. **Dorschel, C.,** Analysis of sugars. I. Retention times on Sugar-Pak I, in *Waters Lab Highlights*, LAH 0210, Waters Associates, Milford, 1984, 8.
3. **Baust, J. G., Lee, R. E., Jr., Rojas, R. R., Hendrix, D. L., Friday, D., and James, H.,** Comparative separation of low-molecular-weight carbohydrates and polyols by high-performance liquid chromatography: radially compressed amine-modified silica *versus* ion exchange, *J. Chromatogr.*, 261, 65, 1983.
4. **Owens, J. A. and Robinson, J. S.,** Isolation and quantitation of carbohydrates in sheep plasma by high-performance liquid chromatography, *J. Chromatogr.*, 338, 303, 1985.
5. **Brons, C. and Olieman, C.,** Study of the high-performance liquid chromatographic separation of reducing sugars, applied to the determination of lactose in milk, *J. Chromatogr.*, 259, 79, 1983.
6. Bio-Rad Catalog L (Bio-Rad Laboratories, Richmond, CA), 87, 1986.
7. **Sasaki, K., Hicks, K. B., and Nagahashi, G.,** Separation of eight inositol isomers by liquid chromatography under pressure using a calcium-form, cation-exchange column, *Carbohydr. Res.*, 183, 1, 1988.
8. **Baker, J. O. and Himmel, M. E.,** Separation of sugar anomers by aqueous chromatography on calcium- and lead-form ion-exchange columns. Application to anomeric analysis of enzyme reaction products, *J. Chromatogr.*, 357, 161, 1986.

TABLE LC 31
HPLC of Higher Oligosaccharides and Cyclodextrins on Ca^{2+}-Form Cation-Exchange Resins

Packing	P1	P1	P1	P2	P2	P2
Column						
Length, cm	61[a]	50[a]	25	90[b]	90[b]	90[b]
Diameter (I.D.), cm	0.7	0.62	0.90	0.78	0.78	0.78
Material	SS	SS	SS	SS	SS	SS
Solvent	H_2O[c]	H_2O[d]	S1[d]	H_2O[e]	H_2O[e]	H_2O[e]
Flow rate (ml/min)	0.5	0.3	0.5	0.5	0.5	0.5
Temperature (°C)	80	90	90	85	85	85
Detection	RI	RI	RI	RI	RI	RI
Reference	1	2	3	4	4	4
	Malto-				**Cello-**	**Xylo-**
Linear oligosaccharides	r[f]	r[f]	r[f]	r[f]	K_{av}	K_{av}
DP						
1	1.00	1.00	1.00	1.00	0.42	0.45
2	0.83	0.82	0.85	0.84	0.29	0.36
3	0.72	0.74	0.76	0.75	0.20	0.27
4	0.64	0.67	0.68	0.68	0.14	0.21
5	0.57	0.61	0.60	0.62	0.10	0.16
6	0.52	0.57	0.54	0.57	0.07	0.12
7	0.47	0.55	—	0.54	0.04	0.09
8	0.43	—	—	0.51	—	0.07
9	—	—	—	—	—	0.05
Cyclodextrins						
Cyclomaltohexaose (α-CD)	—	0.58	0.54	—	—	—
-heptaose (β-CD)	—	1.15	1.01	—	—	—
-octaose (γ-CD)	—	0.84	0.81	—	—	—

[a] Two columns of equal length (30.5 cm,[1] 25 cm[2]) connected in series.
[b] Three columns, each 30 cm long, connected in series.
[c] Distilled, deionized water from Millipore Milli-Q® purification system (Waters).
[d] Distilled water filtered through 0.2-μm filters and degassed at 90°C.
[e] Double-distilled water, degassed at 80°C.
[f] t_r relative to D-glucose (30 min,[1] 27 min,[2] 18.5 min,[3] 45 min[4]).

Packing P1 = Aminex® 50W-X4 (Bio-Rad), particle size 20—30 μm, 4% cross-linked cation-exchange resin, converted to Ca^{2+}-form by washing with $CaCl_2$ solutions.

P2 = Aminex® HPX-42C (Bio-Rad), 25-μm, 4% cross-linked cation-exchange resin in Ca^{2+}-form; precolumn (4 × 0.46 cm) of Aminex Q-15OS (Bio-Rad).

Solvent S1 = water containing 50 ppm Ca-EDTA (Ca-Titriplex; Merck).

REFERENCES

1. **Fitt, L. E., Hassler, W., and Just, D. E.,** A rapid and high-resolution method to determine the composition of corn syrups by liquid chromatography, *J. Chromatogr.*, 187, 381, 1980.
2. **Hokse, H.,** Analysis of cyclodextrins by high-performance liquid chromatography, *J. Chromatogr.*, 189, 98, 1980.
3. **Brunt, K.,** Rapid separation of linear and cyclic glucooligosaccharides on a cation-exchange resin using a calcium ethylenediaminetetraacetate solution as eluent, *J. Chromatogr.*, 246, 145, 1982.
4. **Schmidt, J., John, M., and Wandrey, C.,** Rapid separation of malto-, xylo- and cello-oligosaccharides (DP 2—9) on cation-exchange resin using water as eluent, *J. Chromatogr.*, 213, 151, 1981.

TABLE LC 32
Ion-Moderated Partitioning: HPLC of Sugars on Pb^{2+}-Form Cation-Exchange Resins

Packing	P1	P1	P1	P1	P1
Column					
Length, cm	30	30	30	30	30
Diameter (I.D.), cm	0.78	0.78	0.78	0.78	0.78
Material	SS	SS	SS	SS	SS
Solvent	H_2O[a]	H_2O[b]	H_2O[c]	H_2O[a]	H_2O[c]
Flow rate (ml/min)	0.6	0.6	0.5	0.25	0.6
Temperature (°C)	80	85	85	45	65
Detection	RI	RI	RI	D1,D2	RI
Reference	1	2	3	4	5
Sugar			r[d]		
D-Xylose	1.08	1.07	—	1.08	1.09
L-Arabinose	1.25	1.23	—	—	1.31
D-Ribose	—	—	—	3.93	—
L-Rhamnose	1.19	—	—	1.21	1.19
L-Fucose	—	—	—	1.23	1.25
D-Glucose	1.00	1.00	1.00	1.00	1.00
D-Galactose	1.14	1.15	1.16	1.14	1.18
D-Mannose	1.32	1.31	1.31	1.38	1.35
D-Allose	—	—	1.70	—	—
D-Fructose	—	1.38	—	—	1.45
Sucrose	—	0.80	—	0.86	0.83
Gentiobiose	—	—	—	—	0.78
Cellobiose	—	0.84	—	0.86	0.82
Maltose	—	—	—	0.91	0.89
Lactose	—	0.88	—	0.93	0.91
Melibiose	—	—	—	0.99	0.93
Lactulose	—	—	—	—	1.11
Melezitose	—	—	—	0.79	0.74
Raffinose	—	—	—	0.82	0.78
Maltotriose	—	—	—	—	0.84
Stachyose	—	—	—	—	0.73

[a] Distilled and degassed.
[b] Doubly distilled and deionised.
[c] Distilled, deionised water from Millipore Milli-Q® purification system.
[d] t_r relative to D-glucose (12.68 min,[1] 11.7 min,[2] 14.1 min,[3] 57 min,[4] 12.37 min[5]).

Packing P1 = Aminex® HPX-87P (Bio-Rad), 9-μm, 8% cross-linked cation-exchange resin, Pb^{2+}-form, with precolumn of Aminex® Q-150S (Bio-Rad),[1] deashing system (Bio-Rad) with cation- and anion-exchange micro-guard cartridges in series,[3] or mixed cation- and anion-exchange resins (Waters).[5]

Detection D1 = scintillation counting of radioactively labeled sugars.
D2 = automated orcinol-H_2SO_4 colorimetric assay.

REFERENCES

1. **Blaschek, W.,** Complete separation and quantification of neutral sugars from plant cell walls and mucilages by high-performance liquid chromatography, *J. Chromatogr.*, 256, 157, 1983.
2. Bio-Rad Catalog L (Bio-Rad Laboratories, Richmond, CA) 87, 1986.

TABLE LC 32 (continued)
Ion-Moderated Partitioning: HPLC of Sugars on Pb^{2+}-Form Cation-Exchange Resins

3. **Lenherr, A., Mabry, T. J., and Gretz, M. R.,** Differentiation between glucose, mannose, allose and galactose in plant glycosides by high-performance liquid chromatographic analysis, *J. Chromatogr.*, 388, 455, 1987.
4. **Lohmander, L. S.,** Analysis by high-performance liquid chromatography of radioactively labeled carbohydrate components of proteoglycans, *Anal. Biochem.*, 154, 75, 1986.
5. **Van Riel, J. A. M. and Olieman, C.,** High-performance liquid chromatography of sugars on a mixed cation-exchange resin column, *J. Chromatogr.*, 362, 235, 1986.

TABLE LC 33
Ion-Moderated Partitioning: HPLC of Sugars and Higher Oligosaccharides on Ag^+-Form Cation-Exchange Resins

Packing	P1	P1	P1	P2	P2
Column					
Length, cm	30	30	30	30	30
Diameter (I.D.), cm	0.78	0.78	0.78	0.78	0.78
Material	SS	SS	SS	SS	SS
Solvent	H_2O[a]	H_2O[a]	H_2O[b]	H_2O[b]	H_2O[c]
Flow rate (ml/min)	0.6	0.6	1.2	0.4	0.5
Temperature (°C)	25	45	85	85	85
Detection	RI	RI	RI	RI	RI
Reference	1	1	2	2	3
			Malto-	**Malto-**	**Cello-**
Compound	**r**[d]	**r**[d]	**r**[d]	**r**[d]	**r**[d]
Sugars					
D-Xylose	1.03	1.03	—	—	—
L-Arabinose	1.27	1.22	—	—	—
L-Rhamnose	1.06	1.07	—	—	—
L-Fucose	1.31	1.30	—	—	—
D-Glucose	1.00	1.00	1.00	1.00	1.00
D-Galactose	1.21	1.16	—	—	—
D-Mannose	1.20	1.17	—	—	—
D-Tagatose	0.97	0.97	—	—	—
D-Fructose	1.23	1.17	—	—	—
Sucrose	0.82	0.82[e]	—	—	—
Gentiobiose	0.85	0.83	—	—	—
Cellobiose	0.87	0.85	—	—	0.86
Maltose	0.89	0.86	0.84	0.85	—
Melibiose	1.01	0.97	—	—	—
Lactose	1.02	0.98	—	—	—
Lactulose	1.23	1.12	—	—	—
Melezitose	0.72	0.71[e]	—	—	—
Raffinose	0.80	0.77[e]	—	—	—
Maltotriose	0.82	0.77	0.71	0.75	—
Cellotriose	—	—	—	—	0.75
Stachyose	0.76	0.72[e]	—	—	—
D-Gluco-oligosaccharides					
DP 4	—	—	0.64	0.65	0.66
5	—	—	0.58	0.58	0.60
6	—	—	0.52	0.53	0.54
7	—	—	—	0.48	0.49

TABLE LC 33 (continued)
Ion-Moderated Partitioning: HPLC of Sugars and Higher Oligosaccharides on Ag^+-Form Cation-Exchange Resins

Compound	r[d]	r[d]	Malto- r[d]	Malto- r[d]	Cello- r[d]
8	—	—	—	0.45	0.45
9	—	—	—	0.42	—
10	—	—	—	0.40	—
11	—	—	—	0.38	—

[a] Doubly distilled water filtered through 0.45-μm filter (Millipore).
[b] Doubly distilled and deionized.
[c] Distilled and degassed.
[d] t_r relative to D-glucose (13.88 min at 25°C,[1] 13.76 min at 45°C,[1] 7.75 min on P1,[2] 24.76 min on P2,[2] 22.33 min[3]).
[e] Some decomposition on column at raised temperatures.

Packing P1 = Aminex® HPX-65A (Bio-Rad), 11-μm, 6% cross-linked cation-exchange resin in Ag^+-form, with precolumn (Waters) of mixed cation- and anion-exchange resins (2:3, meq)[1] or deashing system consisting of cation- and anion-exchange cartridges in series (Bio-Rad).[2]
P2 = Aminex® HPX-42A (Bio-Rad), 25-μm, 4% cross-linked cation-exchange resin in Ag^+-form, with deashing system of guard cartridges (Bio-Rad).

REFERENCES

1. **Van Riel, J. A. M. and Olieman, C.,** High-performance liquid chromatography of sugars on a mixed cation-exchange resin column, *J. Chromatogr.*, 362, 235, 1986.
2. Bio-Rad Catalog L, Bio-Rad Laboratories, Richmond, CA, 87, 1986.
3. **Bonn, G., Pecina, R., Burtscher, E., and Bobleter, O.,** Separation of wood degradation products by high-performance liquid chromatography, *J. Chromatogr.*, 287, 215, 1984.

TABLE LC 34
Use of Mixed-Counterion Resin and Coupled Column Systems in Ion-Moderated Partitioning of Mono- and Oligosaccharides

Packing	P1	P1 + P2	P3 + P4	P3 + P4	P5 + P4
Column					
Length, cm	30	60	60[a]	60[a]	40[b]
Diameter (I.D.), cm	0.78	0.78	0.78	0.78	0.78
Material	SS	SS	SS	SS	SS
Solvent	H_2O[c]	H_2O[c]	H_2O[d]	H_2O[d]	H_2O[d]
Flow rate (ml/min)	0.6	0.6	0.6	1.2	1.2
Temperature (°C)	65	70	85	95	95
Detection	RI	RI	RI	RI	RI
Reference	1	1	2	2	2
Compound	r[e]				
Sugars					
D-Xylose	1.04	—	1.07	1.06	1.02
L-Arabinose	1.21	—	1.15	1.15	1.09
L-Rhamnose	1.11	—	—	—	—

TABLE LC 34 (continued)
Use of Mixed-Counterion Resin and Coupled Column Systems in Ion-Moderated Partitioning of Mono- and Oligosaccharides

Compound	r[e]				
L-Fucose	1.26	—	—	—	—
D-Glucose	1.00	1.00	1.00	1.00	1.00
D-Galactose	1.16	1.16	—	—	—
D-Mannose	1.24	—	—	—	—
D-Fructose	1.24	1.25	1.10	1.09	1.09
D-Tagatose	1.42	—	—	—	—
Sucrose	0.84	0.83	—	—	—
Gentiobiose	0.85	—	—	—	—
Cellobiose	0.87	—	0.82	—	—
Maltose	0.89	0.89	0.84	0.84	0.85
Lactose	0.97	0.94	—	—	—
Melibiose	0.98	—	—	—	—
Lactulose	1.13	—	—	—	—
Melezitose	0.76	0.74	—	—	—
Raffinose	0.81	—	—	—	—
Maltotriose	0.84	—	0.76	0.75	0.78
Cellotriose	—	—	0.73	—	—
Stachyose	0.78	—	—	—	—
D-Gluco-oligosaccharides					
DP 4	—	—	0.68[f](0.63)[g]	0.68[f]	0.70[f]
5	—	—	0.62[f](0.56)[g]	0.62[f]	0.63[f]
6	—	—	0.58[f](0.52)[g]	0.59[f]	0.60[f]
7	—	—	0.55[f]	0.56[f]	0.56[f]

[a] Two columns, each 30 cm long, connected in series.
[b] P5 column, 10 cm, connected in series with P4, 30 cm.
[c] Doubly distilled, filtered thorugh 0.45-μm filter (Millipore).
[d] Doubly distilled.
[e] t_r relative to D-glucose (14.5 min on P1[1], 26.6 min on P1 + P2[1], 25.2 min on P3 + P4 at 85°C and 0.6 ml/min[2], 12.3 min on P3 + P4 at 95°C and 1.2 ml/min[2], 11.3 min on P5 + P4[2]).
[f] Malto-oligosaccharides.
[g] Cello-oligosaccharides.

Packing P1 = Aminex® HPX-65A (Bio-Rad) 11-μm, 6% cross-linked cation-exchange resin in Ag^+-form, with all residual sulfonic acid groups converted to Pb^{2+}-form by passage of solution containing both lead and silver nitrate (mol Pb/mol Ag 1.5%), total concentration 0.1 *M*; precolumn (Waters) of mixed-bed cation- and anion-exchange resins (2:3, meq).
P2 = Aminex HPX-87P (Bio-Rad), 9-μm, 8% cross-linked cation-exchange resin in Pb^{2+}-form.
P3 = μ-Spherogel® Carbohydrate N (Beckman); 8-μm, 7.5% cross-linked cation-exchange resin in Ca^{2+}-form; deashing system containing cation- and anion-exchange resins in series (Bio-Rad).
P4 = Aminex HPX-42A (Bio-Rad), 25-μm, 4% cross-linked cation-exchange resin in Ag^+-form; guard columns as for P3.
P5 = Fast Carbohydrate Analysis Column (Bio-Rad), Pb^{2+}-loaded cation-exchange resin, 8% cross-linked.

REFERENCES

1. **Van Riel, J. A. M. and Olieman, C.,** High-performance liquid chromatography of sugars on a mixed cation-exchange resin column, *J. Chromatogr.*, 362, 235, 1986.
2. **Bonn, G.,** High-performance liquid chromatographic elution behaviour of oligosaccharides, monosaccharides and sugar degradation products on series-connected ion-exchange resin columns using water as the mobile phase, *J. Chromatogr.*, 322, 411, 1985.

TABLE LC 35
Ion-Moderated Partitioning: HPLC of Neutral and Amino Sugars and Alditols on Cation-Exchangers with Aqueous-Organic Eluents

Packing	P1	P2	P3	P4	P5	P6	P6	P7	P7	P8	P9	P10
Column												
Length, cm	30	25	60	25	50	30	25	25	25	25	15	15
Diameter (I.D.), cm	0.78	0.4	0.4	0.4	0.8	0.78	0.4	0.46	0.46	0.4	0.6	0.6
Material	SS	SS	SS	SS	SS	SS	SS	SS	SS	SS	SS	SS
Solvent	S1	S2	S3	S4	S5	S6	S7	S8	S9	S10	S11	S11
Flow rate (ml/min)	0.5	0.6	0.3	0.2	0.5	0.6	0.15	1	1	0.8	0.5	0.5
Temperature (°C)	40	30	60	80	80	75	65	20	20	20	4	4
Detection	D1	D2	D3	RI, D1	RI, D1	RI	UV,[a] D1	RI	RI	RI	D2	D2
Reference	1	2	3	4	5	6	1	7	7	8	9	9
Compound	r^b				r^c			r^b			k′	
Sugars												
L-Rhamnose	—	0.45	—	—	—	—	—	0.69	0.74	0.34	1.61(α); 2.69	1.65(α); 2.65
L-Fucose	—	0.61	0.62	—	—	—	—	—	—	0.48	2.93(α); 2.69	2.88; 3.60(α)
D-Lyxose	—	—	—	—	—	—	—	—	—	—	2.43(α); 3.26	2.26(α); 6.73
D-Xylose	—	0.55	0.76	—	—	—	—	0.74	0.81	0.53	2.86[d]	2.20; 2.68(α)
D-Ribose	—	—	—	—	—	—	—	—	—	0.57	2.75; 5.04(α)	—[e]
L-Arabinose	—	0.67	0.84	—	—	—	—	—	—	0.70	3.48; 4.36(α)	4.46(α); 5.96
D-Idose	—	—	—	—	—	—	—	—	—	—	3.32(α); 3.44	4.12[d]
D-Allose	—	—	—	—	—	—	—	—	—	—	4.84; 5.14(α)	4.26(β); —[e](α)
D-Altrose	—	—	—	—	—	—	—	—	—	—	3.66; 4.08(α)	3.70(α); 5.36
D-Gulose	—	—	—	—	—	—	—	—	—	—	4.54[d]	4.74(β); —[e](α)
D-Glucose	1.00	1.00	1.00	1.00	—	0.37	—	1.00	1.00	1.00	4.90(α); 5.00	3.56; 4.50(α)
D-Mannose	—	1.07	0.93	—	—	—	—	—	—	0.64	3.86(α); 6.73	3.62(α); 5.76
D-Talose	—	—	—	—	—	—	—	—	—	—	5.40[d]	—[e]
D-Galactose	1.21	1.31	1.10	—	—	—	—	1.21	1.29	1.18	5.66(α); 6.67	5.42; 7.20(α)
L-Sorbose	—	—	—	—	—	—	—	—	—	0.70	—	—
D-Fructose	—	—	—	1.07	—	—	—	1.23	1.49	0.88	—	—
2-Acetamido-2-deoxy-D-glucose	1.70	2.77	—	—	—	—	—	—	—	—	—	—
2-Acetamido-2-deoxy-D-galactose	1.91	3.35	—	—	—	—	—	—	—	—	—	—

N-Glycolylneuraminic acid	—	1.10[f]	—	—	—	—	—	—	—	—	—	—
N-Acetylneuraminic acid	0.67	1.95[f]	—	—	—	—	—	—	—	—	—	—
Sucrose	—	—	—	0.80	—	—	—	1.25	1.26	0.66	—	—
Trehalose	—	—	—	—	—	—	—	—	—	0.91	—	—
Cellobiose	—	—	—	—	—	—	—	—	—	0.92	—	—
Maltose	—	—	—	—	—	—	—	1.49	1.48	0.97	—	—
Lactose	—	—	—	—	—	—	—	1.85	1.90	1.16	—	—
Maltotriose	—	—	—	0.64	—	—	—	2.30	2.21	—	—	—
Raffinose	—	—	—	—	—	—	—	2.48	2.46	1.82	—	—
Alditols												
Glycerol	—	—	—	—	0.44	—	—	—	—	—	—	—
2-Deoxy-D-*erythro*-pentitol	—	—	—	—	0.40	—	—	—	—	—	—	—
Rhamnitol	—	—	—	—	0.68	—	—	—	—	—	—	—
Fucitol	—	—	—	—	0.80	—	0.81	—	—	—	—	—
6-Deoxy-D-glucitol	—	—	—	—	0.94	—	—	—	—	—	—	—
Ribitol	—	—	—	—	0.52	0.59	0.61	—	—	—	—	—
Arabinitol	—	—	—	—	0.65	0.66	0.72	—	—	—	—	—
Xylitol	1.42	—	—	—	0.82	0.86	0.88	—	—	—	—	—
Allitol	—	—	—	—	0.55	—	—	—	—	—	—	—
D-Mannitol	—	—	—	—	0.68	0.73	0.76	—	—	0.83	—	—
Talitol	—	—	—	—	0.74	—	—	—	—	—	—	—
Galactitol	—	—	—	—	0.91	0.94	0.95	—	—	—	—	—
D-Glucitol	—	—	—	—	1.00	1.00	1.00	—	—	1.01	—	—
Iditol	—	—	—	—	1.24	—	—	—	—	—	—	—
2-Acetamido-2-deoxy-D-mannitol	—	—	—	—	0.50	—	—	—	—	—	—	—
2-Acetamido-2-deoxy-D-galactitol	—	—	—	—	0.50	—	—	—	—	—	—	—
2-Acetamido-2-deoxy-D-glucitol	—	—	—	—	0.58	—	—	—	—	—	—	—

Note: Earlier data on partition chromatography of sugars and alditols on cation-exchange resins in Li^+- and $(CH_3)_3N^+H$-forms in aqueous ethanol will be found in *CRC Handbook of Chromatography*, Volume I., Tables LC 13 and LC 14.

[a] Absorbance at 195 nm.
[b] t_r relative to D-glucose (29 min,[2] 98 min,[3] 12.5 min,[4] 8.6 min with S8, 7.3 min with S9,[7] 14.8 min[8]).
[c] t_r relative to D-glucitol (43.8 min,[1] 67.6 min,[5] 26.7 min[6]).
[d] Anomers not resolved.
[e] Peaks not detected owing to strong retention on column.
[f] As N-acyl mannosamines, from cleavage by neuraminic acid lyase.

TABLE LC 35 (continued)
Ion-Moderated Partitioning: HPLC of Neutral and Amino Sugars and Alditols on Cation-Exchangers with Aqueous-Organic Eluents

Packing	P1	= Aminex® HPX-87H (Bio-Rad), 9-μm, 8% cross-linked cation-exchange resin, H^+-form; precolumn (4 × 0.46 cm) of Cation-H^+ (Bio-Rad).
	P2	= Shodex® DC-613 (Showa Denko, Tokyo), 6-μm, 55% cross-linked resin, H^+-form.
	P3	= Aminex A-6 (Bio-Rad), 18-μm, 8% cross-linked cation-exchange resin, converted to $(CH_3)_3N^+H$-form by treatment with 2 *M* solution of trimethylammonium chloride, pH 9.
	P4	= Ostion® LGKS 0802 (Spolchemie, Ústí, Czechoslovakia), strongly acidic cation-exchange resin, particle diameter 12 to 15-μm, 8% cross-linked, converted to Li^+-form by treatment with *M* LiCl solution.
	P5	= Shodex® SUGAR SP-1010 (Showa Denko, Tokyo), cation-exchange resin in Pb^{2+}-form, with Shodex SP-1010 P precolumn.
	P6	= Aminex® HPX-87C (Bio-Rad), 9-μm, 8% cross-linked cation-exchange resin in Ca^{2+}-form; pre-column (4 × 0.46 cm) of Carbo-C (Bio-Rad).
	P7	= RSiL-Cat® (Alltech), irregularly shaped silica gel (5-μm) with bonded cation-exchanger phase, used in Ca^{2+}-form.
	P8	= HCO95AA (Showa Denko, Tokyo), 9-μm, 10% cross-linked cation-exchange resin, converted to Na^+-form by washing with 0.3 *M* NaOH solution, (reduced-pressure filtration used), then washed with water.
	P9	= prepacked Shodex RSPak DC-613; as P2, but resin in Na^+-form.
	P10	= as P9, but Ca^{2+}-form.
Solvent	S1	= acetonitrile-3 m*M* sulfuric acid (3:17).
	S2	= acetonitrile-water (23:2).
	S3	= ethanol-water (89:11).
	S4	= ethanol-water (3:2).
	S5	= ethanol-water (1:4).
	S6	= acetonitrile-water (1:3).
	S7	= acetonitrile-water (3:7).
	S8	= acetonitrile-water (3:1), containing 0.075% of triethylamine.
	S9	= acetonitrile-water-methanol (34:11:5), containing 0.075% of triethylamine.
	S10	= acetonitrile-water (3:1).
	S11	= acetonitrile-water (4:1).
Detection	D1	= scintillation counting of radioactively labeled compounds.
	D2	= automated post-column reaction with 2-cyanoacetamide; absorbance of products measured at 280 nm.
	D3	= automated tetrazolium blue colorimetric assay.

REFERENCES

1. **Lohmander, L. S.,** Analysis by high-performance liquid chromatography of radioactively labeled carbohydrate components of proteoglycans, *Anal. Biochem.*, 154, 75, 1986.

2. **Honda, S. and Suzuki, S.,** Common conditions for high-performance liquid chromatographic microdetermination of aldoses, hexosamines, and sialic acids in glycoproteins, *Anal. Biochem.*, 142, 167, 1984.
3. **Van Eijk, H. G., Van Noort, W. L., Dekker, C., and Van der Heul, C.,** A simple method for the separation and quantitation of neutral carbohydrates of glycoproteins in the one nanomole range on an adapted amino acid analyzer, *Clin. Chim. Acta,* 139, 187, 1984.
4. **Plicka, J., Kleinmann, I., and Svoboda, V.,** Preparative isoaltion of ^{14}C-labelled saccharose from a partly purified fraction of *Chlorella vulgaris, J. Chromatogr.*, 442, 237, 1988.
5. **Takeuchi, M., Takasaki, S., Inoue, N., and Kobata, A.,** Sensitive method for carbohydrate composition analysis of glycoproteins by high-performance liquid chromatography, *J. Chromatogr.*, 400, 207, 1987.
6. **Wood, R., Cummings, L., and Jupille, T.,** Recent developments in ion-exchange chromatography, *J. Chromatogr. Sci.*, 18, 551, 1980.
7. **Verzele, M., Simoens, G., and Van Damme, F.,** A critical review of some liquid chromatography systems for the separation of sugars, *Chromatographia,* 23, 292, 1987.
8. **Kawamoto, T. and Okada, E.,** Separation of mono- and disaccharides by high-performance liquid chromatography with a strong cation-exchange resin and an acetonitrile-rich eluent, *J. Chromatogr.*, 258, 284, 1983.
9. **Honda, S., Suzuki, S., and Kakehi, K.,** Improved analysis of aldose anomers by high-performance liquid chromatography on cation-exchange columns, *J. Chromatogr.*, 291, 317, 1984.

Data from Reference 9 reproduced with permission of authors and Elsevier Science Publishers.

TABLE LC 36
HPLC of Unsaturated Disaccharides from Glycosaminoglycuronans on Cation-Exchange Resin with Aqueous-Organic Eluent

Packing	P1	P1
Column		
Length, cm	15	30[a]
Diameter (I.D.), cm	0.6	0.6
Material	SS	SS
Solvent	S1	S2
Flow rate (ml/min)	1	1
Temperature (°C)	70	70
Detection	UV[b]	UV[b]
Reference	1	2
Compound[c]	**k'**	**k'**
ΔDi-OS_{HA}	1.65	1.69
ΔDi-OS_{ChS}	2.12	2.31
ΔDi-4S	—	3.22
ΔDi-6S	—	1.97
ΔDi-diS_B	—	2.55
ΔDi-diS_D/ΔDi-diS_G[d]	—	1.46
ΔDi-diS_E/ΔDi-diS_H[d]	—	4.25
ΔDi-triS	—	2.84

Note: TLC data for these compounds will be found in Table TLC 7.

[a] Coupled columns, each 15 cm long.

[b] Absorbance at 232 nm.

[c] ΔDi-OS_{HA} = 2-acetamido-2-deoxy-3-*O*-(β-D-gluco-4-enepyranosyluronic acid)-D-glucose; ΔDi-OS_{ChS} = 2-acetamido-2-deoxy-3-*O*-(β-D-gluco-4-enepyranosyluronic acid)-D-galactose; ΔDi-4S = 2-acetamido-2-deoxy-3-*O*-(β-D-gluco-4-enepyranosyluronic acid)-4-*O*-sulfo-D-galactose; ΔDi-6S = 2-acetamido-2-deoxy-3-*O*-(β-D-gluco-4-enepyranosyluronic acid)-6-*O*-sulfo-D-galactose; ΔDi-diS_B = 2-acetamido-2-deoxy-3-*O*-(2-*O*-sulfo-β-D-gluco-4-enepyranosyluronic acid)-4-*O*-sulfo-D-galactose; ΔdiS_D = 2-acetamido-2-deoxy-3-*O*-(2-*O*-sulfo-β-D-gluco-4-enepyranosyluronic acid)-6-*O*-sulfo-D-galactose; ΔDi-diS_E = 2-acetamido-2-deoxy-3-*O*-(β-D-gluco-4-enepyranosyluronicacid)-4,6-bis-*O*-sulfo-D-galactose; ΔDi-triS = 2-acetamido-2-deoxy-3-*O*-(2-*O*-sulfo-β-D-gluco-4-enpyranosyluronic acid)-4,6-bis-*O*-sulfo-D-galactose.

[d] ΔDi-diS_G and ΔDi-diS_H differ from ΔDi-diS_D and ΔDi-diS_E, respectively, only in having epimeric configurations in hexuronic acid moiety, derived from L-iduronic acid residues in parent polymer (dermatan sulfate).

Packing P1 = Shodex® RS (Type DC-613) column (Showa Denko, Tokyo), 10-μm, fully porous (55% cross-linked) cation-exchange resin, Na^+-form.

Solvent S1 = acetonitrile-methanol-0.5 *M* ammonium formate buffer, pH 4.5 (13:3:4).

S2 = acetonitrile-methanol-0.8 *M* ammonium formate buffer, pH 4.5 (13:3:4).

REFERENCES

1. **Murata, K. and Yokoyama, Y.,** Analysis of hyaluronic acid and chondroitin by high-performance liquid chromatography of the constituent disaccharide units, *J. Chromatogr.*, 374, 37, 1986.
2. **Murata, K. and Yokoyama, Y.,** Liquid chromatographic assay for constituent disaccharides of hyaluronic acid and chondroitin sulphate isomers, *J. Chromatogr.*, 415, 231, 1987.

TABLE LC 37
Ion-Moderated Partitioning: HPLC of Mono- and Oligosaccharides on Anion-Exchangers With Aqueous-Organic Eluents

Packing	P1	P2	P3	P4	P5	P6	P6
Column							
Length, cm	15	50	50	50	25	10	10
Diameter (I.D.), cm	0.43	0.26	0.26	0.26	0.4	0.4	0.4
Material	SS	SS	SS	SS	SS	SS	SS
Solvent	S1	S2	S2	S2	S3	S2	S2
Flow rate (ml/min)	1	1	1	1	0.9	1.6	1.2
Temperature (°C)	60	70	70	70	20	20	0
Detection	D1	RI	RI	RI	RI	RI	RI
Reference	1	2	2	2	3	4	4
Sugar					k′		
2-Deoxy-D-*erythro*-pentose	—	—	—	—	1.29	—	—
L-Rhamnose	—	—	—	—	1.44(α); 1.80	2.1(α); 3.3	3.5(α); 5.5
L-Fucose	2.16	1.71	1.74	1.69	1.55(α); 2.10	3.0; 4.0(α)	4.6(α); 5.9
D-Ribose	2.16	2.45	2.56	2.26	1.80	2.7	3.0
L-Arabinose	3.74	3.21	3.89	3.41	1.90(α); 2.74	4.1; 5.6(α)	6.5; 8.5(α)
D-Xylose	3.32	4.49	4.96	4.40	1.94(α); 2.48	3.4(α); 4.4	4.7(α); 5.9
D-Lyxose	—	—	—	—	—	3.5	5.0(α); 6.4
D-Mannose	—	4.91	6.38	5.60	2.07(α); 2.85	6.0(α); 9.5	8.0(α); 12.6
D-Galactose	7.66	6.20	8.33	7.25	2.42(α); 3.54	9.1(α); 11.7	12.0(α); 14.3
D-Glucose	8.82	7.29	10.0	8.28	2.50(α); 3.25	7.6(α); 8.8	10.2(α); 11.3
D-Tagatose	—	—	—	—	—	5.3	5.6(α); 8.3
D-Fructose	4.58	3.92	4.71	3.93	2.17	6.1	7.0(α); 9.6
L-Sorbose	—	4.47	5.57	4.82	2.42	5.9	7.7
2-Acetamido-2-deoxy-D-glucose	—	—	—	—	1.83(α); 2.56	—	—
Sucrose	—	8.83	13.6	10.2	4.92	—	—
Lactose	26.0	7.93	13.6	10.7	3.90(α); 4.63	20.6(α); 24.1	—
Maltose	—	11.5	18.1	14.0	4.04(α); 5.67	25.9(α); 29.2	—
Raffinose	—	—	—	—	10.8	—	—

Note: Earlier data on HPLC of mono- and oligosaccharides on anion-exchange resins in SO_4^{2-}-form in aqueous ethanol will be found in *CRC Handbook of Cromatography: Carbohydrates*, Volume I, Tables LC 13—16.

TABLE LC 37 (continued)
Ion-Moderated Partitioning: HPLC of Mono- and Oligosaccharides on Anion-Exchangers With Aqueous-Organic Eluents

Packing	P1 =	3013 N (Hitachi, Danbury, CT), macroporous anion-exchange resin; converted to phosphate form by treatment with orthophosphoric acid.
	P2 =	*N*-butylpyridinium bromide polymer (10—15 μm); converted to phosphate form by treatment with 0.5 *M* NaH_2PO_4 buffer, pH 4.3.
	P3 =	*N*-methyl-4-vinylpyridinium bromide polymer (10—15 μm), converted to phosphate form as for P2.
	P4 =	as P3, but polymer converted to sulfate form by treatment with 0.5 *M* sodium sulfate solution.
	P5 =	CDR-10 (Mitsubishi, Tokyo), strongly basic macroreticular anion-exchange resin, 5- to 7-μm particle diameter. 35% cross-linked; converted to sulfate form by treatment with 1% sodium sulfate solution.
	P6 =	Silasorb® (Lachema, Brno, Czechoslovakia), irregularly shaped silica gel, particle diameter 10 μm; treated with 3-aminopropyltriethoxysilane to introduce bonded amino phase, then with 50 m*M* sulfuric acid to convert aminopropyl groups to sulfate form.
Solvent	S1 =	acetonitrile-water (83:17).
	S2 =	acetonitrile-water (4:1).
	S3 =	ethanol-water (4:1).
Detection	D1 =	Automated tetrazolium blue colorimetric method.

REFERENCES

1. **D'Amboise, M., Hanai, T., and Noël, D.,** Liquid-chromatographic measurement of urinary monosaccharides, *Clin. Chem.*, 26, 1348, 1980.
2. **Sugii, A., Harada, K., and Tomita, Y.,** Separation of carbohydrates by high-performance liquid chromatograpny on porous pyridinium polymer columns, *J. Chromatogr.*, 366, 412, 1986.
3. **Oshima, R., Takai, N., and Kumanotani, J.,** Improved separation of anomers of saccharides by high-performance liquid chromatography on macroreticular anion-exchange resin in the sulphate form, *J. Chromatogr.*, 192, 452, 1980.
4. **Kahle, V., and Tesařík, K.,** Separation of saccharides and their anomers by high-performance liquid chromatography, *J. Chromatogr.*, 191, 121, 1980.

Data from References 2, 3 and 4 reproduced with permission of authors and Elsevier Science Publishers.

TABLE LC 38
High-Performance Ion-Exchange Chromatography: Separation of Uronic Acids on Silica-Based Bonded Anion-Exchange Phases

Packing	P1	P2	P3	P3	P3	P4
Column						
Length, cm	25	25	25	25	25	25
Diameter (I.D.), cm	0.46	0.46	0.46	0.46	0.46	0.46
Material	SS	SS	SS	SS	SS	SS
Solvent	S1	S2	S3	S4	S5	S6
Flow rate (ml/min)	0.7	0.4	1.4	1	1	1.5
Temperature (°C)	Ambient	Ambient	40	40	40	40
Detection	RI	UV[a]	RI	RI, UV[a]	UV[a]	UV[a]
Reference	1	2	3	3	3	3
Compound		r[b]			r[c]	
D-Glucuronic acid	1.00; 1.49	1.00	1.00	—	—	—
D-Mannuronic acid	—	1.04	0.84	—	—	—
D-Galacturonic acid	—	1.15	0.64	—	—	—
L-Guluronic acid	—	1.23	0.68	—	—	—
L-Iduronic acid	1.15; 1.29;[d] 1.55	—	—	—	—	—
Oligogalacturonic acids[e]						
DP 2	—	—	—	0.90	—	—
3	—	—	—	1.21	—	—
4	—	—	—	1.94	—	—
Unsaturated[f]						
DP 2	—	—	—	1.00	1.00	1.00
3	—	—	—	1.41	1.08	1.42
4	—	—	—	2.00	1.30	2.10
5	—	—	—	3.00	1.56	3.52
6	—	—	—	4.77	1.78	—
7	—	—	—	—	2.17	—
8	—	—	—	—	2.60	—

[a] Absorbance at 210 nm[2] or 235 nm.[3]
[b] t_r relative to D-glucuronic acid (12.25 min for first peak,[1] 43.2 min,[2] 12.6 min[3]).
[c] t_r relative to unsaturated dimer (5.74 min with S4, 4.8 min with S5, 4.3 min on P4).
[d] Major peak.
[e] Normal oligogalacturonic acids obtained by degradation of pectic acid with endo-polygalacturonase.
[f] Double bond between C-4 and C-5 of terminal nonreducing GalA unit; products of degradation of pectic acid with endo-pectic acid lyase.

Packing P1 = APS-Hypersil® (Shandon, U.K.); weakly basic primary amino phase bonded to silica.
P2 = Partisil® 10-SAX (Whatman); quaternary ammonium phase bonded to 10-μm silica.
P3 = Nucleosil® 10 SB (Chrompack, Middelburg, Netherlands); quaternary ammonium phase bonded to 10-μm silica; Zorbax® SAX (Du Pont) gives identical results.
P4 = LiChrosorb® 10 NH_2 (Merck); weakly basic primary amine phase bonded to 10-μm silica.

Solvent S1 = 0.02 *M* sodium acetate, pH adjusted to 3.4 with acetic acid.
S2 = 5 m*M* KH_2PO_4 (pH 4.6) containing 5% of methanol.
S3 = 0.7 *M* acetic acid.
S4 = 0.3 *M* sodium acetate buffer, pH 5.4.
S5 = 0.4 *M* sodium acetate buffer, pH 5.4.
S6 = 0.11 *M* sodium acetate buffer, pH 7.5.

REFERENCES

1. **Hjerpe, A., Antonopoulos, C. A., Classon, B., Engfeldt, B., and Nurminen, M.,** Uronic acid analysis by high-performance liquid chromatography after methanolysis of glycosaminoglycans, *J. Chromatogr.*, 235, 221, 1982.
2. **Gacesa, P., Squire, A., and Winterburn, P. J.,** The determination of the uronic acid compositon of alginates by anion-exchange liquid chromatography, *Carbohydr. Res.*, 118, 1, 1983.
3. **Voragen, A. G. J., Schols, H. A., De Vries, J. A., and Pilnik, W.,** High-performance liquid chromatographic analysis of uronic acids and oligogalacturonic acids, *J. Chromatogr.*, 244, 327, 1982.

TABLE LC 39
High-Performance Ion-Exchange Chromatography of Oligosaccharides and Reduced Oligosaccharides from Glycosaminoglycuronans on Silica-Based Bonded Anion-Exchange Phases

Packing	P1	P1	P1	P2	P3	P4	P5
Column							
Length, cm	25	25	25	25	25	25	25
Diameter (I.D.), cm	0.45	0.45	0.45	0.46	0.46	0.46	0.46
Material	SS	SS	SS	SS	SS	SS	SS
Solvent	S1	S2	S3	S4	S5	S6	S7
Flow rate (ml/min)	1	1	1	1	0.9	2.2	1.5
Temperature (°C)	Ambient					50	Ambient
Detection	D1	D1	UV,[a] D1	UV[a]	UV[a]	UV[a]	UV[a]
Reference	1	2	2	3	4	5	6
Compound	k′			r[b]	r[c]	r[c]	k′
Saturated oligosaccharides							
GlcAβ1-4(6-S)AMan-ol[d]	5.9	—	—	—	—	—	—
IdAα1-4(6-S)AMan-ol[d]	7.0	—	—	—	—	—	—
(2-S)IdAα1-4Man-ol[d]	8.2	—	—	—	—	—	—
(2-S)IdAα1-4(6-S)AMan-ol[d]	12.3	—	—	—	—	—	—
GlcAβ1-3(4-S)GalNAc-ol (Ch4S-2)[e]	—	0.34	4.75	—	—	—	—
Dimer (Ch4S-4)[e]	—	1.23	—	—	—	—	—
Trimer (Ch4S-6[e]	—	4.17	—	—	—	—	—
Tetramer (Ch4S-8)[e]	—	5.96	—	—	—	—	—
Pentamer (Ch4S-10)[e]	—	7.21	—	—	—	—	—
GlcAβ1-3(6-S)GalNAc-ol (Ch6S-2)[e]	—	0.51	5.87	—	—	—	—
Dimer (Ch6S-4)[e]	—	1.32	—	—	—	—	—
Trimer (Ch6S-6)[e]	—	4.00	—	—	—	—	—
Tetramer (Ch6S-8)[e]	—	5.43	—	—	—	—	—
Pentamer (Ch6S-10)[e]	—	6.50	—	—	—	—	—
GlcAβ1-3GlcNAc (HA-2)[f]	—	—	—	1.28	—	—	—
(GlcAβ1-3GlcNAcβ1-4)$_2$ (HA-4)[f]	—	—	—	1.79	—	—	—
Trimer (HA-6)[f]	—	—	—	2.80	—	—	—
Tetramer (HA-8)[f]	—	—	—	4.87	—	—	—

(GlcNAcβ1-4GlcAβ1-3)$_2$ (HA-4′)[f]	—	—	—	1.79	—	—	—
Trimer (HA-6′)[f]	—	—	—	2.80	—	—	—
Tetramer (HA-8′)[f]	—	—	—	4.87	—	—	—
GlcAβ1-3GlcNAcβ1-4GlcA (HA-3)[f]	—	—	—	2.08	—	—	—
GlcAβ1-3(GlcNAcβ1-4GlcAβ1-3)$_2$ (HA-5)[f]	—	—	—	3.40	—	—	—
GlcAβ1-3(GlcNAcβ1-4GlcAβ1-3)$_3$ (HA-7)[f]	—	—	—	6.25	—	—	—
GlcNAcβ1-4GlcAβ1-3GlcNAc (HA-3′)[f]	—	—	—	1.18	—	—	—
GlcNAcβ1-4(GlcAβ1-3GlcNAcβ1-4)$_2$ (HA-5′)[f]	—	—	—	1.55	—	—	—
GlcNAcβ1-4(GlcAβ1-3GlcNAcβ1-4)$_3$ (HA-7′)[f]	—	—	—	2.30	—	—	—
Unsaturated oligosaccharides							
ΔDi-4S (reduced)[g]	—	—	5.38	—	—	—	—
ΔDi-6S (reduced)[g]	—	—	8.12	—	—	—	—
ΔDi-0S[g]	—	—	—	—	1.0	1.00	—
ΔDi-4S[g]	—	—	—	—	2.6	2.04	—
ΔDi-6S[g]	—	—	—	—	2.2	1.77	—
ΔUA1-3GlcNAcβ1-4GlcAβ1-3GlcNAc[h]	—	—	—	—	1.4	—	—
ΔUA1-3GlcNAcβ1-4(GlcAβ1-3GlcNAcβ1-4)$_2$[h]	—	—	—	—	2.1	—	—
(2-S)ΔUA1-4(2-S)GlcN[i]	—	—	—	—	—	—	4.19
(2-S)ΔUA1-4(2-S)(6-S)GlcN[i]	—	—	—	—	—	—	10.2
(2-S)ΔUA1-4(2-S)GlcNα1-4(2-S)IdAα1-4(2-S)(6-S)GlcN[i]	—	—	—	—	—	—	17.8
(2-S)ΔUA1-4(2-S)(6-S)GlcNα1-4GlcAβ1-4(2-S)(6-S)GlcN[i]	—	—	—	—	—	—	19.6
(2-S)ΔUA1-4(2-S)(6-S)GlcNα1-4(2-S)IdAα1-4(2-S)(6-S)GlcN[i]	—	—	—	—	—	—	22.2
(2-S)ΔUA1-4(2-S)(6-S)GlcNα1-4IdAα1-4(6-S)GlcNAcα1-4GlcAβ1-4-(2-S)(3-S)(6-S)GlcN[i]	—	—	—	—	—	—	26.6

Note: GlcA = D-glucuronic acid, IdA = L-iduronic acid, GlcN = 2-amino-2-deoxy-D-glucose, GlcNAc = 2-acetamido-2-deoxy-D-glucose, GalNAc = 2-acetamido-2-deoxy-D-galactose.
AMan = 2,5-anhydro-D-mannose (from deamination of GlcN), ΔUA = 4-deoxy-α-L-*threo*-hex-4-enopyranosyluronic acid, from action of lyase on GlcA.
Positions of sulfate groups indicated in brackets. For structures of ΔDi-0S, ΔDi-4S, and ΔDi-6S, see footnotes to Table LC 36.

[a] Absorbance at 195 nm,[2] 206 nm,[3] 232 nm.[4-6]
[b] t_r relative to GlcNAc (4.16 min).
[c] t_r relative to ΔDi-0S (5 min,[4] 1.81 min[5]).
[d] From nitrous acid deamination of heparin, followed by reduction with sodium boro[^{3}H]hydride.
[e] From digestion of chondroitin 4- and 6-sulfates with hyaluronidase; labeled alditol at reducing end after reduction with sodium boro[^{3}H]hydride.

TABLE LC 39 (continued)
High-Performance Ion-Exchange Chromatography of Oligosaccharides and Reduced Oligosaccharides from Glycosaminoglycuronans on Silica-Based Bonded Anion-Exchange Phases

[f] From digestion of hyaluronic acid with bovine (HA series) or leech (HA′ series) hyaluronidase; odd-numbered oligosaccharides obtained from even-numbered by removal of nonreducing GlcA end-groups with β-glucuronidase or nonreducing GlcNAc end-groups with β-*N*-acetylglucosamidase A; TLC data for these oligosaccharides will be found in Table TLC 7.

[g] From digestion of chondroitin and dermatan sulfates with chondroitinase ABC or of chondroitin sulfates with chondroitinase AC; in Reference 2 products reduced with sodium boro[^{3}H]hydride.

[h] From digestion of hyaluronic acid with *Streptomyces* hyaluronidase (lyase).

[i] From depolymerization of heparin with heparin lyase (EC4.2.2.7) from *Flavobacterium heparinum.*

Packing	P1 =	Partisil®-10 SAX (Whatman); quaternary ammonium phase bonded to 10-μm silica.
	P2 =	Ultrasil®-NH_2 (Beckman; 5-μm silica), with precolumn (4 × 0.46 cm) of LiChrosorb-NH_2 (Merck; 10-μm silica); both with bonded weakly basic primary amine phase.
	P3 =	Zorbax® NH_2 (DuPont); weakly basic primary amine phase bonded to 6-μm silica.
	P4 =	Separon®-6 NH_2 (Laboratorní Přístroje, Prague, Czechoslovakia), weakly basic aminopropyl phase bonded to 10-μm silica.
	P5 =	Spherisorb® SAX (Phase Separations); strongly basic quaternary ammonium phase bonded to 5-μm silica.
Solvent	S1 =	40 m*M* KH_2PO_4 (pH 4.6) for 30 min, then convex gradient, 40 → 400 m*M* KH_2PO_4 over 40 min to elute disulfated disaccharide.
	S2 =	linear gradient, 250 → 550 m*M* KH_2PO_4 over 20 min, then isocratic for 20 min to elute octa- and deca-saccharides.
	S3 =	40 m*M* KH_2PO_4-methanol (17:3).
	S4 =	0.1 *M* KH_2PO_4, pH 4.75.
	S5 =	0.5 *M* ammonium formate, pH 5.5:water:methanol (15:9:1).
	S6 =	water containing sodium sulfate (10 m*M*) and acetic acid (1 m*M*).
	S7 =	linear gradient, 0.2 → 0.8 *M* sodium chloride (pH 3.5) over 50 min.
Detection	D1 =	scintillation counting of oligosaccharides labeled by reduction with sodium boro[^{3}H]hydride.

REFERENCES

1. **Delaney, S. R., Leger, M., and Conrad, H. E.,** Quantitation of the sulfated disaccharides of heparin by high-performance liquid chromatography, *Anal. Biochem.*, 106, 253, 1980.
2. **Delaney, S. R., Conrad, H. E., and Glaser, J. H.,** A high-performance liquid chromatography approach for isolation and sequencing of chondroitin sulfate oligosaccharides, *Anal. Biochem.*, 108, 25, 1980.
3. **Nebinger, P., Koel, M., Franz, A., and Werries, E.,** High-performance liquid chromatographic analysis of even- and odd-numbered hyaluronate oligosaccharides, *J. Chromatogr.*, 265, 19, 1983.

4. **Gherezghiher, T., Koss, M. C., Nordquist, R. E., and Wilkinson, C. P.,** Rapid and sensitive method for measurement of hyaluronic acid and isomeric chondroitin sulfates using high-performance liquid chromatography, *J. Chromatogr.*, 413, 9, 1987.
5. **Macek, J., Krajíčková, J., and Adam, M.,** Rapid determination of disaccharides from chondroitin and dermatan sulfates by high-performance liquid chromatography, *J. Chromatogr.*, 414, 156, 1987.
6. **Linhardt, R. J., Rice, K. G., Kim, Y. S., Lohse, D. L., Wang, H. M., and Loganathan, D.,** Mapping and quantification of the major oligosaccharide components of heparin, *Biochem. J.*, 254, 781, 1988.

TABLE LC 40
Ion-Exchange Chromatography of Oligosaccharides and Reduced Oligosaccharides from Glycoproteins on Silica- and Resin-Based Anion-Exchangers

Packing	P1	P2	P2	P2	P2	P2	P3	P3
Column								
Length, cm	30	25	25	25	25	25	5	5
Diameter (I.D.), cm	0.4	0.46	0.46	0.46	0.46	0.46	0.5	0.5
Material	SS	SS	SS	SS	SS	SS	SS	SS
Solvent	S1	S2	S3	S4	S5	S6	S7	S8
Flow rate (ml/min)	1	0.5	1	0.5	0.5	0.5	2	2
Temperature (°C)	Ambient	55	55	55	55	55	Ambient	Ambient
Detection	D1	UV,[a] D2, D3	UV,[a] D3	UV,[a] D2-D4	UV,[a] D3	UV,[a] D2, D3	UV[a]	UV[a]
Reference	1	2	2	2	2	2	3	3
Compound	r[b]	k′						
NeuAc2-3Gal1-4Glc	—	32.8	—	—	—	—	—	—
NeuAc2-6Gal1-4Glc	—	31.4	—	—	—	—	—	—
NeuAc2-6Gal1-4GlcNAc	—	30.7	—	—	—	—	—	—
NeuAc2-6GalNAc-ol	—	—	40.5	—	—	—	—	—
NeuGc2-6GalNAc-ol	—	—	40.5	35.0	—	—	—	—
GlcNAc1-3(NeuAc2-6)GalNAc-ol	—	—	36.0	—	—	—	—	—
NeuGc2-6Gal1-3GalNAc-ol	—	—	—	32.0	—	—	—	—
NeuGc2-6(Fuc1-2)Gal1-3GalNAc-ol	—	—	—	29.2	—	—	—	—
NeuAcα2-3Galβ1-3GalNAc-ol	—	—	—	—	27.6	—	3.01	5.61
Galβ1-3(NeuAcα2-6)GalNAc-ol	—	—	—	—	28.9	—	—	—
NeuAcα2-3Galβ1-3(NeuAcα2-6)GalNAc-ol	—	—	—	—	64.5	—	5.01	—
N1[c,d]	—	—	—	—	—	—	2.60	5.00
N2[c,d]	—	—	—	—	—	—	3.81	—
N3[c,d]	—	—	—	—	—	—	5.21	—
PTG1-a[c,e]	—	—	—	—	—	17.7	—	—
PTG1-b[c,e]	—	—	—	—	—	19.0	—	—
CPN-S-1[c,f]	0.67	—	—	—	—	—	—	—
CPN-S-2[c,f]	1.87	—	—	—	—	—	—	—
CPN-S-3[c,f]	2.62	—	—	—	—	—	—	—

ORS-S-IV[c,g]	2.91	—	—	—	—	—	—	—
OV-V[c,h]	0.37	—	—	—	—	—	—	—
PO_4-6Manα1-3Manα1-3Manα1-3Manα1-2-Man[i]	1.00	—	—	—	—	—	—	—

Note: NeuAc = *N*-acetylneuraminic acid, NeuGc = *N*-glycolylneuraminic acid.

[a] Absorbance at 210 nm[2] or 214 nm.[3]
[b] t_r relative to pentamannosylphosphate (10.3 min).
[c] See appendix for structure.
[d] From *N*-glycosylated proteins isolated from urine of sialidosis patient.
[e] Asparagine-linked oligosaccharides released from porcine thyroglobulin by hydrazinolysis.
[f] From hydrazinolysis of human ceruloplasmin.
[g] From orosomucoid.
[h] From endo-β-*N*-acetylglucosaminidase H digestion of ovalbumin glycopeptide V.
[i] From partial acid hydrolysis of yeast mannan.

Packing	P1	=	MicroPak® AX-10 (Varian).
	P2	=	Custom Resin No. 2630 (Hitachi).
	P3	=	Mono Q® (Pharmacia, Uppsala, Sweden); strong-base anion-exchange resin, 9.8-μm particles.
Solvent	S1	=	25 m*M* KH_2PO_4, pH adjusted to 4.0 with H_3PO_4; isocratic for 15 min, then linear gradient 25 → 500 m*M* KH_2PO_4 over 30 min.
Solvent	S2	=	concave gradient, 0 → 150 m*M* NaCl in 6 h.
	S3	=	concave gradient, 0 → 150 m*M* NaCl in 4 h.
	S4	=	concave gradient, 0 → 100 m*M* NaCl in 6 h.
	S5	=	concave gradient, 0 → 250 m*M* NaCl in 8 h.
	S6	=	concave gradient, 0 → 80 m*M* NaCl in 8 h.
	S7	=	linear gradient, 0 → 100 m*M* NaCl over 5 min, then isocratic for 2.5 min.
	S8	=	linear gradient, 0 → 20 m*M* NaCl over 11 min.
Detection	D1	=	scintillation counting for oligosaccharides labeled by reduction with sodium boro[^{3}H]hydride.
	D2	=	phenol-sulpuric acid colorimetric assay.
	D3	=	sodium periodate-resorcinol assay for sialic acid.
	D4	=	thioglycolate-sulfuric acid assay for 6-deoxyhexose.

REFERENCES

1. **Baenziger, J. U. and Natowicz, M.,** Rapid separation of anionic oligosaccharide species by high-performance liquid chromatography, *Anal. Biochem.*, 112, 357, 1981.
2. **Tsuji, T., Yamamoto, K., Konami, Y., Irimura, T., and Osawa, T.,** Separation of acidic oligosaccharides by liquid chromatography: application to analysis of sugar chains of glycoproteins, *Carbohydr. Res.*, 109, 259, 1982.
3. **Van Pelt, J., Damm, J. B. L., Kamerling, J. P., and Vliegenthart, J. F. G.,** Separation of sialyl-oligosaccharides by medium pressure anion-exchange chromatography on Mono Q, *Carbohydr. Res.*, 169, 43, 1987.

APPENDIX TO TABLE LC40: STRUCTURES OF OLIGOSACCHARIDES (continued)

N1 NeuAcα2-6Galβ1-4GlcNAcβ1-2Manα1-3Manβ1-4GlcNAc

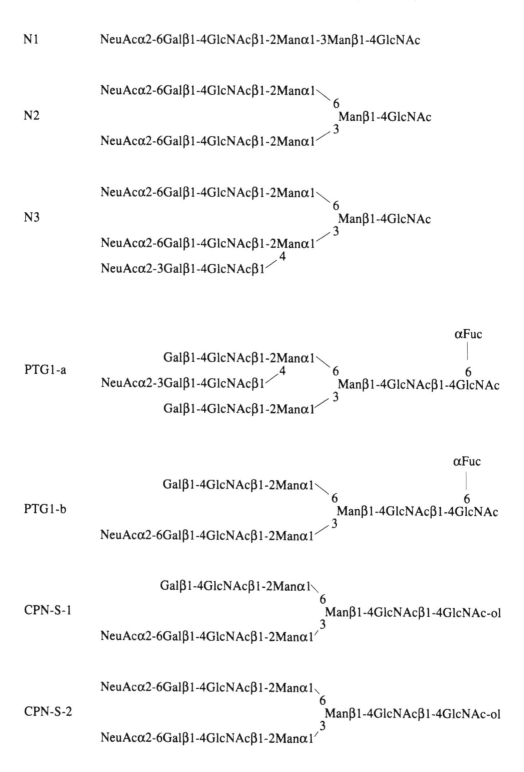

N2

```
NeuAcα2-6Galβ1-4GlcNAcβ1-2Manα1\
                                6
                                 Manβ1-4GlcNAc
                                3
NeuAcα2-6Galβ1-4GlcNAcβ1-2Manα1/
```

N3

```
NeuAcα2-6Galβ1-4GlcNAcβ1-2Manα1\
                                6
                                 Manβ1-4GlcNAc
                                3
NeuAcα2-6Galβ1-4GlcNAcβ1-2Manα1/
                          4
NeuAcα2-3Galβ1-4GlcNAcβ1/
```

PTG1-a

```
                                                     αFuc
                                                      |
     Galβ1-4GlcNAcβ1-2Manα1\                          6
                          4 6
NeuAcα2-3Galβ1-4GlcNAcβ1/    Manβ1-4GlcNAcβ1-4GlcNAc
                            3
     Galβ1-4GlcNAcβ1-2Manα1/
```

PTG1-b

```
                                                       αFuc
                                                        |
     Galβ1-4GlcNAcβ1-2Manα1\                            6
                            6
                             Manβ1-4GlcNAcβ1-4GlcNAc
                            3
NeuAcα2-6Galβ1-4GlcNAcβ1-2Manα1/
```

CPN-S-1

```
     Galβ1-4GlcNAcβ1-2Manα1\
                            6
                             Manβ1-4GlcNAcβ1-4GlcNAc-ol
                            3
NeuAcα2-6Galβ1-4GlcNAcβ1-2Manα1/
```

CPN-S-2

```
NeuAcα2-6Galβ1-4GlcNAcβ1-2Manα1\
                                6
                                 Manβ1-4GlcNAcβ1-4GlcNAc-ol
                                3
NeuAcα2-6Galβ1-4GlcNAcβ1-2Manα1/
```

APPENDIX TO TABLE LC40: STRUCTURES OF OLIGOSACCHARIDES (continued)

CPN-S-3

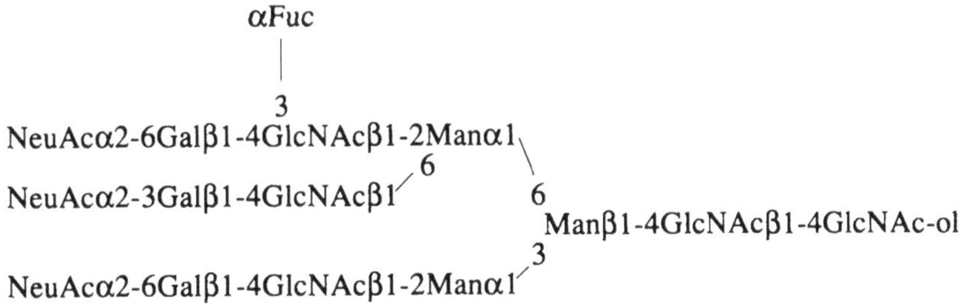

ORS-S-IV

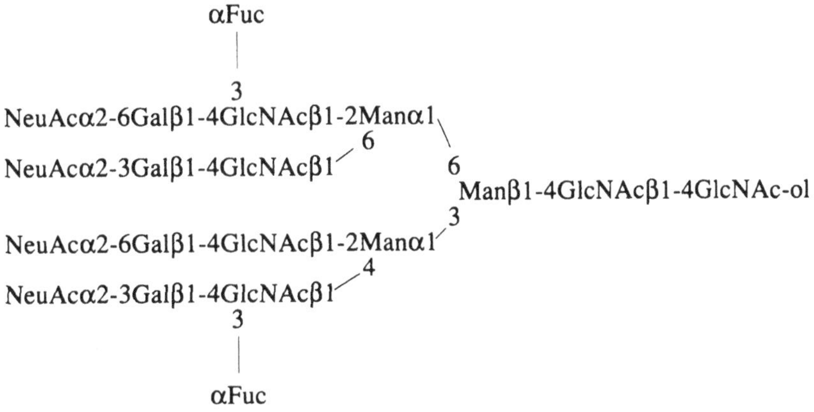

OV-V

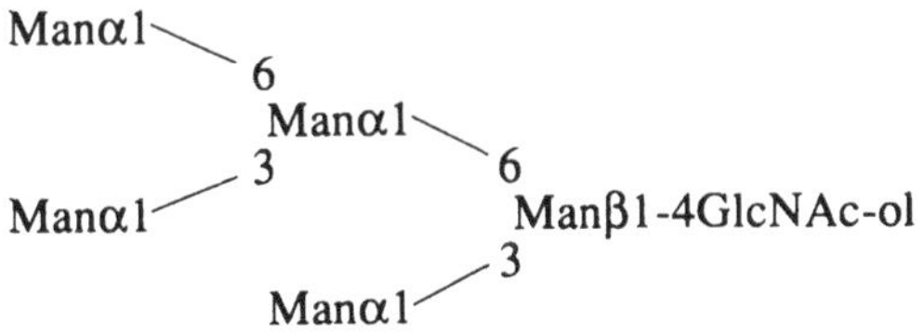

TABLE LC 41
Anion-Exchange Chromatography of Aldonic and Aldaric Acids on High-Performance Resins

Packing	P1	P2	P3	P4
Column				
Length, cm	30	25	25	25
Diameter (I.D.), cm	0.4	0.46	0.46	0.46
Material	SS	SS	SS	SS
Solvent	S1	S2	S3	S4
Flow rate (ml/min)	0.5	2	1	0.3
Temperature (°C)	60	85	45	45
Detection	RI	UV[a]	RI	RI, UV[a]
Reference	1	2	3	3
Compound	k'			
D-Glucono-1,5-lactone	1.03	—	—	—
D-Glucono-1,4-lactone	2.15	—	—	—
D-Gluconic acid	6.00	2.34	4.92	4.40
D-Galactonic acid	6.07	—	—	—
L-Idonic acid	—	—	4.88	4.16
L-Gulonic acid	—	—	5.32	4.68
D-Mannonic acid	—	—	6.32	—
D-Arabinonic acid	—	2.84	—	—
5-Keto-D-gluconic acid	9.31	—	9.80	7.88
2-Keto-D-gluconic acid	10.9	—	14.5	9.2
2-Keto-L-gulonic acid	—	—	15.8	10.0
2,5-Diketo-D-gluconic acid	13.5	—	18.2	11.2
Glucaric acid	—	13.8	—	

[a] Absorbance at 210—212 nm.

Packing P1 = Aminex® A-28 (Bio-Rad); strongly basic anion-exchange resin, particle diameter 9—13 μm, 8% cross-linked.
P2 = BA-X8 (Alltech); strongly basic anion-exchange resin, particle diameter 7—10 μm, 8% cross-linked.
P3 = Aminex A-27 (Bio-Rad); strongly basic anion-exchange resin, particle diameter 13—17 μm, 8% cross-linked.
P4 = Aminex A-29 (Bio-Rad); strongly basic anion-exchange resin, particle diameter 5—8 μm, 8% cross-linked.

Solvent S1 = ammonium formate solution; 0.3 *M* HCOOH adjusted to pH 3.75 with aqueous ammonia.
S2 = 0.16 *M* sodium chloride solution containing magnesium chloride (0.02 *M*).
S3 = 0.2 *M* ammonium formate, pH 3.2.
S4 = 0.2 *M* KH_2PO_4, pH 3.35.

REFERENCES

1. **Blake, J. D., Clarke, M. L., and Richards, G. N.,** Determination of D-gluconic, 5-keto-D-gluconic, 2-keto-D-gluconic and 2,5-diketo-D-gluconic acids by high-performance liquid chromatography, *J. Chromatogr.*, 312, 211, 1984.
2. **Dijkgraaf, P. J. M., Verhaar, L. A. Th., Groenland, W. P. T., and Van der Wiele, K.,** High speed liquid chromatography of sugar acids on anion-exchange resins, *J. Chromatogr.*, 329, 371, 1985.
3. **Lazarus, R. A. and Seymour, J. L.,** Determination of 2-keto-L-gulonic and other ketoaldonic and aldonic acids produced by ketogenic bacterial fermentation, *Anal. Biochem.*, 157, 360, 1986.

TABLE LC 42
Anion-Exchange Chromatography of Sialic Acids, KDO, and Derivatives on High-Performance Resins

Packing	P1	P1	P2	P2
Column				
Length, cm	40	40	40	40
Diameter (I.D.), cm	0.46	0.46	0.46	0.46
Material	SS	SS	SS	SS
Solvent	S1	S2	S1	S3
Flow rate (ml/min)	0.5	0.5	0.5	0.5
Temperature (°C)	20	20	20	20
Detection	UV[a]	UV[a]	UV[a]	UV[a]
Reference	1	1	2	2
Compound	r[b]			r[c]
2-Deoxy-2-acetamido-D-mannose[d]	0.12	—	—	—
2-Deoxy-2-glycolylamino-D-mannose[d]	0.14	—	—	—
N-Acetylneuraminic acid[e]	1.00	1.00	1.00	—
4-mono-*O*-acetyl-	1.41	—	—	—
9-mono-*O*-acetyl-	1.47	1.31	—	—
2-deoxy-2,3-dehydro-	1.67	1.65	—	—
7,9-di-*O*-acetyl-	1.76	—	—	—
7,8,9-tri-*O*-acetyl-	2.05	—	—	—
methyl β-glycoside	—	—	0.74	—
methyl α-glycoside	—	—	1.22	—
N-Glycolylneuraminic acid	1.33	1.06	1.33	—
4-mono-*O*-acetyl-	1.69	—	—	—
9-mono-*O*-acetyl-	1.70	—	—	—
7,9-di-*O*-acetyl-	2.02	—	—	—
7,8,9-tri-*O*-acetyl-	2.52	—	—	—
3-Deoxy-D-*manno*-2-octulosonic acid[f]	—	—	1.49	—
methyl α-glycoside	—	—	1.09	1.00
methyl β-glycoside	—	—	1.39	1.23
Oligosaccharides				
Di-*N*-acetylneuraminyllactose	0.27	0.17	—	—
Neu5Acα2-6Galβ1-4GlcNAc	0.32	—	0.32	—
Neu5Acα2-6Galβ1-4Glc	0.37	0.23	0.38	—
Neu5Acα2-3Galβ1-4Glc	0.48	0.34	0.49	—
Neu5Acβ2-6Galβ1-4GlcNAc	—	—	0.50	—
(Neu5,9Ac$_2$)α2-3Galβ1-4Glc	0.57	—	—	—
Heptoseα1-5KDO	—	—	1.10	—
KDOα2-4KDO	—	—	—	7.42
methyl α-glycoside	—	—	—	4.78
KDOβ2-4KDO, methyl β-glycoside	—	—	—	10.6

[a] Absorbance at 195 or 215 nm,[1] 200 nm.[2]
[b] t_r relative to *N*-acetylneuraminic acid (6.0 min with S1,[1,2] 15.5 min with S2[1]).
[c] t_r relative to KDO methyl α-glycoside (2.3 min).
[d] Products of cleavage of *N*-acetyl- and *N*-glycolylneuraminic acid by neuraminic acid aldolase.
[e] Designated Neu5Ac in oligosaccharide structures.
[f] KDO.

Packing P1 = Aminex® A-28 (Bio-Rad); strongly basic anion-exchange resin, particle diameter 9—13 μm, 8% cross-linked.

P2 = Aminex A-29 (Bio-Rad); similar to Aminex A-28, particle diameter 5—8 μm.

TABLE LC 42 (continued)
Anion-Exchange Chromatography of Sialic Acids, KDO, and Derivatives on High-Performance Resins

Solvent	S1	=	0.75 *M* sodium sulfate solution.
	S2	=	40 m*M* sodium acetate buffer, pH 5.5.
	S3	=	10 m*M* sodium sulfate solution.

REFERENCES

1. **Shukla, A. K. and Schauer, R.,** Analysis of *N,O*-acylated neuraminic acids by high-performance liquid anion-exchange chromatography, *J. Chromatogr.*, 244, 81, 1982.
2. **Shukla, A. K., Schauer, R., Unger, F. M., Zähringer, U., Rietschel, E. T., and Brade, H.,** Determination of 3-deoxy-D-*manno*-octulosonic acid (KDO), *N*-acetylneuraminic acid, and their derivatives by ion-exchange liquid chromatography, *Carbohydr. Res.*, 140, 1, 1985.

TABLE LC 43
Chromatography of Aminodeoxyhexoses, Aminodeoxyhexitols, and Sugars as Derived 1-Amino-1-Deoxyglycitols on Cation-Exchange Resins

Packing	P1	P2	P3	P4	P4	P4
Column						
Length, cm	15	25	25	32	32	32
Diameter (I.D.), cm	0.28	0.32	0.4	0.32	0.32	0.20
Material	Glass	Glass	SS	Glass	Glass	Glass
Solvent	S1	S2	S3	S4	S4	S4
Flow rate (ml/min)	0.15	0.17	0.25	0.20	0.20	0.20
Temperature (°C)	65	T1	72	50	60	70
Dection	D1	D2	D3, D4	D5	D5	D5
Reference	1	2	3	4	4	4
Compound			r[a]			
2-Amino-2-deoxy-D-mannose	0.75	1.11	—	0.80	0.85	0.90
2-Amino-2-deoxy-D-glucose	1.00	1.00	1.00	1.00	1.00	1.00
2-Amino-2-deoxy-D-galactose	1.28	1.25	1.39	1.38	1.32	1.27
2-Amino-2-deoxy-D-galactitol	—	—	2.31	0.62	0.83	1.04
2-Amino-2-deoxy-D-glucitol	—	—	2.58	0.68	0.87	1.10
1-Amino-1-deoxy-D-glucitol[b]	—	—	—	0.53	0.73	0.88
1-Amino-1-deoxygalactitol[b]	—	—	—	0.60	0.85	1.06
1-Amino-1-D-mannitol[b]	—	—	—	0.82	1.14	1.37
1-Amino-1-deoxyfucitol[b]	—	—	—	0.90	1.31	1.64
1-Amino-1-deoxy-L-rhamnitol[b]	—	—	—	0.99	—	—
1-Amino-1-deoxyxylitol[b]	—	—	—	1.14	1.59	1.95
1-Amino-1-deoxyarabinitol[b]	—	—	—	1.34	—	—
1-Amino-1-deoxyribitol[b]	—	—	—	2.21	—	—

Note: Earlier data on chromatography of aminodeoxyhexoses and -hexitols on an amino acid analyzer will be found in *CRC Handbook of Chromatography: Carbohydrates*, Volume I, Table LC 25.

[a] t_r relative to 2-amino-2-deoxy-D-glucose (34.5 min,[1] 45.0 min,[2] 21.6 min,[3] in Reference 4, 46.81 min at 50°C, 36.98 min at 60°C, 30.17 min at 70°C).

[b] 1-Amino-1-deoxyglycitol (glycamine) derivatives prepared by reductive amination of sugars with sodium cyanoborohydride (0.2 *M*, in 1 *M* ammonium sulfate, pH 7.0) at 100°C for 90 min; reaction terminated by addition of 0.2 *M* HCl and samples evaporated to dryness under nitrogen; residues dissolved in sodium citrate buffer (67 m*M*, pH 3.25) for injection.

TABLE LC 43 (continued)
Chromatography of Aminodeoxyhexoses, Aminodeoxyhexitols, and Sugars as Derived 1-Amino-1-Deoxyglycitols on Cation-Exchange Resins

Packing	P1	= W-2 resin (Beckman); cation-exchange resin (Na^+-form) used in amino acid analyzer.
	P2	= BT2710 resin (Biotronik®, Frankfurtan-Mair, West Germany); cation-exchange resin (Na^+-form) used in amino acid analyzer.
	P3	= Aminex® A9 (Bio-Rad); strongly acidic cation-exchange resin, particle diameter 11 μm, 10% cross-linked, K^+-form.
	P4	= DC-4A resin (Durrum, Palo Alto, CA); cation-exchange resin (Na^+-form) used in amino acid analyzer.
Solvent	S1	= 0.07 *M* Na^+ borate-citrate buffer, pH 7.24.
	S2	= 0.1 *M* sodium citrate, pH 7.2.
	S2	= 88.4 m*M* KH_2PO_4, adjusted to pH 7.00 with KOH and containing acetonitrile (10%) and methanol (5%).
	S4	= 0.02 *M* sodium tetraborate, pH adjusted to 8.00 with HCl.
Temperature	T1	= 40°C for 15 min then 63°C.
Detection	D1	= automated ninhydrin colorimetric assay; absorbance measured at 570 nm.
	D2	= automated copper bicinchoninate colorimetric assay; absorbance measured at 570 nm.
	D3	= scintillation counting of radioactively labeled compounds.
	D4	= automated 2-cyanoacetamide method; absorbance measured at 276 nm.
	D5	= fluorometric detection after postcolumn reaction with *o*-phthalaldehyde; excitation wavelength 340 nm, emission 455 nm.

REFERENCES

1. **Madden, D. E., Alpenfels, W. F., Mathews, R. A., and Newsom, A. E.,** Improved method for the determination of hexosamines using the Beckman 121-M amino acid analyzer, *J. Chromatogr.*, 248, 476, 1982.
2. **Josić, Dj., Hofermaas, R., Bauer, Ch., and Reutter, W.,** Automatic amino acid and sugar analysis of glycoproteins, *J. Chromatogr.*, 317, 35, 1984.
3. **Lohmander, L. S.,** Analysis by high-performance liquid chromatography of radioactively labeled carbohydrate components of proteoglycans, *Anal. Biochem.*, 154, 75, 1986.
4. **Perini, F. and Peters, B. P.,** Fluorometric analysis of amino sugars and derivatized neutral sugars, *Anal. Biochem.*, 123, 357, 1982.

TABLE LC 44
Anion-Exchange Chromatography of Alditols, Mono- and Oligosaccharides, as Borate Complexes, on High-Performance Resins

Packing	P1	P1	P1	P1	P2	P3	P3	P3	P3	P4	P4	P5 + P4[a]
Column												
Length, cm	8	8	15	8	48.7	25	25	6.5	6.5	20	20	25 + 25[a]
Diameter (I.D.), cm	0.8	0.8	0.4	0.8	0.6	0.6	0.6	0.6	0.6	0.8	0.8	0.8 + 0.8[a]
Material	SS	SS	SS	SS	Glass	Glass	Glass	Glass	Glass	Glass	Glass	Glass
Solvent	S1	S2	S3	S4	S5	S6	S7	S8	S9	S10	S11	S10
Flow rate (ml/min)	1	1	0.5	1	0.83	0.67	0.67	0.33	0.33	0.83	0.83	0.83
Temperature (°C)	65	65	65	65	60	63	63	63	63	60	60	60
Detection	D1	D1	D2	D2, D3	D4	D5	D5, D6	D5	D5	D4	D4	D4
Reference	1	1	2	2, 3	4	5	5	6	6	7	7	7
Compound	r[b]		r[c]							r[d]	r[e]	
Xylitol	0.57	0.59	—	—	—	—	—	—	—	—	—	—
Ribitol	0.66	0.65	—	—	—	—	—	—	—	—	—	—
Arabinitol	0.74	0.76	—	—	—	—	—	—	—	—	—	—
Rhamnitol	0.76	0.80	—	—	—	—	—	—	—	—	—	—
D-Glucitol	1.00	1.00	—	—	—	—	—	—	—	—	—	—
Fucitol	1.33	1.32	—	—	—	—	—	—	—	—	—	—
D-Mannitol	1.38	1.37	—	—	—	—	—	—	—	—	—	—
Galactitol	2.66	1.44	—	—	—	—	—	—	—	—	—	—
D-Lyxose	—	—	0.59	0.52	0.57	—	—	—	—	—	—	—
D-Ribose	—	—	0.59	0.55	0.59	—	—	—	—	—	—	—
L-Arabinose	—	—	0.76	0.73	0.75	—	—	—	—	—	0.76	—
D-Xylose	—	—	0.91	0.90	0.87	0.91	0.92	—	—	—	0.91	—
2-Deoxy-D-*erythro*-pentose	—	—	—	—	—	—	—	0.54	—	—	—	—
L-Rhamnose	—	—	0.22	0.31	0.49	—	—	0.61	—	—	—	—
L-Fucose	—	—	0.67	0.68	0.77	0.83	0.72	0.79	—	—	—	—
D-*altro*-Heptulose	—	—	—	—	0.55	—	—	—	—	—	—	—
D-Psicose	—	—	—	—	0.70	—	—	—	—	—	—	—
D-Fructose	—	—	—	—	0.72	—	—	0.77	—	—	—	—

D-Tagatose	—	—	—	—	0.82	—	—	—	—	—	—	—
L-Sorbose	—	—	—	—	0.90	—	—	—	—	—	—	—
D-Mannose	—	—	0.59	0.60	0.69	0.75	0.56	0.71	—	—	0.69	—
D-Allose	—	—	—	—	0.80	—	—	—	—	—	—	—
D-Galactose	—	—	0.85	0.79	0.80	0.86	0.83	0.83	1.00	—	0.82	—
D-Altrose	—	—	—	—	0.86	—	—	—	—	—	—	—
D-Glucose	—	—	1.00	1.00	1.00	1.00	1.00	1.00	1.00	—	1.00	1.00
L-Ascorbic acid	—	—	—	—	—	—	—	0.52	—	—	—	—
2-Acetamido-2-deoxy-D-galactose	—	—	—	—	—	—	—	0.55	—	—	—	—
2-Acetamido-2-deoxy-D-glucose	—	—	—	—	—	—	—	0.55	—	—	—	—
N-Acetylneuraminic acid	—	—	—	—	—	—	—	1.80	1.65	—	—	—
9-mono-*O*-acetyl-	—	—	—	—	—	—	—	1.08	1.38	—	—	—
7,9-di-*O*-acetyl-	—	—	—	—	—	—	—	—	1.30	—	—	—
7,8,9-tri-*O*-acetyl-	—	—	—	—	—	—	—		1.15	—	—	—
N-Glycolylneuraminic acid	—	—	—	—	—	—	—	3.16	2.15	—	—	—
9-mono-*O*-acetyl—	—	—	—	—	—	—	—	—	1.67	—	—	—
Oligosaccharides												
Sucrose	—	—	—	—	0.11	—	—	—	—	—	—	—
Trehalose	—	—	—	—	0.12	—	—	—	—	—	—	—
Cellobiose	—	—	—	—	0.22	—	—	—	—	1.00	0.35	0.83
Maltose	—	—	—	—	0.32	—	—	—	—	—	—	—
Lactose	—	—	—	—	0.46	—	—	—	—	—	—	—
Turanose	—	—	—	—	0.72	—	—	—	—	—	—	—
Melibiose	—	—	—	—	1.04	—	—	—	—	—	—	—
Melezitose	—	—	—	—	0.16	—	—	—	—	—	—	—
Raffinose	—	—	—	—	0.18	—	—	—	—	—	—	—
Cellotriose	—	—	—	—	—	—	—	—	—	0.84	0.30	0.75
Stachyose	—	—	—	—	0.43	—	—	—	—	—	—	—
Cellotetraose	—	—	—	—	—	—	—	—	—	0.76	0.25	0.68
Cellopentaose	—	—	—	—	—	—	—	—	—	0.68	0.22	0.63
Cellohexaose	—	—	—	—	—	—	—	—	—	0.63	0.19	—

Note: Earlier data on anion-exchange chromatography of carbohydrates as borate complexes will be found in *CRC Handbook of Chromatography: Carbohydrates*, Volume I, Tables LC 18—21.

[a] Dual column system; after injection of sample columns coupled for 20 min, until cellodextrins have passed into P4 column, with glucose retained on P5; P5 column then excluded for 50 min, until elution of cellodextrins; P4 column then excluded and P5 switched in until elution of glucose.

TABLE LC 44 (continued)
Anion-Exchange Chromatography of Alditols, Mono- and Oligosaccharides, as Borate Complexes, on High-Performance Resins

[b] t_r relative to D-glucitol (42 min with S1, 58.5 min with S2).
[c] t_r relative to D-glucose (89.5 min with S3,[2] 64.8 min with S4,[2,3] 208 min,[4] 70 min with S6,[5] 96 min with S7,[5] 22.1 min with S8,[6] 13.3 min with S9,[6] 206 min with S11,[7] 102 min with P5 + P4[7]).
[d] t_r relative to cellobiose (63 min).

Packing	P1	= Hitachi 2633 resin (Hitachi, Tokyo); strongly basic anion-exchange resin (quaternary ammonium type) with spherical particles, diameter 11 μm.
	P2	= DEAE-Spheron® 300, capacity 2.25 meq/g, mean particle diameter 27 μm; diethylaminoethyl derivative prepared in authors' laboratory from Spheron P-300 (Lachema, Brno, Czechoslovakia), poly(hydroxyethylmethacrylate) resin.
	P3	= DA-X8-11 (Durrum), strongly basic anion-exchange resin, 8% cross-linked, particle diameter 11 μm.
	P4	= TEAE-Spheron® 1000, capacity 2 meq/g, particle diameter 13—18 μm; experimental laboratory product from Lachema (Brno, Czechoslovakia), prepared by quaternization of DEAE-Spheron with ethyl halide to form triethylaminoethyl (TEAE) groups.
	P5	= DEAE-Spheron 300, capacity 0.6 meq/g, particle diameter 20—40 μm; prepared from Spheron P-300 as for P2.
Solvent	S1	= 0.50 *M* borate buffer, pH 7.5 (pH of aqueous solution of boric acid adjusted with KOH).
	S2	= stepwise elution program: 0.50 *M* borate buffer, pH 7.1, for 35 min; 0.30 *M* borate buffer, pH 8.1, for 30 min; 0.50 *M* borate buffer, pH 10.5, for 25 min, (all buffers prepared as described for S1).
	S3	= stepwise elution program; 0.20 *M* borate buffer, pH 7.4, for 30 min; 0.35 *M* borate buffer, pH 7.7, for 30 min; 0.50 *M* borate buffer, pH 9.2, for 40 min (buffers prepared as described for S1); column packed in 0.50 *M* buffer initially.
	S4	= gradient elution, using buffers (*A*) 0.25 *M* borate, pH 8.2, (*B*) 0.40 *M* borate, pH 7.4, (*C*) 0.60 *M* borate, pH 9.3: linear gradient, 100% *A* → 100% *B* over 15 min; isocratic elution with *B* for 20 min; linear gradient, 100% *B* → 100% *C* over 10 min; isocratic elution with *C* for 25 min; column packed in *C* initially.
	S5	= stepwise elution program: 30 m*M* borate buffer, pH 7.50, for 95 min; 100 m*M* borate buffer, pH 8.85, for 30 min; 250 m*M* borate buffer, pH 8.88, for 2 h.
	S6	= 0.7 *M* potassium borate, pH 9.2.
	S7	= 0.4 *M* potassium borate, pH 7.6.
	S8	= 0.8 *M* borate buffer, pH 8.55.
	S9	= 0.8 *M* borate buffer, pH 8.55, containing sodium acetate (0.3 *M*).
	S10	= 25 m*M* borate buffer, pH 7.5.
	S11	= programmed elution with S10 and (*D*) 350 m*M* borate buffer, pH 8.8: isocratic elution with S10 for 48 min; linear gradient, S10 → 100% *D* over 2 h; isocratic elution with *D* for 1 h; column equilibrated with S10 initially.

Detection D1 = automated periodate-pentane-2,4-dione assay; photometric (412 nm) or fluorimetric detection (excitation wavelength 410 nm, emission 503 nm).
D2 = automated 2-cyanoacetamide method, with fluorimetric detection; excitation wavelength 331 nm, emission 383 nm.
D3 = as D2, but with UV detection at 276 nm.
D4 = automated orcinol-sulfuric acid colorimetric method; abosrbance measured at 420 nm.
D5 = automated copper bincinchoninate colorimetric method; absorbance measured at 570 nm.
D6 = scintillation counting of radioactively labeled sugars.

REFERENCES

1. **Honda, S., Takahashi, M., Shimada, S., Kakehi, K., and Ganno, S.,** Automated analysis of alditols by anion-exchange chromatography with photometric and fluorometric postcolumn derivatization, *Anal. Biochem.*, 128, 429, 1983.
2. **Honda, S., Takahashi, M., Kakehi, K., and Ganno, S.,** Rapid analysis of monosaccharides by high-performance anion-exchange chromatography of borate complexes with fluorometric detection using 2-cyanoacetamide, *Anal. Biochem.*, 113, 130, 1981.
3. **Honda, S. Takahashi, M., Nishimura, Y., Kakehi, K., and Ganno, S.,** Sensitive ultraviolet monitoring of aldoses in automated borate complex anion-exchange chromatography with 2-cyanoacetamide, *Anal. Biochem.*, 118, 162, 1981.
4. **Vrátný, P., Mikeš, O., Farkaš, J., Štrop, P., Čopíková, J., and Nejepínská, K.,** Chromatography of mixtures of oligo- and monosaccharides on DEAE-Spheron, *J. Chromatogr.*, 180, 39, 1979.
5. **Josić, Dj., Hofermaas, R., Bauer, Ch., and Reutter, W.,** Automatic amino acid and sugar analysis of glycoproteins, *J. Chromatogr.*, 317, 35, 1984.
6. **Shukla, A. K., Scholz, N., Reimerdes, E. H., and Schauer, R.,** High-performance liquid chromatography of *N,O*-acylated sialic acids, *Anal. Biochem.*, 123, 78, 1982.
7. **Hostomská-Chytilová, Z., Mikeš, O., Vrátný, P., and Smrž, M.,** Chromatography of cellodextrins and enzymatic hydrolysates of cellulose on ion-exchange derivatives of Spheron, *J. Chromatogr.*, 235, 229, 1982.

TABLE LC 45
Resolution of Polyols, Mono-, and Oligosaccharides by Ion Chromatography, with Amperometric Detection

Packing	P1	P1	P2	P1	P3	P3	P1	P1	P3	P1[a]
Column[b]										
Length, cm	25	25	25	25	25	25	25	25	25	25
Diameter (I.D.), cm	0.4	0.4	0.4	0.4	0.46	0.46	0.4	0.4	0.46	0.4
Solvent	S1	S1	S2	S3	S4	S5	S6	S7	S8	S1
Flow rate (ml/min)	1	1	0.8	1	1	1	0.5	1	1	0.9
Temperature (°C)	36	Ambient								32
Detection	D1	D1	D1	D1	D1	D1	D1	D1	D1	D2
Reference	1	2	3, 4	4	4	4	5—7	6, 7	8	9
Compound	r[c]				k′	k′	r[c]			
Inositol	—	—	0.14	—	—	—	0.13	—	—	—
Xylitol	0.47	0.36	—	—	—	—	0.15	—	—	0.50
D-Glucitol	0.60	0.45	0.21	—	—	—	0.20	—	—	0.60
D-Mannitol	—	—	—	—	—	—	0.22	—	—	—
L-Fucose	—	—	0.38	0.33	—	—	0.34	0.33	0.45	—
2-Deoxy-D-*erythro*-pentose	—	—	0.41	—	—	—	0.40	0.39	—	—
2-Deoxy-D-*arabino*-hexose	—	—	0.62	—	—	—	—	—	0.68	—
L-Rhamnose	0.73	—	0.76	—	—	—	0.60	—	—	0.75
L-Arabinose	0.87	0.89	0.69	—	—	—	0.69	0.69	—	0.87
D-Xylose	—	1.00	1.17	—	—	—	1.30	1.22	—	—
D-Ribose	—	1.29	—	—	—	—	—	—	—	—
D-Galactose	—	1.00	0.83	0.85	—	—	0.90	0.87	0.93	—
D-Glucose	1.00	1.00	1.00	1.00	—	—	1.00	1.00	1.00	1.00
D-Mannose	—	1.00	1.24	1.10	—	—	—	1.32	1.08	—
D-Fructose	1.13	1.19	1.38	—	—	—	1.28	1.52	—	1.15
2-Amino-2-deoxy-D-galactose	—	—	—	0.68	4.75	—	—	—	0.76	—
2-Amino-2-deoxy-D-glucose	—	—	—	0.81	5.75	—	—	—	0.86	—
2-Acetamido-2-deoxy-D-galactose	—	—	—	—	9.00	—	—	—	—	—
N-Acetylneuraminic acid	—	—	—	—	—	1.00	—	—	—	—

D-Glucuronic acid	—	—	—	—	—	3.14	—	—	—	—
N-Glycolylneuraminic acid	—	—	—	—	—	3.85	—	—	—	—
Lactose	1.67	—	—	—	—	—	—	—	—	1.60
Sucrose	1.87	2.14	—	—	—	—	1.50	—	—	1.75
Melibiose	—	—	1.58	—	—	—	—	—	—	—
Isomaltose	—	1.78	1.79	—	—	—	—	—	—	—
Gentiobiose	—	2.75	2.13	—	—	—	—	—	—	—
Cellobiose	—	2.86	2.27	—	—	—	—	—	—	—
Turanose	—	—	2.41	—	—	—	—	—	—	—
Maltose	3.73	4.46	3.10	—	—	—	—	—	—	3.65
Raffinose	3.13	—	—	—	—	—	—	—	—	2.70
Stachyose	3.47	—	—	—	—	—	—	—	—	—

[a] Used in modular system, with components from various manufacturers.
[b] Composed of polymer (details not available) to obviate contamination from metal, or ions in glass.
[c] t_r relative to D-glucose (3.75 min,[1] 5.6 min,[2] 12.1 min,[3] 15 min,[4] 10.4 min with S6,[5-7] 11.5 min with S7,[6,7] 9.5 min,[8] 4.44 min[9]).

Packing
P1 = HPIC-AS6 (Dionex, Sunnyvale, CA) in Dionex System 2011i Ion Chromatograph; pellicular anion-exchange resin, particle diameter 10 μm, used with HPIC-AG6 guard column (5 × 0.4 cm).
P2 = HPIC-AS6A (Dionex); as P1 but particle diameter 5 μm; used with HPIC-AG6A guard column (5 × 0.4 cm).
P3 = CarboPac® PA-1 (Dionex) in Dionex BioLC System Ion Chromatograph; pellicular anion-exchange resin, particle diameter 5 μm, used with HPIC-AG6 guard column.

Solvent
S1 = 0.15 *M* sodium hydroxide.
S2 = (*A*) Distilled, deionized water, (*B*) 50 m*M* sodium hydroxide solution containing acetic acid (1.5 m*M*); isocratic elution with *A* + *B* (93:7) for 15 min, then linear gradient, 7 → 100% *B* in 10 min; isocratic elution with *B* for 15 min; 0.3 *M* NaOH introduced as postcolumn reagent to raise pH to optimum for detection.
S3 = 15 m*M* sodium hydroxide; 0.3 *M* NaOH added post column as for S2.
S4 = 10 m*M* sodium hydroxide; postcolumn addition of 0.3 *M* NaOH.
S5 = 0.1 *M* sodium hydroxide solution containing sodium acetate (0.15 m*M*).
S6 = 8 m*M* barium hydroxide solution containing acetic acid (0.5 m*M*).
S7 = 1 m*M* barium hydroxide solution containing acetic acid (0.125 m*M*).
S8 = 22 m*M* sodium hydroxide for 15 min; column washed with 0.2 *M* NaOH for 10 min, then equilibrated with 22 m*M* NaOH for 15 min.

TABLE LC 45 (continued)
Resolution of Polyols, Mono-, and Oligosaccharides by Ion Chromatography, with Amperometric Detection

Detection D1 = pulsed amperometric detector (Dionex); gold working electrode, Ag/AgCl reference electrode, glassy carbon counterelectrode.

D2 = amperometric detector based on oxidation at active nickel(III) oxide electrode.

REFERENCES

1. **Rocklin, R. D. and Pohl, C. A.,** Determination of carbohydrates by anion-exchange chromatography with pulsed amperometric detection, *J. Liq. Chromatogr.*, 6, 1577, 1983.
2. **Rocklin, R. D.,** Ion chromatography: a versatile technique for the analysis of beer, *LC, Liq. Chromatogr. HPLC Mag.*, 1, 504, 1983.
3. Dionex Technical Note No. 20, Dionex Corporation, Sunnyvale, CA, March, 1987.
4. **Olechno, J. D., Carter, S. R., Edwards, W. T., and Gillen, D. G.,** Developments in the chromatographic determination of carbohydrates, *Am. Biotechnol. Lab.*, 5(5), 38, 1987.
5. **Johnson, D. C. and Polta, T. Z.,** Amperometric detection in liquid chromatography with pulsed cleaning and reaction of noble metal electrodes, *Chromatogr. Forum*, 1, 37, 1986.
6. **Johnson, D. C.,** Carbohydrate detection gains potential, *Nature (London)*, 321, 451, 1986.
7. **Edwards, W. T., Pohl, C. A., and Rubin, R.,** Determination of carbohydrates using pulsed amperometric detection combined with anion exchange separations, *Tappi*, 70, 138, 1987.
8. **Hardy, M. R., Townsend, R. R., and Lee, Y. C.,** Monosaccharide analysis of glycoconjugates by anion exchange chromatography with pulsed amperometric detection, *Anal. Biochem.*, 170, 54, 1988.
9. **Reim, R. E. and Van Effen, R. M.,** Determination of carbohydrates by liquid chromatography with oxidation at a nickel(III) oxide electrode, *Anal. Chem.*, 58, 3203, 1986.

TABLE LC 46
Resolution of Malto-Oligosaccharides, DP 2—43, by Ion Chromatography with Pulsed Amperometric Detection

Packing	P1	P1	P1	P1
Column				
Length, cm	25	25	25	25
Diameter (I.D.), cm	0.4	0.4	0.4	0.4
Solvent	S1	S2	S3	S4
Flow rate (ml/min)	1	1	1	1
Temperature (°C)	34	Ambient		
Detection	PAD[a]	PAD	PAD	PAD
Reference	1	2	3	4
DP	r[b]			
2	1.00	1.00	1.00	1.00
3	1.12	1.30	1.12	1.95
4	1.25	1.60	1.35	2.62
5	1.50	2.20	1.60	3.21
6	1.81	3.00	2.00	3.80
7	2.25	4.10	2.45	4.38
8	2.68	5.31	3.00	4.91
9	3.38	7.31	3.55	5.45
10	4.18	—	3.81	5.84
11	—	—	4.16	6.23
12	—	—	4.46	6.62
13	—	—	4.71	7.00
14	—	—	4.96	7.21
15	—	—	5.21	7.40
16	—	—	5.46	7.59
17	—	—	5.71	7.78
18	—	—	5.91	7.98
19	—	—	6.11	8.17
20	—	—	6.28	8.37
21	—	—	6.46	8.57
22	—	—	6.61	8.76
23	—	—	—	8.95
24[c]	—	—	—	9.10
25	—	—	—	9.23
26	—	—	—	9.35
27	—	—	—	9.48
28	—	—	—	9.61
29	—	—	—	9.74
30	—	—	—	9.87
31	—	—	—	10.00
32	—	—	—	10.08
33	—	—	—	10.16
34	—	—	—	10.24
35[c]	—	—	—	10.32
36	—	—	—	10.40
37	—	—	—	10.47
38	—	—	—	10.54
39	—	—	—	10.61
40	—	—	—	10.68
41	—	—	—	10.74
42	—	—	—	10.80
43	—	—	—	10.85

TABLE LC 46 (continued)
Resolution of Malto-Oligosaccharides, DP 2—43, by Ion Chromatography with Pulsed Amperometric Detection

[a] Pulsed amperometric detector, gold working electrode.
[b] t_r relative to maltose (2.0 min,[1] 2.3 min,[2] 2.1 min,[3] 3.2 min[4]).
[c] Sensitivity of detector increased threefold for oligomers of DP 24—34; further tenfold increase required for detection of oligomers of DP 35—43.

Packing P1 = HPIC-AS6 (Dionex); pellicular anion-exchange resin, particle diameter 10 μm, with AG6 guard column (5 × 0.4 cm), in Dionex System 2011i Ion Chromatograph.
Solvent S1 = 0.2 *M* sodium hydroxide solution containing sodium acetate (0.2 *M*).
S2 = 0.15 *M* sodium hydroxide solution containing sodium acetate (0.15 *M*).
S3 = (*A*) 0.15 M sodium hydroxide, (*B*) 0.15 *M* sodium hydroxide containing sodium acetate (0.5 *M*); isocratic elution with *A* + *B* (1:1) for 1 min, then linear gradient, 50 → 100% *B* in 8 min; isocratic elution with *B* for 6 min.
S4 = (*C*) 0.1 *M* sodium hydroxide, (*D*) 0.1 *M* sodium hydroxide containing sodium acetate (0.6 *M*); linear gradient 0 → 100% *D* in 30 min, then isocratic elution with *D* for 5 min.

REFERENCES

1. **Rocklin, R. D. and Pohl, C. A.,** Determination of carbohydrates by anion exchange chromatography with pulsed amperometric detection, *J. Liq. Chromatogr.*, 6, 1577, 1983.
2. **Rocklin, R. D.,** Ion chromatography: a versatile technique for the analysis of beer, *LC, Liq. Chromatogr. HPLC Mag.*, 1, 504, 1983.
3. Dionex Technical Note No. 20, Dionex Corporation, Sunnyvale, CA, March, 1987.
4. **Olechno, J. D., Carter, S. R., Edwards, W. T., and Gillen, D. G.,** Developments in the chromatographic determination of carbohydrates, *Am. Biotechnol. Lab.*, 5(5), 38, 1987.

TABLE LC 47
Resolution of Cyclic Glucans by Ion Chromatography with Pulsed Amperometric Detection

Packing	P1	P1
Column		
Length, cm	25	25
Diameter (I.D.), cm	0.4	0.4
Solvent	S1	S2
Flow rate (ml/min)	1	1
Temperature (°C)	25	25
Detection	PAD[a]	PAD
Reference	1	1

TABLE LC 47 (continued)
Resolution of Cyclic Glucans by Ion Chromatography with Pulsed Amperometric Detection

Compound	k′	k′
Cyclodextrins		
Cyclomaltohexaose	0.76	—
α-D-glucosyl-	0.84	—
α-maltosyl-	1.18	—
α-maltotriosyl-	1.84	—
α-maltotetraosyl-	2.34	—
α-maltopentaosyl-	3.35	—
Cyclomaltoheptaose	5.31	—
α-D-glucosyl-	3.75	—
α-maltosyl-	4.95	—
α-maltotriosyl-	7.50	—
α-maltotetraosyl-	10.2	—
Cyclomaltooctaose	3.26	—
α-D-glucosyl-	3.75	—
α-maltosyl-	4.85	—
α-maltotriosyl-	6.20	—
α-maltotetraosyl-	8.45	—
Cyclosophoraoses		
DP 17	1.45	6.29
18	1.68	7.68
19	1.78	8,38
20	2.65	15.0
21	2.29	12.9
22	2.87	17.8
23	2.99	19.5
24	3.01	22.3
25	4.17	31.6
	(8.08 min)[b]	(52.2 min)[b]

Note: See data for HPLC of these oligosaccharides (Tables LC 4 and LC 13) and SEC (Table LC 52).

[a] Pulsed amperometric detection at gold electrode.
[b] t_r for highest oligomer given in brackets at foot of column.

Packing P1 = HPIC-AS6 (Dionex); pellicular anion-exchange resin, particle size 10 μm, with AG6 guard column (5 × 0.4 cm), in Dionex BioLC Model 4000i Ion Chromatograph.

Solvent S1 = 0.15 *M* sodium hydroxide solution containing sodium acetate (0.20 *M*).

S2 = 0.15 *M* sodium hydroxide solution containing sodium acetate (0.14 *M*).

REFERENCE

1. **Koizumi, K., Kubota, Y., Tanimoto, T., and Okada, Y.,** Determination of cyclic glucans by anion-exchange chromatography with pulsed amperometric detection, *J. Chromatogr.*, 454, 303, 1988.

TABLE LC 48
Resolution of Oligosaccharides, Neutral, Sialylated, and Phosphorylated, From Glycoproteins by Ion Chromatography with Pulsed Amperometric Detection

Packing	P1	P1	P1	P1	P1	P1	P2	P2
Column								
Length, cm	25	25	25	25	25	25	25	25
Diameter (I.D.), cm	0.46	0.46	0.46	0.46	0.46	0.46	0.4	0.4
Solvent	S1	S2	S3	S4	S5	S6	S7	S8
Flow rate (ml/min)	1	1	1	1	1	1	0.5	0.5
Temperature	Ambient							
Detection	PAD[a]	PAD	PAD	PAD	PAD	PAD	PAD	PAD
Reference	1, 2	1	2	2	2	2	3	3
Compound	k'						r[b]	r[c]
Neutral oligosaccharides								
Galβ1-4GlcNAc	2.52	—	—	—	—	—	—	—
Galβ1-4GlcNAcβ1-6Man	3.35	—	—	—	—	—	—	—
Fucα1-3GlcNAcβ1-2Man	0.30	—	—	—	—	—	—	—
Galβ1-4(Fucα1-3)GlcNAcβ1-2Man	1.04	—	—	—	—	—	—	—
Galβ1-4GlcNAcβ1-3Galβ1-4Glc	6.87	—	—	—	—	—	—	—
Galβ1-3GlcNAcβ1-3Galβ1-4Glc	10.5	—	—	—	—	—	—	—
Galβ1-3GalNAcβ1-4Galβ1-4Glc	3.52	—	—	—	—	—	—	—
Galβ1-4GlcNAcβ1-2(Galβ1-4GlcNAcβ1-4)Man-ol	2.39	—	—	—	—	—	—	—
Galβ1-4GlcNAcβ1-2(Galβ1-4GlcNAcβ1-4)Man	9.09	—	—	—	—	—	—	—
Galβ1-4GlcNAcβ1-2(Galβ1-4GlcNAcβ1-6)Man	9.52	—	—	—	—	—	—	—
Galβ1-4Glcβ1-2(Galβ1-4Glcβ1-6)Man	13.2	—	—	—	—	—	—	—
Galβ1-4GlcNAcβ1-3(Galβ1-4GlcNAcβ1-6)Manα1-2Man	11.5	—	—	—	—	—	—	—
Galβ1-4GlcNAcβ1-2Manα1-3 (Galβ1-4GlcNAcβ1 -2Manα1-6)Man	13.3	—	—	—	—	—	—	—
Fucα1-3GlcNAcβ1-2Manα1-3(Fucα1-3GlcNAcβ1-2Manα1-6)Man	1.43	—	—	—	—	—	—	—
Galβ1-4(Fucα1-3)GlcNAcβ1-2Manα1-3[Galβ1-4(Fucα1-3)GlcNAc-β1-2Manα1-6]Man	3.00	—	—	—	—	—	—	—

Triantennary oligosaccharide I[d]	14.4	—	—	—	—	—	—	—
Triantennary oligosaccharide II[d]	15.2	—	—	—	—	—	—	—
Triantennary glycopeptide III[d]	—	18.5	—	—	—	—	—	—
Triantennary glycopeptide IV[d]	—	21.2	—	—	—	—	—	—
Triantennary glycopeptide V[d]	—	18.5	—	—	—	—	—	—
Triantennary glycopeptide VI[d]	—	21.0	—	—	—	—	—	—
Tetraantennary oligosaccharide VII[d]	16.1	—	—	—	—	—	—	—
	(39.4 min)[e]	(51.0 min)[e]						
Sialylated oligosaccharides[f]								
Neu5Acα2-3Galβ1-4Glc	—	—	7.56	14.8	—	—	1.39	—
Neu5Acα2-6Galβ1-4Glc	—	—	6.78	16.9	—	—	1.26	—
Neu5Acα2-3Galβ1-4(Fucα1-3)Glc	—	—	—	—	—	—	0.90	—
Neu5Acα2-3Galβ1-3(Fucα1-4)GlcNAc	—	—	—	—	—	—	0.61	—
Neu5Acα2-3Galβ1-3GlcNAcβ1-3Galβ1-4Glc	—	—	12.8	—	—	—	0.85	—
Neu5Acα2-6Galβ1-4GlcNAcβ1-3Galβ1-4Glc	—	—	6.00	13.1	—	—	0.62	—
Galβ1-3(Neu5Acα2-6)GlcNAcβ1-3Galβ1-4Glc	—	—	14.2	—	—	—	1.15	—
Galβ1-3GalNAcβ1-4(Neu5Acα2-3)Galβ1-4Glc	—	—	6.35	17.9	—	—	—	1.00
Neu5Acα2-3Galβ1-3(Fucα1-4)GlcNAcβ1-3Galβ1-4Glc	—	—	—	—	—	—	0.61	—
Fucα1-2(Neu5Acα2-3)Galβ1-3GlcNAcβ1-3Galβ1-4Glc	—	—	—	—	—	—	0.69	—
Neu5Acα2-3Galβ1-3(Neu5Acα2-6)GlcNAcβ1-3Galβ1-4Glc	—	—	41.7	—	—	—	7.05	—
Neu5Acα2-3Galβ1-3GalNAcβ1-3(Neu5Acα2-3)Galβ1-4Glc	—	—	—	—	—	—	—	2.54
Galβ1-3GalNAcβ1-3(Neu5Acα2-8Neu5Acα2-3)Galβ1-4Glc	—	—	—	—	—	—	—	3.27
Neu5Acα2-6Galβ1-4GlcNAcβ1-3Gal(Galβ1-4GlcNAcβ1-6)Galβ1-4Glc	—	—	—	—	—	—	0.62	—
Neu5Acα2-3Galβ1-3GalNACβ1-3(Neu5Acα2-8Neu5Acα2-3)Galβ1-4Glc	—	—	—	—	—	—	—	13.8
Neu5Acα2-6Galβ1-4GlcNAcβ1-3[Galβ1-4(Fucα1-3)GlcNAcβ1-6]Galβ1-4Glc	—	—	—	—	—	—	0.47	—
Neu5Acα2-6Galβ1-3GlcNAcβ1-3[Galβ1-4(Fucα1-3)GlcNAcβ1-6]Galβ1-4Glc	—	—	—	—	—	—	0.56	—
Glycopeptide VIII[d]	—	—	—	—	—	—	2.17	—
			(98.2 min)[e]	(43.5 min)[e]				
Phosphorylated oligosaccharides[g]								
Manα1-2Manα-OR	—	—	—	—	7.39	0.14	—	—
ManPα1-2Manα-OR	—	—	—	—	19.6	11.4	—	—
Manα1-2ManPα-OR	—	—	—	—	19.6	11.4	—	—
ManPα1-2ManPα-OR	—	—	—	—	28.1	23.2	—	—

TABLE LC 48 (continued)
Resolution of Oligosaccharides, Neutral, Sialylated, and Phosphorylated, From Glycoproteins by Ion Chromatography with Pulsed Amperometric Detection

Compound	k'						r[b]	r[c]
ManPα1-3Manα-OR	—	—	—	—	19.6	11.1	—	—
ManPα1-6Manα-OR	—	—	—	—	20.0	11.1	—	—
ManPα1-2Manα1-2Manα-OR	—	—	—	—	18.3	10.2	—	—
ManPα1-2Manα1-3Manα-OR	—	—	—	—	18.6	10.2	—	—
ManPα1-2Manα1-6Manα-OR	—	—	—	—	18.4	9.30	—	—
Manα1-2ManPα1-2Manα-OR	—	—	—	—	18.0	9.30	—	—
Manα1-2ManPα1-3Manα-OR	—	—	—	—	18.0	9.30	—	—
Manα1-2ManPα1-6Manα-OR	—	—	—	—	18.1	9.30	—	—
Manα1-2Manα1-3(Manα1-2Manα1-6)Manα-OR	—	—	—	—	7.65	0.14	—	—
Manα1-2Manα1-3(ManPα1-2Manα1-6)Manα-OR	—	—	—	—	16.0	6.78	—	—
ManPα1-2Manα1-3(Manα1-2Manα1-6)Manα-OR	—	—	—	—	15.9	15.5	—	—
ManPα1-2Manα1-3(ManPα1-2Manα1-6)Manα-OR	—	—	—	—	22.6	19.1	—	—
					(67.0 min)[e]	(55.8 min)[e]		

[a] Pulsed amperometric detection at gold electrode.
[b] t_r relative to *N*-acetylneuraminic acid (10.24 min).
[c] t_r relative to Galβ1-3GalNAcβ1-4(Neu5Acα2-3)Galβ1-4Glc (4.91 min).
[d] Structures appended.
[e] Retention time of latest eluting oligosaccharide in series.
[f] Neu5Ac denotes *N*-acetylneuraminic acid.
[g] ManP denotes Man-6-PO_4, R = $(CH_2)_8COOCH_3$.

Packing P1 = CarboPac® PA-1 (Dionex); pellicular anion-exchange resin, particle diameter 5 μm, used with CarboPac PA guard column (2.5 × 0.3 cm) in Dionex BioLC System Ion Chromatograph.
P2 = HPIC-AS6 (Dionex); pellicular anion-exchange resin, particle diameter 10 μm.

Solvent S1 = (*A*) 0.10 *M* sodium hydroxide, (*B*) 0.10 *M* sodium hydroxide solution containing sodium acetate (0.15 *M*); isocratic elution with *A* for 10 min, then linear gradient 0 → 80% *B* over 60 min; postcolumn addition of 0.3 *M* NaOH to raise pH to optimum for detection.
S2 = isocratic elution with *A* for 2 min, then linear gradient 0 → 100% *B* over 65 min; postcolumn addition of 0.3 *M* NaOH.
S3 = isocratic elution with 0.1 *M* sodium hydroxide solution, containing 50 m*M* sodium acetate, for 25 min, then linear gradient in sodium acetate concentration, 50 → 90 m*M* over 60 min, 90 → 150 m*M* over 10 min; reequilibration to initial conditions over 7 min; postcolumn addition of 0.3 *M* NaOH.
S4 = pH gradient applied, 0.2 → 0.5 *M* sodium hydroxide over 60 min.

S5 = isocratic elution with 0.1 *M* sodium hydroxide for 5 min, then linear gradient to 0.1 *M* NaOH, 0.6 *M* sodium acetate over 67 min; isocratic for 5 min then reequilibration to initial conditions; postcolumn addition of 0.3 *M* NaOH.

S6 = 20 m*M* sodium acetate solution, pH adjusted to 6.0 with 43 m*M* acetic acid, to equilibrate column, then linear gradient in acetate concentration, 20 → 600 m*M* over 80 min, achieved by addition of *M* sodium acetate, pH 6.0; postcolumn addition of 0.3 *M* NaOH.

S7 = 50 m*M* sodium hydroxide containing 0.10 *M* sodium acetate.

S8 = 50 m*M* sodium hydroxide containing 0.15 *M* sodium acetate.

REFERENCES

1. **Hardy, M. R. and Townsend, R. R.,** Separation of positional isomers of oligosaccharides and glycopeptides by high-performance anion-exchange chromatography with pulsed amperometric detection, *Proc. Natl. Acad. Sci. U.S.A.*, 85, 3289, 1988.
2. **Townsend, R. R., Hardy, M. R., Hindsgaul, O., and Lee, Y. C.,** High-performance anion-exchange chromatography of oligosaccharides using pellicular resins and pulsed amperometric detection, *Anal. Biochem.*, 174, 459, 1988.
3. **Wang, W.-T., Lindh, F., Lundgren, T., and Zopf, D.,** HPLC of sialic acid-containing oligosaccharides and acidic monosaccharides, Abstr. A57, *Proc. 14th Int. Symp. Carbohydrate Chemistry, Stockholm*, 1988.

APPENDIX TO TABLE LC48:
STRUCTURES OF NUMBERED OLIGOSACCHARIDES AND GLYCOPEPTIDES

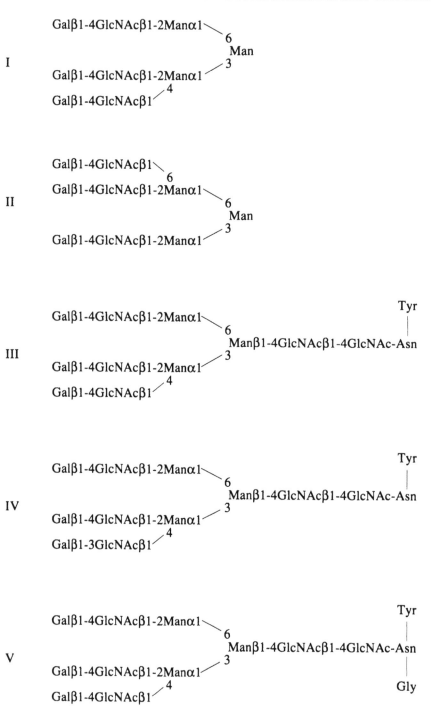

I

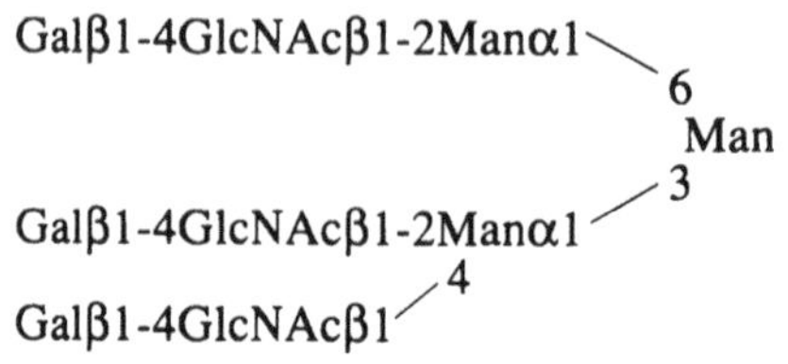

II

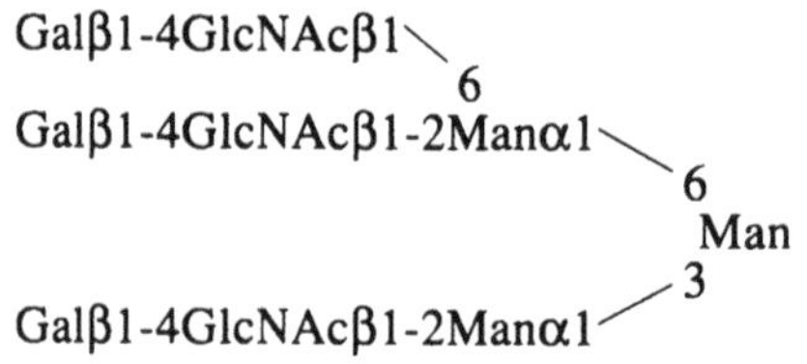

III

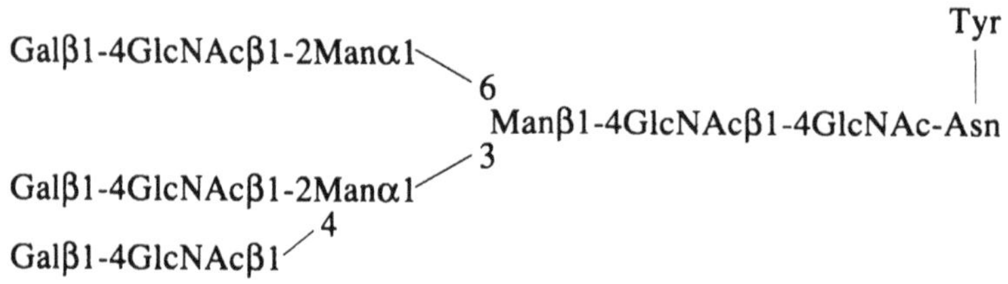

IV

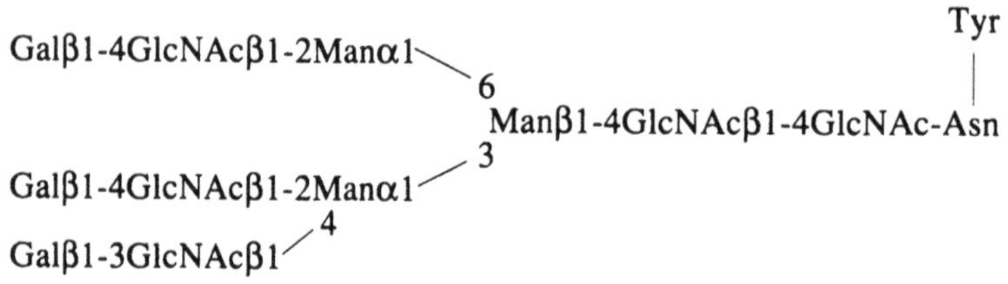

V

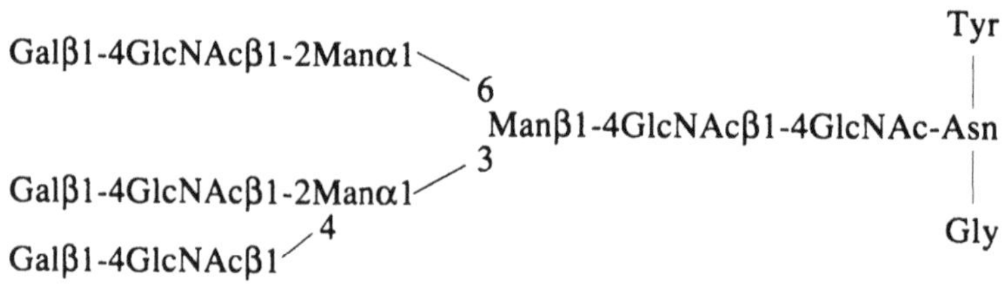

APPENDIX TO TABLE LC48: STRUCTURES OF NUMBERED OLIGOSACCHARIDES AND GLYCOPEPTIDES (continued)

VI

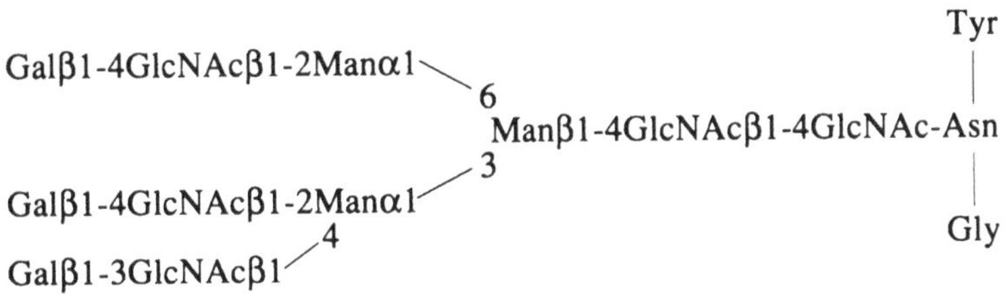

VII

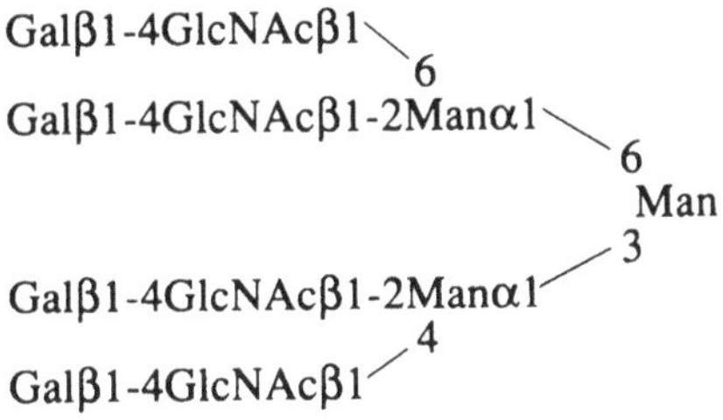

VIII

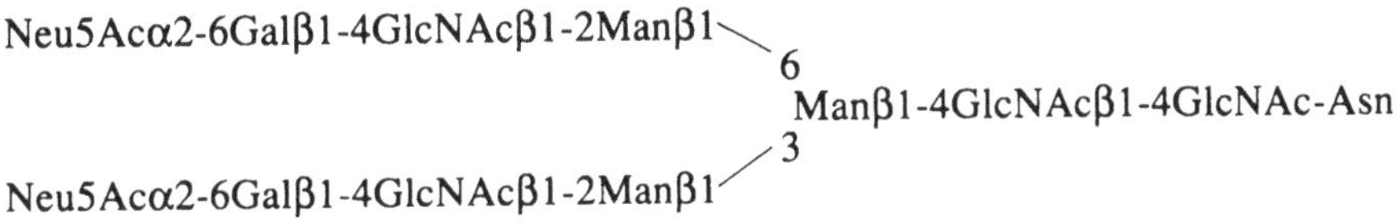

TABLE LC 49
Effect of Methyl and Carboxylate Groups on SEC of Mono- and Oligosaccharides on Low-Porosity Gels

Packing	P1	P2	P3	P3
Column				
Length, cm	200	142	95	95
Diameter (I.D.), cm	2	1.2	1.5	1.5
Material	Glass	Glass	Glass	Glass
Solvent	H_2O	H_2O	S1	S2
Flow rate (ml/h)	7	3	8	8
Temperature	Ambient			
Detection	D1	D1	D1, D2	D1, D2
Reference	1	2	3	3
Compound	r[a]	k′		
D-Glucose	1.00	—	1.24	—
6-*O*-methyl	0.93	—	—	—
4,6-di-*O*-methyl	0.87	—	—	—
2,3,6-tri-*O*-methyl	0.81	—	—	—
2,3,4,6-tetra-*O*-methyl	0.78	—	—	—
D-Mannose	—	1.27	—	—
Methyl-α-D-mannopyranoside	—	0.95	—	—
L-Fucose	—	1.14	—	—
L-Rhamnose	—	1.14	—	1.14
2-*O*-methyl	—	—	—	0.94
D-Glucuronic acid	—	—	0.65	1.24[b]
4-*O*-methyl	—	—	—	0.94
Cellobiose	—	—	1.02	—
Mannobiose	—	1.05	—	—
Fucosylmannose	—	0.95	—	—
$(GlcA)_2$	—	—	—	0.96
$(Rha)_2$	—	—	—	0.73
Rha-(4-*O*-methyl-GlcA)	—	—	—	0.73
Raffinose	—	—	0.81	—
Mannotriose	—	0.82	—	—
Glc-$(GlcA)_2$	—	—	—	0.79
$(Rha)_2$-(4-*O*-methyl-GlcA)	—	—	—	0.51
Mannotetraose	—	0.68	—	—
$(Glc)_2$-$(GlcA)_2$	—	—	—	0.63
$(Rha)_3$-(4-*O*-methyl-GlcA)	—	—	—	0.42
$(Rha)_2$-(2-*O*-methyl-Rha)-(4-*O*-methyl-GlcA)	—	—	—	0.37
$(Glc)_3$-$(GlcA)_2$	—	—	—	0.47
$(Glc)_4$-$(GlcA)_2$	—	—	—	0.22

Note: This table supplements Table LC 29 in *CRC Handbook of Chromatography: Carbohydrates*, Volume I.

[a] t_r relative to D-glucose (ca. 76 h).
[b] At pH of eluent S2, which approximates to pK_a for uronic acid, glucuronic acid elutes at same position as glucose.

Packing	P1	= Bio-Gel® P-4 (Bio-Rad; polyacrylamide gel), −400 mesh.
	P2	= Bio-Gel® P-2 (200—400 mesh).
	P3	= Bio-Gel® P-2 (−400 mesh).
Solvent	S1	= 0.2 *M* trimethylammonium formate buffer, pH 7.0.
	S2	= 0.2 *M* trimethylammonium formate buffer, pH 3.3.
Detection	D1	= phenol-sulfuric acid colorimetric assay of fractions.
	D2	= assay of uronic acid in fractions by carbazole method.

TABLE LC 49 (continued)
Effect of Methyl and Carboxylate Groups on SEC of Mono- and Oligosaccharides on Low-Porosity Gels

REFERENCES

1. **Grellert, E. and Ballou, C. E.,** Separation of methyl ethers of sugars by gel filtration, *Carbohydr. Res.*, 30, 218, 1973.
2. **Yamada, H., Ohshima, Y., Tamura, K., and Miyazaki, T.,** Separation of a 6-deoxyhexose and a hexose by gel filtration, *Carbohydr. Res.*, 83, 377, 1980.
3. **Djordjevic, S. P., Batley, M., and Redmond, J. W.,** Preparative gel chromatography of acidic oligosaccharides using a volatile buffer, *J. Chromatogr.*, 354, 507, 1986.

TABLE LC 50
Effect of Glycosidic Linkage Distribution on SEC of D-Gluco-Oligosaccharides on Low-Porosity Gels

Packing	P1	P1	P1	P1	P2
Column					
Length, cm	170	170	170	170	35
Diameter (I.D.), cm	0.9	0.9	0.9	0.9	1
Material	Glass	Glass	Glass	Glass	Glass
Solvent	H_2O	S1	H_2O	S1	H_2O
Flow rate (ml/h)	28	28	28	28	2
Temperature (°C)	60	60	60	60	25
Detection	RI	RI	RI	RI	RI
Reference	1	1	2	2	3
Compound	r^a	r^a	K_{mf}	K_{mf}	K_d
Kojibiose (α1-2)	—	—	—	—	0.511
Sophorose (β1-2)	—	—	—	—	0.506
Nigerose(α1-3)	—	—	—	—	0.519
Laminaribiose (β1-3)	0.90	0.86	0.521	0.429	0.535
Maltose (α1-4)	—	—	—	—	0.536
Cellobiose (β1-4)	0.91	0.94	0.525	0.503	0.523
Isomaltose (α1-6)	—	—	—	—	0.493
Gentiobiose (β1-6)	—	—	—	—	0.466
Laminaritriose	0.80	0.77	0.430	0.351	—
Maltotriose	—	—	—	—	0.441
Cellotriose	0.81	0.84	0.441	0.414	0.409
Glcβ1-3Glcβ1-4Glc	—	—	0.437	0.424	—
Glcβ1-4Glcβ1-3Glc	0.75	—	0.435	0.392	—
Isomaltotriose	—	—	—	—	0.375
Laminaritetraose	0.73	0.70	0.357	0.289	—
Maltotetraose	—	—	—	—	0.361
Cellotetraose	0.74	0.78	0.370	0.361	—
Glcβ1-3Glcβ1-4Glcβ1-4Glc	0.74	—	0.362	0.341	—
Glcβ1-4Glcβ1-4Glcβ1-3Glc	0.70	—	0.360	0.336	—
Isomaltotetraose	—	—	—	—	0.286
Laminaripentaose	—	—	0.295	0.253	—
Maltopentaose	—	—	—	—	0.299
Cellopentaose	—	—	0.306	0.295	—
Glcβ1-3(Glcβ1-4Glc)$_2$	—	—	0.303	—	—
(Glcβ1-4Glc)$_2$β1-3Glc	—	—	0.315	0.285	—
Isomaltopentaose	—	—	—	—	0.224
Cellohexaose	—	—	0.254	0.233	—
Glcβ1-3(Glcβ1-4Glc)$_2$ β1-4Glc	—	—	0.255	—	—
(Glcβ1-4Glc)$_2$β1-4Glcβ1-3Glc	—	—	0.262	0.244	—

Note: This table supplements Table LC 29 in *CRC Handbook of Chromatography: Carbohydrates*, Volume I.

TABLE LC 50 (continued)
Effect of Glycosidic Linkage Distribution on SEC of D-Gluco-Oligosaccharides on Low-Porosity Gels

[a] t_r relative to D-glucose (160 min with H_2O, 156 min with S1).

Packing	P1	=	Bio-Gel® P-2 (Bio-Rad; polyacrylamide gel).
	P2	=	Sephadex® G-15 (Pharmacia; cross-linked dextran gel).
Solvent	S1	=	0.1 *M* boric acid.

REFERENCES

1. **Luchsinger, W. W. and Stone, B. A.,** Linkage sequencing of oligosaccharides by their rates of alkaline degradation, *Carbohydr. Res.*, 46, 1, 1976.
2. **Luchsinger, W. W., Luchsinger, S. W., and Luchsinger, D. W.,** Gel chromatography of (1 → 3), (1 → 4), and mixed-linkage (1 → 3), (1 → 4)-β-D-gluco-oligosaccharides, *Carbohydr. Res.*, 104, 153, 1982.
3. **Haglund, Å. C., Marsden, N. V. B., and Östling, S. G.,** Partitioning of oligoglucans in Sephadex G-15 in relation to their conformational structure, *J. Chromatogr.*, 318, 57, 1985.

TABLE LC 51
SEC of Oligosaccharides (DP2—60) of Various Series on Polyacrylamide Gels

Packing	P1	P1	P1	P2	P1	P1	P2	P3	P1	P1	P3
Column											
Length, cm	210[a]	201[b]	201[b]	198[c]	200[d]	200[c]	200	200	201[b]	200[d]	203[c]
Diameter (I.D.), cm	5	2.54	2.54	1.8	0.6	2	2	2	2.54	0.6	2
Material	Glass[e]	Glass[e]	Glass[e]	Glass[e]	Glass	Glass	Glass	Glass	Glass[e]	Glass	Glass
Solvent	H_2O[f]	H_2O[g]	H_2O[g]	H_2O[g]	H_2O[f]	H_2O	H_2O	H_2O	H_2O[g]	H_2O[f]	S1
Flow rate (ml/h)	220	55	55	25	18	60	60	60	55	18	50
Temperature (°C)	65	60	60	60	55	55	55	55	60	55	65
Detection	RI	D1	D1	D1	D2	RI	RI	RI	D1	D2	D3
Reference	1	2	2	2	3	4	4	4	2	3	5
Series	Cello-	Malto-	Pullulan[h]		Isomalto-				Xylo-	Chito-	$(GalA)_n$[i]
DP	K_G[j]	r[k]	r[l]	K_{av}	K_{av}	r[k]	r[k]	r[k]	r[m]	K_{av}	K_{av}
2	0.91	0.95	—	—	0.90	0.93	1.00	0.75	0.96	0.76	0.52
3	0.83	0.91	1.00	0.87	0.82	0.89	1.00	0.69	0.92	0.64	0.40
4	0.77	0.87	—	—	0.76	0.85	1.00	0.64	0.88	0.54	0.31
5	0.72	0.83	—	—	0.70	0.81	0.98	0.59	0.84	0.45	0.24
6	0.66	0.80	0.87	0.78	0.64	0.77	0.95	0.54	0.80	—	0.19
7	0.60	0.77	—	—	0.59	0.74	0.92	0.50	0.77	—	0.14
8	0.56	0.74	—	—	0.54	0.71	0.89	0.46	0.74	—	0.11
9	0.52	0.71	0.76	0.69	0.49	0.68	0.86	0.43	0.71	—	0.08
10	0.48	0.68	—	—	0.45	0.65	0.84	0.41	0.68	—	—
11	0.44	0.65	—	—	0.42	0.63	0.82	0.40	0.65	—	—
12	0.40	0.63	0.67	0.62	0.39	0.61	0.80	0.39	0.63	—	—
13	0.36	0.61	—	—	0.36	0.59	0.78	0.38	0.61	—	—
14	—	0.59	—	—	0.33	0.58	0.75	—	0.59	—	—
15	—	0.57	0.62	0.55	0.30	0.56	0.73	—	0.57	—	—
16	—	0.55	—	—	0.27	0.54	0.71	—	—	—	—
17	—	0.53	—	—	0.24	0.53	0.69	—	—	—	—
18	—	0.52	0.57	0.48	—	0.52	0.67	—	—	—	—
19	—	0.51	—	—	—	0.51	0.66	—	—	—	—
20	—	0.50	—	—	—	0.50	0.65	—	—	—	—

TABLE LC 51 (continued)
SEC of Oligosaccharides (DP2—60) of Various Series on Polyacrylamide Gels

Series	Cello-	Malto-	Pullulan[h]		Isomalto-				Xylo-	Chito-	(GalA)$_n$[i]
DP	KG[j]	r[k]	r[l]	Kav	Kav	r[k]	r[k]	r[k]	r[m]	K$_{av}$	K$_{av}$
21	—	0.49	0.52	0.42	—	0.49	0.63	—	—	—	—
22	—	0.48	—	—	—	0.48	0.62	—	—	—	—
23	—	0.47	—	—	—	0.47	—	—	—	—	—
24	—	0.46	0.48	0.36	—	0.46	—	—	—	—	—
27	—	—	0.45	0.33	—	—	—	—	—	—	—
30	—	—	0.43	0.30	—	—	—	—	—	—	—
33	—	—	0.42	0.27	—	—	—	—	—	—	—
36	—	—	0.40	0.24	—	—	—	—	—	—	—
39	—	—	0.39	0.21	—	—	—	—	—	—	—
42	—	—	—	0.19	—	—	—	—	—	—	—
45	—	—	—	0.17	—	—	—	—	—	—	—
48	—	—	—	0.15	—	—	—	—	—	—	—
51	—	—	—	0.13	—	—	—	—	—	—	—
54	—	—	—	0.12	—	—	—	—	—	—	—
57	—	—	—	0.11	—	—	—	—	—	—	—
60	—	—	—	0.10	—	—	—	—	—	—	—

Note: Earlier data on SEC of oligosaccharides of malto-, isomalto-, cello-, pullulan, and xylo- series will be found in Tables LC 31—38 in *CRC Handbook of Chromatography: Carbohydrates,* Volume I.

[a] Precolumn, 30 × 5 cm, and two columns, each 100 × 5 cm, connected in series, with elution in downward-upward-downward direction.
[b] Two columns, each 109 × 2.54 cm, connected in series with zero dead volume union, with elution in upward-downward direction.
[c] Single column, eluted in upward direction.
[d] Two coupled columns, each 100 × 0.6 cm.
[e] Coated with dichlorodimethylsilane.
[f] Doubly distilled and continuously pre-filtered (0.22-μm filter unit) and supplied to degassing reservoir kept at 80—90°C.
[g] Deionized and degassed (80°C).
[h] Maltotriose units joined through α1-6 linkages.
[i] Oligogalacturonic acids.
[j] $KG = V_e - V_0/V_{Glc} - V_0$.

[k] t_r relative to D-glucose (19.8 h;[2] in ref. 4, 9.6 h on P1, 9.9 h on P2, 8.25 h on P3).
[l] t_r relative to maltotriose (18 h).
[m] t_r relative to D-xylose (19.8 h).

Packing	P1 =	Bio-Gel® P-4 (Bio-Rad), −400 mesh; narrow range of particle sizes obtained by wind-sieving dry gel[1] and removal of fines from hydrated particles by several decantations,[1,3] or elutriation in water (50°C), yielding mean particle diameter of 40 ± 4 μm; gel slurry degassed at 55°C under vacuum[2,4] before packing of column.
	P2 =	Bio-Gel® P-6, −400 mesh; elutriation as for P1 gave mean particle diameter of 47 ± 4 μm; slurry degassed as for P1 before
	P3 =	Bio-Gel® packing of column.[34] P-2, −400 mesh; slurry degassed before packing[4] and column equilibrated with eluent by passage of 2—3 column volumes.[5]
Solvent	S1 =	0.1 *M* acetate buffer, pH 3.6, prepared by titration of sodium acetate solution with acetic acid.
Detection	D1 =	automated orcinol-sulfuric acid colorimetric assay.
	D2 =	scintillation counting of samples labeled at reducing end by treatment with $NaB[^3H]_4$.
	D3 =	automated *m*-hydroxydiphenyl assay for uronic acid.

REFERENCES

1. **Hamacher, K., Schmid, G., Sahm, H., and Wandrey, Ch.,** Structural heterogeneity of cellooligomers homogeneous according to high-resolution size-exclusion chromatography, *J. Chromatogr.*, 319, 311, 1985.
2. **John, M., Schmidt, J., Wandrey, C., and Sahm, H.,** Gel chromatography of oligosaccharides up to DP 60, *J. Chromatogr.*, 247, 281, 1982.
3. **Natowicz, M. and Baenziger, J. U.,** A rapid method for chromatographic analysis of oligosaccharides on Bio-Gel P-4, *Anal. Biochem.*, 105, 159, 1980.
4. **Yamashita, K., Mizuochi, T., and Kobata, A.,** Analysis of oligosaccharides by gel filtration, *Methods Enzymol.*, 83, 105, 1982.
5. **Thibault, J.-F.,** Separation of α-D-galacturonic acid oligomers by chromatography on polyacrylamide gel, *J. Chromatogr.*, 194, 315, 1980.

TABLE LC 52
Chromatography of Cyclodextrins and Branched Cyclodextrins on Vinyl Alcohol Polymer Gel

Packing	P1	P1	P1	P1
Column				
Length, cm	50	50	50	50
Diameter (I.D.), cm	0.76	0.76	0.76	0.76
Material	SS	SS	SS	SS
Solvent	H_2O	S1	S2	S3
Flow rate (ml/h)	60	60	60	60
Temperature	Ambient			
Detection	RI	RI	RI	RI
Reference	1	1	1	1
Compound	r[a]			
Cyclomaltohexaose (α-CD)	1.40	1.19	1.07	1.07
6-*O*-α-D-glucosyl-	1.26	1.09	1.02	0.99
6-*O*-α-maltosyl-	1.20	1.04	0.98	0.96
6-*O*-α-maltotriosyl-	1.14	1.01	0.96	0.94
Cyclomaltoheptaose (β-CD)	2.25	2.04	1.85	1.75
6-*O*-α-D-glucosyl-	1.78	1.67	1.56	1.49
6-*O*-α-maltosyl-	1.66	1.54	1.45	1.39
6,6‴-di-*O*-α-D-glucosyl-	1.53	1.43	1.35	1.30
6-*O*-γ-maltotriosyl-	1.57	1.48	1.41	1.33
Cyclomaltooctaose (γ-CD)	0.95	0.95	0.96	0.97
6-*O*-α-D-glucosyl-	0.89	0.89	0.90	0.91
6-*O*-α-maltosyl-	0.86	0.87	0.88	0.89
6-*O*-α-maltotriosyl-	0.85	0.86	0.87	0.88

Note: See other chromatographic data for these cyclodextrins in Tables LC 4, LC 13, LC 47, and TLC 6.

[a] t_r relative to water (17.22 min).

Packing P1 = Asahipak® GS-320 (Asahi Kasei, Tokyo).
Solvent S1 = water-methanol (19:1).
S2 = water-methanol (9:1).
S3 = water-methanol (17:3).

REFERENCE

1. **Koizumi, K., Utamura, T., Kuroyanagi, T., Hizukuri, S., and Abe, J.-I.,** Analyses of branched cyclodextrins by high-performance liquid and thin-layer chromatography, *J. Chromatogr.*, 360, 397, 1986.

TABLE LC 53
SEC of Reduced Oligosaccharides Derived from Glycoproteins, and of Asparagine-Linked Glycopeptides, on Polyacrylamide and Dextran Gels

Packing	P1	P1	P2
Column			
Length, cm	200[a]	200[b]	135
Diameter (I.D.), cm	0.6	2	1
Material	Glass	Glass	Glass
Solvent	H_2O	H_2O	S1
Flow rate (ml/h)	18	60	3
Temperature (°C)	55	55	Ambient
Detection	D1	D1	D2, D3
Reference	1	2	3
Compound	**K_{av}[c]**	**r[d]**	**K_{av}**
Milk oligosaccharides			
Galβ1-4Glc-ol	—	0.92	—
Fucα1-2Galβ1-4Glc-ol	—	0.87	—
Galβ1-4(Fucα1-3)Glc-ol	—	0.88	—
GlcNAcβ1-3Galβ1-4Glc-ol	—	0.82	—
Galβ1-3GlcNAcβ1-3Galβ1-4Glc-ol	—	0.78	—
Galβ1-4GlcNAcβ1-3Galβ1-4Glc-ol	—	0.78	—
Galβ1-3(Fucα1-4)GlcNAcβ1-3Galβ1-4Glc-ol	—	0.76	—
Galβ1-4(Fucα1-3)GlcNAcβ1-3Galβ1-4Glc-ol	—	0.76	—
Fucα1-2Galβ1-3GlcNAcβ1-3Galβ1-4Glc-ol	—	0.75	—
Galβ1-4GlcNAcβ1-3(Galβ1-4GlcNAcβ1-6)Galβ1-4Glc-ol	—	0.70	—
Galβ1-3(4)GlcNAcβ1-3[Galβ1-4(Fucα1-3)GlcNAcβ1-6]Galβ1-4Glc-ol	—	0.68	—
Fucα1-2Galβ1-3GlcNAcβ1-3(Galβ1-4GlcNAcβ1-6)Galβ1-4Glc-ol	—	0.67	—
Galβ1-3(4)GlcNAcβ1-3Galβ1-4(Fucα1-3)GlcNAcβ1-3Galβ1-4Glc-ol	—	0.66	—
Galβ1-3(4)[Fucα1-4(3)]GlcNAcβ1-3[Galβ1-4(Fucα1-3)GlcNAcβ1-6]Galβ1-4Glc-ol	—	0.66	—
Fucα1-2Galβ1-3GlcNAcβ1-3[Galβ1-4(Fucα1-3)GlcNAcβ1-6]Galβ1-4Glc-ol	—	0.65	—
Galβ1-3(4)[Fucα1-4(3)]GlcNAcβ1-3Galβ1-4(Fucα1-3)GlcNAcβ1-3Galβ1-4Glc-ol	—	0.64	—
Galβ1-4(3)GlcNAcβ1-3[Galβ1-3(4)GlcNAcβ1-3Galβ1-4GlcNAcβ1-6]Galβ1-4Glc-ol	—	0.62	—
Galβ1-4(3)GlcNAcβ1-3[Galβ1-3(4)GlcNAcβ1-3Galβ1-4(Fucα1-3)GlcNAcβ1-6]Galβ1-4Glc-ol	—	0.60	—
Galβ1-4(3)[Fucα1-3(4)]GlcNAcβ1-3[Galβ1-3(4)GlcNAcβ1-3Galβ1-4(Fucα1-3)GlcNAcβ1-6]Galβ1-4Glc-ol	—	0.59	—

TABLE LC 53 (continued)
SEC of Reduced Oligosaccharides Derived from Glycoproteins, and of Asparagine-Linked Glycopeptides, on Polyacrylamide and Dextran Gels

Compound	K_{av}^c	r^d	K_{av}
Galβ1-4(3)GlcNAcβ1-3[Galβ1-3(4){Fucα1-4(3)}GlcNAcβ1-3Galβ1-4(Fucα1-3)GlcNAcβ1-6]Galβ1-4Glc-ol	—	0.59	—
Galβ1-4(3)[Fucα1-3(4)]GlcNAcβ1-3[Galβ1-3(4){Fucα1-4(3)}GlcNAcβ1-3Galβ1-4(Fucα1-3)GlcNAcβ1-6]Galβ1-4Glc-ol	—	0.58	—
Oligosaccharides related to asparagine-linked sugar chains			
Manβ1-4(GlcNAc-ol	0.84	0.88	—
Manα1-6Manβ1-4GlcNAc-ol	—	0.85	—
Manα1-3Manβ1-4GlcNAc-ol	—	0.84	—
Manα1-3(Manα1-6)Manβ1-4GlcNAc-ol	0.72	0.81	—
Manα1-2Manα1-3Manβ1-4GlcNAc-ol	—	0.80	—
Manα1-3(Manα1-3Manα1-6)Manβ1-4GlcNAc-ol	—	0.79	—
GlcNAcβ1-2Manα1-6Manβ1-4GlcNAc-ol	—	0.78	—
Manα1-2Manα1-2Manα1-3Manβ1-4GlcNAc-ol	—	0.77	—
GlcNAcβ1-2(4)Manα1-3Manβ1-4GlcNAc-ol	—	0.77	—
Manα1-3[Manα1-3(Manα1-6)Manα1-6]Manβ1-4GlcNAc-ol	0.60	0.75	—
Manα1-3(GlcNAcβ1-2Manα1-6)Manβ1-4GlcNAc-ol	0.60	0.75	—
Galβ1-4GlcNAcβ1-2Manα1-6Manβ1-4GlcNAc-ol	—	0.74	—
Galβ1-4GlcNAcβ1-2Manα1-3Manβ1-4GlcNAc-ol	—	0.73	—
Galβ1-4(Fucα1-3)GlcNAcβ1-2Manα1-6Manβ1-4GlcNAc-ol	—	0.73	—
Galβ1-4(Fucα1-3)GlcNAcβ1-2(4)Manα1-3Manβ1-4GlcNAc-ol	—	0.72	—
GlcNAcβ1-2Manα1-3(GlcNAcβ1-4)(Manα1-6)Manβ1-4GlcNAc-ol	—	0.71	—
Manα1-3(Galβ1-4GlcNAcβ1-2Manα1-6)Manβ1-4GlcNAc-ol	0.56	0.71	—
GlcNAcβ1-2Manα1-3(GlcNAcβ1-2Manα1-6)Manβ1-4GlcNAc-ol	—	0.68	—
GlcNAcβ1-2Manα1-3[Manα1-3(Manα1-6)Manα1-6]Manβ1-4GlcNAc-ol	—	0.67	—
GlcNAcβ1-2Manα1-3(GlcNAcβ1-4)(GlcNAcβ1-2Manα1-6)Manβ1-4GlcNAc-ol	—	0.67	—
Triantennary oligosaccharide I[e]	—	0.65	—
GlcNAcβ1-2(GlcNAcβ1-4)Manα1-3(GlcNAcβ1-4)(Manα1-3Manα1-6)Manβ1-4GlcNAc-ol	—	0.65	—
GlcNAcβ1-2Manα1-3[GlcNAcβ1-2(GlcNAcβ1-6)Manα1-6] Manβ1-4GlcNAc-ol	—	0.64	—
GlcNAcβ1-2(GlcNAcβ1-4)Manα1-3(GlcNAcβ1-2Manα1-6)Manβ1-4GlcNAc-ol	—	0.63	—
Triantennary oligosaccharide II[e]	—	0.62	—
Galβ1-4GlcNAcβ1-2Manα1-3(Galβ1-4GlcNAcβ1-2Manα1-6)Manβ1-4GlcNAc-ol	—	0.62	—
GlcNAcβ1-2(GlcNAcβ1-4)Manα1-3(GlcNAcβ1-4)(GlcNAcβ1-2Manα1-6)Manβ1-4GlcNAc-ol	—	0.61	—
Tetraantennary oligosaccharide III[e]	—	0.60	—
Galβ1-4(Fucα1-3)GlcNAcβ1-2Manα1-3[Galβ1-4(Fucα1-3)GlcNAcβ1-2Manα1-6]Manβ1-4GlcNAc-ol	—	0.60	—

Triantennary oligosaccharide IV[e]	—	0.58	—
Triantennary oligosaccharide V[e]	—	0.58	—
Tetraantennary oligosaccharide VI[e]	—	0.54	—
Manβ1-4GlcNAcβ1-4(Fucα1-6)GlcNAc-ol series:			
Fucα1-6GlcNAc-ol	—	0.87	—
GlcNAcβ1-4GlcNAc-ol	—	0.83	—
GlcNAcβ1-4(Fucα1-6)GlcNAc-ol	—	0.79	—
Manβ1-4GlcNAcβ1-4GlcNAc-ol	0.69	0.79	—
Manβ1-4GlcNAcβ1-4(Fucα1-6)GlcNAc-ol	—	0.76	—
Manα1-3(Manα1-6)Manβ1-4GlcNAcβ1-4GlcNAc-ol	0.59	0.73	—
Manα1-6Manβ1-4GlcNAcβ1-4(Fucα1-6)GlcNAc-ol	—	0.73	—
Manα1-3(Manα1-6)Manβ1-4GlcNAcβ1-4(Fucα1-6)GlcNAc-ol	—	0.70	—
Manα1-3[Manα1-3(Manα1-6)Manα1-6]Manβ1-4GlcNAcβ1-4GlcNAc-ol	—	0.68	—
GlcNAcβ1-2Manα1-3(Manα1-6)Manβ1-4GlcNAcβ1-4GlcNAc-ol	—	0.68	—
Galβ1-4GlcNAcβ1-2Manα1-6Manβ1-4GlcNAcβ1-4GlcNAc-ol	—	0.67	—
GlcNAcβ1-2Manα1-3[Manα1-3(6)Manα1-6]Manβ1-4GlcNAcβ1-4GlcNAc-ol	—	0.65	—
Galβ1-4GlcNAcβ1-2Manα1-3(Manα1-6)Manβ1-4GlcNAcβ1-4GlcNAc-ol	—	0.65	—
Manα1-3(6)[Galβ1-4GlcNAcβl-2Manα1-6(3)]Manβ1-4GlcNAcβ1-4GlcNAc-ol	—	0.64	—
GlcNAcβ1-2Manα1-3[Manα1-3(Manα1-6)Manα1-6]β1-4GlcNAcβ1-4GlcNAc-ol	—	0.63	—
GlcNAcβ1-2Manα1-3(GlcNAcβ1-2Manα1-6)Manβ1-4GlcNAcβ1-4GlcNAc-ol	0.42	0.62	—
GlcManβ1-2Manα1-3(GlcNAcβ1-2Manα1-6)Manβ1-4GlcNAcβ1-4(Fucα1-6)GlcNAc-ol	—	0.61	—
GlcNAcβ1-2Manα1-3[GlcNAcβ1-2(GlcNAcβ1-4)Manα1-6]Manβ1-4GlcNAcβ1-4GlcNAc-ol	0.37	—	—
Triantennary oligosaccharide Ia[e]	—	0.60	—
Galβ1-3GlcNAcβ1-2Manα1-3(Galβ1-3GlcNAcβ1-2Manα1-6)Manβ1-4GlcNAcβ1-4GlcNAc-ol	—	0.59	—
Galβ1-3GlcNAcβ1-2Manα1-3(Galβ1-4GlcNAcβ1-2Manα1-6)Manβ1-4GlcNAcβ1-4GlcNAc-ol	—	0.59	—
Galβ1-4GlcNAcβ1-2Manα1-3(Galβ1-4GlcNAcβ1-2Manα1-6)Manβ1-4GlcNAcβ1-4GlcNAc-ol	0.36	0.58	—
Galβ1-4GlcNAcβ1-2Manα1-3(Galβ1-4GlcNAcβ1-2Manα1-6)Manβ1-4GlcNAcβ1-4(Fucα1-6)GlcNAc-ol	—	0.57	—
Galβ1-4GlcNAcβ1-2Manα1-3(GlcNAcβ1-4)(Galβ1-4GlcNAcβ1-2Manα1-6)Manβ1-4GlcNAcβ1-4(Fucα1-6)GlcNAc-ol	—	0.56	—
Galβ1-4(Fucα1-3)GlcNAcβ1-2Manα1-3(Galβ1-4GlcNAcβ1-2Manα1-6)Manβ1-4GlcNAcβ1-4(Fucα1-6)GlcNAc-ol	—	0.56	—
Galβ1-4GlcNAcβ1-2Manα1-3(Galβ1-3Galβ1-4GlcNAcβ1-2Manα1-6)Manβ1-4GlcNAcβ1-4(Fucα1-6)GlcNAc-ol	—	0.55	—
Galβ1-4(Fucα1-3)GlcNAcβ1-2Manα1-3[Galβ1-4(Fucα1-3)GlcNAcβ1-2Manα1-6]Manβ1-4GlcNAcβ1-4(Fucα1-6)-GlcNAc-ol	—	0.54	—
Galβ1-3Galβ1-4GlcNAcβ1-2Manα1-3(Galβ1-3Galβ1-4GlcNAcβ1-2Manα1-6)Manβ1-4GlcNAcβ1-4(Fucα1-6)GlcNAc-ol	—	0.54	—
Triantennary oligosaccharide Va[e]	—	0.54	—
Triantennary oligosaccharide VII[e]	0.28	—	—
Triantennary oligosaccharide VIIa[e]	—	0.53	—
Triantennary oligosaccharide VIII[e]	—	0.52	—
Triantennary oligosaccharide IX[e]	—	0.51	—
Tetraantennary oligosaccharide X[e]	—	0.51	—

TABLE LC 53 (continued)
SEC of Reduced Oligosaccharides Derived from Glycoproteins, and of Asparagine-Linked Glycopeptides, on Polyacrylamide and Dextran Gels

Compound	K_{av}[c]	r[d]	K_{av}
Triantennary oligosaccharide XI[e]	—	0.50	—
Tetraantennary oligosaccharide XII[e]	—	0.49	—
Tetraantennary oligosaccharide XIII[e]	—	0.48	—
Tetraantennary oligosaccharide XIV[e]	—	0.47	—
Tetraantennary oligosaccharide XV[e]	—	0.45	—
Tetraantennary oligosaccharide XVI[e]	—	0.43	—
Asparagine-linked glycopeptides			
Manα1-3[Manα1-3(Manα1-6)Manα1-6]Manβ1-4GlcNAcβ1-4GlcNAc-Asn	—	—	0.64
GlcNAcβ1-2Manα1-3(GlcNAcβ1-4)[Manα1-3(Manα1-6)Manα1-6]Manβ1-4GlcNAcβ1-4GlcNAc-Asn	—	—	0.61
Galβ1-4GlcNAcβ1-2(GlcNAcβ1-4)Manα1-3(GlcNAcβ1-4)[Manα1-3(Manα1-6)Manα1-6]Manβ1-4GlcNAcβ1-4GlcNAc-Asn	—	—	0.55
GlcNAcβ1-2Manα1-3(Galβ1-4GlcNAcβ1-2Manα1-6)Manβ1-4GlcNAcβ1-4GlcNAc-Asn	—	—	0.49
Galβ1-4GlcNAcβ1-2Manα1-3(Neu5Acα2-6Galβ1-4GlcNAcβ1-2Manα1-6)Manβ1-4GlcNAcβ1-4GlcNAc-Asn	—	—	0.31
Neu5Acα2-6Galβ1-4GlcNAcβ1-2Manα1-3(Neu5Acα2-6Galβ1-4GlcNAcβ1-2Manα1-6)Manβ1-4GlcNAcβ1-4GlcNAc-Asn	—	—	0.20

[a] Two coupled columns, each 100 × 0.6 cm.
[b] Single column or two coupled columns, each 100 × 2 cm.
[c] For equivalent number of D-glucose units per molecule compare K_{av} with values for isomalto-oligosaccharides; see data for Reference 3 in Table LC 51.
[d] t_r relative to D-glucose (9.6 h); for equivalent number of D-glucose units per molecule see data for P1 in Reference 4, Table LC 51.
[e] Structure appended.

Packing P1 = Bio-Gel® P-4 (Bio-Rad; polyacrylamide gel), −400 mesh; narrow range of particle sizes obtained by repeated suspension in water and decantation of fines;[1] slurry degassed at 55°C prior to packing of column.
P2 = Sephadex® G-50 (Pharmacia; dextran gel), superfine grade.
Solvent S1 = alkaline borate buffer: 45.5 m*M* boric acid—4.5 m*M* sodium tetraborate, pH 8.2, containing EDTA disodium salt (2 m*M*) and sodium azide (0.02%).
Detection D1 = scintillation counting of oligosaccharides labeled by reduction with $NaB[^3H]_4$.
D2 = scintillation counting of glycopeptides labeled with $[^{14}C]$- or $[^3H]$-GlcNAc.
D3 = phenol-sulfuric acid colorimetric assay of fractions.

REFERENCES

1. **Natowicz, M. and Baenziger, J. U.,** A rapid method for chromatographic analysis of oligosaccharides on Bio-Gel P-4, *Anal. Biochem.*, 105, 159, 1980.
2. **Yamashita, K., Mizuochi, T., and Kobata, A.,** Analysis of oligosaccharides by gel filtration, *Methods Enzymol.*, 83, 105, 1982.
3. **Rothman, R. J. and Warren, L.,** Analysis of IgG glycopeptides by alkaline borate gel filtration chromatography, *Biochim. Biophys. Acta*, 955, 143, 1988.

APPENDIX TO TABLE LC 53: STRUCTURES OF NUMBERED COMPOUNDS

I

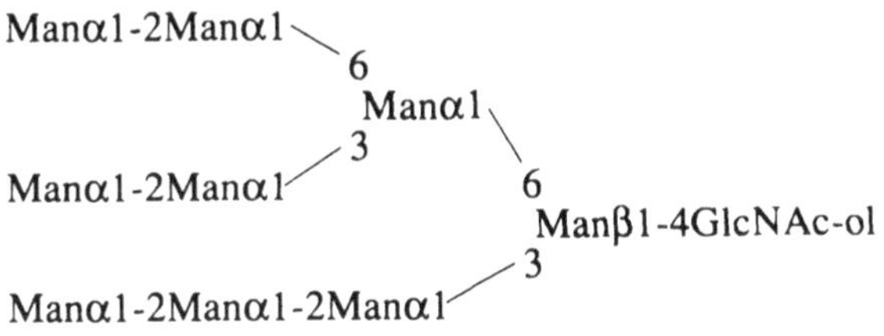

Ia As I but

II

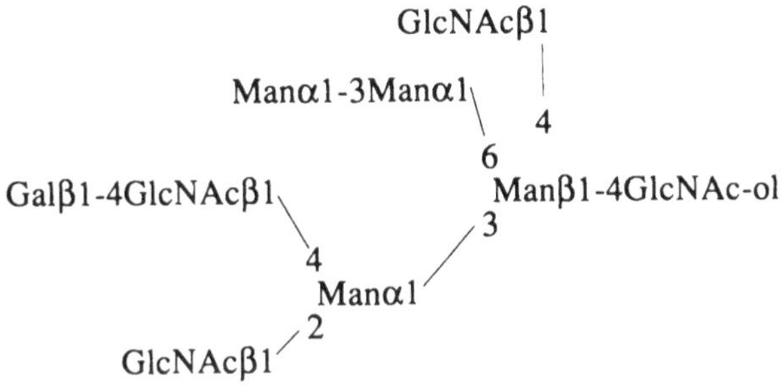

III

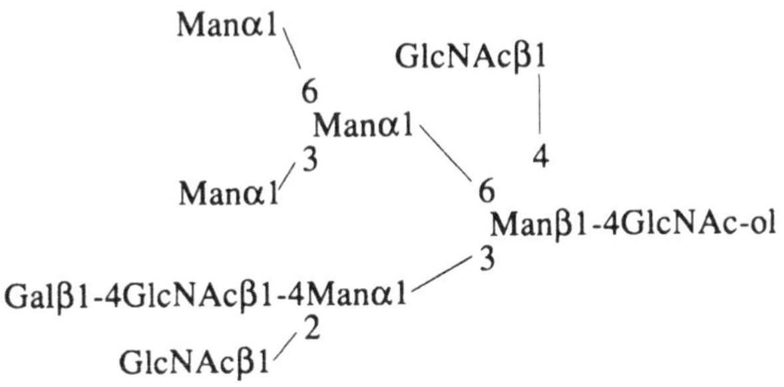

IV

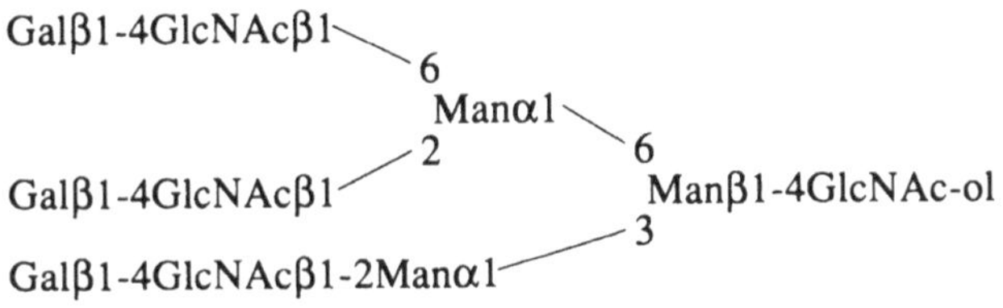

APPENDIX TO TABLE LC 53: STRUCTURES OF NUMBERED COMPOUNDS (continued)

V

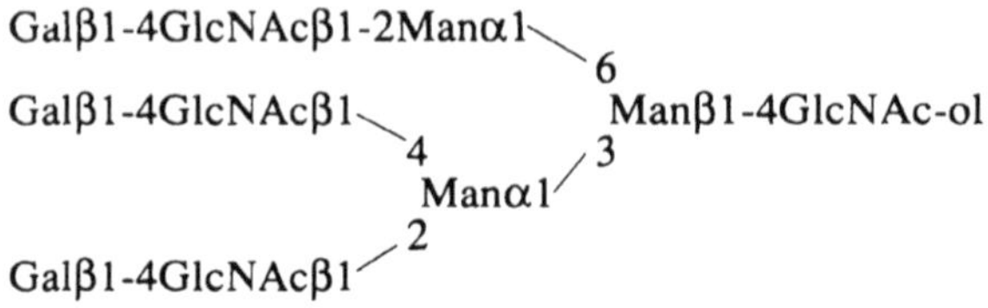

Va As V but

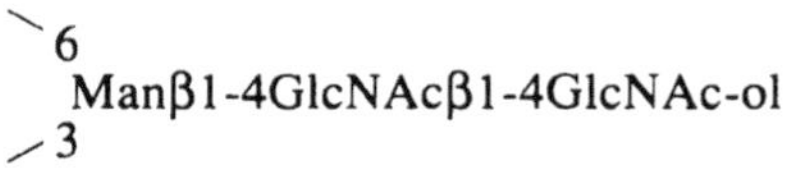

VI

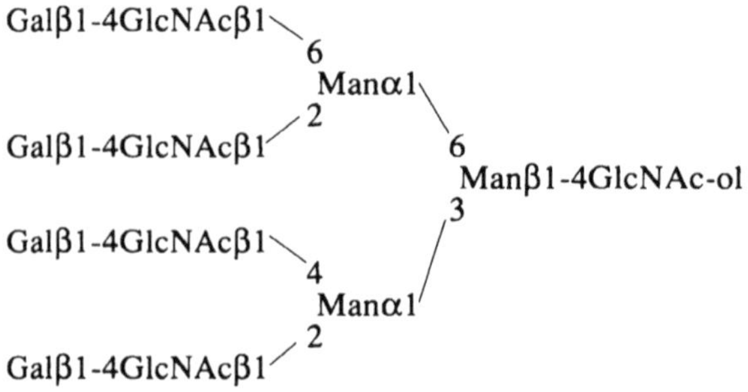

VII

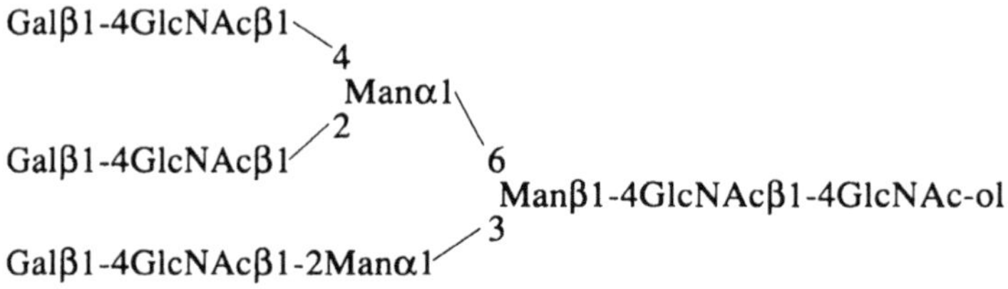

VIIa As VII but

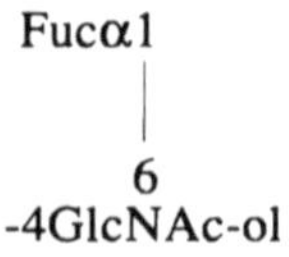

VIII

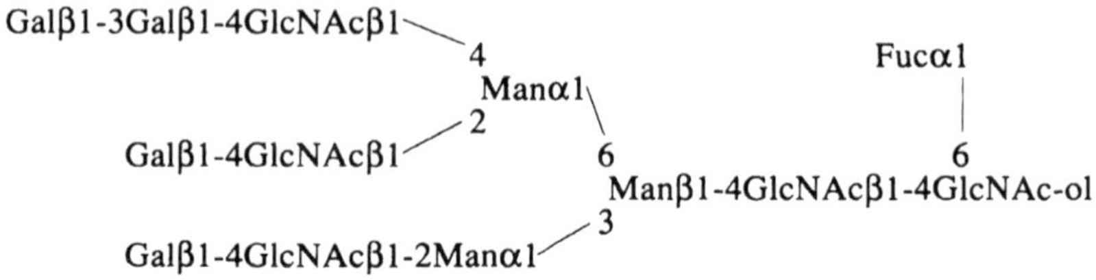

APPENDIX TO TABLE LC 53: STRUCTURES OF NUMBERED COMPOUNDS (continued)

IX

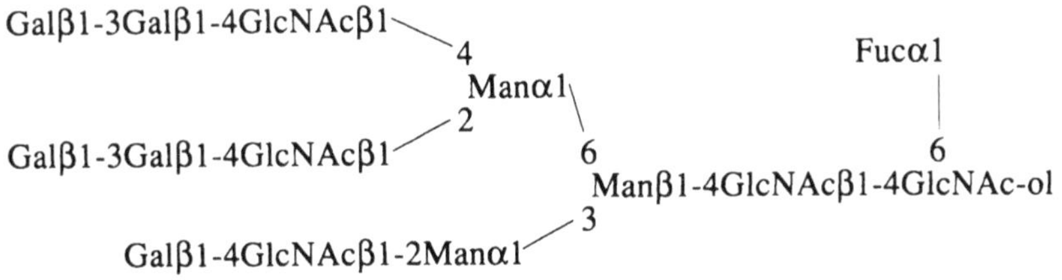

X

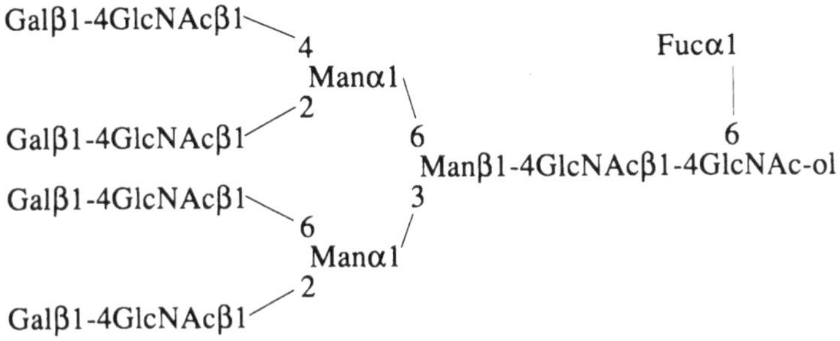

XI

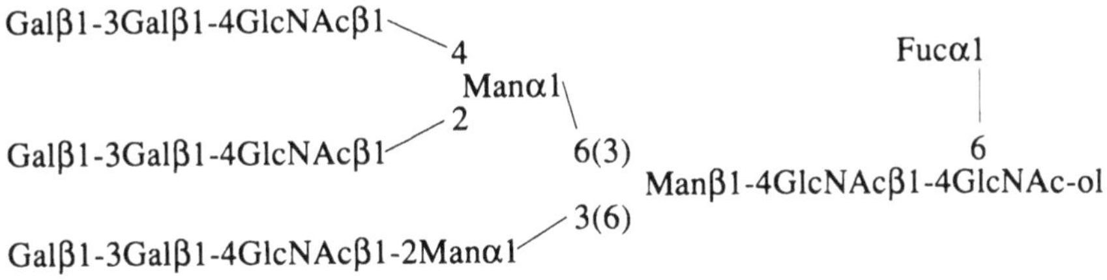

XII

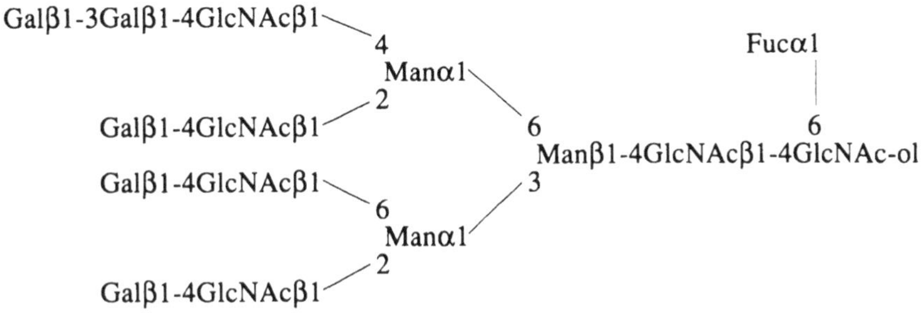

APPENDIX TO TABLE LC 53: STRUCTURES OF NUMBERED COMPOUNDS (continued)

XIII

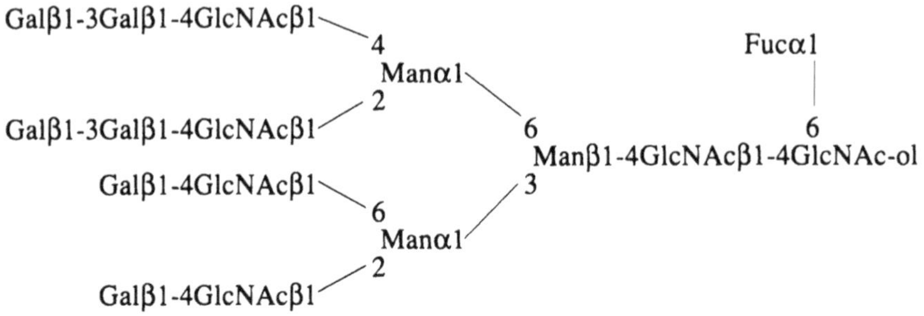

XIV

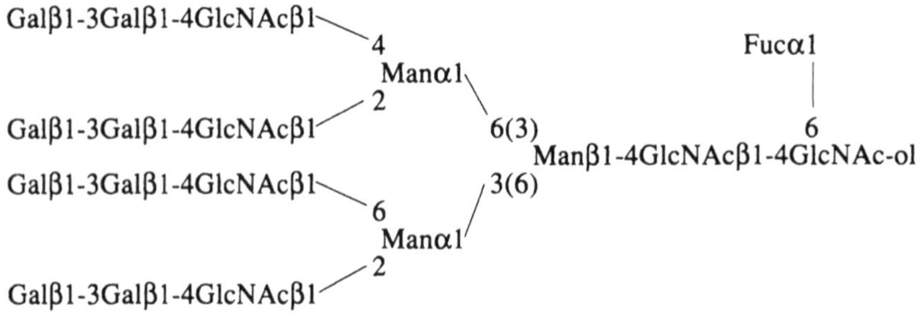

XV

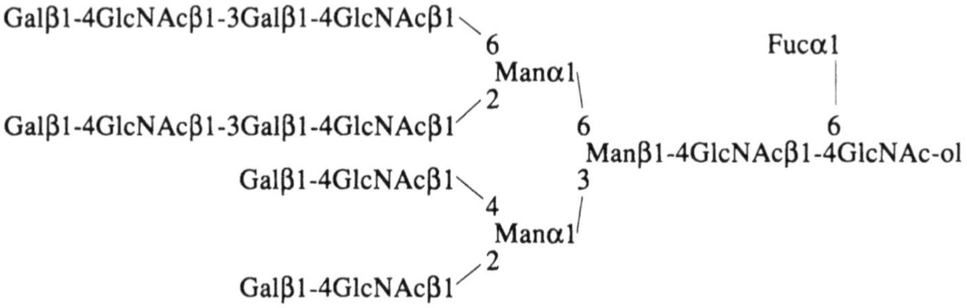

XVI

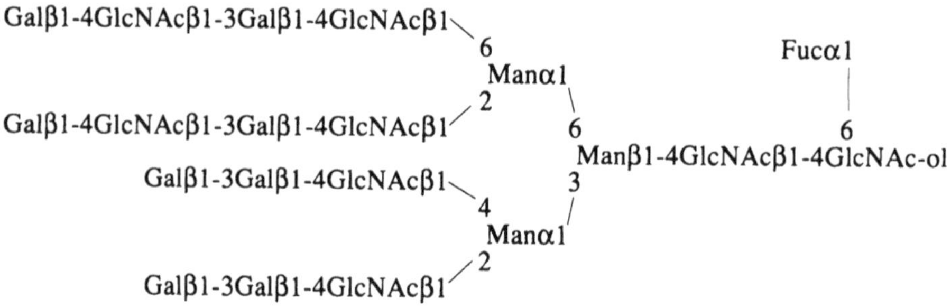

TABLE LC 54
Molecular-Weight Distribution Analysis of Dextrans by SEC on Agarose-Polyacrylamide and Cross-Linked Agarose Gels

Packing	P1	P2	P2	P3	P4	P5	P6
Column							
Length, cm	45	45	70	70	70	30	30
Diameter (I.D.), cm	1.7	1.7	1.6	1.6	1.6	1	1
Material	Glass	Glass	Glass	Glass	Glass	SS	SS
Solvent	S1	S1	S2	S2	S2	S3	S3
Flow rate (ml/h)	25	30	5	5	5	30	30
Temperature (°C)	20	20	25	25	25	Ambient	
Detection	D1	D1	RI	RI	RI	RI	RI
Reference	1	1	2	2	2	3	3
Dextran standards $\overline{M}_w$	K_d		K_{av}				
1,080,000	—	—	0.036	0.059	0.19	—	—
965,000	—	—	—	—	—	0.051	—
450,000	0.27	—	—	—	—	—	—
409,000	—	—	0.091	0.20	0.35	0.19	—
278,000	—	—	—	—	—	0.27	0.10
175,000	—	—	0.20	0.31	0.46	—	—
158,000	0.47	0.10	—	—	—	0.32	0.18
114,000	—	—	—	—	—	—	0.24
89,000	—	—	0.34	0.41	0.57	—	—
68,500	0.61	0.42	—	—	—	—	—
63,850	—	—	—	—	—	0.49	—
44,400	0.69	0.58	—	—	—	—	—
41,200	—	—	—	—	—	—	0.35
39,800	—	—	0.54	0.52	0.67	—	—
25,600	—	—	—	—	—	—	0.44
21,600	0.82	0.86	—	—	—	—	—
14,700	—	—	0.75	0.67	0.81	0.73	—
11,700	—	—	—	—	—	—	0.60
10,400	0.96	—	—	—	—	—	—
3,940	—	—	—	—	—	0.84	0.74
3,500	—	—	0.90	0.83	0.91	—	—

Note: This table supplements Tables LC 39—41 in *CRC Handbook of Chromatography: Carbohydrates,* Volume I, which show calibration data for SEC of dextrans on polyacrylamide, agarose, and Sephadex gels.

Packing
- P1 = Ultrogel® AcA-22 (LKB, Bromma, Sweden), agarose-polyacrylamide composite gel.
- P2 = Ultrogel® AcA-34.
- P3 = Sepharose® CL-6B (Pharmacia), cross-linked agarose gel (6% agarose, average particle diameter 110 μm).
- P4 = Sepharose® CL-4B (4% agarose).
- P5 = Superose® 6 HR 10/30 (Pharmacia), highly cross-linked agarose gel (6% agarose, average particle diameter 13.5 μm), in prepacked column.
- P6 = Superose® 12 HR 10/30, highly cross-linked agarose gel (12% agarose, average particle diameter 10.5 μm), in prepacked column.

Solvent
- S1 = 0.01 *M* acetate buffer solution, pH 5.5, containing sodium azide (0.02%) as preservative.
- S2 = 0.51 *M* sodium chloride solution, containing trichlorobutanol as preservative.
- S3 = 0.05 *M* phosphate buffer, pH 7.0, containing 0.15 *M* NaCl.

Detection
- D1 = phenol-sulfuric acid colorimetric assay of fractions.

TABLE LC 54 (continued)
Molecular-Weight Distribution Analysis of Dextrans by SEC on Agarose-Polyacrylamide and Cross-Linked Agarose Gels

REFERENCES

1. **Dreher, T. W., Hawthorne, D. B., and Grant, B. R.,** Comparison of open-column and high-performance gel permeation chromatography in the separation and molecular-weight estimation of polysaccharides, *J. Chromatogr.*, 174, 443, 1979.
2. **Hagel, L.,** Comparison of some soft gels for the molecular-weight distribution analysis of dextran at enhanced flow rates, *J. Chromatogr.*, 160, 59, 1978.
3. **Andersson, T., Carlsson, M., Hagel, L., Pernemalm, P.-Å., and Janson, J.-C.,** Agarose-based media for high-resolution gel filtration of biopolymers, *J. Chromatogr.*, 326, 33, 1985.

TABLE LC 55
Molecular-Weight Distribution Analysis of Dextrans by High-Performance SEC on Semirigid Gels

Packing	P1 + P2	P1 + P2
Column		
Length, cm	180[a]	120[b]
Diameter (I.D.), cm	0.75	0.75
Material	SS	SS
Solvent	S1	S1
Flow rate (ml/h)	60	49
Temperature (°C)	35	33
Detection	RI	RI[c]
Reference	1	2

Dextran standards $\overline{M}_w$	K_{av}	
490,000	0.27	—
239,800	—	0.24
231,000	0.31	—
154,000	0.39	—
145,000	—	0.29
104,500	—	0.32
73,600	—	0.37
72,000	0.49	—
44,400	0.54	—
42,100	—	0.45
41,800	0.54	—
39,200	—	0.45
31,900	—	0.48
22,300	0.62	—
14,700	—	0.57
11,500	—	0.62
9,300	0.71	—
8,800	—	0.66
7,900	—	0.65
5,200	—	0.71
4,500	—	0.71
4,100	—	0.71
2,500	—	0.77
1,170	—	0.83

TABLE LC 55 (continued)
Molecular-Weight Distribution Analysis of Dextrans by High-Performance SEC on Semirigid Gels

Note: Data for SEC of dextrans on Spheron and Hydrogel packings (the latter is no longer available) can be found in *CRC Handbook of Chromatography: Carbohydrates,* Volume I, Table LC 42.

[a] Two G5000 PW columns and one G3000 PW column, each 60 × 0.75 cm, connected in series in sequence of descending pore size, enclosed in larger glass column (70 × 7.5 cm) filled with water at 35°C.

[b] Two water-jacketed columns each 60 × 0.75 cm, connected in series.

[c] Interference refractometer, using wavelength 546 nm.

Packing P1 = TSK® G5000 PW (Toyo Soda, Tokyo); hydrophilic polymer-based, semirigid gel, average particle diameter 10 μm, SEC exclusion limit $\bar{M}_w$ 7 × 10^6.

P2 = TSK G3000 PW; average particle diameter 13 μm, SEC exclusion limit $\bar{M}_w$ 60,000.

Solvent S1 = water containing sodium azide (0.02%) as preservative.

REFERENCES

1. **Alsop, R. M. and Vlachogiannis, G. J.,** Determination of the molecular weight of clinical dextran by gel permeation chromatography on TSK PW type columns, *J. Chromatogr.*, 246, 227, 1982.
2. **Cullen, M. P., Turner, C., and Haycock, G. B.,** Gel permeation chromatography with interference refractometry for the rapid assay of polydisperse dextrans in biological fluids, *J. Chromatogr.*, 337, 29, 1985.

TABLE LC 56
Molecular-Weight Distribution Analysis of Dextrans by High-Performance SEC on Silica-Based Packings

Packing	P1	P2	P3	P4	P5	P6 + P7	P6 + P8	P6 + P9
Column								
Length, cm	30	30	120[a]	120[a]	120[a]	50[b]	50[b]	50[b]
Diameter (I.D.), cm	0.39	0.39	0.75	0.75	0.75	0.41	0.41	0.41
Material	SS	SS	SS	SS	SS	SS	SS	SS
Solvent	S1	S1	H_2O	H_2O	H_2O	S2	S2	S2
Flow rate (ml/h)	60	60	60	60	60	30	30	30
Temperature (°C)	25	25	25	25	25	24	24	24
Detection	RI	RI	RI	RI	RI	RI	RI	RI
Reference	1	1	2	2	2	3	3	3

TABLE LC 56 (continued)
Molecular-Weight Distribution Analysis of Dextrans by High-Performance SEC on Silica-Based Packings

Dextran standards $\overline{M}_w$	K_{av}						r^c	
2,000,000	—	—	—	—	—	0.087	0.61	0.73
518,000	—	—	0.076	—	—	0.14	0.63	0.74
450,000	0.24	0.024	—	—	—	—	—	—
237,000	—	—	0.21	—	—	0.21	0.68	0.75
167,000	—	—	0.32	0.013	—	0.24	0.69	0.76
154,000	0.41	0.24	—	—	—	—	—	—
110,000	—	—	—	—	—	0.27	0.72	0.77
68,500	0.55	0.42	0.50	0.085	—	0.34	0.74	0.79
44,400	0.62	0.50	—	—	—	—	—	—
39,500	—	—	0.59	0.20	—	0.43	0.76	0.80
22,000	0.74	0.65	0.70	0.41	0.12	0.55	0.81	0.84
10,400	0.86	0.80	0.75	0.54	0.26	0.71	0.86	0.89
8,000	—	—	0.82	0.72	0.41	—	—	—

[a] Two columns, each 60 × 0.75 cm, connected in series.
[b] Two columns, each 25 × 0.41 cm, connected in series, in order of increasing pore size.
[c] t_r relative to D-glucose (10.16 min on P6 + P8, 10.24 min on P6 + P9).

Packing P1 = μ-Bondagel® E-linear (Waters); polyether phase bonded to 10-μm silica gel; SEC exclusion limit $\overline{M}_w$ 2 × 10^6.
P2 = μ-Bondagel® E-300; as P1 but exclusion limit $\overline{M}_w$ 500,000.
P3 = TSK® G 4000 SW (Toyo Soda, Tokyo); inert phase bonded to silica gel; average particle diameter 13 μm, SEC exclusion limit $\overline{M}_w$ 600,000.
P4 = TSK® G 3000 SW; as P3 but average particle diameter 10 μm, SEC exclusion limit $\overline{M}_w$ 200,000.
P5 = TSK® G 2000 SW; as P4 but exclusion limit $\overline{M}_w$ 30,000.
P6 = SynChropak® 100 Å (SynChrom, Linden, IN); glycerolpropyl phase bonded to 10-μm silica; SEC exclusion limit $\overline{M}_w$ 100,000.
P7 = SynChropak® 500 Å; as P6 but exclusion limit $\overline{M}_w$ 5 × 10^6.
P8 = SynChropak® 1000 Å; as P6 but exclusion limit $\overline{M}_w$ > 10^7.
P9 = SynChropak® 4000 Å; as P6 but very high exclusion limit.

Solvent S1 = 0.1 *M* sodium acetate-acetic acid buffer solution, pH 5.5, containing sodium azide (0.02%) as preservative.
S2 = 0.11 *M* sodium acetate-acetic acid buffer solution, pH 3.7, containing sodium sulfate (total ionic strength 0.71 *M*).

REFERENCES

1. **Dreher, T. W., Hawthorne, D. B., and Grant, B. R.,** Comparison of open-column and high-performance gel permeation chromatography in the separation and molecular-weight estimation of polysaccharides, *J. Chromatogr.*, 174, 443, 1979.
2. **Kato, Y., Komiya, K., Sasaki, H., and Hashimoto, T.,** Separation range and separation efficiency in high-speed gel filtration on TSK-gel SW columns, *J. Chromatogr.*, 190, 297, 1980.
3. **Barth, H. G. and Regnier, F. E.,** High-performance gel permeation chromatography of water-soluble cellulosics, *J. Chromatogr.*, 192, 275, 1980.

TABLE LC 57
Molecular-Weight Distribution Analysis of Starch Polysaccharides and Pullulans by High-Performance and Medium-Pressure SEC

Packing	P1 + P2 + P3	P4 + P5 + P6	P7
Column			
Length, cm	180[a]	262[b]	30
Diameter (I.D.) cm	0.75	1.6	1
Material	SS	Glass	SS
Solvent	S1	S2	H_2O
Flow rate (ml/h)	31	25	36
Temperature (°C)	35	Ambient	Ambient
Detection	D1 + RI	RI + D2	RI
Reference	1	2	3
Polymer	**r**[c]	**K_{av}**	
Amylopectin			
M_w 8 × 10⁶ [d]	—	0.029	—
9 × 10⁶ [e]	—	0	—
Amylose[f]			
M_w 2,400	—	—	0.81
7,200	—	0.83	—
20,000	—	0.75	—
50,000	—	0.67	—
80,000	—	—	0.36
95,000	1.14	0.61	—
135,000	1.10	0.59	0.30
200,000	1.08	—	—
230,000	—	0.55	—
300,000	1.04	—	—
348,000	—	0.43	0.19
400,000	1.03	—	—
590,000	0.99	0.40	—
700,000	0.98	—	—
730,000	—	0.35	—
950,000	0.95	0.33	0.044
1.1×10^6	0.93	0.29	—
1.6×10^6	0.90	0.23	—
2.5×10^6	0.88	0.20	—
Pullulan[g]			
M_w 5,300	—	—	0.74
10,400	1.28	—	0.63
22,800	—	—	0.53
45,500	1.17	—	—
50,000	—	—	0.40
100,000	—	—	0.32
187,000	1.04	—	0.22
348,000	1.00	—	—
400,000	—	—	0.10
750,000	0.95	—	—
850,000	—	—	0.044

[a] Three water-jacketed columns, each 60 × 0.75 cm, connected in series in order of increasing pore size; elution in downward-upward-downward direction.

[b] Three columns, 60 × 1.6 cm, 67 × 1.6 cm, and 135 × 1.6 cm, connected in series.

[c] t_r relative to pullulan standard $\overline{M}_w$ 348,000 (79.8 min).

[d] Component of maize starch.

TABLE LC 57 (continued)
Molecular-Weight Distribution Analysis of Starch Polysaccharides and Pullulans by High-Performance and Medium-Pressure SEC

[e] Standard used to determine V_o.
[f] Standards, isolated from various starches and purified by repeated recrystallization from 1-butanol-saturated water,[1] or synthesized enzymically with potato phosphorylase,[2,3] characterized by viscometry[2] or static low-angle laser-light-scattering.[3]
[g] Standards prepared by Hayashibara Biochemical Institute (Japan)[1] or Shodex (Showa Denka, Tokyo).[3]

Packing	P1	=	TSK® G3000 PW (Toyo Soda, Tokyo); hydrophilic vinyl polymer gel, average particle diameter 13 μm, SEC exclusion limit $\overline{M}_w$ 60,000.
	P2	=	TSK® G4000 PW; average particle diameter 13 μm, SEC exclusion limit $\overline{M}_w$ 700,000.
	P3	=	TSK® G6000 PW; average particle diameter 17 μm, SEC exclusion limit $\overline{M}_w$ 3 × 10^7.
	P4	=	Sephacryl® S-400 (Pharmacia); cross-linked allyl-dextran gel, SEC fractionation range $\overline{M}_w$ 10,000—500,000 (dextrans).
	P5	=	Sephacryl® S-500, fractionation range $\overline{M}_w$ 40,000—2,000,000 (dextrans).
	P6	=	Sephacryl® S-1000, fractionation range $\overline{M}_w$ 100,000—10,000,000 (dextrans).
	P7	=	Superose® 6 HR 10/30 (Pharmacia); cross-linked agarose gel (6% agarose), average particle diameter 13.5 μm, SEC exclusion limit $\overline{M}_w$ 1 × 10^6; in prepacked column.
Solvent	S1	=	50 m*M* sodium phosphate buffer, pH 6.1, containing sodium azide (0.02%) as preservative.
	S2	=	5 m*M* sodium hydroxide solution, containing sodium azide (0.002%).
Detection	D1	=	low-angle laser light scattering.
	D2	=	total carbohydrate in fractions determined by anthrone colorimetric method; amylose and amylopectin distinguished by iodine staining.

REFERENCES

1. **Hizukuri, S. and Takagi, T.,** Estimation of the distribution of molecular weight for amylose by the low-angle laser-light-scattering technique combined with high-performance gel chromatography, *Carbohydr. Res.*, 134, 1, 1984.
2. **Praznik, W., Burdicek, G., and Beck, R. H. F.,** Molecular weight analysis of starch polysaccharides using cross-linked allyl-dextran gels, *J. Chromatogr.*, 357, 216, 1986.
3. **Praznik, W., Beck, R. H. F., and Eigner, W. D.,** New high-performance gel permeation chromatographic system for the determination of low molecular-weight amyloses, *J. Chromatogr.*, 387, 467, 1987.

TABLE LC 58
Molecular-Weight Distribution Analysis of Carrageenans, Inulins, and Pectins by Medium-Pressure SEC

Packing	P1	P1	P2	P3
Column				
Length, cm	100	40	30	71
Diameter (I.D.), cm	2.5	1.6	1	1.6
Material	Glass	Glass	SS	Glass

TABLE LC 58 (continued)
Molecular-Weight Distribution Analysis of Carrageenans, Inulins, and Pectins by Medium-Pressure SEC

Solvent	S1	S1	H_2O	S2
Flow rate (ml/h)	48	18	n.a.	1
Temperature (°C)	60	60	Ambient	Ambient
Detection	D1	RI	RI	D2
Reference	1	2	3	4

Polymer	Carrageenans Iota-	Carrageenans Kappa-		Inulins	Pectins
$\overline{M}_w$	K_{av}				r[a]
1,600	—	—	—	—	0.68
2,450	—	—	—	0.74	—
4,770	—	—	—	—	0.48
6,010	—	—	—	0.66	—
8,460	—	—	—	—	0.41
9,250	—	—	—	0.60	—
10,050	—	—	—	—	0.37
11,300	—	—	0.83	—	—
12,700	—	0.82	—	—	—
14,400	—	—	0.75	—	—
16,200	—	—	—	0.53	—
20,700	—	0.69	—	—	—
25,000	0.72	0.67	—	—	—
35,700	—	—	0.64	—	—
40,300	—	0.58	—	—	—
48,300	—	—	0.56	—	—
50,000	0.59	0.54	—	—	—
78,500	—	0.42	—	—	—
100,000	0.42	0.37	—	—	—
143,800	—	0.30	—	—	—
152,800	—	—	0.27	—	—
200,000	0.29	0.24	—	—	—
234,000	—	—	0.20	—	—
298,000	—	0.16	—	—	—

[a] t_r relative to D-galacturonic acid (144 min).

Packing P1 = Sepharose® CL-4B (Pharmacia), cross-linked agarose gel (4% agarose); SEC exclusion limit M_w 5 × 10^6.

P2 = Superose® 12 HR 10/30 (Pharmacia), highly cross-linked agarose gel (12% agarose); average particle diameter 10.5 μm, SEC exclusion limit M_w 500,000; in prepacked column.

P3 = Sephacryl® S-200 (Pharmacia), cross-linked allyl-dextran gel, average particle diameter 40—50 μm, SEC exclusion limit M_w 50,000.

Solvent S1 = 0.2 *M* lithium chloride.

S2 = 0.1 *M* acetate buffer, pH 3.6, containing benzoic acid (0.05%) as preservative.

Detection D1 = enzymic assay for D-galactose in hydrolyzed fractions; assay based on oxidation of β-D-galactose to D-galactonic acid by NAD^+ in presence of β-D-galactose dehydrogenase, NADH formed being determined spectrophotometrically at 340 nm.

D2 = fractions assayed for D-galacturonic acid by Blumenkrantz method and reducing end-groups determined by Cu^{2+}-arsenomolybdate method (Milner and Avigad).

TABLE LC 58 (continued)
Molecular-Weight Distribution Analysis of Carrageenans, Inulins, and Pectins by Medium-Pressure SEC

REFERENCES

1. **Ekström, L.-G., Kuivinen, J., and Johansson, G.,** Molecular weight distribution and hydrolysis behaviour of carrageenans, *Carbohydr. Res.*, 116, 89, 1983.
2. **Ekström, L.-G.,** Molecular-weight distribution and the behavior of kappa-carrageenan on hydrolysis, *Carbohydr. Res.*, 135, 283, 1985.
3. **Beck, R. H. F. and Praznik, W.,** Molecular characterization of fructans by high-performance gel chromatography, *J. Chromatogr.*, 369, 208, 1986.
4. **Tuerena, C. E., Taylor, A. J., and Mitchell, J. R.,** Evaluation of a method for determining the free carboxyl group distribution in pectins, *Carbohydr. Polym.*, 2, 193, 1982.

TABLE LC 59
Molecular-Weight Distribution Analysis of Glycosaminoglycuronans and Derived Oligosaccharides by High-Performance SEC

Packing	P1	P2 + P3	P4	P4 + P5	P6 + P7	
Column						
Length, cm	60	90[a]	50	100[b]	60[c]	
Diameter (I.D.), cm	0.75	0.4	0.8	0.8	0.75	
Material	SS	SS	SS	SS	SS	
Solvent	S1	S2	S3	S3	S4	
Flow rate (ml/h)	60	24	30	30	21	
Temperature (°C)	Ambient	40	40	40	Ambient	
Detection	UV[d], DI	RI	RI	RI, UV[d]	UV[d], D2	
Reference	1	2	3	4	5	
Polymer	Hep[e]	HA[e]	HA	HA	HA	ChS, DS[e]
$\overline{M}_n$	r[f]	K_{av}	r[g]	r[g]	K_d	K_d
380	—	0.57	—	—	—	—
760	—	0.43	—	—	—	—
1,050	—	0.34	—	—	—	—
1,520	—	0.28	—	—	—	—
2,800	—	0.23	—	—	—	—
11,250	—	0.14	—	—	—	—
24,700	—	0.11	—	—	—	—
$\overline{M}_w$						
2,230	0.91	—	—	—	—	—
3,010	—	—	—	—	—	0.78
4,300	—	—	—	—	—	0.70
4,450	0.85	—	—	—	—	—
4,610	—	—	—	—	—	0.66
6,680	0.79	—	—	—	—	—
7,040	—	—	—	—	—	0.58
7,540	—	—	—	—	0.57	—
8,250	—	—	—	—	—	0.55
8,910	0.74	—	—	—	—	—
11,300	—	—	—	—	0.45	—
12,600	—	—	—	—	—	0.43
15,000	—	—	—	—	—	0.40
16,700	—	—	—	—	0.35	—
17,300	—	—	—	—	—	0.34
23,500	—	—	—	—	—	0.28

TABLE LC 59 (continued)
Molecular-Weight Distribution Analysis of Glycosaminoglycuronans and Derived Oligosaccharides by High-Performance SEC

$\overline{M}_n$	r[f]	K_{av}	r[g]	r[g]	K_d	K_d
26,500	—	—	—	—	0.26	—
38,500	—	—	—	—	0.18	—
110,000	—	—	1.00	1.00	—	—
120,000	—	—	0.98	—	—	—
257,000	—	—	—	0.91	—	—
335,000	—	—	—	0.89	—	—
450,000	—	—	0.88	—	—	—
470,000	—	—	0.87	—	—	—
610,000	—	—	—	0.82	—	—
980,000	—	—	0.81	—	—	—

Note: Earlier data on SEC of glycosaminoglycuronans will be found in *CRC Handbook of Chromatography: Carbohydrates*, Volume I, Table LC 45.

[a] Three columns, each 30 × 0.4 cm, connected in series, P2 followed by two P3 columns.
[b] Two columns, each 50 × 0.8 cm, connected in series, P4 followed by P5.
[c] Two columns, each 30 × 0.75 cm, connected in series, P6 followed by P7.
[d] Absorbance measured at 204,[5] 206,[1] or 280 nm.[4]
[e] Hep = heparin, HA = hyaluronic acid, ChS = chondroitin 4-sulfate, DS = dermatan sulfate.
[f] t_r relative to D-glucuronic acid (21.2 min).
[g] t_r relative to hog skin hyaluronic acid, $\overline{M}_w$ 110,000 (32 min[3], 57.4 min[4]).

Packing P1 = TSK® G3000 SW (Toyo Soda, Tokyo); inert phase bonded to silica gel, average particle diameter 10 μm, SEC exclusion limit $\overline{M}_w$ 200,000 (dextrans); used with TSK G SWP precolumn (7.5 × 0.75 cm).
P2 = μ-Bondagel® E-linear (Waters); polyether phase bonded to 10-μm silica gel, SEC exclusion limit $\overline{M}_w$ 2 × 10^6 (dextrans).
P3 = μ-Porasil® GPC 60 Å (Waters); porous silica, SEC exclusion limit $\overline{M}_w$ 10,000.
P4 = Shodex® OHpak B-806 (Showa Denko, Tokyo); semirigid poly(hydroxyalkylmethacrylate) gel, SEC exclusion limit $\overline{M}_w$ 2.5 × 10^7 (dextrans); used with Shodex OHpak 800P precolumn (5 × 0.6 cm).
P5 = Shodex® OHpak B-805; as P4 but exclusion limit $\overline{M}_w$ 5 × 10^6.
P6 = TSK® G4000 SW (Toyo Soda); as P1 but exclusion limit $\overline{M}_w$ 600,000 (dextrans).
P7 = TSK® G2000 SW; as P1 but exclusion limit $\overline{M}_w$ 20,000 (dextrans).

Solvent S1 = 0.1 *M* sodium chloride solution (degassed).
S2 = 20 m*M* sodium acetate buffer, pH 4.0, containing trace (1.5 mg/l) of hyaluronic acid to block adsorption sites on silica.
S3 = 0.02 *M* sodium chloride solution (degassed).
S4 = 0.15 *M* sodium chloride.

Detection D1 = scintillation counting of oligosaccharides (produced by treatment of heparin with nitrous acid), labeled by reduction with $NaB[^3H]_4$.
D2 = polyacrylamide gel electrophoresis of fractions and densitometry of bands, stained with alcian blue and silver, to determine mass ratio of components.

REFERENCES

1. **Harenberg, J. and De Vries, J. X.**, Characterisation of heparins by high-performance size exclusion liquid chromatography, *J. Chromatogr.*, 261, 287, 1983.
2. **Knudsen, P. J., Eriksen, P. B., Fenger, M., and Florentz, K.**, High-performance liquid chromatography of hyaluronic acid and oligosaccharides produced by bovine testes hyaluronidase, *J. Chromatogr.*, 187, 373, 1980.
3. **Motohashi, N. and Mori, I.**, Molecular weight determination of hyaluronic acid and its separation from mouse skin extract by high-performance gel permeation chromatography using a precision differential refractometer, *J. Chromatogr.*, 299, 508, 1984.
4. **Motohashi, N., Nakamichi, Y., Mori, I., Nishikawa, H., and Umemoto, J.**, Analysis by high-performance gel permeation chromatography of hyaluronic acid in animal skins and rabbit synovial fluid, *J. Chromatogr.*, 435, 335, 1988.
5. **Hittner, D. M. and Cowman, M. K.**, High-performance gel-permeation chromatography of glycosaminoglycans. Column calibration by gel electrophoresis, *J. Chromatogr.*, 402, 149, 1987.

TABLE LC 60
Low- and High-Pressure SEC of Glycoproteins

		P1		P2	
Packing		P1		P2	
Column					
Length, cm		102		60	
Diameter (I.D.), cm		0.9		0.75	
Material		Glass		SS	
Solvent		S1		S2	
Flow rate (ml/h)		n.a.		30	
Temperature (°C)		4		Ambient	
Detection		UV[a]		UV[a]	
Reference		1		2	
Glycoprotein	**M[b]**	**K_{av}**	**M'[c]**	**K_{av}**	**M'[d]**
Ribonuclease B[e]	14,900	—	—	0.487	14,700
Ovomucoid[e]	28,000	—	—	0.350	26,900
α_1-Acid glycoprotein[e]	44,000[f]	0.467	386,000	0.312[f]	31,500
Asialo-(α_1-acid glycoprotein)[g]	39,000	0.587	191,000	—	—
Ovalbumin[e]	44,300	—	—	0.238	42,800
Fetuin[e]	48,000[f]	0.477	368,000	0.220	46,300
Asialofetuin[g]	44,000	1.07	<10,000	—	—
Ovoinhibitor[e] (Japanese quail)	48,300	—	—	0.202	50,000
Ovoinhibitor[e] (chicken)	48,300	—	—	0.197	51,000
Taka-amylase A[e]	51,000	—	—	0.192	52,100
Bovine γ-globulin,[e] H chain	51,500	—	—	0.195	51,400
Acid carboxypeptidase[e] (*Aspergillus niger*)	64,000	—	—	0.170	57,600
Transferrin[e]	76,000	—	—	0.107	77,700
Thyroglobulin	669,000	0.154	>570,000	—	—
Asialo-thyroglobulin[g]	660,000	0.887	34,000	—	—
Bovine submaxillary mucin	1.3×10^6	0.014	ca. 2×10^6	—	—
Asialo-(bovine submaxillary mucin)[g]	1.27×10^6	0.892	33,000	—	—

[a] Absorbance measured at 280 nm.
[b] True molecular weight.
[c] Apparent molecular weight, from calibration plot of log M vs. K_{av} for standard proteins.
[d] Apparent molecular weight, from calibration plot of $M^{0.555}$ vs. $(K_{av})^{1/3}$ for standard proteins.
[e] In reference 2, disulfide bonds in glycoproteins were reduced by dithiothreitol and resultant SH groups carboxyamidomethylated by reaction with iodoacetamide.
[f] In Reference 2, sample had mol wt 36,500.
[g] In Reference 1, glycoproteins were desialylated by mild acid hydrolysis (12.5 m*M* H_2SO_4, 80°C, 1 h or 10 m*M* HCl, 80°C, 100 min) or incubation with neuraminidase attached to beaded agarose (37°C, 5 h, in 0.1 *M* citrate buffer, pH 5.0).

Packing P1 = Sepharose® 6B (Pharmacia), 6% agarose gel; SEC fractionation range mol wt 10,000—4,000,000 for globular proteins, 10,000—1,000,000 for dextrans.
P2 = TSK® G3000 SW (Toyo Soda, Tokyo), chemically bonded silica gel, average particle diameter 10 μm, SEC exclusion limits mol wt 400,000 (proteins) or 200,000 (dextrans): used with TSK® GSWP precolumn (5 × 0.75 cm).

Solvent S1 = 10 m*M* NaH_2PO_4 buffer, pH 5.5, containing sodium azide (0.02%) as preservative.
S2 = 6 *M* guanidine hydrochloride in 10 m*M* phosphate buffer, pH 6.5, containing EDTA (1 m*M*).

REFERENCES

1. **Alhadeff, J. A.,** Gel filtration of sialoglycoproteins, *Biochem. J.*, 173, 315, 1978.
2. **Ui, N.,** High-speed filtration of glycopeptides in 6 *M* guanidine hydrochloride, *J. Chromatogr.*, 215, 289, 1981.

TABLE LC 61
Application of "Universal Calibration" to Determination of Molecular Weights of Polysaccharides by SEC

Packing	P1 + P2 + P3	P2 + P4	P5	P1 + P2 + P3	P2 + P4	P2 + P4	P5	P6	P7	P8
Column										
Length, cm	240[a]	120[b]	60	240[a]	120[b]	120[b]	60	58	57	25
Diameter (I.D.), cm	0.75	0.75	0.9	0.75	0.75	0.75	0.9	2.6	2.6	0.4
Material	SS	SS	Glass	SS	SS	SS	Glass	Glass	Glass	SS
Solvent	S1	S2	S3	S1	S2	S2	S3	S3	S3	S4
Flow rate (ml/h)	60	42	15	60	42	42	15	131	133	12
Temperature (°C)	25	40	20	25	40	40	20	25	25	50
Detection	RI	RI	D1	RI	RI	RI	D1	UV[c]	UV[c]	RI
Reference	1	2	3	1	2	2	3	4	4	5
Polymer	**Dextran**			**Pullulan**		**Starch**	**Gums**[d]	**Gum arabic**[e]		**Carra-geenan**
log $[\eta]\cdot\bar{M}_w$[f]	V_e **(ml)**		K_{av}	V_e **(ml)**			K_{av}			
4.973	66.7	—	—	—	—	—	—	—	—	—
5.079	—	—	—	—	32.9	—	—	—	—	—
5.835	61.0	—	—	—	—	—	—	—	—	—
5.934	60.5	—	—	—	—	—	—	—	—	—
6.055	—	—	—	—	30.4	—	—	—	—	—
6.144	—	—	—	59.5	—	—	—	—	—	—
6.181	—	—	0.80	—	—	—	—	—	—	—
6.246	—	—	—	—	—	—	—	0.54	—	—
6.316	—	—	—	—	—	—	—	0.53	—	—
6.323	58.1	—	—	—	—	—	—	—	—	—
6.352	—	—	—	—	—	—	—	—	0.75	—
6.355	—	29.8	—	—	—	—	—	—	—	0.72
6.460	—	—	0.76	—	—	—	—	—	—	—
6.478	—	—	—	57.0	—	—	—	—	—	—
6.515	—	—	—	—	—	28.9[g]	—	—	—	—
6.535	—	—	—	—	—	—	—	0.40	0.68	—
6.553	—	—	—	—	—	—	—	—	0.66	—

TABLE LC 61 (continued)
Application of "Universal Calibration" to Determination of Molecular Weights of Polysaccharides by SEC

Polymer	Dextran			Pullulan		Starch	Gums[d]	Gum arabic[e]		Carra-geenan
log $[\eta]\cdot\bar{M}_w$[f]	V_e (ml)		K_{av}	V_e (ml)			K_{av}			
6.584	—	—	—	—	29.4	—	—	—	—	—
6.690	—	—	—	—	—	—	—	0.38	—	—
6.697	—	29.2	—	—	—	—	—	—	—	—
6.719	55.5	—	—	—	—	—	—	—	—	—
6.750	—	—	0.70	—	—	—	—	—	—	—
6.808	—	—	—	55.2	—	—	—	—	—	—
6.845	—	—	—	—	—	—	—	0.30	0.50	—
6.856	—	—	—	—	—	—	—	—	—	0.52
6.995	—	—	—	—	—	—	0.65	—	—	—
7.025	—	28.4	—	—	—	28.4[g]	—	—	—	—
7.048	—	—	—	53.9	—	—	—	—	—	—
7.149	53.2	—	—	—	—	—	—	—	—	—
7.162	—	—	—	—	—	28.2[g]	—	—	—	—
7.215	—	—	—	—	—	—	0.61	—	—	—
7.275	—	—	—	—	—	—	—	0.10	—	—
7.320	—	—	—	—	—	—	0.59	—	—	—
7.376	—	—	—	—	—	—	—	—	—	0.32
7.378	—	—	—	52.3	—	—	—	—	—	—
7.388	—	—	—	—	—	27.8[h]	—	—	—	—
7.502	51.3	—	—	—	—	—	—	—	—	—
7.539	—	—	0.54	—	—	—	—	—	—	—
7.547	—	—	—	—	27.3	—	—	—	—	—
7.579	—	—	—	—	—	—	—	—	0.42	—
7.615	—	—	—	—	—	27.7[g]	—	—	—	—
7.616	—	27.5	—	—	—	—	—	—	—	—
7.621	—	—	—	—	—	—	—	—	—	0.25
7.658	—	—	—	—	—	—	0.53	—	—	—
7.692	—	—	—	50.3	—	—	—	—	—	—
7.734	—	—	—	—	—	—	—	0	—	—

7.749	—	—	—	—	—	—	—	—	0.33	—
7.770	—	—	—	—	—	—	0.50	—	—	—
7.840	—	—	—	—	—	—	—	—	—	0.18
7.844	49.7	—	—	—	—	—	—	—	—	—
7.870	—	26.6	—	—	—	—	—	—	—	—
7.944	—	—	—	48.6	—	—	—	—	—	—
7.955	—	—	—	—	—	—	—	0	—	—
7.960	—	—	—	—	—	26.8[g]	—	—	—	—
7.964	—	—	—	—	—	—	—	—	0.30	—
8.025	—	—	—	—	—	26.2[h]	—	—	—	—
8.074	—	—	—	—	26.2	—	—	—	—	—
8.168	—	—	—	—	—	—	—	—	0.27	—
8.200	—	—	—	—	—	—	0.39	—	—	—
8.274	—	—	—	47.3	—	—	—	—	—	—
8.289	—	—	—	—	—	—	—	—	0.23	—
8.295	—	—	0.35	—	—	—	—	—	—	—
8.331	—	25.8	—	—	—	—	—	—	—	—
8.338	—	—	—	—	—	25.3[g]	—	—	—	—
8.349	—	—	—	—	—	—	0.33	—	—	—
8.448	—	—	—	—	—	—	—	0	—	—
8.554	—	—	—	—	—	—	—	—	0.18	—
8.603	—	—	—	45.4	—	—	—	—	—	—
8.623	45.3	—	—	—	—	—	—	—	—	—
8.673	—	25.0	—	—	—	—	—	—	—	—
8.696	—	—	—	—	—	—	—	—	0.16	—
8.793	—	24.8	—	—	—	—	—	—	—	—

[a] Four columns, each 60 × 0.75 cm, connected in series, P1 followed by P2 then two of P3.
[b] Two columns, each 60 × 0.75 cm, connected in series, P4 followed by P2.
[c] Absorbance measured at 214 nm.
[d] Arabinogalactan gum exudates of *Acacia longifolia*, *A. mearnsii*, *A. robusta*, *A. hebeclada*, and *A. senegal*, glucuronomannoglycan gum of *Grevillea robusta*, and partially hydrolyzed samples of *A. senegal* and *G. robusta* gum (both had log $[\eta]\cdot\overline{M}_w$ = 6.995); all fitted universal calibration plot established with dextran standards (column 3).
[e] $\overline{M}_w$ of fractions from gel columns determined by low-angle laser-light scattering.
[f] Intrinsic viscosity $[\eta]$ determined using Ubbelohde viscometer, at same temperature and ionic strength as SEC conditions; $\overline{M}_w$ by light scattering. Parameter $[\eta]\cdot\overline{M}$ is proportional to hydrodynamic volume of polymer molecule.
[g] Amylopectin.
[h] Amylose

TABLE LC 61 (continued)
Application of "Universal Calibration" to Determination of Molecular Weights of Polysaccharides by SEC

Packing	P1 =	TSK® G 2000 PW (Toyo Soda, Tokyo); cross-linked hydrophilic vinyl polymer, average particle diameter 10 μm, SEC exclusion limit $\overline{M}_w$ 5000.
	P2 =	TSK® G3000 PW; average particle diameter 13 μm, exclusion limit $\overline{M}_w$ 60,000.
	P3 =	TSK® G5000 PW; average particle diameter 17 μm, exclusion limit $\overline{M}_w$ 7×10^6.
	P4 =	TSK® G6000 PW; average particle diameter 17 μm, exclusion limit $\overline{M}_w$ 3×10^7.
	P5 =	Sepharose® 4B (Pharmacia); 4% agarose gel, SEC fractionation range $\overline{M}_w$ 300,000—5,000,000 (dextrans).
	P6 =	Sephacryl® S-400 (Pharmacia); allyl-dextran gel, fractionation range $\overline{M}_w$ 10,000—500,000 (dextrans).
	P7 =	Sephacryl® S-500; as P6 but fractionation range $\overline{M}_w$ 40,000—2,000,000 (dextrans).
	P8 =	LiChrospher® 1000 DIOL (Merck); porous silica, average particle diameter 10 μm, pore size 1,000 Å, surface-modified with 1,2-dihydroxy-3-propoxypropyl groups, fractionation range 10,000—10,000,000 (dextrans).
Solvent	S1 =	0.1 *M* sodium chloride.
	S2 =	0.25 *M* potassium chloride, degassed by heating and sonication.
	S3 =	1 *M* sodium chloride.
	S4 =	0.1 *M* sodium sulfate.
Detection	D1 =	phenol-sulfuric acid assay of fractions.

REFERENCES

1. **Kato, T., Tokuya, T., and Takahashi, A.,** Comparison of poly(ethylene oxide), pullulan and dextran as polymer standards in aqueous gel chromatography, *J. Chromatogr.*, 256, 61, 1983.
2. **Kuge, T., Kobayashi, K., Tanahashi, H., Igushi, T., and Kitamura, S.,** Gel permeation chromatography of polysaccharides: universal calibration curve, *Agric. Biol. Chem.*, 48, 2375, 1984.
3. **Churms, S. C. and Stephen, A. M.,** Application of "universal calibration" to estimation of $\overline{M}_w$ of plant-gum polysaccharides by steric-exclusion chromatography, *S. Afr. J. Sci.*, 84, 855, 1988.
4. **Vandevelde, M.-C. and Fenyo, J.-C.,** Macromolecular distribution of *Acacia senegal* gum (gum arabic) by size-exclusion chromatography, *Carbohydr. Polym.*, 5, 251, 1985.
5. **Sworn, G., Marrs, W. M., and Hart, R. J.,** Characterisation of carrageenans by high-performance size-exclusion chromatography using a LiChropher 1000 DIOL column, *J. Chromatogr.*, 403, 307, 1987.

TWO-DIMENSIONAL HPLC MAPPING OF OLIGOSACCHARIDES DERIVED FROM GLYCOPROTEINS

Two recent papers have reported two-dimensional HPLC mapping of many of the oligosaccharides produced by sequential digestion of glycoproteins with pepsin and *N*-oligosaccharide glycopeptidase (almond),[1] or by hydrazinolysis followed by *N*-acetylation.[2] In both cases oligosaccharides were derivatized at the reducing end by reductive amination with 2-aminopyridine in the presence of sodium cyanoborohydride, which makes possible the use of fluorometric detection of sensitivity such that the detection limit for carbohydrate can be as low as 10 pmol. The retention data for these derivatized oligosaccharides on two different columns, one operating by reversed-phase partition, the other containing an amino or amide phase that fractionates oligosaccharides mainly according to size, are combined to give a two-dimensional ''oligosaccharide map''. Subsequent identification of unknown oligosaccharides is greatly facilitated by reference to this map after HPLC on the same two columns, since oligosaccharides which overlap on elution from one of the columns are usually separated on the other and, thus, on the map. The retention parameters used to establish the data bases from which these maps have been drawn are shown in Table LC 62.

In addition to the two major papers cited, a note on studies of the structure and heterogeneity of the carbohydrate chains of the glycoproteins immunoglobulin G, thyroglobulin, fetuin, and α_1-acid glycoprotein by use of similar oligosaccharide maps, in this case without derivatization of the oligosaccharides prior to HPLC, has recently been published.* Although details of the oligosaccharide structures given in this publication are insufficient to enable the corresponding chromatographic data to be included in Table LC 62, the paper should nevertheless be noted as a further example of the power of the two-dimensional mapping technique in structural studies of complex oligosaccharides.

* Arbatsky, N. P., Martynova, M. D., Zheltova, A. O., Derevitskaya, V. A., and Kochetkov, N. K., Studies on structure and heterogeneity of carbohydrate chains of *N*-glycoproteins by use of liquid chromatography. ''Oligosaccharide maps'' of glycoproteins, *Carbohydr. Res.*, 187, 165, 1989.

TABLE LC 62
HPLC Data Used to Establish Two-Dimensional Oligosaccharide Maps in Structural Studies of Glycoproteins

Packing	P1		P2		P3	P4
Column						
Length, cm	15		25		15	15
Diameter (I.D.), cm	0.6		0.46		0.46	0.46
Material	SS		SS		SS	SS
Solvent	S1		S2		S3	S4
Flow rate (ml/min)	1		1		1.5	1
Temperature (°C)	55		40		24	23
Detection	D1		D1		D1	D1
Reference	1		1		2	2
Oligosaccharides	**r**[a]	**Glc**[b]	**r**[a]	**Glc**[b]	**r**[c]	**Man**[d]
Oligomannose type						
GlcNAcβ1-4GlcNAc	—	—	—	—	0.50	0
Manβ1-4GlcNAcβ1-4GlcNAc	1.52	6.6	n.a.[e]	2.0	0.75	0.9
Manα1-3Manβ1-4GlcNAcβ1-4GlcNAc	—	—	—	—	0.74	1.8
Manα1-6Manβ1-4GlcNAcβ1-4GlcNAc	1.77	7.4	n.a.[e]	3.3	0.95	1.8
Manα1-6(Manα1-3)Manβ1-4GlcNAcβ1-4GlcNAc	1.77	7.4	1.07	4.3	0.97	2.8
Manα1-2Manα1-3Manβ1-4GlcNAcβ1-4GlcNAc	—	—	—	—	0.67	3.0
Manα1-3Manα1-6(Manα1-3)Manβ1-4GlcNAcβ1-4GlcNAc	—	—	—	—	1.00	3.8
Manα1-2Manα1-2Manα1-3Manβ1-4GlcNAcβ1-4GlcNAc	—	—	—	—	0.62	4.0
Manα1-3(Manα1-6)Manα1-6(Manα1-3)Manβ1-4GlcNAcβ1-4GlcNAc	1.70	7.2	1.63	6.2	1.00	5.0
Manα1-3(Manα1-2Manα1-6)Manα1-6(Manα1-3)Manβ1-4GlcNAcβ1-4GlcNAc	—	—	—	—	0.81	6.0
Manα1-3(Manα1-6)Manα1-6(Manα1-2Manα1-3)Manβ1-4GlcNAcβ1-4GlcNAc	1.39	6.1	1.98	7.1	0.86	6.0
Manα1-2Manα1-3(Manα1-6)Manα1-6(Manα1-3)Manβ1-4GlcNAcβ1-4GlcNAc	—	—	—	—	1.12	6.0
Manα1-3(Manα1-2Manα1-6)Manα1-6(Manα1-2Manα1-3)Manβ1-4GlcNAcβ1-4GlcNAc	1.20	5.1	2.37	8.1	0.67	7.0
Manα1-3(Manα1-6)Manα1-6(Manα1-2Manα1-2Manα1-3)Manβ1-4GlcNAcβ1-4GlcNAc	1.32	5.8	2.33	8.0	0.83	7.0
Manα1-2Manα1-3(Manα1-2Manα1-6)Manα1-6(Manα1-3)Manβ1-4GlcNAcβ1-4GlcNAc	—	—	—	—	0.94	7.0
Manα1-2Manα1-3(Manα1-6)Manα1-6(Manα1-2Manα1-3)Manβ1-4GlcNAcβ1-4GlcNAc	—	—	—	—	1.02	7.0
Manα1-3(Manα1-2Manα1-6)Manα1-6(Manα1-2Manα1-2Manα1-3)Manβ1-4GlcNAcβ1-4GlcNAc	1.16	4.9	2.71	9.0	0.67	8.0
Manα1-2Manα1-3(Manα1-2Manα1-6)Manα1-6(Manα1-2Manα1-3)Manβ1-4GlcNAcβ1-4GlcNAc	—	—	—	—	0.81	8.0
Manα1-2Manα1-3(Manα1-6)Manα1-6(Manα1-2Manα1-2Manα1-3)Manβ1-4GlcNAcβ1-4GlcNAc	—	—	—	—	0.95	8.0
Manα1-2Manα1-3(Manα1-2Manα1-6)Manα1-6(Manα1-2Manα1-2Manα1-3)Manβ1-4GlcNAc-β1-4GlcNAc	1.44	6.3	3.28	10.3	0.75	9.0
Hybrid type						
GlcNAcβ1-2Manα1-3(GlcNAcβ1-4)(Manα1-3Manα1-6)Manβ1-4GlcNAcβ1-4GlcNAc	2.75	9.5	1.52	5.9	—	—
GlcNAcβ1-2(GlcNAcβ1-4)Manα1-3(GlcNAcβ1-4)(Manα1-3Manα1-6)Manβ1-4GlcNAcβ1-4GlcNAc	5.23	13.1	1.63	6.2	—	—

TABLE LC 62 (continued)
HPLC Data Used to Establish Two-Dimensional Oligosaccharide Maps in Structural Studies of Glycoproteins

Oligosaccharides	r[a]	Glc[b]	r[a]	Glc[b]	r[c]	Man[d]
GlcNAcβ1-2(Galβ1-4GlcNAcβ1-4)Manα1-3(GlcNAcβ1-4)(Manα1-3Manα1-6)Manβ1-4GlcNAc-β1-4GlcNAc	5.45	13.6	2.02	7.2	—	—
GlcNAcβ1-2Manα1-3(GlcNAcβ1-4)[Manα1-3(Manα1-6)Manα1-6]Manβ1-4GlcNAcβ1-4GlcNAc	2.32	8.7	1.90	6.9	1.20	6.2
GlcNAcβ1-2(GlcNAcβ1-4)Manα1-3(GlcNAcβ1-4)[Manα1-3(Manα1-6)Manα1-6]Manβ1-4GlcNAc-β1-4GlcNAc	4.37	11.9	2.02	7.2	—	—
GlcNAcβ1-2(Galβ1-4GlcNAcβ1-4)Manα1-3(GlcNAcβ1-4)[Manα1-3(Manα1-6)Manα1-6]Manβ1-4-GlcNAcβ1-4GlcNAc	4.74	12.4	2.40	8.2	—	—
N-Acetyllactosamine type						
GlcNAcβ1-2Manα1-6(Manα1-3)Manβ1-4GlcNAcβ1-4GlcNAc	2.75	9.5	1.17	4.7	1.27	3.7
GlcNAcβ1-2Manα1-3(Manα1-6)Manβ1-4GlcNAcβ1-4GlcNAc	1.89	7.7	1.17	4.7	0.93	3.8
Manα1-3(Manα1-6)Manβ1-4GlcNAcβ1-4(Fucα1-6)GlcNAc	3.17	10.2	1.17	4.7	—	—
GlcNAcβ1-2Manα1-6(Manα1-3)Manβ1-4GlcNAcβ1-4(Fucα1-6)GlcNAc	4.96	12.7	1.27	5.1	—	—
GlcNAcβ1-2Manα1-3(Manα1-6)Manβ1-4GlcNAcβ1-4(Fucα1-6)GlcNAc	3.17	10.2	1.27	5.1	—	—
Galβ1-4GlcNAcβ1-2Manα1-6(Manα1-3)Manβ1-4GlcNAcβ1-4GlcNAc	2.99	9.9	1.43	5.6	1.30	4.5
Galβ1-4GlcNAcβ1-2Manα1-3(Manα1-6)Manβ1-4GlcNAcβ1-4GlcNAc	2.00	8.0	1.43	5.7	1.00	4.7
Galβ1-4GlcNAcβ1-2Manα1-6(Manα1-3)Manβ1-4GlcNAcβ1-4(Fucα1-6)GlcNAc	5.45	13.6	1.59	6.1	—	—
Galβ1-4GlcNAcβ1-2Manα1-3(Manα1-6)Manβ1-4GlcNAcβ1-4(Fucα1-6)GlcNAc	3.57	10.9	1.67	6.3	—	—
GlcNAcβ1-2Manα1-3(GlcNAcβ1-4)(Manα1-6)Manβ1-4GlcNAcβ1-4GlcNAc	3.28	10.4	1.27	5.1	—	—
GlcNAcβ1-2Manα1-3(GlcNAcβ1-4)(Manα1-6)Manβ1-4GlcNAc(Fucα1-6)GlcNAc	5.79	14.3	1.40	5.5	—	—
Galβ1-4GlcNAcβ1-2Manα1-3(GlcNAcβ1-4)(Manα1-6)Manβ1-4GlcNAcβ1-4GlcNAc	3.63	11.0	1.49	5.8	—	—
Galβ1-4GlcNAcβ1-2Manα1-3(GlcNAcβ1-4)(Manα1-6)Manβ1-4GlcNAcβ1-4(Fucα1-6)GlcNAc	6.39	15.5	1.71	6.4	—	—
GlcNAcβ1-2Manα1-3(GlcNAcβ1-2Manα1-6)Manβ1-4GlcNAcβ1-4GlcNAc	2.40	8.9	1.27	5.1	1.20	4.5
GlcNAcβ1-2Manα1-3(GlcNAcβ1-2Manα1-6)Manβ1-4GlcNAcβ1-4(Fucα1-6)GlcNAc	4.67	12.3	1.40	5.5	—	—
Galβ1-4GlcNAcβ1-2Manα1-6(GlcNAcβ1-2Manα1-3)Manβ1-4GlcNAcβ1-4GlcNAc	2.69	9.4	1.52	5.9	—	—
Galβ1-4GlcNAcβ1-2Manα1-3(GlcNAcβ1-2Manα1-6)Manβ1-4GlcNAcβ1-4GlcNAc	2.81	9.6	1.59	6.1	—	—
Galβ1-4GlcNAcβ1-2Manα1-6(GlcNAcβ1-2Manα1-3)Manβ1-4GlcNAcβ1-4(Fucα1-6)GlcNAc	5.23	13.1	1.67	6.3	—	—
Galβ1-4GlcNAcβ1-2Manα1-3(GlcNAcβ1-2Manα1-6)Manβ1-4GlcNAcβ1-4(Fucα1-6)GlcNAc	5.32	13.3	1.71	6.4	—	—

TABLE LC 62 (continued)
HPLC Data Used to Establish Two-Dimensional Oligosaccharide Maps in Structural Studies of Glycoproteins

Oligosaccharides	r[a]	Glc[b]	r[a]	Glc[b]	r[c]	Man[d]
Galβ1-4GlcNAcβ1-2Manα1-3(Galβ1-4GlcNAcβ1-2Manα1-6)Manβ1-4GlcNAcβ1-4GlcNAc	3.17	10.2	1.94	7.0	1.28	6.0
Galβ1-4GlcNAcβ1-2Manα1-3(Galβ1-4GlcNAcβ1-2Manα1-6)Manβ1-4GlcNAcβ1-4(Fucα1-6)-GlcNAc	5.69	14.1	2.10	7.4	—	—
Galβ1-4GlcNAcβ1-2Manα1-3(Galα1-3Galβ1-4GlcNAcβ1-2Manα1-6)Manβ1-4GlcNAcβ1-4GlcNAc	3.22	10.3	2.14	7.5	—	—
GlcNAcβ1-2Manα1-3(Galα1-3Galβ1-4GlcNAcβ1-2Manα1-6)Manβ1-4GlcNAcβ1-4(Fucα1-6)-GlcNAc	5.23	13.1	1.98	7.1	—	—
Galβ1-4GlcNAcβ1-2Manα-1-3(Galα1-3Galβ1-4GlcNAcβ1-2Manα1-6)Manβ1-4GlcNAcβ1-4-(Fucα1-6)GlcNAc	5.69	14.1	2.37	8.1	—	—
Galα1-3Galβ1-4GlcNAcβ1-2Manα1-3(Galα1-3Galβ1-4GlcNAcβ1-2Manα1-6)Manβ1-4GlcNAc-β1-4GlcNAc	3.40	10.6	2.56	8.6	—	—
Galα1-3Galβ1-4GlcNAcβ1-2Manα1-3(Galα1-3Galβ1-4GlcNAcβ1-2Manα1-6)Manβ1-4GlcNAc-β1-4(Fucα1-6)GlcNAc	5.84	14.4	2.71	9.0	—	—
Galβ1-4GlcNAcβ1-2(Galβ1-4GlcNAcβ1-6)Manα1-6(Manα1-3)Manβ1-4GlcNAcβ1-4(Fucα1-6)-GlcNAc	3.22	10.3	2.21	7.7	—	—
Galβ1-4GlcNAcβ1-2Manα1-3(Galβ1-4GlcNAcβ1-6Manα1-6)Manβ1-4GlcNAcβ1-4(Fucα1-6)-GlcNAc	4.12	11.6	2.13	7.5	—	—
Galβ1-4GlcNAcβ1-2(Galβ1-4GlcNAcβ1-4)Manα1-3(Manα1-6)Manβ1-4GlcNAcβ1-4(Fucα1-6)-GlcNAc	5.59	13.9	2.13	7.5	—	—
GlcNAcβ1-2Manα1-3(GlcNAcβ1-4)(GlcNAcβ1-2Manα1-6)Manβ1-4GlcNAcβ1-4GlcNAc	4.67	12.3	1.34	5.4	—	—
GlcNAcβ1-2Manα1-3(GlcNAcβ1-4)(GlcNAcβ1-2Manα1-6)Manβ1-4GlcNAcβ1-4(Fucα1-6)GlcNAc	7.55	17.8	1.49	5.8	—	—
GlcNAcβ1-2Manα1-3(GlcNAcβ1-4)(Galβ1-4GlcNAcβ1-2Manα1-6)Manβ1-4GlcNAcβ1-4GlcNAc	5.32	13.3	1.63	6.2	—	—
GlcNAcβ1-2Manα1-3(GlcNAcβ1-4)(Galβ1-4GlcNAcβ1-2Manα1-6)Manβ1-4GlcNAcβ1-4-(Fucα1-6)GlcNAc	7.96	18.7	1.78	6.6	—	—
Galβ1-4GlcNAcβ1-2Manα1-3(GlcNAcβ1-4)(GlcNAcβ1-2Manα1-6)Manβ1-4GlcNAcβ1-4GlcNAc	5.50	13.7	1.67	6.3	—	—
Galβ1-4GlcNAcβ1-2Manα1-3(GlcNAcβ1-4)(GlcNAcβ1-2Manα1-6)Manβ1-4GlcNAcβ1-4-(Fucα1-6)GlcNAc	8.05	18.9	1.82	6.7	—	—
Galβ1-4GlcNAcβ1-2Manα1-3(GlcNAcβ1-4)(Galβ1-4GlcNAcβ1-2Manα1-6)Manβ1-4GlcNAcβ1-4GlcNAc	5.89	14.5	2.02	7.2	—	—
Galβ1-4GlcNAcβ1-2Manα1-3(GlcNAcβ1-4)(Galβ1-4GlcNAcβ1-2Manα1-6)Manβ1-4GlcNAcβ1-4(Fucα1-6)GlcNAc	8.64	20.2	2.13	7.5	—	—
GlcNAcβ1-2(GlcNAcβ1-4)Manα1-3(GlcNAcβ1-2Manα1-6)Manβ1-4GlcNAcβ1-4GlcNAc	3.57	10.9	1.40	5.5	1.45	4.7

TABLE LC 62 (continued)
HPLC Data Used to Establish Two-Dimensional Oligosaccharide Maps in Structural Studies of Glycoproteins

Oligosaccharides	r[a]	Glc[b]	r[a]	Glc[b]	r[c]	Man[d]
GlcNAcβ1-2(GlcNAcβ1-4)Manα1-3(GlcNAcβ1-2Manα1-6)Manβ1-4GlcNAcβ1-4(Fucα1-6)GlcNAc	6.04	14.8	1.49	5.8	—	—
GlcNAcβ1-2Manα1-3[GlcNAcβ1-2(GlcNAcβ1-6)Manα1-6]Manβ1-4GlcNAcβ1-4GlcNAc	1.73	7.3	1.49	5.8	—	—
GlcNAcβ1-2Manα1-3[GlcNAcβ1-2(GlcNAcβ1-6)Manα1-6]Manβ1-4GlcNAcβ1-4(Fucα1-6)GlcNAc	3.22	10.3	1.71	6.4	—	—
GlcNAcβ1-2(Galβ1-3GlcNAcβ1-4)Manα1-3(GlcNAcβ1-2Manα1-6)Manβ1-4GlcNAcβ1-4GlcNAc	4.12	11.6	1.63	6.2	—	—
GlcNAcβ1-2(Galβ1-3GlcNAcβ1-4)Manα1-3(GlcNAcβ1-2Manα1-6)Manβ1-4GlcNAcβ1-4-(Fucα1-6)GlcNAc	6.39	15.5	1.90	6.9	—	—
GlcNAcβ1-2(GlcNAcβ1-4)Manα1-3(Galβ1-4GlcNAcβ1-2Manα1-6)Manβ1-4GlcNAcβ1-4-(Fucα1-6)GlcNAc	6.49	15.7	1.86	6.8	—	—
Galβ1-4GlcNAcβ1-2(GlcNAcβ1-4)Manα1-3(GlcNAcβ1-2Manα1-6)Manβ1-4GlcNAcβ1-4-(Fucα1-6)GlcNAc	6.96	16.6	1.90	6.9	—	—
Galβ1-4GlcNAcβ1-2(GlcNAcβ1-4)Manα1-3(Galβ1-4GlcNAcβ1-2Manα1-6)Manβ1-4GlcNAcβ1-4(Fucα1-6)GlcNAc	7.27	17.2	2.25	7.8	—	—
Triantennary oligosaccharide I[f]	2.05	8.1	2.56	8.6	—	—
Triantennary oligosaccharide Ia[f]	3.79	11.2	2.71	9.0	—	—
Triantennary oligosaccharide II[f]	4.96	12.7	2.44	8.3	1.70	7.4
Triantennary oligosaccharide III[f]	5.27	13.2	2.41	8.2	—	—
Triantennary oligosaccharide IIa[f]	7.46	17.7	2.60	8.7	—	—
Triantennary oligosaccharide IIIa[f]	7.64	18.0	2.56	8.6	—	—
Triantennary oligosaccharide IV[f]	5.03	12.8	2.71	9.0	1.62	8.5
GlcNAcβ1-2(GlcNAcβ1-4)Manα1-3[GlcNAcβ1-2(GlcNAcβ1-6)Manα1-6]Manβ1-4GlcNAcβ1-4-GlcNAc	2.45	9.0	1.67	6.3	1.18	5.5
GlcNAcβ1-2(GlcNAcβ1-4)Manα1-3[GlcNAcβ1-2(GlcNAcβ1-6)Manα1-6]Manβ1-4GlcNAcβ1-4-(Fucα1-6)GlcNAc	4.74	12.4	1.78	6.6	—	—
GlcNAcβ1-2(GlcNAcβ1-4)Manα1-3[Galβ1-4GlcNAcβ1-2(GlcNAcβ1-6)Manα1-6]Manβ1-4-GlcNAcβ1-4(Fucα1-6)GlcNAc	4.67	12.3	2.13	7.5	—	—
GlcNAcβ1-2(Galβ1-4GlcNAcβ1-4)Manα1-3[GlcNAcβ1-2(GlcNAcβ1-6)Manα1-6]Manβ1-4GlcNAcβ1-4(Fucα1-6)GlcNAc	5.03	12.8	2.13	7.5	—	—
GlcNAcβ1-2(GlcNAcβ1-4)Manα1-3[GlcNAcβ1-2(Galβ1-4GlcNAcβ1-6)Manα1-6]Manβ1-4GlcNAc-β1-4(Fucα1-6)GlcNAc	5.03	12.8	2.13	7.5	—	—
Galβ1-4GlcNAcβ1-2(GlcNAcβ1-4)Manα1-3[GlcNAcβ1-2(GlcNAcβ1-6)Manα1-6]Manβ1-4GlcNAc-β1-4(Fucα1-6)GlcNAc	5.55	13.8	2.17	7.6	—	—
Galβ1-4GlcNAcβ1-2(GlcNAcβ1-4)Manα1-3[Galβ1-4GlcNAcβ1-2(GlcNAcβ1-6)Manα1-6]-Manβ1-4GlcNAcβ1-4(Fucα1-6)GlcNAc	5.45	13.6	2.48	8.4	—	—
Galβ1-4GlcNAcβ1-2(GlcNAcβ1-4)Manα1-3[GlcNAcβ1-2(Galβ1-4GlcNAcβ1-6)Manα1-6]-Manβ1-4GlcNAcβ1-4(Fucα1-6)GlcNAc	5.69	14.1	2.52	8.5	—	—

TABLE LC 62 (continued)
HPLC Data Used to Establish Two-Dimensional Oligosaccharide Maps in Structural Studies of Glycoproteins

Oligosaccharides	r[a]	Glc[b]	r[a]	Glc[b]	r[c]	Man[d]
Galβ1-4GlcNAcβ1-2(Galβ1-4GlcNAcβ1-4)Manα1-3[GlcNAcβ1-2(GlcNAcβ1-6)Manα1-6]-Manβ1-4GlcNAcβ1-4(Fucα1-6)GlcNAc	5.84	14.4	2.52	8.5	—	—
GlcNAcβ1-2(GlcNAcβ1-4)Manα1-3[Galβ1-4GlcNAcβ1-2(Galβ1-4GlcNAcβ1-6)Manα1-6]-Manβ1-4GlcNAcβ1-4(Fucα1-6)GlcNAc	5.03	12.8	2.48	8.4	—	—
Galβ1-4GlcNAcβ1-2(Galβ1-4GlcNAcβ1-4)Manα1-3[Galβ1-4GlcNAcβ1-2(GlcNAcβ1-6)Manα1-6]Manβ1-4GlcNAcβ1-4(Fucα1-6)GlcNAc	5.69	14.1	2.93	9.5	—	—
Galβ1-4GlcNAcβ1-2(GlcNAcβ1-4)Manα1-3[Galβ1-4GlcNAcβ1-2(Galβ1-4GlcNAcβ1-6)-Manα1-6]Manβ1-4GlcNAcβ1-4(Fucα1-6)GlcNAc	5.55	13.8	2.86	9.3	—	—
Tetraantennary oligosaccharide V[f]	3.40	10.6	3.11	9.9	1.33	8.8
Tetraantennary oligosaccharide Va[f]	5.74	14.2	3.28	10.3	—	—
Tetraantennary oligosaccharide VI[f]	3.45	10.7	3.02	9.7	—	—
Tetraantennary oligosaccharide VIa[f]	5.84	14.4	3.27	10.2	—	—
Tetraantennary oligosaccharide VII[f]	3.34	10.5	3.37	10.5	1.25	9.8
GlcNAcβ1-2(Galβ1-3GlcNAcβ1-4)Manα1-3[GlcNAcβ1-2(GlcNAcβ1-6)Manα1-6]Manβ1-4-GlcNAcβ1-4GlcNAc	3.09	10.1	1.94	7.0	—	—
GlcNAcβ1-2(Galβ1-3GlcNAcβ1-4)Manα1-3[GlcNAcβ1-2(GlcNAcβ1-6)Manα1-6]Manβ1-4-GlcNAcβ1-4(Fucα1-6)GlcNAc	5.32	13.3	2.10	7.4	—	—
GlcNAcβ1-2(GlcNAcβ1-4)Manα1-3[GlcNAcβ1-3Galβ1-4GlcNAcβ1-2(GlcNAcβ1-6)Manα1-6]-Manβ1-4GlcNAcβ1-4GlcNAc	2.99	9.9	2.02	7.2	—	—
GlcNAcβ1-2(GlcNAcβ1-4)Manα1-3[GlcNAcβ1-3Galβ1-4GlcNAcβ1-2(GlcNAcβ1-6)Manα1-6]-Manβ1-4GlcNAcβ1-4(Fucα1-6)GlcNAc	5.50	13.7	2.21	7.7	—	—
GlcNAcβ1-3Galβ1-4GlcNAcβ1-2(GlcNAcβ1-4)Manα1-3[GlcNAcβ1-2(GlcNAcβ1-6)Manα1-6]-Manβ1-4GlcNAcβ1-4GlcNAc	3.34	10.5	2.02	7.2	—	—
GlcNAcβ1-3Galβ1-4GlcNAcβ1-2(GlcNAcβ1-4)Manα1-3[GlcNAcβ1-2(GlcNAcβ1-6)Manα1-6]-Manβ1-4GlcNAcβ1-4(Fucα1-6)GlcNAc	5.94	14.6	2.25	7.8	—	—
GlcNAcβ1-2(GlcNAcβ1-4)Manα1-3[GlcNAcβ1-3Galβ1-4GlcNAcβ1-2(GlcNAcβ1-3Galβ1-4-GlcNAcβ1-6)Manα1-6]Manβ1-4GlcNAcβ1-4GlcNAc	4.12	11.6	2.63	8.8	—	—
GlcNAcβ1-2(GlcNAcβ1-4)Manα1-3[GlcNAcβ1-3Galβ1-4GlcNAcβ1-2(GlcNAcβ1-3Galβ1-4-GlcNAcβ1-6)Manα1-6]Manβ1-4GlcNAcβ1-4(Fucα1-6)GlcNAc	6.54	15.8	2.71	9.0	—	—
Tetraantennary oligosaccharide VIII[f]	4.37	11.9	3.67	11.2	—	—
Tetraantennary oligosaccharide VIIIa[f]	6.75	16.2	3.84	11.6	—	—
Tetraantennary oligosaccharide IX[f]	4.60	12.2	3.63	11.1	—	—
Tetraantennary oligosaccharide IXa[f]	7.02	16.7	3.84	11.6	—	—
Tetraantennary oligosaccharide X[f]	5.41	13.5	4.16	12.4	—	—
Tetraantennary oligosaccharide Xa[f]	7.78	18.3	4.25	12.6	—	—
Tetraantennary oligosaccharide XI[f]	5.64	14.0	3.06	9.8	—	—
Tetraantennary oligosaccharide XIa[f]	8.01	18.8	3.15	10.0	—	—
Tetraantennary oligosaccharide XII[f]	6.49	15.7	4.59	13.6	—	—
Tetraantennary oligosaccharide XIIa[f]	8.84	20.8	4.65	13.8	—	—

TABLE LC 62 (continued)
HPLC Data Used to Establish Two-Dimensional Oligosaccharide Maps in Structural Studies of Glycoproteins

Oligosaccharides	r[a]	Glc[b]	r[a]	Glc[b]	r[c]	Man[d]
Small fucosylated oligosaccharides						
GlcNAcβ1-4(Fucα1-6)GlcNAc	—	—	—	—	0.84	0.6
GlcNAcβ1-4(Fucα1-3)GlcNAc	—	—	—	—	0.28	0.9
Manβ1-4GlcNAcβ1-4(Fucα1-3)GlcNAc	—	—	—	—	0.48	2.0
Manα1-6Manβ1-4GlcNAcβ1-4(Fucα1-3)GlcNAc	—	—	—	—	0.62	2.8
Manα1-3(Manα1-6)Manβ1-4GlcNAcβ1-4(Fucα1-3)GlcNAc	—	—	—	—	0.64	3.8
Xylose-containing oligosaccharides						
Xylβ1-2Manβ1-4GlcNAcβ1-4GlcNAc	—	—	—	—	0.97	1.3
Xylβ1-2Manβ1-4GlcNAcβ1-4(Fucα1-3)GlcNAc	—	—	—	—	0.62	2.2
Xylβ1-2(Manα1-6)Manβ1-4GlcNAcβ1-4GlcNAc	—	—	—	—	1.12	2.7
Xylβ1-2(Manα1-6)Manβ1-4GlcNAcβ1-4(Fucα1-3)GlcNAc	1.73	7.3	1.10	4.4	0.80	3.5
Xylβ1-2(Manα1-3)(Manα1-6)Manβ1-4GlcNAcβ1-4GlcNAc	—	—	—	—	0.95	3.7
Xylβ1-2(Manα1-3)(Manα1-6)Manβ1-4GlcNAcβ1-4(Fucα1-3)GlcNAc	1.29	5.6	1.46	5.7	0.62	4.8
Xylβ1-2(Manα1-3)(Manα1-3Manα1-6)Manβ1-4GlcNAcβ1-4GlcNAc	—	—	—	—	1.00	4.3
Xylβ1-2(Manα1-3)(GlcNAcβ1-2Manα1-6)Manβ1-4GlcNAcβ1-4(Fucα1-3)GlcNAc	1.32	5.8	1.63	6.2	—	—
Xylβ1-2(GlcNAcβ1-2Manα1-3)(Manα1-6)Manβ1-4GlcNAcβ1-4(Fucα1-3)GlcNAc	1.66	7.1	1.63	6.2	—	—
Xylβ1-2(GlcNAcβ1-2Manα1-3)(GlcNAcβ1-2Manα1-6)Manβ1-4GlcNAcβ1-4(Fucα1-3)GlcNAc	1.62	7.0	1.75	6.5	—	—
Xylβ1-2(GlcNAcβ1-2Manα1-3)[Galβ1-4(Fucα1-6)GlcNAcβ1-2Manα1-6]Manβ1-4GlcNAcβ1-4-(Fucα1-3)GlcNAc	1.70	7.2	2.44	8.3	—	—
Xylβ1-2[Galβ1-4(Fucα1-6)GlcNAcβ1-2Manα1-3][Galβ1-4(Fucα1-6)GlcNAcβ1-2Manα1-6]-Manβ1-4GlcNAcβ1-4(Fucα1-3)GlcNAc	1.89	7.7	3.28	10.3	—	—

[a] t_r relative to pyridylaminated isomaltotetraose (6.5 min on P1, 8.5 min on P2); columns calibrated with pyridylaminated hydrolysate of dextran.
[b] Equivalent DP in glucose units, from comparison of t_r for oligosaccharide with calibration curve given by dextran hydrolysate.
[c] t_r relative to pyridylaminated Manα1-3(Manα1-6)Manα1-6(Manα1-3)Manβ1-4GlcNAcβ1-4GlcNAc (24 min).
[d] Retention data on P4 expressed in terms of molecular size as number of D-mannosyl residues attached to $(GlcNAc)_2$ core; actual retention times not given.
[e] Retention times for D-glucose oligomers of DP below 4 not given in Reference 1.
[f] Structure appended.

Packing P1 = Shimpack® CLC-ODS (Shimadzu, Japan); C_{18} phase bonded to microparticulate silica.
P2 = TSK®-gel Amide 80 (Toyo Soda, Japan); carbamoyl groups bonded to microparticulate silica.
P3 = Cosmosil® 5C_{18}-P (Nacalai Tesque, Kyoto, Japan); C_{18} phase bonded to 5-μm silica.
P4 = MicroPak® AX-5 (Varian); amine phase bonded to 5-μm silica.

Solvent S1 = (*A*) 10 m*M* sodium phosphate buffer, pH 3.8;
(*B*) *A* containing 1-butanol (0.5%); column equilibrated with *A*-*B* (4:1); after injection linear gradient, 20 → 50% *B* in 60 min.

Solvent S2 = (*C*) acetonitrile-3% acetic acid in water containing triethylamine, pH 7.3 (13:7);
(*D*) acetonitrile-3% acetic acid in water containing triethylamine, pH 7.3 (1:1); column equilibrated with *C*; after injection linear gradient, 100% *C* → 100% *D* in 50 min.

TABLE LC 62 (continued)
HPLC Data Used to Establish Two-Dimensional Oligosaccharide Maps in Structural Studies of Glycoproteins

S3 = (*E*) 0.1 *M* ammonium acetate buffer, pH 4.0;
(*F*) *E* containing 1-butanol (0.5%); column equilibrated with *E-F* (19:1); after injection linear gradient, 5 → 100% *F* in 55 min.

S4 = (*G*) acetonitrile-3% acetic acid in water containing triethylamine, pH 7.3 (4:1);
(*H*) 3% aqueous acetic acid containing triethylamine, pH 7.3; column equilibrated with *G*; after injection linear gradient, 0 → 20% *H* in 2 min, then 20 → 45% *H* in 18 min.

Detection D1 = fluorometric detection of pyridylamino-oligosaccharides; excitation wavelength 320 nm, emission 400 nm.

REFERENCES

1. **Tomiya, N., Awaya, J., Kurono, M., Endo, S., Arata, Y., and Takahashi, N.,** Analyses of N-linked oligosaccharides using a two-dimensional mapping technique, *Anal. Biochem.*, 171, 73, 1988.
2. **Hase, S., Ikenaka, K., Mikoshiba, K., and Ikenaka, T.,** Analyses of tissue glycoprotein sugar chains by two-dimensional high-performance liquid chromatographic mapping, *J. Chromatogr.*, 434, 51, 1988.

APPENDIX TO TABLE LC 62: STRUCTURES OF NUMBERED OLIGOSACCHARIDES

I

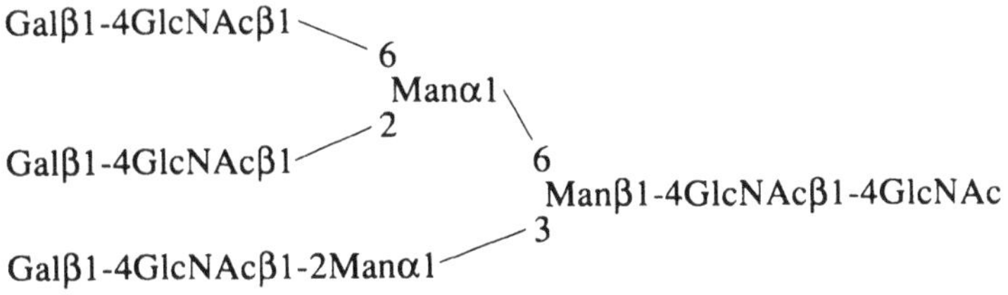

Ia As I but

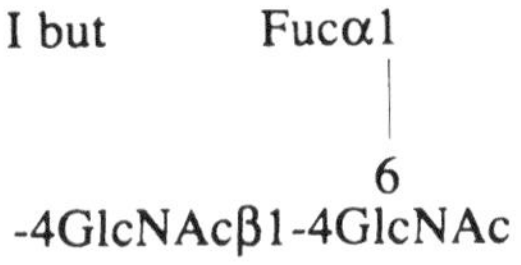

II

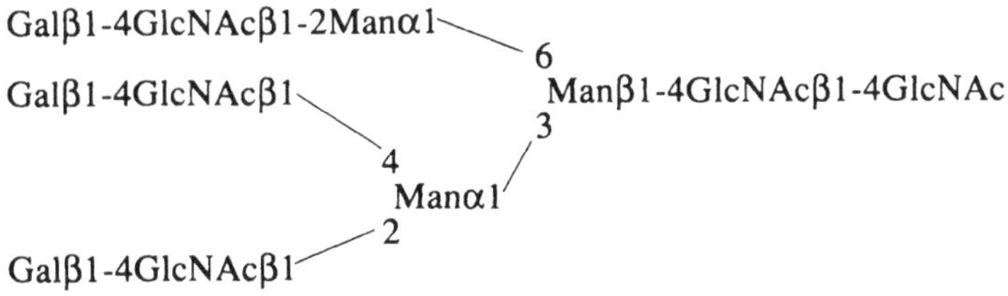

IIa As II but

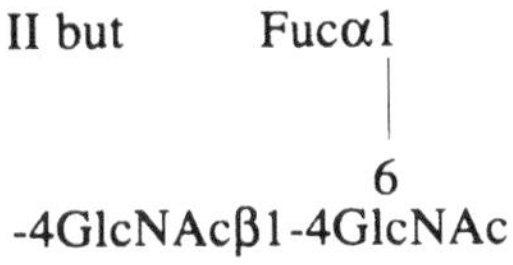

III

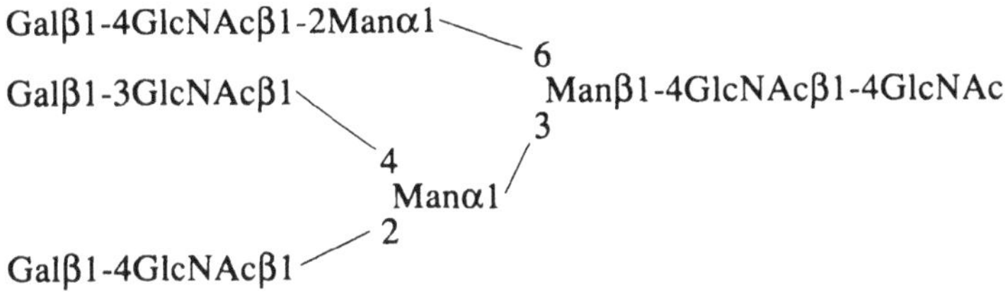

IIIa As III but

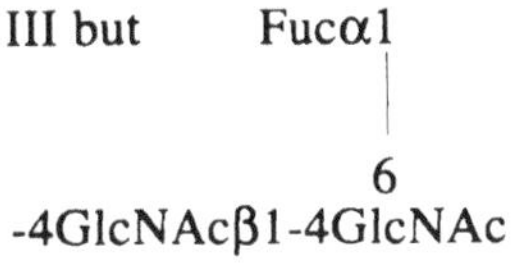

APPENDIX TO TABLE LC 62: STRUCTURES OF NUMBERED OLIGOSACCHARIDES (continued)

IV

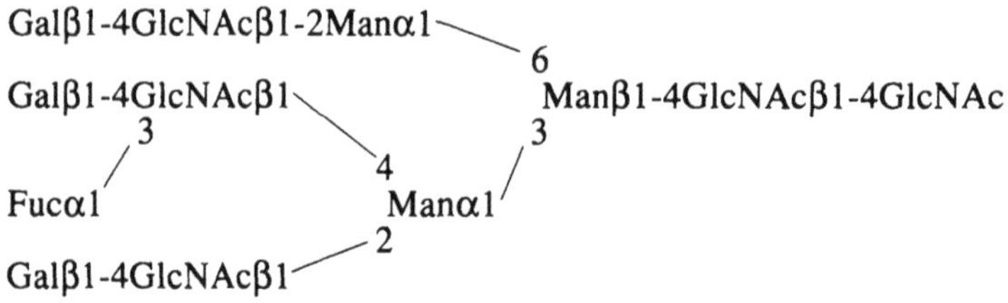

V

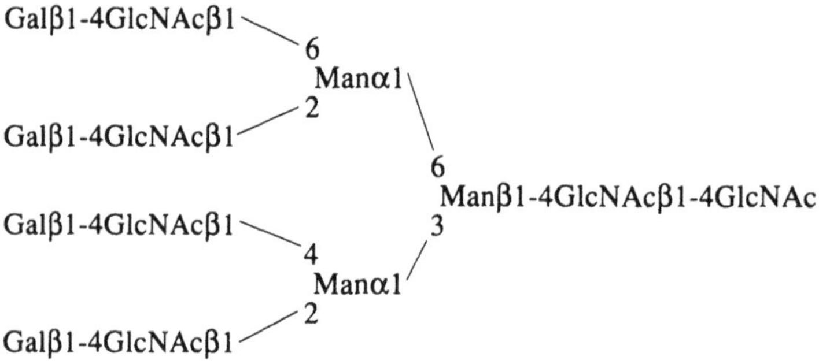

Va As V but

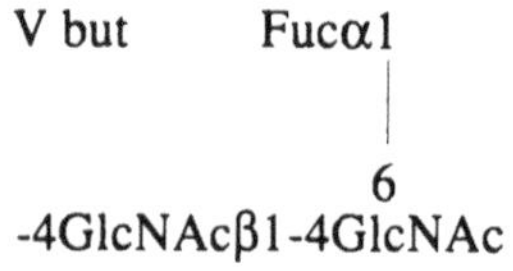

VI

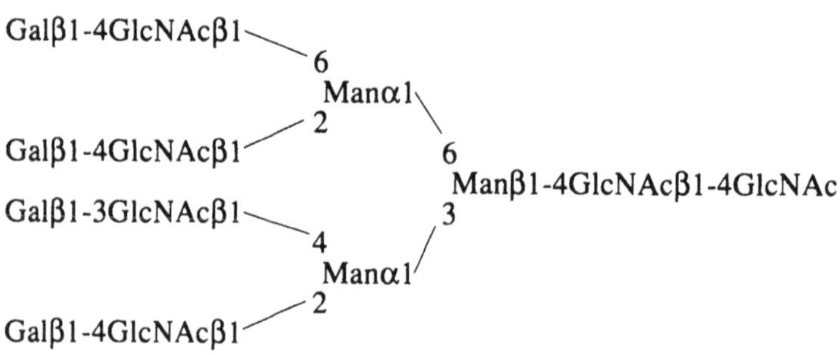

VIa As VI but

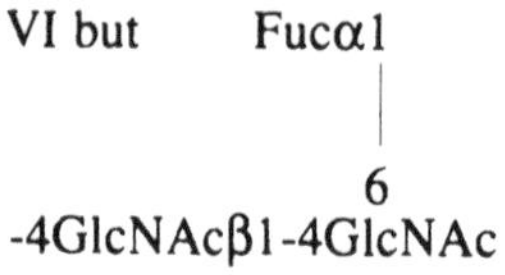

APPENDIX TO TABLE LC 62: STRUCTURES OF NUMBERED OLIGOSACCHARIDES (continued)

VII

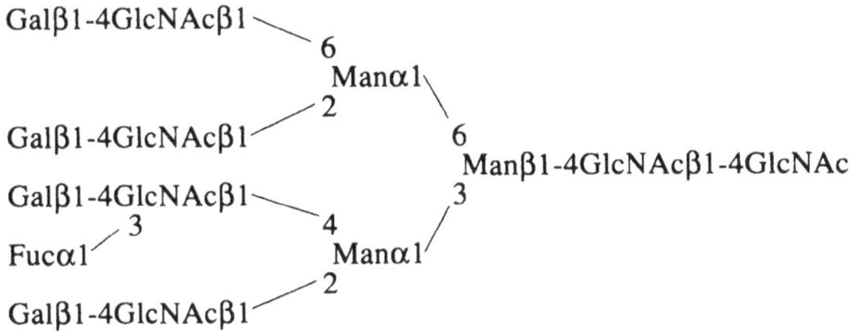

VIII

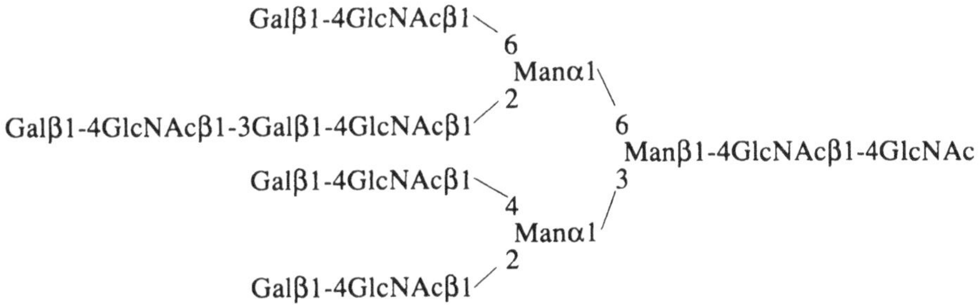

VIIIa As VIII but

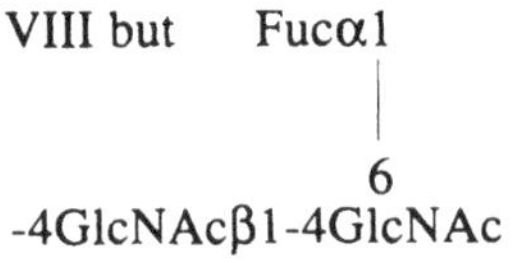

IX

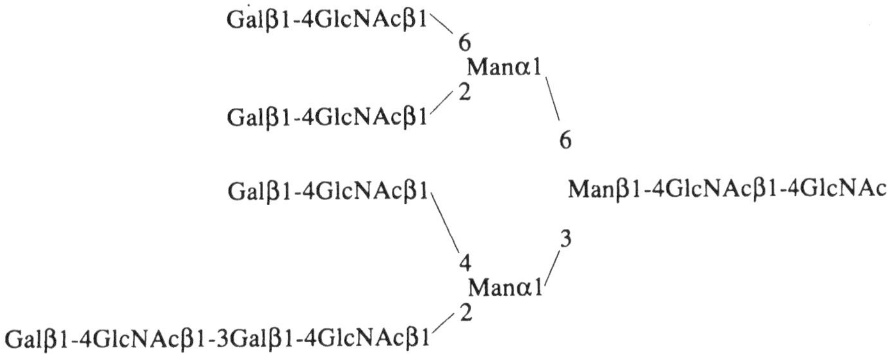

IXa As IX but

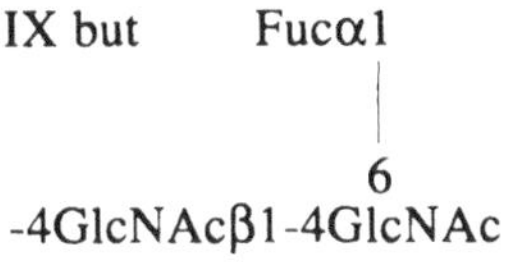

APPENDIX TO TABLE LC 62: STRUCTURES OF NUMBERED OLIGOSACCHARIDES (continued)

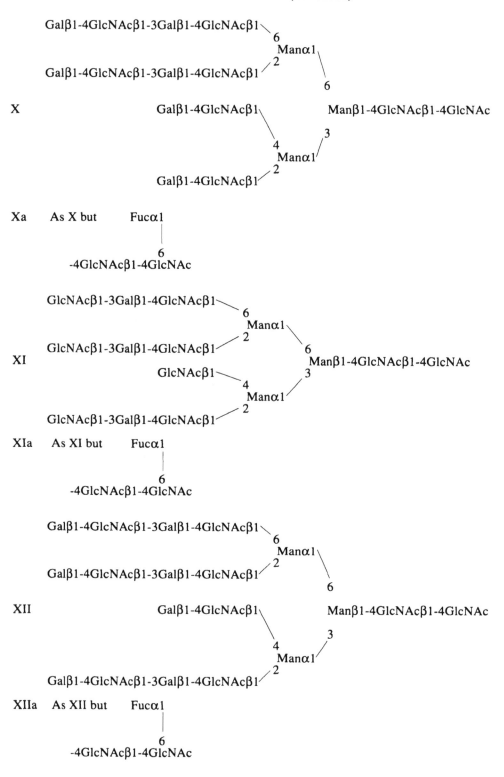

```
     Galβ1-4GlcNAcβ1-3Galβ1-4GlcNAcβ1\
                                      6
                                      Manα1\
                                      2     \
     Galβ1-4GlcNAcβ1-3Galβ1-4GlcNAcβ1/       \
                                              6
X                    Galβ1-4GlcNAcβ1\          Manβ1-4GlcNAcβ1-4GlcNAc
                                     \        3
                                      4      /
                                      Manα1/
                                      2
                     Galβ1-4GlcNAcβ1/

Xa   As X but      Fucα1
                     |
                     6
       -4GlcNAcβ1-4GlcNAc

     GlcNAcβ1-3Galβ1-4GlcNAcβ1\
                               6
                               Manα1\
                               2     \
     GlcNAcβ1-3Galβ1-4GlcNAcβ1/       6
XI                                     Manβ1-4GlcNAcβ1-4GlcNAc
                     GlcNAcβ1\        3
                               4     /
                               Manα1/
                               2
     GlcNAcβ1-3Galβ1-4GlcNAcβ1/

XIa  As XI but     Fucα1
                     |
                     6
       -4GlcNAcβ1-4GlcNAc

     Galβ1-4GlcNAcβ1-3Galβ1-4GlcNAcβ1\
                                      6
                                      Manα1\
                                      2     \
     Galβ1-4GlcNAcβ1-3Galβ1-4GlcNAcβ1/       \
                                              6
XII                  Galβ1-4GlcNAcβ1\          Manβ1-4GlcNAcβ1-4GlcNAc
                                     \        3
                                      4      /
                                      Manα1/
                                      2
     Galβ1-4GlcNAcβ1-3Galβ1-4GlcNAcβ1/

XIIa As XII but    Fucα1
                     |
                     6
       -4GlcNAcβ1-4GlcNAc
```

OTHER NOVEL APPROACHES TO ANALYSIS OF CARBOHYDRATES BY LIQUID CHROMATOGRAPHY

A. ION PAIR CHROMATOGRAPHY ON REVERSED-PHASE COLUMNS

In analysis of acidic carbohydrates, an alternative to the use of ion-exchange columns (see Table LC 38) is afforded by the introduction of ion-pairing reagents into the eluent, which permits analysis of the resulting neutral complexes on reversed-phase packings. An effective organic counter-ion for this purpose is the tetrabutylammonium ion; alkylammonium cations with shorter alkyl chains do not confer sufficient hydrophobic character on the complex formed with the anionic carbohydrate molecules to promote resolution on reversed-phase columns.[1]

Application of this method to HPLC of sugar phosphates gave the first reported resolution of D-glucose- and D-fructose-6-phosphate, on a 25 × 0.46 cm stainless steel column packed with 5-μm Spherisorb® ODS-2 (Phase Separations, Norwalk, CT), eluted at 38°C (flow rate 2ml/min) with an aqueous solution containing 0.02 *M* formic acid and 8.5 m*M* tetrabutylammonium hydroxide.[1] Addition of a europium complex, tris (2,2,6,6-tetramethyl-3,5-heptanedionato)europium (III), in trace amounts (2×10^{-5} *M*) to the eluent resulted in the formation of an adduct between this complex and the sugar phosphates which absorbed in the ultraviolet (maximum at 240 nm), making possible the use of UV detection in HPLC of the sugar phosphates, with a detection limit of 0.2 μg. At this low concentration the europium complex did not affect the retention of the solutes on the reversed-phase column. Under these conditions the 6-phosphates of D-glucose and D-fructose eluted with capacity factors of approximately 4.0 and 4.9, respectively. Resolution of the isomeric 1,6- and 2,6-diphosphates of D-fructose with this system was not successful, but at higher ionic strength and lower pH, achieved by the addition of hydrochloric acid (0.02 *M*) and sodium chloride (0.04 *M*) to the eluent, and a higher concentration of tetrabutylammonium hydroxide (0.03 *M*), the two diphosphates were partially resolved. Resolution of both mono- and diphosphates in a single analysis may be possible using gradient elution.

Ion-pair chromatography on reversed-phase columns has been applied successfully to HPLC analyses of the normal and unsaturated oligogalacturonic acids produced by degradation of pectic acid with endo-polygalacturonase or endo-pectic acid lyase, respectively,[2] and of the products of digestion of a 1 → 4-β-D-mannuronan from alginate by the alginate lyase isolated from tissue of *Sargassum fluitans*.[3] The method has also been used in the quantification of hyaluronic acid in biological tissues and fluids, the tetra- and hexasaccharide, each carrying 4,5-unsaturated D-glucuronosyl residues at their nonreducing ends, produced by degradation of hyaluronic acid with *Streptomyces* hyaluronidase being resolved completely and quantitatively.[4] The chromatographic conditions used in these analyses are summarized in the table that follows. For the unsaturated oligomers resulting from lyase degradation of the uronic acid-containing polysaccharides, UV detection (at 230 to 235 nm) can be used. The addition of the tetrabutylammonium ion to the eluent does not interfere with detection by this sensitive method.

Solutes	Column	Eluent	Temperature (°C)	Flow rate (ml/min)
Oligogalacturonic acids (normal), DP 2—4	LiChrosorb® 10 RP-18, 25 × 0.46 cm (Merck)	Methanol-0.05 *M* phosphate buffer, pH 7.0 (1:9) containing 25 m*M* tetrabutylammonium bromide	40	1
Oligogalacturonic acids (unsaturated), DP 2—7	As above	As above but methanol:buffer ratio 3:7	40	1

Solutes	Column	Eluent	Temperature (°C)	Flow rate (ml/min)
Unsaturated D-mannuronans, DP 2—7	C_{18} μ-Bondapak® cartridge, 10 × 0.8 cm (Waters), used under radial compression (Waters Z-module)	Acetonitrile-0.1 *M* phosphate buffer, pH 6.5 (1:9), containing 10 m*M* tetrabutylammonium hydroxide	Ambient	1
Unsaturated tetra- and hexasaccharide from *Streptomyces* hyaluronidase digestion of hyaluronic acid	Ultrasphere® ODS (5 μm), 25 × 0.46 cm (Beckman)	Gradient elution (18 min), 20 → 22% acetonitrile in 8 → 7.9 m*M* orthophosphoric acid, pH 7.5, containing 10 m*M* tetrabutylammonium hydroxide	Ambient	n.a.

B. HYDROPHOBIC-INTERACTION CHROMATOGRAPHY OF CARBOHYDRATES AND GLYCOCONJUGATES

Hydrophobic-interaction chromatography *per se* is applicable only to glycoconjugates in which the noncarbohydrate moiety is capable of binding to the hydrophobic groups attached to the column packing. A notable example is found in lignin-carbohydrate complexes: those isolated from milled wood of *Pinus densiflora* and fractionated by SEC (Sepharose® 4B) were further fractionated by chromatography on hydrophobic packings Phenyl- and Octyl-Sepharose® CL-4B (Pharmacia) into several components, differing in lignin content.[5] This was achieved by application of the complexes to the columns in 25 m*M* sodium phosphate buffer, pH 6.8, containing 0.8 *M* ammonium sulfate (under which conditions 90% of the material was adsorbed on Phenyl-Sepharose® CL-4B, 80% on Octyl-Sepharose® CL-4B), and desorption by stepwise elution with the same buffer containing (*A*) 0.6 *M* ammonium sulfate and 15% of 2-ethoxyethanol, (*B*) 0.4 *M* ammonium sulfate and 30% of 2-ethoxyethanol, (*C*) 0.2 *M* ammonium sulfate and 45% of 2-ethoxyethanol, and (*D*) 50% 2-ethoxyethanol. The lignin content of the 5 fractions thus obtained was found to vary from 8—14% (unadsorbed fraction) to 50—60% (fraction eluted with solvent *D*). Although no marked difference in the neutral sugar compositions of the fractions was observed, the arabinose content of the unadsorbed fraction was highest in all cases, and the proportion of either glucose or galactose increased concomitantly from the unadsorbed fraction to that of highest lignin content, suggesting that any covalent bonding between lignin and carbohydrate was likely to involve these hexosyl residues.

That adsorption was due to hydrophobic interaction rather than hydrogen bonding was indicated by the finding that neither urea nor guanidine hydrochloride (both good splitting agents for hydrogen bonds) desorbs the lignin-carbohydate complexes to the same extent as does 2-ethoxyethanol. Another solvent producing desorption of the complexes is 1,4-dioxane, which could possibly replace 2-ethoxyethanol in the desorbing eluents. The stronger adsorption by Phenyl-Sepharose® CL-4B is believed to be due to the participation of interactions between π electrons in the phenyl groups and the lignin moieties of the complexes, over and above the hydrophobic interaction. The carbohydrate components of the complexes are clearly not involved in binding, and the method thus affords a powerful technique for purifying lignin-carbohydrate complexes from polysaccharides not linked to lignin, as well as for fractionation of the complexes on the basis of their lignin content.

The fact that carbohydrate as such has no effect on binding of glycoconjugates to supports for hydrophobic-interaction chromatography is further demonstrated by a study of the behavior of various glycoproteins on different forms of chromatography,[6] including chromatography on the hydrophobic packing TSK® Phenyl 5 PW (Toyo Soda), with a gradient of decreasing salt concentration (molarity of ammonium sulfate 1.7 → 0) in 0.1 *M* sodium phosphate buffer, pH 7.0. Despite the known microheterogeneity of some of these glycoproteins, due to variation in the number and length of the oligosaccharide side chains attached to the polypeptide core, the only one giving more than one peak on hydrophobic chromatography was β_2-glycoprotein I, a lipid-associated protein, which gave two peaks, corresponding to differences in lipid content. The method is possibly one of the most suitable for isolation and purification of glycoproteins if fractionation on the basis of microheterogeneity of the carbohydrate moiety is to be avoided.

Observed adsorption of heparin on Phenyl-Sepharose® CL-4B and even on Sepharose CL-4B itself in the presence of 3.8 *M* ammonium sulfate in 0.01 *M* hydrochloric acid, with subsequent desorption by stepwise elution with eluents containing progressively lower concentrations of ammonium sulfate (3.4 → 2.0 *M*) in the same medium, has been ascribed to hydrophobic interaction, involving the cross-linking structure $-O-CH_2-CH(OH)-CH_2-O-$ present in Sepharose® CL-4B (introduced by reaction of Sepharose 4B with epichlorohydrin). This seems to be refuted by later work, in which the chromatographic behavior of various glycosaminoglycuronans on Phenyl-Sepharose® CL-4B in the presence of high concentrations of ammonium sulfate has been correlated with their solubility in such salt solutions.[8] When fractions of chondroitin 4- or 6-sulfate and dermatan sulfate, matched in molecular weight distribution, were adsorbed on a column of Phenyl-Sepharose CL-4B equilibrated in 4 *M* ammonium sulfate in 10 m*M* hydrochloric acid (2.8 *M* ammonium sulfate in the case of dermatan sulfate) and eluted with a reverse linear gradient (4.0 → 2.0 *M* or 2.8 → 1.0 *M* ammonium sulfate in 10 mM HCl), chondroitin 6-sulfate fractions were retained more strongly than chondroitin 4-sulfate of similar $\overline{M}_w$, and dermatan sulfate was retained less strongly than chondroitin sulfates of similar size. Temperature had no effect on adsorption of dermatan sulfate, but the chondroitin sulfates were retained much more strongly at 4°C than at room temperature, as was found also with binding of heparin to Sepharose® CL-4B in the presence of 3.8 *M* ammonium sulfate.[7] Since hydrophobic interaction is generally strengthened with increasing temperature, this greater binding at low temperature indicates another mechanism. The observation[8] that the order of binding of dermatan sulfate and chondroitin 4- and 6-sulfates was the reverse of the order of solubility of the glycosaminoglycuronans in the eluents used, solubility decreasing with temperature for the chondroitin sulfates but not dermatan sulfate, suggested that retention of the glycosaminoglycuronans under these conditions is probably due to interfacial precipitation on the gel matrix arising from changes in the hydration of the molecules in the presence of high concentrations of ammonium sulfate.

Thus, chromatography of glycosaminoglycuronans on hydrophobic packings, or of heparin on Sepharose® CL-4B, in media containing ammonium sulfate in high concentration is not hydrophobic-interaction chromatography. Nevertheless, the method may well afford a valuable fractionation technique, since it has been found[7] that the heparin fraction retained most strongly by Sepharose® CL-4B, and eluted only when the concentration of ammonium sulfate was decreased to 2.5 *M*, was that of highest $\overline{M}_w$ and *N*-acetyl content, lowest sulfate content and, significantly, highest anticoagulant activity. This fraction comprised only 8% of the total heparin fractionated, and isolation by other means would be difficult.

C. APPLICATION OF HYDRODYNAMIC CHROMATOGRAPHY (FIELD FLOW FRACTIONATION) TO MOLECULAR SIZE DETERMINATION OF LARGE POLYSACCHARIDES

In recent years the potential of hydrodynamic chromatography, also known as field flow fractionation, as a means of estimating the molecular size of polymers so large that they are totally excluded from porous packings or gels of even the largest available pore sizes has been recognized. In contrast to SEC packings, those used in hydrodynamic chromatography are impenetrable solid spheres, the presence of which produces a nonuniform velocity field in the column through which eluent is flowing, the fluid in the center of the tube flowing faster than that near the walls. Solute particles are fractionated on the basis of size because steric, electrical, and hydrodynamic exclusion of larger particles from the wall region results in their being confined mainly to the faster-moving fluid in the center, while smaller particles, which can be present in either the slow-moving fluid near the wall or the faster central region, on average move more slowly through the column. Comparison of the retention time of a polymer with those of latex spheres of known diameter (0.05 to 1 μm) gives an estimate of the hydrodynamic volume of the polymer.

This method has been applied to xanthan,[9] which has $\overline{M}_w$ (>2 × 10^6) beyond the fractionation range of most SEC packings. Two coupled stainless steel columns, each 100 × 0.46 cm, one packed with spherical silica beads (DuPont Zipax®) having an average particle diameter of 27 μm, the other with fractionated glass beads, average diameter 30 μm were used. An aqueous solution consisting of 0.05 *M* sodium sulfate, 0.2% sodium dodecyl sulfate, and 2 m*M* sodium azide (bactericide), which produced maximal resolution for latex spheres, gave an elution volume for xanthan equal to that of a latex sphere of diameter 0.153 μm, from which the hydrodynamic volume of the xanthan molecule could be estimated. To avoid aggregation of the xanthan very low concentrations (ca. 70 ppm in 20 μl) were injected, and therefore a detection method more sensitive than RI or UV detection was required: the xanthan was labeled with 5-amino-fluorescein, which permitted the use of fluorometric detection (excitation wavelength 480 nm, emission above 500 nm). A small amount of fluorescein was added to the xanthan solution to act as a marker, and retention times were measured relative to that of this small molecule. That retention time was indeed a function of molecular size was demonstrated by the effect of sonication for various periods from 5 to 60 min, which resulted in some cleavage of the polysaccharide chains; as sonication progressed, the xanthan retention time moved closer to that of the fluorescein marker peak. The apparent molecular volume, 1.03 × 10^7 dl/g mol, given by this method is intermediate between that calculated if the rod-like xanthan molecule were freely rotating and the actual volume of the molecule based on a model of a prolate ellipsoid of minor-axis diameter 2 nm and contour length ca. 2 μm, and thus indicates considerable orientation of the molecule in the field of flow. These results were obtained at a flow rate of 1 ml/min; the effect of flow rate remains to be investigated, as does the potential application of the technique to polysaccharides in general.

REFERENCES

1. **Henderson, S. K. and Henderson, D. E.,** Reversed phase ion pair HPLC analysis of sugar phosphates, *J. Chromatogr. Sci.*, 24, 198, 1986.
2. **Voragen, A. G. J., Schols, H. A., De Vries, J. A., and Pilnik, W.,** High-performance liquid chromatographic analysis of uronic acids and oligogalacturonic acids, *J. Chromatogr.*, 244, 327, 1982.
3. **Romeo, T. and Preston, J. F.,** Liquid chromatographic analysis of the depolymerisation of (1 → 4)-β-D-mannuronan by an extracellular alginate lyase from a marine bacterium, *Carbohydr. Res.*, 153, 181, 1986.
4. **Chun, L. E., Koob, T. J., and Eyre, D. R.,** Quantitation of hyaluronic acid in tissues by ion-pair reverse-phase high-performance liquid chromatography of oligosaccharide cleavage products, *Anal. Biochem.*, 171, 197, 1988.
5. **Takahashi, N., Azuma, J.-I., and Koshijima, T.,** Fractionation of lignin-carbohydrate complexes by hydrophobic-interaction chromatography, *Carbohydr. Res.*, 107, 161, 1982.
6. **Putnam, F. W. and Takahashi, N.,** Structural characterisation of glycoproteins, *J. Chromatogr.*, 443, 267, 1988.
7. **Uchiyama, H., Fujimoto, N., Sakurai, K., and Nagasawa, K.,** Separation of heparin on Sepharose CL-4B in the presence of high concentrations of ammonium sulfate, *J. Chromatogr.*, 287, 55, 1984.
8. **Uchiyama, H., Okouchi, K., and Nakasawa, K.,** Chromatography of glycosaminoglycans on hydrophobic gel. Correlation between chromatographic behavior of glycosaminoglycans on Phenyl-Sepharose CL-4B and their solubility in the presence of high concentrations of ammonium sulfate, *Carbohydr. Res.*, 140 239, 1985.
9. **Prud'homme, R. K., Froiman, G., and Hoagland, D. A.,** Molecular size determination of xantha polysaccharide, *Carbohydr. Res.*, 106, 225, 1982.

AFFINITY CHROMATOGRAPHY

Affinity chromatography, the application of which to fractionation and purification of polysaccharides and glycoproteins was reviewed briefly in *CRC Handbook of Chromatography: Carbohydrates,* Volume I, has assumed great importance during the past decade, particularly in biochemistry and clinical chemistry. The ability of lectins to complex exclusively or preferably with particular sugars or oligosaccharide groupings has proved invaluable in such applications as isolation and purification of cell membrane glycoproteins, typing of lymphocytes and erythrocytes, characterization of neural glycopeptides and glycoproteins, and differentiation between normal and malignant cells.[1-4] Lectins were classified initially with respect to their binding specificity for simple sugars, sometimes with pronounced anomeric specificity, and tables were compiled on this basis.[1,5] Sugar-lectin complementarity was established mainly by the Landsteiner hapten-inhibition technique,[6] sugars being compared with respect to the minimal concentration required to inhibit either the precipitin reaction between the lectin and a reactive macromolecule or the hemagglutination reaction. The use of α- and β-glycosides (alkyl or *p*-nitrophenyl) as hapten inhibitors yielded information on anomeric specificity. However, in their comprehensive review published in 1978 Goldstein and Hayes[1] expressed the *caveat* that the true nature of the binding of lectins to specific glycoproteins would be interpreted correctly only by using these data in conjunction with the results of studies of the binding of the lectins to structurally significant oligosaccharides derived from the glycoproteins. Subsequent investigations, involving affinity chromatography of such oligosaccharides on immobilized lectins, have revealed many instances of binding strength being determined by the presence or absence of a particular sequence of sugar residues in the oligosaccharide chain. Several examples are included in Table LC 63, which lists the carbohydrate-binding specificities of lectins that have proved important in affinity chromatography and the other applications cited above.

TABLE LC 63
Affinity Chromatography: Carbohydrate-Binding Specificities of Lectins

Source of lectin[a]	Carbohydrate specificity[b]	Ref.
Abrus precatorius (jequirity bean)	β-D-Gal*p* > α-D-Gal*p*	1, 5
Agaricus bisporus (meadow	Galβ1-3GalNAcα	7
mushroom)	NeuAcα2-3Galβ1-3(NeuAcα2-6)GalNAcα	7
Anquilla anguilla (eel)	α-L-Fuc*p*, 3-*O*-Me-D-Gal*p*	1
Arachis hypogaea (peanut)	Galβ1-3GalNAcα	1, 7
	⪢ Galβ1-4(3)GlcNAc	8
Artocarpus integrifolia (jackfruit)	α-D-Gal*p*	9
	Galβ1-3GalNAcα, Galβ1-4(3)GlcNAc	8
Bandeiraea simplicifolia		
BSI isolectins	α-D-Gal*p* > α-D-Gal*p*NAc	1, 10, 11
BSI-A_4	GalNAcα1-3Galβ > GalNAcα1-3(Fucα1-2)Galβ	8
BSI-B_4	α-D-Gal*p*	10, 11
BSII	D-Glc*p*NAc (α = β)	1
Bauhinia purpurea	Galβ1-3GalNAcα	7
	> Galβ1-4(3)GlcNAc,α-D-GalNAc	8
Bauhinia variegata, var. *candida*	β-D-Gal*p*, α-D-Gal*p*NAc	5
Canavalia ensiformis or Concanavalin A (jackbean)	α-D-Man*p* > α-D-Glc*p* > α-DGlc*p*NAc	1, 5
	Manα1-2Manα1-2Man > Manα1-2Man > α-D-Man*p*	1
	Glycopeptides:[c] 1 > 2 ⪢ 3; 4 does not bind.	12
	Also some binding of Fru*p*(β > α), α-D-Ara*f*, 2,5-anhydro-D-mannitol and -D-glucitol, and 1,4-anhydro-D-arabinitol	1
Coronilla varia	α-D-Gal*p*, α-D-Gal*p*NAc	5
Crotalaria juncea (sunn hemp)	Galβ1-4Glc > β-D-Gal > α-D-Gal	1, 5, 13
Cytisus sessilifolius	GlcNAcβ1-4GlcNAc > Glcβ1-4Glc > Galβ1-4Glc	1, 5
Datura stramonium (thorn apple)	Galβ1-4GlcNAcβ1-2(Galβ1-4GlcNAcβ1-6)Man	8
	> Galβ1-4GlcNAcβ1-2(Galβ1-4GlcNAcβ1-4)Man	14
	Substitution destroys binding capacity, but bisecting GlcNAc has no effect.	
Dolichos biflorus (horse gram)	α-D-Gal*p*NAc ⪢ α-D-Gal*p*	1, 5
	GalNAcα1-3GalNAc > GalNAcα1-3(Fucα1-2)Gal > GalNAcα1-3Gal	8
Echinocystis lobata (wild cucumber)	β-D-Gal*p* > α-D-Gal*p*	13
Erythrina cristagalli (coral tree)	Galβ1-4(3)GlcNAc in multiple antennary glycoproteins	8
Geodia cydonium	Galβ1-4(3)GlcNAc, α-D-Gal*p*NAc	8
Glycine max (soybean)	α-D-Gal*p*NAc > β-D-Gal*p*NAc ⪢ α-D-Gal*p*	1, 5
	GalNAcα1-3Galβ1-3GlcNAcβ1-3Galβ1-4Glc; substitution destroys binding capacity	7, 8
Helix pomatia (edible snail)	α-D-Gal*p*NAc ⪢ α-D-Glc*p*NAc ⪢ α-D-Gal*p*	1, 5
	GalNAcα1-3GalNAc > GalNAcα1-3Gal > GalNAcα1-3(Fucα1-2)Gal > GalNAc	8
Laburnum alpinum	GlcNAcβ1-4GlcNAc > Glcβ1-4Glc > Galβ1-4Glc	1, 5
Lens culinaris (lentil)	α-D-Man*p* > α-D-Glc*p*, α-D-Glc*p*NAc	1, 5
Limulus polyphemus (horseshoe crab), hemolymph	NeuAc	1, 5
Lotus tetragonolobus (asparagus pea)	α-L-Fuc*p* ⪢ L-Gal*p*	1, 5
Maclura pomifera (osage orange)	D-Gal*p*, D-Gal*p*NAc	1
	Galβ1-3GalNAc > α-D-Gal*p*NAc	8
Momordia charantia (bitter gourd)	β-D-Gal*p* > α-D-Gal*p*	13
Phaseolus lunatus, syn. *limensis*	α-D-Gal*p*NAc > α-D-Gal*p*	1, 5
(lima bean)	GalNAcα1-3(Fucα1-2)Gal > GalNAcα1-3Gal	8
Phaseolus vulgaris (kidney bean)	Galβ1-4GlcNAcα1-2Man in multiple antennary glycoproteins	1
Pisum sativum (garden pea)	α-D-Man*p* > α-D-Glc*p* > α-DGlc*p*NAc	1, 5
Ricinus communis (castor bean)		1, 5

TABLE LC 63 (continued)
Affinity Chromatography: Carbohydrate-Binding Specificities of Lectins

Source of lectin[a]	Carbohydrate specificity[b]	Ref.
RCA1 (agglutinin, mol wt 120,000)	Galβ1-4Glc > β-D-Galp- > α-D-Galp	8
	Galβ1-4(3)GlcNAc > Galβ1-3GalNAc	
RCA2 (ricin, toxin, mol wt 60,000)	Galβ1-4Glc > β-D-Galp > α-D-Galp > GalpNAc	1
	Galβ1-3GalNAc > Galβ1-4(3)GlcNAc	8
Salvia sclarea	α-D-GalpNAc, linked to Ser or Thr in peptide chain; presence of two greatly increases binding	8
Sambucus nigra L. (elderberry), bark	NeuAcα2-6Gal > NeuAcα2-3Gal > GalNAcβ1-6Gal > GalNAcβ1-4Glc > GalNAcα1-6Gal > Galβ1-4Glc > D-GalpNAc(β > α) > D-Galp(β > α)	15
Solanum tuberosum (potato), tuber	$(GlcNAc\beta1\text{-}4GlcNAc)_2$ ≫ GlcNAcβ1-4GlcNAcβ1-4GlcNAc > GlcNAcβ1-4GlcNAc ≫ GlcNAcβ1-4MurNAc[d]	1, 5
Sophora japonica (Japanese pagoda tree)	β-D-GalpNAc > β-D-Galp > α-D-Galp	1, 5
	Galβ1-3GalNAc, Galβ1-4(3)GlcNAc	8
Tridacna maxima (Röding) (giant clam), hemolymph	β-D-GalpNAc > β-D-Galp ≫ α-D-Galp	1, 16, 17
Triticum vulgare (wheat germ)	GlcNAcβ1-4GlcNAcβ1-4GlcNAc > GlcNAcβ1-4GlcNAc	1, 18
	> GlcNAcβ1-4MurNAc[d]	5
	Also binds NeuAc	
Ulex europeus (gorse)		
Lectin I	α-L-Fucp	1, 5
Lectin II	GlcNAcβ1-4GlcNAc	1, 5
Ulex galli, *Ulex nanus*	GlcNAcβ1-4GlcNAc	5
Ulex parviflorus	α-L-Fucp	5
Vicia faba (broad bean)	α-D-Manp > α-D-Glcp ≫ α-D-GlcpNAc	1
Vicia villosa		
Isolectin B_4	α-D-GalpNAc, linked to Ser or Thr in peptide chain; two required for strong binding.	7, 8
Mixture of A_4, A_2 B_2 and B_4	GalNAcα1-3Gal > GalNAcα1-3(Fucα1-2)Gal; α-D-GalpNAc as for B_4	8
Wisteria floribunda	GalNAcα1-3Gal > GalNAcα1-3(Fucα1-2)Gal; GalNAcα1-3GalNAc > α-D-GalpNAc linked to Ser/Thr, Galβ1-4(3)GlcNAc	8

[a] Lectins from plant sources occur in seed, unless otherwise indicated.
[b] Sugars are nonreducing end-groups, unless otherwise indicated.
[c] Structures of glycopeptides examined are appended.
[d] MurNAc denotes *N*-acetylmuramic acid.

APPENDIX TO TABLE LC 63: STRUCTURES OF NUMBERED COMPOUNDS

Glycopeptide 1

Fucα1
GlcNAcβ1-2Manα1 → 6
Manβ1-4GlcNAcβ1-4GlcNAc-Asn-peptide (Fucα1 → 6 of the GlcNAc)
R_1β1-2Manα1 → 3

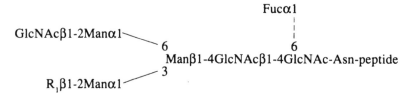

R_1 = GlcNAc, Galβ1-4GlcNAc, or NeuAc-Galβ1-4GlcNAc

Glycopeptide 2

Fucα1
Manα1 → 6
Manβ1-4GlcNAcβ1-4GlcNAc-Asn-peptide (Fucα1 → 6 of the GlcNAc)
R_2β1-2Manα1 → 3

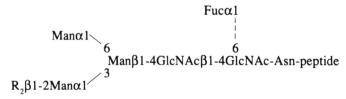

R_2 = H, GlcNAc, Galβ1-4GlcNAc, or NeuAc-Galβ1-4GlcNAc

Glycopeptide 3

Fucα1
R_3-Galβ1-4GlcNAcβ1-2Manα1 → 6
Manβ1-4GlcNAcβ1-4GlcNAc-Asn-peptide (Fucα1 → 6 of the GlcNAc)
R_3-Galβ1-4GlcNAcβ1-2Manα1 → 3

R_3 = H or NeuAc

Glycopeptide 4

Galβ1-4GlcNAcβ1 → 2
Manα1 → 6
Galβ1-4GlcNAcβ1 → 4
Manβ1-4GlcNAcβ1-4GlcNAc-Asn-peptide
Galβ1-4GlcNAcβ1-2Manα1 → 3

Dotted line indicates that terminal GlcNAc may or may not carry α-L-Fuc at O-6; this does not affect binding capacity.

HIGH-PERFORMANCE AFFINITY CHROMATOGRAPHY

Initially, packings for affinity chromatography were prepared by coupling of the appropriate lectin to agarose or polyacrylamide gels, but recently coupling to microparticulate silica has made possible more rapid affinity chromatography, which can be performed using HPLC equipment. Examples include purification of the enzymes peroxidase and glucose oxidase within 30 min on columns packed with porous silica (Merck LiChrospher® 1000, 10 μm diameter) to which concanavalin A (Con A) was coupled.[19] Con A-silica columns remain stable for long periods when used at ambient temperature. Human serum glycoproteins have been fractionated within 20 min on a system in which a column packed with the same silica, to which the *Phaseolus vulgaris* lectin was coupled, was connected in series to a TSK® G3000 SW column,[20] and columns packed with LiChrosorb® DIOL coupled to Con A and to wheat germ lectin have been used for the rapid differentiation of a series of glycopeptides derived from ovalbumin.[21] An interesting variation is the immobilization of ConA on macroporous, microparticulate silica (Nucleosil, 300 Å pore size, 5 μm particle diameter, Macherey-Nagel) via complexing interactions with Cu(II), chelated with iminodiacetic acid groups coupled to the silica.[22] This results in a column exhibiting dual behavior, glycoproteins such as peroxidase and α_1-acid glycoprotein being retained by the Con A moiety and eluted by displacement with methyl α-D-mannopyranoside (linear gradient 0 → 50 m*M* in 10 min, in 25 m*M* phosphate buffer, pH 6.0, containing 0.2 *M* NaCl), and proteins (β-lactoglobulin, α-chymotrypsinogen A, lysozyme, and ribonuclease A) being retained by the chelated copper and eluted with a sodium chloride gradient (0 → 1.0 *M* NaCl in 30 min) in the same buffer. The presence of the Con A substantially reduces the retentive capacity of the column for solutes binding to the copper, but some glycoproteins, such as glucose oxidase and ribonuclease B, are retained so strongly that sodium borate (0.3 *M*, in pH 6.0 buffer) is required for the elution of the latter, and the former is eluted only at low pH, under which conditions the Con A is stripped from the Cu(II) complex. Thus, this method is not suitable for chromatography of glycoproteins binding very strongly to Con A, but otherwise offers a means of simultaneous fractionation of proteins and glycoproteins.

The thermodynamics and kinetics of chromatography of carbohydrates on silica-bound Con A have been studied in depth by Muller and Carr,[23,24] who have concluded that the large plate heights observed, with resultant peak broadening, are due mainly to slow dissociation of the solutes from the immobilized ligands, rate constants being about 30 times smaller than those found in solution for methyl and *p*-nitrophenyl glycosides of α-D-mannose and -glucose. For glycoproteins, being multivalent with respect to sugar residues binding to the lectin, the rate constants for dissociation are even smaller, almost one tenth of those for the monovalent carbohydrates, so that very low flow rates, of the order of 0.02 ml/min, are required to achieve acceptably small plate heights and peak widths.[25] A stopped-flow elution technique, in which the column, once the eluent containing competing solute (methyl α-D-mannopyranoside) has been introduced, is isolated from the HPLC system for a period sufficient to enable desorption of the glycoprotein to take place, has been advocated by these authors.[25]

AFFINITY CHROMATOGRAPHY WITH NONLECTIN IMMOBILIZED LIGANDS

Although affinity chromatography of carbohydrates has been largely confined to supports containing immobilized lectins, other ligands have been used successfully in some instances. A notable example is the use of serotonin (5-hydroxytryptamine), coupled to Sepharose® 4B, in affinity chromatography of oligosaccharides and glycoproteins containing *N*-acetylneuraminic acid.[26] Sequential elution with water, a phosphate or Tris buffer (100 m*M*, pH 7.0), and the same buffer containing sodium chloride (500 m*M*) or by the use of a buffer gradient gave satisfactory fractionation and purification of a number of important sialoglycoproteins on such columns. These included fetuin, ovomocoid, and sialoglycoproteins from tumor extracts such as carcinoembryonic antigen (CEA), a marker protein in primary colonic cancer. Asialofetuin and -ovomucoid were not retained by the column, nor were *N*-glycolylneuraminic acid or products of periodate oxidation and borohydride reduction of *N*-acetylneuraminic acid residues, which resulted in cleavage of the C-7—C-9 chain; it was evident, therefore, that both this portion of the molecule and the *N*-acetyl group were essential for binding. This method of affinity chromatography of sialoglycoproteins has advantages over the use of wheat germ lectin, which is not specific for *N*-acetylneuraminic acid only (see Table LC 63), or the lectin from *Limulus polyphemus*, which is specific but is difficult to obtain in amounts large enough for use in preparative chromatography.

Another novel approach[27] was the separation of pyruvylated polysaccharides, such as those from *Xanthomonas campestris* and *Enterobacter aerogenes*, into fractions differing in pyruvate content by chromatography on an affinity matrix prepared by coupling to Sepharose 6B the globulin fraction of antiserum against the pyruvylated polysaccharide of *Rhizobium* strain TA 1. Pyruvate-rich fractions were retained on elution with 0.1 *M* acetate buffer (pH 4.0) containing 0.5 *M* sodium chloride, which released those poorer in pyruvate; the same buffer containing 0.25 *M* sodium chloride, with 0.25 *M* sodium pyruvate as a competing solute, subsequently eluted the pyruvate-rich material.

The use of antibodies as ligands has recently been introduced into high-performance affinity chromatography as a means of overcoming the problems arising from strong interaction between carbohydrates and immobilized lectins, resulting in the low dissociation constants mentioned in the previous section.[23-25] Coupling weakly interactive monoclonal antibodies to 10-μm silica gel gives affinity chromatography packings for which the dissociation constants of the solute-ligand complexes are higher,[28] closer to the range associated with other modes of chromatography, such as ion-exchange and reversed-phase partition, that have been successfully adapted to HPLC. Solute-ligand interactions are in dynamic equilibrium, with solutes bound by the antibodies merely being retarded on the column, not totally immobilized until displaced by a competing solute or under conditions that disrupt the complex, as is the case with lectin-carbohydrate interactions; therefore, fractionation can be achieved by isocratic elution under mild conditions. Dissociation constants, and thence peak widths, can be controlled by varying the temperature,[28] sharp peaks being obtained at 37 to 50°C. The method has been used successfully in analysis of serum and urine samples for the clinically significant tetrasaccharide Glcα1-6Glcα1-4Glcα1-4Glc, using a column (10 × 0.5 cm) packed with 10-μm silica gel (Selectispher®-10 activated tresyl column; Perstorp Biolytica AB, Lund, Sweden) to which was coupled the monoclonal antibody 39.5 (IgG2b).[28,29] Isocratic elution with 0.02 *M* sodium phosphate buffer, pH 7.5, containing 0.2 *M* sodium chloride or 0.1 *M* sodium sulfate, at 0.2 ml/min gave a sharp peak for the oligosaccharide in less than 20 min, and the use of this system in conjunction with a pulsed amperometric detector (Dionex) made possible detection in the 10 ng (ca. 2 pmol) region, with a linear response up to at least 100 ng.[29] This new technique has great potential, especially in clinical analysis, for which highly specific, sensitive methods are required.

REFERENCES

1. **Goldstein, I. J. and Hayes, C. E.,** The lectins: carbohydrate-binding proteins of plants and animals, *Adv. Carbohydr. Chem. Biochem.*, 35, 127, 1978.
2. **Lotan, R. and Nicolson, G. L.,** Purification of cell membrane glycoproteins by lectin affinity chromatography, *Biochim. Biophys. Acta*, 559, 329, 1979.
3. **Gabius, H.-J.,** Tumor lectinology: at the intersection of carbohydrate chemistry, biochemistry, cell biology, and oncology, *Angew. Chem. Int. Ed. Engl.*, 27, 1267, 1988.

4. **Koyama, I., Miura, M., Matsuzaki, H., Sakagishi, Y., and Komoda, T.**, Sugar-chain heterogeneity of human alkaline phosphatases: differences between normal and tumour-associated isozymes, *J. Chromatogr.*, 413, 65, 1987.
5. **Kristiansen, T.**, Group-specific separation of glycoproteins, *Methods Enzymol.*, 34, 331, 1974.
6. **Landsteiner, K.**, *The Specificity of Serological Reactions*, Dover Publications, New York, 1962.
7. **Sueyoshi, S., Tsuji, T., and Osawa, T.**, Carbohydrate-binding specificities of five lectins that bind to *O*-glycosyl linked carbohydrate chains. Quantitative analysis by frontal affinity chromatography, *Carbohydr. Res.*, 178, 213, 1988.
8. **Wu, A. M., and Sugii, S.**, Code and classification of Gal/GalNAc specific lectins, Abstr. C34, in *Proc. 14th Int. Symp. Carbohydr. Chem.*, Stockholm, 1988; see also Wu, A. M. and Sugii, S., Differential binding properties of Gal and GalNAc specific lectins, in *The Molecular Immunology of Complex Carbohydrates*, Wu, A. M., Ed., Plenum Press, New York, 1988.
9. **Chowdhury, S., Ray, S., and Chatterjee, B. P.**, Single step purification of polysaccharides using immobilized jackfruit lentin as affinity adsorbent, *Glycoconjugate J.*, 5, 27, 1988.
10. **Blake, D. A. and Goldstein, I. J.**, Resolution of nucleotide sugars and oligosaccharides by lectin affinity chromatography, *Anal. Biochem.*, 102, 103, 1980.
11. **Blake, D. A. and Goldstein, I. J.**, Resolution of carbohydrates by affinity chromatography, *Methods Enzymol.*, 83, 127, 1982.
12. **Narasimhan, S., Wilson, J. R., Martin, E., and Schachter, H.**, A structural basis for four distinct elution profiles on concanavalin A-Sepharose affinity chromatography of glycopeptides, *Can. J. Biochem.*, 57, 83, 1979.
13. **Majumdar, T. and Surolia, A.**, A general method for isolation of galactopyranosyl-specific lectins, *Indian J. Biochem. Biophys.*, 16, 200, 1979.
14. **Yamashita, K., Totani, K., Ohkura, T., Takasaki, S., Goldstein, I. J., and Kobata, A.**, Carbohydrate binding properties of complex-type oligosaccharides on immobilized *Datura stramonium* lectin, *J. Biol. Chem.*, 262, 1602, 1987.
15. **Shibuya, N., Goldstein, I. J., Broekaert, W. F., Nsimba-Lubaka, M., Peeters, B., and Peumans, W. J.**, The elderberry (*Sambucus nigra* L.) bark lectin recognizes the Neu5Ac(α2-6)Gal/GalNAc sequence, *J. Biol. Chem.*, 262, 1596, 1987.
16. **Baldo, B. A., Sawyer, W. H., Stick, R. V., and Uhlenbruck, G.**, Purification and characterization of a galactan-reactive agglutinin from the clam *Tridacna maxima* (Röding) and a study of its combining site, *Biochem. J.*, 175, 467, 1978.
17. **Gleeson, P. A., Jermyn, M. A., and Clarke, A. E.**, Isolation of an arabinogalactan protein by lectin affinity chromatography on tridacnin-Sepharose 4B, *Anal. Biochem.*, 92, 41, 1979.
18. **Mintz, G. and Glaser, L.**, Glycoprotein purification on a high-capacity wheat germ lectin affinity column, *Anal. Biochem.*, 97, 423, 1979.
19. **Borchert, A., Larsson, P.-O., and Mosbach, K.**, High-performance liquid affinity chromatography on silica-bound concanavalin A, *J. Chromatogr.*, 244, 49, 1982.
20. **Borrebaeck, C. A. K., Soares, J., and Mattiasson, B.**, Fractionation of glycoproteins according to lectin affinity and molecular size using a high-performance liquid chromatography system with sequentially coupled columns, *J. Chromatogr.*, 284, 187, 1984.
21. **Honda, S., Suzuki, S., Nitta, T., and Kakehi, K.**, Analytical high-performance affinity chromatography of ovalbumin-derived glycopeptides on columns of concanavalin A- and wheat germ agglutinin-immobilized gels, *J. Chromatogr.*, 438, 73, 1988.
22. **El Rassi, Z., Truei, Y., Maa, Y.-F., and Horváth, C.**, High-performance liquid chromatography with concanavalin A immobilized by metal interactions on the stationary phase, *Anal. Biochem.*, 169, 172, 1988.
23. **Muller, A. J. and Carr, P. W.**, Chromatographic study of the thermodynamic and kinetic characteristics of silica-bound concanavalin A, *J. Chromatogr.*, 284, 33, 1984.
24. **Muller, A. J. and Carr, P. W.**, Examination of the thermodynamic and kinetic characteristics of microparticulate affinity chromatography supports. Application to concanavalin A, *J. Chromatogr.*, 357, 11, 1986.
25. **Muller, A. J. and Carr, P. W.**, Examination of kinetic effects in the high-performance liquid affinity chromatography of glycoproteins by stopped-flow and pulsed elution methods, *J. Chromatogr.*, 294, 235, 1984.
26. **Sturgeon, R. J. and Sturgeon, C. M.**, Affinity chromatography of sialoglycoproteins, utilising the interaction of serotonin with *N*-acetylneuraminic acid and its derivatives, *Carbohydr. Res.*, 103, 213, 1982.
27. **Sutherland, I. W.**, Affinity column for separation of pyruvylated and non-pyruvylated polysaccharides, *J. Chromatogr.*, 213, 301, 1981.
28. **Ohlson, S., Lundblad, A., and Zopf, D.**, Novel approach to affinity chromatography using "weak" monoclonal antibodies, *Anal. Biochem.*, 169, 204, 1988.
29. **Wang, W.-T., Kumlien, J., Lundblad, A., Ohlson, S., and Zopf, D.**, Analysis of a glucose containing tetrasaccharide by high-performance liquid affinity chromatography, Abstr. A56, in *Proc. 14th Int. Symp. Carbohydrate Chemistry*, Stockholm, 1988.

PREPARATIVE LIQUID CHROMATOGRAPHY

Although the emphasis in Section I.III has been on the analytical results obtainable with the various forms of liquid chromatography now available, many of these methods can be used on a preparative scale to isolate pure compounds in quantities sufficient to permit their further analysis by chemical and spectroscopic means. The adaptability to preparative liquid chromatography of the diverse types of HPLC packings employed in analysis of carbohydrates is discussed in the excellent, comprehensive review of HPLC of carbohydrates recently published by Hicks,[1] who emphasizes the point that the conditions appertaining to most of the methods, *viz.* no derivatization of the sample, high-resolution separations, and nondestructive detection techniques, are ideal for the isolation of pure compounds. If quantities of only a few milligrams are required, conventional, analytical-scale columns can often be used,[1] but the isolation of compounds in higher amounts (from ca. 100 mg up to gram quantities) requires larger columns, with packings of high capacity that permit fast flow rates. Use of HPLC equipment for this purpose necessitates modifications such as larger sample loops and pumps capable of producing flow rates in excess of 100 ml/min, which give rise to considerable band spreading in the column and consequent loss of resolution.[2] Therefore, for large-scale preparative chromatography, conventional, low-pressure methods are preferable despite their slowness. For example, for fractionation of small polysaccharides (mol wt range 4,000 to 20,000) and isolation of discrete components the author has found that chromatography on soft gels, rather than the newer SEC packings, remains the technique of choice (see References 39 and 40 in Table LC 64). At the University of Cape Town (Cape Town, South Africa) the time-honored method of partition chromatography on a large column packed with cellulose has been used successfully to isolate pure methylated sugars, in amounts sufficient to permit their use as standards for chromatography and spectroscopy, from hydrolysates of methylated polysaccharides (Table LC 64, References 5 and 6).

Another low-pressure method that continues to prove useful in preparative liquid chromatography of charged oligosaccharides and polysaccharides is that involving ion exchange on dextran or cellulose gels bearing anionic (DEAE or QAE) or cationic (CM or SP) groups. This technique was reviewed in *CRC Handbook of Chromatography: Carbohydrates*, Volume I (pages 101 to 104); some more recent applications are listed in Table LC 64 (References 29 to 31). Attachment of such groups to more rigid matrices has made possible separations combining the resolution of the low-pressure method with faster flow rates, and the use of these packings is often called medium-pressure liquid chromatography (MPLC). Resin-based ion-exchangers for MPLC are also available (e.g., Mono P, Mono Q; Pharmacia) and have been applied to acidic polysaccharides, gangliosides, and charged oligosaccharides derived from glycoproteins (see Table LC 64, References 35 to 38).

Tne HPLC methods that have been used for preparative purposes have been assessed and compared by Hicks and co-workers,[1,3,4] who concluded that C_{18}-bonded silica packings and cation-exchange resins, with water or (for H^+-form resins) dilute acid, preferably volatile, as eluents are preferable to aminopropyl-bonded silica gels, which are easily degraded by reactive reducing sugars, contaminants, and aqueous mobile phases. Where such packings have been used in preparative liquid chromatography they have functioned mainly as weak ion-exchangers, with eluents containing aqueous buffer solutions in high proportions; adaptation of normal-phase partition, with eluents rich in acetonitrile, to a preparative scale is hazardous and expensive. Some successful applications of HPLC packings in preparative chromatography of a diversity of carbohydrates are listed in Table LC 64 (References 3, 4, and 11 to 24).

Adsorption chromatography on silica gel is another classical method that has been applied in preparative chromatography of carbohydrates in both low-pressure and HPLC systems (Table LC 64, References 7 to 10), but it has the disadvantage that derivatization is required, except in the case of gangliosides.[10] Affinity chromatography, where applicable, is an extremely efficient preparative method currently being adapted to HPLC; examples of the use of this technique have been cited in the preceding subsection.

TABLE LC 64
Fractionation and Isolation of Carbohydrates by Preparative Liquid Chromatography

Packing	Column size: Length, cm	Column size: I.D., cm	Eluent	Flow rate ml/min	Temperature °C	Type of carbohydrate	Quantity (total sample)	Ref.
Cellulose (Whatman)	90	4.2—6.5	Light petroleum (b.p. 100—120°C)-water-saturated butan-1-ol (7:3, changed in 9 steps to 7:50), then butan-1-ol half-saturated with water	1	30	Methylated sugars	1—5 g	5
Cellulose	65	4	Butanone-water azeotrope	1	28	Methylated sugars	500 mg	6
Silica gel L (100—160 μm)	45	3	Light petroleum (b.p. 70—100°C)-acetone (9:1 → 0:100 in 7 steps)	7	Ambient	Methyl ethers of methyl-L-rhamnopyranoside	14 g	7
Silica gel (70—230 mesh)	100	3	Benzene-ethyl acetate (2:1)	7	Ambient	D-Glucose and α-(1 → 4)-linked oligomers to DP5, peracetylated	10 g	8
Silica gel (3 μm)	25	0.8	Hexane-1,4-dioxane-dichloromethane (8:2:1, changed to 4:2:1 in 3 steps)	2.5	Ambient	Benzoylated oligosaccharide-alditols from partial acid hydrolysate of plant gum	200 mg	9
Silica gel (5 μm)	25	0.46	2-Propanol-hexane-water (linear gradient, 55:42:3 → 55:25:20 in 2 h)	0.5	Ambient	Gangliosides	<1 mg	10
C_{18}-bonded silica (3 μm)	15	0.45	Acetonitrile-water (9:1 for disaccharide-alditols, 19:1 for trisaccharide-alditols)	1.25	Ambient	Benzoylated oligosaccharide-alditols, previously fractionated on silica gel (see above)	ca. 40 mg	9
Prep-pak® C_{18} cartridge (Waters)	30	5.8	Distilled water	300	Ambient	Methyl α- and β-maltosides	3 g	11
				250		Methyl ethers of sucrose	8 g	12
Dextropak® cartridges (Waters), 2 in series	20	0.8	Distilled water	2	Ambient	Human milk oligosaccharides	n.a.	13

Alltech® 600 RP	25	0.46	Distilled, deionized water	0.5	Ambient	Human milk oligosaccharides	n.a.	14
			Phosphate buffer, pH 6.0			Glycopeptides from ovalbumin	n.a.	15
Prep-Pak®C_{18} cartridges (Waters), 4 in series	120	5.8	Degassed, distilled water	100	Ambient	Trehalulose isolated from isomaltulose, sucrose, D-glucose, D-fructose, and trisaccharide	ca. 35 g	16
Rainin® Dynamax C_{18}	25	2.14	Distilled water	3	Ambient	Malto-oligosaccharides to DP 11	12 mg	4
Rainin Dynamax® (NH_2 (aminopropyl-silica)	25	2.14	Acetonitrile-water (11:9)	12	Ambient	Malto-oligosaccharides to DP 7	500 mg	1, 4
MicroPak® AX-5 (Varian)	30	0.4	Acetonitrile-1.0 m*M* KH_2PO_4 (3:2)	1.3	Ambient	Oligosaccharide-alditols from mucins	n.a.	17
			Acetonitrile-1.0 m*M* KH_2PO_4 (11:9)	1	Ambient	Human milk oligosaccharides	n.a.	18
	50	0.8	Water-500 m*M* KH_2PO_4, pH 4.0, stepwise and gradient elution, buffer 0 → 40% over 2 h	2	Ambient	Sialyl-oligosaccharides from hydrazinolysis of ovomucoid	210 mg	19
Chromatem Amino AS 5A (5 μm), 2 in series	25	0.4	Acetonitrile-50 m*M* KH_2PO_4 with 0.01% 1,4-diaminobutane (7:3) for 30 min, linear gradient to solvent-water (17:3) over 1 h, isocratic for 1 h	1	Ambient	Monosialyl-oligosaccharides from ovomucoid	18 mg	19
AS 3A (3 μm), in series with AS 5A columns	15 7.5	0.4 0.4			40—50	Mono-, di-, tri-, and tetrasialyl-oligosaccharides from hydrazinolysis of α_1-acid glycoprotein	2—4 mg	20
LiChrosorb®-NH_2 (Merck)	25	0.25	Acetonitrile-5 m*M* phosphate buffer, pH 5.6 (83:17) for 7 min, then gradient elution (83:17 → 36:64 over 75 min)	39	20	Gangliosides	1—5 mg	21
Aminex® HPX-87H (Bio-Rad), 9 μm	30	0.78	35 m*M* formic acid	0.3	25	GlcA, GalA	6 mg	3
				0.6	25	GlcA, GlcNAc	30 mg	3

TABLE LC 64 (continued)
Fractionation and Isolation of Carbohydrates by Preparative Liquid Chromatography

Packing	Column size: Length, cm	Column size: I.D., cm	Eluent	Flow rate ml/min	Temperature °C	Type of carbohydrate	Quantity (total sample)	Ref.
Hitachi 3019-s resin, H^+-form, 30—40 μm)	60	2.2	0.5% formic acid	100	Ambient	Dextrans (mol wt 19,000 and 2,900), raffinose, lactose, D-galactose, GlcA, GalA	80 mg	22
						Neutral and acidic oligosaccharides from hydrolysate of plant gum	100 mg	22
Aminex® Q-15S, Ca^{2+}-form (Bio-Rad), 20—25 μm	30	2	Degassed, distilled water	1.5	85	D-glucose, D-fructose, sucrose	200 mg	3
						Lactulose isolated from contaminating lactose, D-galactose, D-fructose, D-tagatose	130 mg	3
	60	0.9	Degassed, distilled water	0.33	80	Cellobiulose isolated from cellobiose, D-glucose, D-fructose	200 mg	23
						Maltulose isolated from maltose, D-glucose, D-fructose	200 mg	23
Aminex® HPX-42A (Bio-Rad), Ag^+-form, 20—30 μm	30	2	Degassed, distilled water	1	85	Malto-oligosaccharides to DP 11	100 mg	4
	30	1				Cello-oligosaccharides to DP 7, D-glucose, D-fructose, 1,6-anhydro-β-D-glucose, from hydrothermolysis of biomass	100 mg	24

Zerolit® 225-X4, K^+-form, −200 mesh, 2 in series With recycling	160	5	Distilled water	1.6	Ambient	Fractionation of human milk oligosaccharides	≤1 g	25
	180	2.2		0.6	Ambient	Isolation of individual tri- to hexaoses from human milk oligosaccharides		25
Ostion® LGKS 0802, Li^+-form, 12—15 μm	50	2.4	Ethanol-water (3:2)	0.8	80	Isolation of sucrose from D-glucose, D-fructose, maltotriose	350 mg	26
AGl-X4 (Bio-Rad), OH^--form, —400 mesh	28	2.6	Distilled water	3	Ambient	Isolation of αβ-trehalose from αα- and ββ-trehalose	250 mg	27
AG MP-1 (Bio-Rad), macroporous resin, 200—400 mesh, $HCOO^-$-form	60	3.8	Water, then stepwise elution, sodium formate, pH 4.7 (0.33 → 0.75 *M* in 8 steps)	2.4	Ambient	Oligogalacturonic acids to DP 6	7 g	28
	40	2.2	Water, then sodium formate, pH 4.7 (0.30→0.65 *M* in 7 steps)	2.4	Ambient	Oligogalacturonic acids to DP 7	2 g	28
DEAE-Sephadex® A-25 (Pharmacia), Cl^--form	135	2.5	100 m*M* potassium chloride, 10 m*M* imidazole, pH 7.0, then concave gradient to 300 m*M* KCl, 10 m*M* imidazole	98	Ambient	Oligogalacturonic acids to DP 16	5 g	29
QAE-Sephadex® A-25-120 (Pharmacia), Cl^--form	45	1.7	0.125 *M* imidazole-HCl, pH 7.0, then linear gradient (0.125 → 0.75 *M*) and step to 0.95 *M*)	n.a.	Ambient	Oligogalacturonic acids, DP 3—13	274 mg	30
DEAE-Sephacel® (Pharmacia), $HCOO^-$-form	20	1.4	Formic acid, linear gradient (0 → 1.0 *M*)	n.a.	Ambient	Even- and odd-numbered oligosaccharides from hyaluronidase digestion of hyaluronic acid, DP 2—10	44 mg	31
DEAE-Ultropac TSK® 545 (LKB), Cl^--form	15	0.75	Sodium chloride gradient (0.05 → 0.75 *M*) in water-methanol (9:1)	0.6	Ambient	Glycosaminoglycuronans hyaluronic acid, heparan sulfate, chondroitin 4- and 6-sulfates, dermatan sulfate	≤1 mg	32
DEAE-silica gel, 120—200 mesh, CH_3COO^--form	60	2	Ammonium acetate gradient (0.2 → 0.5 *M*) in methanol	1.5	Ambient	Mono-, di-, tri-, and tetrasialyl gangliosides separated	40 mg	33

TABLE LC 64 (continued)
Fractionation and Isolation of Carbohydrates by Preparative Liquid Chromatography

Packing	Column size: Length, cm	Column size: I.D., cm	Eluent	Flow rate ml/min	Temperature °C	Type of carbohydrate	Quantity (total sample)	Ref.
DEAE-Spheron® 1000 (Lachema), 17 μm, CH_3COO^--form	25	0.8	Ammonium acetate gradient in methanol (0 → 0.2 *M* over 45 min, then 0.2 → 0.5 *M*)	3.4	25	Mono-, di-, tri-, tetra-, and pentasialyl gangliosides separated	30 mg	34
Mono-Q® (Pharmacia), 10 μm, CH_3COO^--form	5	0.5	Stepwise gradient, potassium acetate (0 → 0.225 *M* in 4 steps) in methanol	2	Ambient	Mono-, di-, tri-, and tetrasialyl gangliosides separated	35 mg	35
Q-Sepharose® (Pharmacia), 10 μm, CH_3COO^--form	75	3	Sodium acetate gradient (0 → 4 M) in chloroform-methanol-water (15:30:4)	1.5	Ambient	Gangliosides; resolution according to structure as well as sialic acid content	5 g	36
Mono-Q® (Pharmacia), Cl^--form	5	0.5	Sodium chloride gradient in water (0 → 10 m*M*, then 10 → 100 m*M*)	2	Ambient	Sialyl-oligosaccharides (reduced) from hydrazinolysis of human serotransferrin	100 mg	37
Mono-P® (Pharmacia), Cl^--form	20	0.5	Sodium chloride gradient (0 → 1.0 *M*) in 15 m*M* phosphate-NaCl buffer, pH 7.0	1	Ambient	Rhamnogalacturonans from elm mucilage	10 mg	38
Bio-Gel® P-10 (Bio-Rad)	70	4	Distilled water	0.8	Ambient	Polysaccharides, $\overline{M}_w$ 8,000 and 16,000, from *Acacia mabellae* gum	1.4 g	39
Sephadex® G-10 (Pharmacia)	55	3.5	Distilled water	1	Ambient	Polysaccharide, $\overline{M}_w$ 4,400 isolated from low-molecular weight products of Smith degradation of *Grevillea robusta* gum	2 g	40
Trisacryl® GF05 (LKB)	70	2.5	0.1 *M* pyridinium acetate buffer, pH 5.0	1	Ambient	Oligosaccharides, mol wt 250—1,250, after prolonged Smith degradation of *Grevillea robusta* gum	25 mg	40

Sephacryl® S-400 and S-500 (Pharmacia)	60	2.6	M sodium chloride	2.25	Ambient	Polysaccharide components, $\overline{M}_w$ 2×10^5—4×10^6, of gum arabic (*Acacia senegal* gum)	200 mg	41
Superose® 6 Prep grade (Pharmacia)	70	1.6	50 m*M* phosphate buffer, pH 7.0	0.8	Ambient	Dextrans, $\overline{M}_w$ 10^3—10^6	≥200 mg	42
HW 55S and 75S (Toyo Soda)	60	10	0.1 *M* sodium nitrate	8	Ambient	Lime pectin, $\overline{M}_w$ 8,000—300,000	500 mg	43
				8	60	Alginate from *Laminaria digitata*, components of $\overline{M}_w$ 2×10^5—2×10^6	720 mg	43
				10	60	κ-Carageenan from *Euchema cottonii*, components of $\overline{M}_w$ 6,000—960,000	1.25 g	43

REFERENCES

1. **Hicks, K. B.**, High-performance liquid chromatography of carbohydrates, *Adv. Carbohydr. Chem. Biochem.*, 46, 17, 1988.
2. **Kaminski, M. and Reusch, J.**, Comparison of methods of sample introduction during scale-up of liquid chromatography, *J. Chromatogr.*, 356, 47, 1986.
3. **Hicks, K. B., Sondey, S. M., and Doner, L. W.**, Preparative liquid chromatography of carbohydrates: mono- and disaccharides, uronic acids, and related derivatives, *Carbohydr. Res.*, 168, 33, 1987.
4. **Hicks, K. B. and Sondey, S. M.**, Preparative high-performance liquid chromatography of malto-oligosaccharides, *J. Chromatogr.*, 389, 183, 1987.
5. **Stephen, A. M.**,*Virgilia oroboides* gum. III. Products of hydrolysis of the methylated gum, *J. Chem. Soc.*, p. 2030, 1962; **Stephen, A. M.**, unpublished data, 1988.
6. **Merrifield, E. H. and Stephen, A. M.**, Structural studies on the capsular polysaccharide from Klebsiella serotype K64, *Carbohydr. Res.*, 74, 241, 1979.
7. **Evtushenko, E. V., Vakhrosheva, N. M., and Ovodov, Yu. S.**, Preparative liquid column chromatography of methyl ethers of methyl α-L-rhamnopyranoside, *J. Chromatogr.*, 196, 331, 1980.
8. **Dziedzic, S. and Kearsley, M. W.**, Preparation of macro quantities of glucose oligomers, *J. Chromatogr.*, 154, 295, 1978.
9. **Oshima, R. and Kumanotani, J.**, Structural studies of plant gum from sap of the lac tree, *Rhus vernicifera, Carbohydr. Res.*, 127, 43, 1984.
10. **Kundu, S. K. and Scott, D. D.**, Rapid separation of gangliosides by high-performance liquid chromatography, *J. Chromatogr.*, 232, 19, 1982.
11. **Cheetham, N. W. H. and Sirimanne, P.**, Synthesis of methyl α- and β-maltosides, *Carbohydr. Res.*, 96, 126, 1981.
12. **Moody, W., Richard, G. N., Cheetham, N. W. H., and Sirimanne, P.**, Isolation and alkaline degradation of some mono-*O*-methylsucroses, *Carbohydr. Res.*, 114, 306, 1983.
13. **Cheetham, N. W. H. and Dube, V. E.**, Preparation of lacto-*N*-neotetraose from human milk by high-performance liquid chromatography, *J. Chromatogr.*, 263, 426, 1983.
14. **Dua, V. K. and Bush, C. A.**, Identification and fractionation of human milk oligosaccharides by proton-nuclear magnetic resonance spectroscopy and reverse-phase high-performance liquid chromatography, *Anal. Biochem.*, 133, 1, 1983.

15. **Dua, V. K. and Bush, C. A.,** Resolution of some glycopeptides of hen ovalbumin by reverse-phase high-pressure liquid chromatography, *Anal. Biochem.*, 137, 33, 1984.
16. **Cookson, D., Cheetham, P. S. J., and Rathbone, E. B.,** Preparative high-performance liquid chromatographic purification and structural determination of 1-*O*-α-D-glucopyranosyl-D-fructose (trehalulose), *J. Chromatogr.*, 402, 265, 1987.
17. **Dua, V. K., Dube, V. E., and Bush, C. A.,** The combination of normal-phase and reverse-phase high-pressure liquid chromatography with NMR for the isolation and characterization of oligosaccharide alditols from ovarian cyst mucins, *Biochim. Biophys. Acta*, 802, 29, 1984.
18. **Dua, V. K., Goso, K., Dube, V. K., and Bush, C. A.,** Characterization of lacto-*N*-hexaose and two fucosylated derivatives from human milk by high-performance liquid chromatography and proton NMR spectroscopy, *J. Chromatogr.*, 328, 259, 1985.
19. **Paz Parente, J. P., Leroy, Y., Montreuil, J., and Fournet, B.,** Separation of sialyl-oligosaccharides by high-performance liquid chromatography. Application to analysis of carbohydrate units of acidic oligosaccharides obtained by hydrazinolysis of hen ovomucoid, *J. Chromatogr.*, 288, 147, 1984.
20. **Cardon, P., Paz Parente, J. P., Leroy, Y., Montreuil, J., and Fournet, B.,** Separation of sialyl-oligosaccharides by high-performance liquid chromatography. Application to the analysis of mono-, di-, tri-, and tetrasialyl-oligosaccharides obtained by hydrazinolysis of α_1-acid glycoprotein, *J. Chromatogr.*, 356, 135, 1986.
21. **Gazzotti, G., Sonnino, S., and Ghidoni, R.,** Normal-phase high-performance liquid chromatographic separation of non-derivatized ganglioside mixtures, *J. Chromatogr.*, 348, 371, 1985.
22. **Kumanotani, J., Oshima, R., Yamauchi, Y., Takai, N., and Kurosu, Y.,** Preparative high-performance gel chromatography for acidic and neutral saccharides, *J. Chromatogr.*, 176, 462, 1979.
23. **Hicks, K. B., Symanski, E. V., and Pfeffer, P. E.,** Synthesis and high-performance liquid chromatography of maltulose and cellobiulose, *Carbohydr. Res.*, 112, 37, 1983.
24. **Bonn, G.,** High-performance liquid chromatographic isolation of ^{14}C-labelled gluco-oligosaccharides, monosaccharides and sugar degradation products on ion-exchange resins, *J. Chromatogr.*, 387, 393, 1987.
25. **Donald, A. S. R. and Feeney, J.,** Separation of human milk oligosaccharides by recycling chromatography. First isolation of lacto-*N*-*neo*-difucohexaose II and 3′-galactosyllactose from this source, *Carbohydr. Res.*, 178, 79, 1988.
26. **Plicka, J., Kleinmann, I., and Svoboda, V.,** Preparative isolation of ^{14}C-labelled saccharose from a partly purified fraction of *Chlorella vulgaris*, *J. Chromatogr.*, 442, 237, 1988.
27. **Parrish, F. W., Meagher, M. M., and Reilly, P. J.,** Chromatographic procedures for isolating αβ-trehalose formed during the preparation of ββ-trehalose, *Carbohydr. Res.*, 168, 129, 1987.
28. **Doner, L. W., Irwin, P.L., and Kurantz, M. J.,** Preparative chromatography of oligogalacturonic acids, *J. Chromatogr.*, 449, 229, 1988.
29. **Jin, D. F. and West, C. A.,** Characteristics of galacturonic acid oligomers as elicitors of casbene synthetase activity in castor bean seedlings, *Plant Physiol.*, 74, 989, 1984.
30. **Davis, K. R., Darvill, A. G., Albersheim, P., and Dell, A.,** Host-pathogen interactions. XXIX. Oligogalacturonides released from sodium polypectate by endopolygalacturonic acid lyase are elicitors of phytoalexins in soybean, *Plant Physiol.*, 80, 568, 1986.
31. **Nebinger, P.,** Comparison of gel permeation and ion-exchange chromatographic procedures for the separation of hyaluronate oligosaccharides, *J. Chromatogr.*, 320, 351, 1985.

32. **Lee, Y. J., Radhakrishnamurthy, B., Dalferes, E. R., Jr., and Berenson, G. S.,** Fractionation of aorta glycosaminoglycans by high-performance liquid chromatography, *J. Chromatogr.*, 419, 275, 1987.
33. **Kundu, S. K., Chakravarty, S. K., Roy, S. K., and Roy, A. K.,** DEAE-silica gel and DEAE-controlled porous glass as ion-exchangers for isolation of Glycolipids, *J. Chromatogr.*, 170, 65, 1979.
34. **Šmid, F., Bradová, V., Mikeš, O., and Sedláčková, J.,** Rapid ion-exchange separation of human brain gangliosides, *J. Chromatogr.*, 377, 69, 1986.
35. **Mansson, J.-E., Rosengren, B., and Svennerholm, L.,** Separation of gangliosides by anion-exchange chromatography on Mono Q, *J. Chromatogr.*, 322, 465, 1985.
36. **Hirabayashi, Y., Nakao, T., Matsumoto, M., Obata, K., and Ando, S.,** Improved method for large-scale purification of brain gangliosides by Q-Sepharose column chromatography. Immunochemical detection of C-series polysialogangliosides in adult bovine brains, *J. Chromatogr.*, 445, 377, 1988.
37. **Van Pelt, J., Damm, J. B. L., Kamerling, J. P., and Vliegenthart, J. F. G.,** Separation of sialyl-oligosaccharides by medium pressure anion-exchange chromatography on Mono Q, *Carbohydr. Res.*, 169, 43, 1987.
38. **Barsett, H. and Paulsen, B. S.,** Separation of acidic polysaccharides from *Ulmus glabra* Huds. on Mono P, *J. Chromatogr.*, 329, 315, 1985.
39. **Churms, S. C., Merrifield, E. H., and Stephen, A. M.,** A comparative examination of two polysaccharide components from the gum of *Acacia mabellae*, *Carbohydr. Res.*, 63, 337, 1978.
40. **Churms, S. C. and Stephen, A. M.,** Studies of the molecular core of *Grevillea robusta* gum, *Carbohydr. Res.*, 167, 239, 1987.
41. **Vandevelde, M.-C. and Fenyo, J.-C.,** Macromolecular distribution of *Acacia senegal* gum (gum arabic) by size-exclusion chromatography, *Carbohydr. Polym.*, 5, 251, 1985.
42. **Hagel, L. and Andersson, T.,** Characteristics of a new agarose medium for high-performance gel filtration chromatography, *J. Chromatogr.*, 285, 295, 1984.
43. **Lecacheux, D. and Brigand, G.,** Preparative fractionation of natural polysaccharides by size exclusion chromatography, *Carbohydr. Polym.*, 8, 119, 1988.

Section I.IV

PLANAR CHROMATOGRAPHY TABLES

Since the publication of Volume I there have been few advances in paper chromatography, and therefore the data included in Tables PC 1 to 5 in that volume are still representative of the results given by standard methods in current use. Only the use of ion-exchanger papers for the PC of sugars, uronic acids, and alditols, and extension of the DP range of higher oligosaccharides resolved by PC with multiple developments, are added in the present volume. For this reason a separate section on PC is not justified here, and the two new tables of PC data have, therefore, been included in Section I.IV together with Tables TLC 1 to 6, which present data obtained as a result of the considerable advances in thin-layer chromatography, particularly in the application of high-performance thin-layer chromatography (HPTLC) plates, since 1978. The name ''planar chromatography'' used in the heading to this section is thus a generic term covering both PC and TLC.

TABLE PC 1
Chromatography of Sugars, Uronic Acids, and Alditols on Ion-Exchanger Papers

Paper	P1	P2	P3	P4
Solvent	S1	S2	S2	S2
Technique	T1	T2	T3	T4
Reference	1	2	2	2
Compound	$R_{Glc} \times 100$			
D-Lyxose	—	190	182	197
D-Xylose	—	180	176	181
D-Ribose	—	153	162	181
L-Arabinose	—	142	142	140
D-Mannose	—	132	135	128
D-Talose	—	116	113	126
D-Glucose	100	100	100	100
D-Allose	—	85	77	98
D-Galactose	—	82	74	76
D-Galacturonic acid	36	—	—	—
D-Glucuronic acid	27	—	—	—
Cellobiouronic acid	12	—	—	—
Glycerol	—	421	405	472
Erythritol	—	237	250	283
D-Threitol	—	210	209	260
Ribitol	—	147	162	173
D-Arabinitol	—	126	142	153
Xylitol	—	86	114	129
Allitol	—	105	105	115
D-Altritol	—	84	88	102
D-Mannitol	—	83	81	96
Galactitol	—	73	71	80
D-Glucitol	—	53	68	79
L-Iditol	—	43	57	66

Paper P1 = Whatman DE 81 (DEAE-cellulose paper).
P2 = Whatman CM 82 (carboxymethylcellulose paper), La^{3+}-form.
P3 = Whatman CM 82, Ca^{2+}-form.
P4 = Whatman CM 82, Ba^{2+}-form.

Solvent S1 = ethyl acetate-acetic acid-water (3:1:1).
S2 = 1-butanol-ethanol-water (10:1:2).

Technique T1 = descending chromatography, 30 h (48 h for aldobiouronic acid).
T2 = descending chromatography; CMC sheet irrigated with 0.1 *M* aqueous lanthanum acetate for 24 h, then with water for 48 h; dried at room temperature before chromatography of alditols or aldoses (2 — 7 d).
T3 = CMC sheet irrigated with 0.15 *M* calcium acetate; other details as in T2.
T4 = CMC sheet irrigated with 0.15 *M* barium acetate; other details as in T2.

REFERENCES

1. **Caldes, G., Prescott, B., and Baker, P. J.,** Use of DEAE-cellulose paper in the paper chromatographic separation of uronic acids, *J. Chromatogr.*, 234, 264, 1982.
2. **Bilisics, L. and Petruš, L.,** Chromatographic separation of alditols and some aldoses on *O*-(carboxymethyl)cellulose paper in the lanthanum, calcium, and barium forms, *Carbohydr. Res.*, 146, 141,

TABLE PC 2
Chromatography of Maltodextrins,[a] DP 5—25

Paper	P1	P1
Solvent	S1 + S2 + S3	S1 + S2 + S3 + S4
Technique	T1	T2
Reference	1	1, 2
DP	**R[b] × 100**	
5	100	100
6	96	96
7	92	92
8	87	88
9	83	84
10	78	80
11	73	76
12	68	72
13	62	68
14	57	64
15	52	60
16	47	56
17	41	52
18	35	48
19	30	43
20	25	38
21	21	32
22	17	27
23	13	22
24	9	17
25	6	12

Note: This table supplements Table PC 4 in *CRC Handbook of Chromatography: Carbohydrates*, Volume I.

[a] Technique also applied to amylodextrins in Reference 2.
[b] Mobility relative to maltopentaose.

Paper	P1 =	Toyo filter paper No. 50 (Toyo Soda, Tokyo).
Solvent	S1 =	1-butanol-pyridine-water (6:4:3).
	S2 =	1-butanol-pyridine-water (1:1:1).
	S3 =	1-butanol-pyridine-water (2:2:2.9).
	S4 =	1-butanol-pyridine-water (1:1:1.9).
Technique	T1 =	descending chromatography: S1 for 40 h at 16° C; three descents (20 h each) with S2; four descents (20 h each) with S3.
	T2 =	as in T1, then two descents (20 h each) with S4.

REFERENCES

1. **Umeki, K. and Kainuma, K.,** Fractionation of maltosaccharides of relatively high degree of polymerization by multiple descending paper chromatography, *J. Chromatogr.*, 150, 242, 1978.
2. **Umeki, K. and Kainuma, K.,** Fine structure of Nägeli amylodextrin obtained by acid treatment of defatted waxy-maize starch — structural evidence to support the double-helix hypothesis, *Carbohydr. Res.*, 96, 143, 1981.

TABLE TLC 1
TLC of Alditols, Mono-, and Oligosaccharides on Cellulose Plates

Layer	L1	L2	L3	L3	L4	L4	L5
Solvent	S1	S2	S3	S4	S5	S6	S7 + S8
Technique	T1	T2	T3	T3	T4	T4	T5
Reference	1	1	2	2	3	3	4
Compound	$R_F \times 100$						$R_{Glc} \times 100$
	*	*	1*	2*	1*	2*	***
Glycerol	46	45	—	—	—	—	—
Erythritol	19	19	—	—	—	—	—
L-Threitol	26	3	—	—	—	—	—
D-Arabinitol	11	5	—	—	—	—	—
1-deoxy-	16	19	—	—	—	—	—
Ribitol	16	9	—	—	—	—	—
Xylitol	12	3	—	—	—	—	—
Allitol	11	9	—	—	—	—	—
D-Altritol	11	3	—	—	—	—	—
1-deoxy-	—	11	—	—	—	—	—
Galactitol	10	3	—	—	—	—	—
1-deoxy-L-	13	12	—	—	—	—	—
D-Glucitol	10	2	—	—	—	—	—
D-Mannitol	11	7	—	—	—	—	—
L-Mannitol, 1-deoxy-	13	14	—	—	—	—	—
D-Talitol, 1-deoxy-	19	10	—	—	—	—	—
D-Lyxose	17	18	—	—	—	—	—
D-Arabinose	20	17	—	—	61	68	—
D-Xylose	24	16	—	—	65	60	—
D-Ribose	24	7	—	—	—	—	—
L-Rhamnose	—	—	—	—	86	74	—
L-Fucose	28	31	—	—	78	79	—
2-Deoxy-D-*arabino*-hexose	39	47	—	—	—	—	—
2-Deoxy-D-*lyxo*-hexose	33	44	—	—	—	—	—
D-Galactose	15	6	—	—	43	58	—
D-Mannose	—	10	—	—	54	58	—
D-Glucose	18	12	—	—	46	52	100
D-Fructose	—	—	—	—	—	—	118
D-Glucuronic acid	—	—	—	—	47	31	—
D-Galacturonic acid	—	—	—	—	44	38	—
2-Amino-2-deoxy-D-glucose	—	—	26	37	—	—	—
2-Amino-2-deoxy-D-galactose	—	—	27	25	—	—	—
2-Amino-2-deoxy-D-mannose	—	—	35	34	—	—	—
2-Amino-2,6-dideoxy-L-mannose	—	—	40	46	—	—	—
2-Amino-2,6-dideoxy-L-galactose	—	—	46	39	—	—	—
2-Amino 2,6-dideoxy-D-glucose	—	—	52	52	—	—	—
Muramic acid	—	—	50	46	—	—	—
Sucrose	—	—	—	—	—	—	89
Nigerose	—	—	—	—	—	—	83
Maltose	—	—	—	—	—	—	78
Isomaltose	—	—	—	—	—	—	65
Raffinose	—	—	—	—	—	—	59
Panose	—	—	—	—	—	—	55
Isomaltotriose	—	—	—	—	—	—	49
Isomaltotetraose	—	—	—	—	—	—	38
Isomaltopentaose	—	—	—	—	—	—	29

TABLE TLC 1 (continued)
TLC of Alditols, Mono-, and Oligosaccharides on Cellulose Plates

Note:	* =	single development.
	*** =	threefold development (monodimensional).
	1* =	two-dimensional development, first run.
	2* =	two-dimensional development, second run orthogonal to first.
Layer	L1 =	Eastman Chromagram® sheets (20 × 20 cm), 6064 cellulose (without fluorescent indicator), impregnated with 5% sodium tungstate dihydrate adjusted with 1 *M* H_2SO_4 to pH 6; air-dried at room temperature, then 60° C for 30 min.
	L2 =	As L1, but pH adjusted to 8.
	L3 =	precoated 0.10-mm cellulose (without fluorescent indicator) on Al plates (20 × 20 cm).
	L4 =	0.25 mm Cellulose MN300 (Machery-Nagel) on glass plates (20 × 20 cm); after drying, plates heated at 100° C for 15 min.
	L5 =	Avicel® SF plates (20 × 20 cm; Asahi, Tokyo).
Solvent	S1 =	1-butanol-ethanol-water (40:11:19).
	S2 =	acetone-1-butanol-water (5:3:2).
	S3 =	2-propanol-90% formic acid-water (20:1:5).
	S4 =	lutidine-water (13:7).
	S5 =	1-butanol-2-butanone-formic acid-water (8:6:3:3).
	S6 =	phenol-water-formic acid (50:49:1, organic phase only).
	S7 =	1-butanol-pyridine-acetic acid-water (10:6:1:3).
	S8 =	phenol-1.5% aqueous ammonia (2:1).
Technique	T1 =	ascending development, 2.5 h.
	T2 =	ascending development, 1.5 h.
	T3 =	two-dimensional development at 30° C; first run with S3 for 5—6 h, second run with S4 for ca. 6 h in orthogonal direction.
	T4 =	two-dimensional development; two developments with S5, then single development with S6 in orthogonal direction.
	T5 =	two successive ascents with S7, then single ascent with S8; each development for 4 h at 20° C.

REFERENCES

1. **Briggs, J., Chambers, I. R., Finch, P., Slaiding, I. R., and Weigel, H.,** Thin-layer chromatography on cellulose impregnated with tungstate: a rapid method of resolving mixtures of some commonly occurring carbohydrates, *Carbohydr. Res.*, 78, 365, 1980.
2. **Ryan, E. A. and Kropinski, A. M.,** Separation of amino sugars and related compounds by two-dimensional thin-layer chromatography, *J. Chromatogr.*, 195, 127, 1980.
3. **Métraux, J. P.,** Thin-layer chromatography of neutral and acidic sugars from plant cell wall polysaccharides, *J. Chromatogr.*, 237, 525, 1982.
4. **Horikoshi, T., Koga, T., and Hamada, S.,** Thin-layer chromatography using multiple development for analysis of reaction products of sucrases, *J. Chromatogr.*, 416, 353, 1987.

TABLE TLC 2
TLC of Mono- and Oligosaccharides, and Pyruvic Acetals, on Unmodified Silica Gel

Layer	L1	L1	L1	L1	L1	L2	L2	L2	L2	L2	L2	L2	L3
Solvent	S1	S2	S3	S4	S5	S6	S7	S8	S9	S10	S11	S12	S13
Technique	T1	T1	T1	T1	T1	T2	T3	T2	T4	T5	T6	T7	T8
Reference	1[a]	1	1	1	1	2	2	2	2	3	4	5	6

TABLE TLC 2 (continued)
TLC of Mono- and Oligosaccharides, and Pyruvic Acetals, on Unmodified Silica Gel

Compound	$R_F \times 100$											$R_{Gal} \times 100$	$R_F \times 100$
	*	*	*	*	*	*	*	*	*	**	***	***	***
DL-Glyceraldehyde	77	[b]	73	68	36	—	—	—	—	—	—	—	—
D-Erythrose	65	74	67	58	35	—	—	—	94	—	—	—	—
D-Threose	71	83	75	63	43	—	—	—	—	—	—	—	—
L-Arabinose	38	43	35	43	17	—	—	40	—	36	—	—	—
D-Ribose	47	55	45	45	32	37	—	79	—	—	—	—	—
D-Xylose	51	59	53	53	23	—	—	53	65	48	68	—	—
D-Lyxose	53	60	54	53	27	—	—	—	—	—	—	—	—
2-Deoxy-D-*erythro*-pentose	69	78	73	62	45	76	85	73	—	71	85	—	—
2-Deoxy-D-*lyxo*-Hexose	59	65	63	59	32	—	—	—	—	—	—	—	—
2-Deoxy-D-*arabino*-Hexose	65	72	71	63	35	—	—	—	—	—	—	—	—
L-Fucose	51	54	48	52	23	57	75	52	—	—	—	—	—
L-Rhamnose	67	78	71	64	38	—	84	—	—	63	79	—	—
6-Deoxy-D-glucose	63	72	67	63	33	—	—	—	—	—	—	—	—
2,6-Dideoxy-D-*ribo*-Hexose	82	89	87	71	64	—	—	—	—	—	—	—	—
D-Galactose	23	23	21	34	10	33	49	32	38	16	37	100	—
D-Allose	30	32	28	38	10	—	—	—	—	—	—	—	—
D-Glucose	29	31	29	41	13	42	61	34	40	22	43	—	—
D-Talose	35	35	32	36	26	—	—	—	—	—	—	—	—
D-Gulose	33	35	33	41	17	—	—	—	—	—	—	—	—
D-Mannose	35	37	37	43	16	—	65	—	50	—	—	—	—
D-Altrose	43	49	45	47	17	—	—	—	—	—	—	—	—
D-Idose	47	55	48	47	9	—	—	—	—	—	—	—	—
D-*glycero*-D-*gluco*-Heptose	20	22	19	33	13	—	—	—	—	—	—	—	—
D-*erythro*-Pentulose	—	—	—	—	—	—	—	—	80	—	—	—	—
D-*threo*-Pentulose	—	—	—	—	—	—	—	—	82	—	—	—	—
D-Fructose	—	—	—	—	—	—	28	42	51	27	49	—	—
L-Sorbose	—	—	—	—	—	13	—	48	—	32	—	—	—
D-Tagatose	—	—	—	—	—	—	—	58	—	—	—	—	—
D-*altro*-Heptulose	—	—	—	—	—	—	—	34	—	—	—	—	—
D-*manno*-Heptulose	—	—	—	—	—	—	—	40	—	—	—	—	—
D-Glucuronic acid	—	—	—	—	—	—	4	—	—	—	—	—	—
D-Galacturonic acid	—	—	—	—	—	—	4	—	—	—	—	—	—
2-Amino-2-deoxy-D-galactose	—	—	—	—	—	—	—	—	—	—	—	—	35
2-Amino-2-deoxy-D-glucose	—	—	—	—	—	—	—	—	—	—	—	—	39
2-Amino-2-deoxy-D-mannose	—	—	—	—	—	—	—	—	—	—	—	—	47
D-Muramic acid	—	—	—	—	—	—	—	—	—	—	—	—	76
-*O*-(1-Carboxyethylidene)-D-galactose													
(*R*)-3,4-	—	—	—	—	—	—	—	—	—	—	—	121	—
(*S*)-3,4-	—	—	—	—	—	—	—	—	—	—	—	136	—
(*R*)-4,6-	—	—	—	—	—	—	—	—	—	—	—	108	—
(*S*)-4,6-	—	—	—	—	—	—	—	—	—	—	—	111	—
-*O*-β-D-Galactopyranosyl-D-galactose													
2-	—	—	—	—	—	—	—	—	—	—	—	40	—
3-	—	—	—	—	—	—	—	—	—	—	—	40	—
4-	—	—	—	—	—	—	—	—	—	—	—	47	—
6-	—	—	—	—	—	—	—	—	—	—	—	30	—
Lactose	—	—	—	—	—	19	29	12	18	4	16	—	—
Lactulose	—	—	—	—	—	—	10	—	—	—	—	—	—
Maltose	—	—	—	—	—	—	44	—	—	—	20	—	—

TABLE TLC 2 (continued)
TLC of Mono- and Oligosaccharides, and Pyruvic Acetals, on Unmodified Silica Gel

	$R_F \times 100$											$R_{Gal} \times 100$	$R_F \times 100$
Compound	*	*	*	*	*	*	*	*	*	**	***	***	***
Isomaltose	—	—	—	—	—	—	22	—	—	—	—	—	—
Sucrose	—	—	—	—	—	—	59	25	34	12	26	—	—
Palatinose	—	—	—	—	—	8	9	20	27	—	—	—	—
Melezitose	—	—	—	—	—	—	—	—	—	—	12	—	—
Raffinose	—	—	—	—	—	—	16	8	12	—	7	—	—

Note: * = single development, ** = double development, *** = threefold development.

[a] Data from Reference 1 supplement earlier data for ketoses: see Table TLC 2 in *CRC Handbook of Chromatography: Carbohydrates,* Volume I.

[b] Streaks.

Layer L1 = 0.25 mm silica gel 60 pre-coated on glass plates (20 × 20 cm; Merck No. 5721).
L2 = silica gel 60 pre-coated on Al sheets (20 × 20 cm; Merck No. 5553).
L3 = K5 silica gel pre-coated plates (Whatman).

Solvent S1 = 1-butanol-acetone-water (4:5:1).
S2 = acetone-water (9:1).
S3 = 2-propanol-ethyl acetate-water (6:3:1).
S4 = 1-butanol-methanol-water (5:3:1).
S5 = 2-butanone-acetic acid-saturated aqueous boric acid solution (9:1:1).
S6 = 2-butanone-2-propanol-acetonitrile-0.5 *M* boric acid + 0.25 *M* 2-propylamine (4:3:1:2)
S7 = 2-butanone-2-propanol-acetonitrile-0.5 *M* boric acid + 0.25 *M* 2-propylamine-acetic acid (4:3:2:1.5:0.04).
S8 = ethyl acetate-2-propanol-0.5 *M* 2-propylammonium acetate-methanol-phenyldihydroxyboronic acid (5:3:1.5:0.5:0.1).
S9 = 2-butanone-2-propanol-diisopropyl ether-pyridine-water-phenyldihydroxyboronic acid (4:3:1:1:1:0.12).
S10 = acetone-chloroform-water (17:2:1).
S11 = acetonitrile-water (17:3).
S12 = ethyl acetate-acetic acid-formic acid-water (12:3:1:4).
S13 = acetonitrile-ethanol-acetic acid-water (13:2:1:4).

Technique T1 = ascending development, 15 cm; developing chamber equilibrated with solvent for 1 h before use.
T2 = ascending development, 10 cm; developing chamber saturated with solvent vapor before use.
T3 = ascending development 60—70 min.
T4 = ascending development, 70—75 min.
T5 = two successive ascents, each 17.5 cm (70 min) in vapor-saturated chamber: plate dried for 2 min at 110° C after each development.
T6 = three successive ascents (each 60 min) at 21° C in vapor-saturated chamber: plate dried for 20 min at 40—45° C after each development.
T7 = three successive ascents (4.5, 9.0, and 18.0 cm): plate dried after each development.
T8 = three successive ascents (each 70—80 min): after each development plate dried in ventilated hood for 15 min, then for 30 min at 37° C in convection oven.

REFERENCES

1. **Papin, J.-P. and Udiman, M.,** Thin-layer chromatography of principal aldoses, *J. Chromatogr.*, 170, 490, 1979.
2. **Ghebregzabher, M., Rufini, S., Sapia, G. M., and Lato, M.,** Improved thin-layer chromatograpic method for sugar separations, *J. Chromatogr.*, 180, 1, 1979.
3. **Žilić, Ž.,** Simple rapid method for the separation and qualitative analysis of carbohydrates in biological fluids, *J. Chromatogr.*, 164, 91, 1979.
4. **Gauch, R., Leuenberger, U., and Baumgartner, E.,** Quantitative determination of mono-, di- and trisaccharides by thin-layer chromatography, *J. Chromatogr.*, 174, 195, 1979.

5. **Fontana, J. D., Duarte, J. H., Iacomini, M., and Gorin, P. A. J.,** Synthesis and chromatographic properties of isomeric *O*-β-D-galactropyranosyl-D-galactoses, and of diastereoisomers of 3,4-*O*- and 4,6-*O*-(1-carboxy-ethylidene)-D-galactose, *Carbohydr. Res.*, 108, 221, 1982.
6. **Rebers, P. A., Wessman, G. E., and Robyt, J. F.,** A thin-layer chromatographic method for analysis of amino sugars in polysaccharide hydrolyzates, *Carbohydr. Res.*, 153, 132, 1986.

TABLE TLC 3
TLC of Alditols, Acids, Lactones, Mono-, and Oligosaccharides on Impregnated Silica Gel and HPTLC Plates

Layer	L1	L2	L3	L4	L5	L6	L7	L7	L7	L7	L7	L8
Solvent	S1	S2	S3	S4	S5	S5	S6	S7	S8	S9	S10	S11
Technique	T1	T2	T2	T2	T3	T3	T4	T4	T5	T4	T4	T5
Reference	1	2	2	2	3	3	4	4	4	4	4	4
	$R_F \times 100$											
Compound	*	*	*	*	*	*	*	*	**	*	*	**
Glycerol	73	—	—	—	—	—	73	60	—	—	—	—
Threitol	61	—	—	—	—	—	—	—	—	—	—	—
Erythritol	64	—	—	—	17	36	69	50	—	—	—	—
Xylitol	38	—	—	—	2	10	—	—	—	—	—	—
D-Arabinitol	47	—	—	—	7	16	—	—	—	30	48	—
Ribitol	53	—	—	—	—	—	—	—	—	34	54	—
D-Glucitol	21	—	—	—	1	5	41	13	—	—	—	—
Galacitol	25	—	—	—	—	—	—	—	—	—	—	—
D-Mannitol	30	—	—	—	4	9	44	18	—	—	—	—
D-*glycero*-D-*galacto*-Heptitol	13	—	—	—	—	—	—	—	—	—	—	—
D-*glycero*-D-*manno*-Heptitol	18	—	—	—	—	—	—	—	—	—	—	—
1,5-Anhydroglucitol	—	—	—	—	44	53	—	—	—	—	—	—
L-Arabinose	37	26	—	45	—	—	61	34	—	29	46	32
D-Ribose	47	43	—	57	—	—	68	32	—	42	46	65
D-Xylose	55	37	—	62	16	35	69	45	—	42	56	41
D-Lyxose	—	—	—	—	—	—	—	—	—	42	56	69
L-Fucose	—	54	—	72	—	—	69	47	77	—	—	—
L-Rhamnose	—	74	—	89	—	—	79	60	83	—	—	—
2-Deoxy-D-*erythro*-pentose	—	—	—	—	—	—	79	62	—	—	—	—
D-Galactose	16	9	—	26	—	—	44	24	—	—	—	—
D-Glucose	28	12	—	37	20	25	53	32	68	—	—	—
D-Mannose	—	20	—	48	—	—	56	35	—	—	—	—
D-Talose	—	—	—	—	—	—	56	24	—	—	—	—
D-Fructose	34	—	—	—	5	14	54	20	—	—	—	—
L-Sorbose	—	—	—	—	—	—	61	16	—	—	—	—
D-Tagatose	—	—	—	—	—	—	66	28	—	—	—	—
D-Ribono-1,4-lactone	—	—	—	—	—	—	—	—	—	69	61	—
L-Arabinono-1,4-lactone	—	—	—	—	—	—	—	—	—	73	76	—
D-Glucono-1,4-lactone	—	—	—	—	—	—	—	—	—	79	76[a]	—
D-Gluconic acid	—	—	—	—	—	—	—	—	—	3	5	—
D-Galacturonic acid	—	—	37	—	—	—	—	—	—	—	—	—
L-Guluronic acid	—	—	49	—	—	—	—	—	—	—	—	—
D-Glucuronic acid	—	—	55	—	—	—	—	—	—	—	—	—
D-Mannuronic acid	—	—	63	—	—	—	—	—	—	—	—	—
D-Mannurono-6,3-lactone	—	—	85	—	—	—	—	—	—	—	—	—
D-Glucurono-6,3-lactone	—	—	89	—	—	—	—	—	—	—	—	—
L-Gulurono-6,3-lactone	—	—	95	—	—	—	—	—	—	—	—	—
Melibiose	—	—	—	—	—	—	16	6	—	—	—	—
Trehalose	—	—	—	—	—	—	27	16	55	—	—	—
Lactulose	—	—	—	—	—	—	27	8	—	—	—	—
Gentiobrose	—	—	—	—	—	—	29	10	—	—	—	—
Lactose	—	—	—	22	—	—	33	13	—	—	—	—
Palatinose	—	—	—	—	—	—	40	8	—	—	—	—

TABLE TLC 3 (continued)
TLC of Alditols, Acids, Lactones, Mono-, and Oligosaccharides on Impregnated Silica Gel and HPTLC Plates

Compound	$R_F \times 100$											
	*	*	*	*	*	*	*	*	**	*	*	**
Cellobiose	—	—	—	—	—	—	43	16	—	—	—	—
Maltose	—	—	—	34	—	—	46	16	—	—	—	—
Sucrose	—	—	—	47	—	—	46	26	62	—	—	—
Raffinose	—	—	—	16	—	—	14	5	46	—	—	—
Melezitose	—	—	—	36	—	—	—	—	—	—	—	—
Maltotriose	—	—	—	—	—	—	—	—	49	—	—	—
Maltotetraose	—	—	—	—	—	—	—	—	43	—	—	—
Maltoheptaose	—	—	—	—	—	—	—	—	34	—	—	—

Note: * = single development, ** = double development.

[a] Slower spot at R_F 0.43 due to 1,5-lactone.

Layer L1 = silica gel 60 precoated on glass plates (20 × 20 cm; Merck No. 5715), impregnated by spraying with 0.5 *M* NaH_2PO_4 in 50% ethanol; allowed to stand for 15 min, then activated by drying at 110° C for 1 h.
L2 = sintered silica gel plates (20 × 5 cm; 0.2 mm silica gel layer), impregnated with 0.3 *M* Na_2HPO_4 by immersion in solution for 5 min.; air-dried in vertical position, then activated at 105° C for 1 h.
L3 = as L2, impregnated with 0.3 *M* NaH_2PO_4.
L4 = as L2, impregnated with 0.5 *M* NaH_2PO_4.
L5 = Silica gel G (Merck), slurried in borate buffer (0.1 *M*) in methanol-water (1:1), coated to thickness of 0.25 mm on glass plates (10 × 20 cm); air-dried at room temperature for 24 h.
L6 = silica gel 60 high-performance thin-layer chromatography (HPTLC) plates (Merck), impregnated by immersion in borate-buffered methanol solution; air-dried at room temperature.
L7 = silica gel 60 HPTLC plates or HPTLC Al sheets (Merck).
L8 = silica gel 60 HPTLC plates, impregnated with phosphate buffer, pH 8, by immersion for 15 s; dried at 120° C for 30 min.

Solvent S1 = 2-propanol-acetone-0.2 *M* lactic acid (6:3:1).
S2 = 1-butanol-acetone-water (4:5:1).
S3 = 1-butanol-ethanol-0.1 *M* orthophosphoric acid (1:10:5).
S4 = 2-propanol-methanol-water (16:1:3).
S5 = 1-butanol-acetone-water (5:4:1).
S6 = ethyl acetate-pyridine-water-acetic acid-propionic acid (10:10:2:1:1).
S7 = 1 butanol-2-propanol-0.08 *M* boric acid (3:5:1).
S8 = 1-propanol-nitromethane-water-glacial acetic acid (5:3:2:0.1).
S9 = ethyl acetate-pyridine-tetrahydrofuran-water-acetic acid (50:22:15:15:4).
S10 = 1-butanol-2-propanol-0.08 *M* boric acid-glacial acetic acid (15:25:5:4).
S11 = 2-butanone-acetic acid 0.08 *M* boric acid (9:1:1).

Technique T1 = ascending development, 15 cm (4 h).
T2 = ascending development in vapor-saturated chamber.
T3 = ascending development, 10 cm (30 — 45 min).
T4 = ascending development in vapor-saturated chamber, to 7 cm.
T5 = two successive ascents, each to 7 cm.

REFERENCES

1. **Kremer, B. P.,** Improved method for the thin-layer chromatographic identification of alditols, *J. Chromatogr.*, 166, 335, 1978.
2. **Iizima, N., Fujihara, M., and Nagumo, T.,** Application of a sintered silica gel plate to the thin-layer chromatography of carbohydrates, *J. Chromatogr.*, 193, 464, 1980.
3. **Lajunen, K., Purokoski, S., and Pitkänen, E.,** Qualitative thin-layer chromatographic separation of 1,5-anhydroglucitol in the presence of other carbohydrates on silica gel impregnated with borate buffer, *J. Chromatogr.*, 187, 455, 1980.
4. **Klaus, R. and Ripphahn, J.,** Quantitative thin-layer chromatographic analysis of sugars, sugar acids and polyalcohols, *J. Chromatogr.*, 244, 99, 1982.

TABLE TLC 4
TLC of Sugars and Derivatives, Oligosaccharides, and Alditols on Bonded-Phase Plates

Layer	L1	L2	L3	L3	L3
Solvent	S1	S2	S2	S3	S4
Technique	T1	T2	T3	T3	T3
Reference	1	2	3	3	3
Compound	$R_F \times 100$				
Glycerol	70	—	—	—	—
Erythritol	56	—	—	—	—
L-Threitol	37	—	—	—	—
Ribitol	56	—	—	—	—
D-Arabinitol	28	—	—	—	—
Xylitol	5	—	—	—	—
Allitol	42	—	—	—	—
Altritol	23	—	—	—	—
1-deoxy-	30	—	—	—	—
D-Mannitol	18	—	—	—	—
L-Mannitol, 1-deoxy-	31	—	—	—	—
Galactitol	13	—	—	—	—
1-deoxy-L-	28	—	—	—	—
D-Glucitol	2	—	—	—	—
D-Talitol, 1-deoxy-	49	—	—	—	—
Maltitol	30	—	—	—	—
Cellobiitol	46	—	—	—	—
D-Arabinose	76	0	30	—	—
D-Xylose	76	0	35	—	—
D-Lyxose	67	0	35	—	—
D-Ribose	30	0	39	—	—
2-Deoxy-D-*erythro*-pentose	—	0	50	—	—
2-Deoxy-D-*arabino*-hexose	—	0	42	—	—
L-Fucose	78	0	37	—	—
L-Rhamnose	69	0	44	—	—
D-Glucose	80	35	22	—	47
D-Galactose	79	0	21	—	—
D-Mannose	63	0	25	—	—
D-Gulose	40—59	—	—	—	—
D-Allose	39	—	—	—	—
D-Talose	14	0	31	—	—
D-*erythro*-Pentulose	—	—	40	—	—
D-Fructose	65	41	27	—	—
D-Psicose	—	47	33	—	—
D-Tagatose	—	42	30	—	—
L-Sorbose	—	40	29	—	—
D-*altro*-Heptulose	—	46	32	—	—
D-Glucuronic acid	2	—	—	—	—
D-Galacturonic acid	0—15	—	—	—	—
D-Mannuronic acid	0	—	—	—	—
L-Ascorbic acid	1	—	—	—	—
2-Amino-deoxy-D-glucose	—	—	4	—	—
2-Acetamido-2-deoxy-D-glucose	—	40	38	—	—
Methyl α-D-glucopyranoside	81	—	—	—	—
Methyl β-D-glucopyranoside	—	54	44	—	—
Methyl α-D-mannopyranoside	75	57	50	—	—
Methyl β-D-galactopyranoside	78	—	—	—	—
Methyl β-L-arabinopyranoside	—	64	55	—	—
D-Glucose					

TABLE TLC 4 (continued)
TLC of Sugars and Derivatives, Oligosaccharides, and Alditols on Bonded-Phase Plates

Compound	$R_F \times 100$				
3-*O*-methyl-	—	—	—	55	—
1,6-anhydro-	—	—	—	49	—
diethyl dithioacetal	—	—	—	31	—
D-Glucopyranose					
4,6-*O*-benzylidene	—	—	—	56	—
4,6-*O*-ethylidene	—	—	—	46	—
α-D-Glucofuranose					
1,2-*O*-isopropylidene	—	66	—	49	—
1,2:5,6-di-*O*-isopropylidene	—	—	—	77	—
Sucrose	—	26	18	—	—
Cellobiose	88	20	15	—	—
Cellobiulose	—	23	17	—	—
Maltose	87	20	14	—	41
Maltulose	—	22	17	—	—
Lactose	87	19	14	—	—
Lactulose	77	22	16	—	—
Kojibiose	87	20	14	—	—
Laminaribiose	87	23	18	—	—
Isomaltose	92	21	12	—	—
Gentiobiose	88	17	12	—	—
Melibiose	—	16	11	—	—
Turanose	—	24	18	—	—
Palatinose	—	24	17	—	—
αα-Trehalose	—	21	14	—	—
Melezitose	—	17	12	—	—
Raffinose	—	14	9	—	—
Maltotriose	—	12	10	—	34
Stachyose	—	6	4	—	—
Maltotetraose	—	—	—	—	29
Maltopentaose	—	—	—	—	24
Maltohexaose	—	—	—	—	20
Maltoheptaose	—	—	—	—	16

Layer	L1	=	Polygram® Ionex-25 SA-Na sheets (Macherey-Nagel), converted to Cu-form by immersion in 5% aqueous cupric acetate solution for 1 h; washed with deionized water, air-dried at room temperature for 24 h.
	L2	=	precoated silica gel plates (0.25 mm layer, 10% of $CaSO_4$ as binder), derivatized by immersion in solution of (3-aminopropyl)triethoxysilane (1% v/v) in dry hexane for 15 min; plates washed by immersion in dry hexane for 15 min, then dried *in vacuo* at 60° C for 15 min.
	L3	=	silica gel HPTLC plates (Whatman HP-KF) derivatized as for L2, then impregnated with NaH_2PO_4 by immersion in 0.2 *M* aqueous solution for 15 min; plates drained on paper towel for 30 min, then dried *in vacuo* at 70° C for 2 h.
Solvent	S1	=	deionized water.
	S2	=	acetonitrile-water (7:3).
	S3	=	acetonitrile-water (9:1).
	S4	=	acetonitrile-water (3:2).
Technique	T1	=	ascending development, 1 — 1.5 h.
	T2	=	ascending development, 10 cm (ca. 22 min).
	T3	=	ascending development.

TABLE TLC 4 (continued)
TLC of Sugars and Derivatives, Oligosaccharides, and Alditols on Bonded-Phase Plates

REFERENCES

1. **Briggs, J., Finch, P., Matulewicz, M. C., and Weigel, H.,** Complexes of copper (II), calcium and other metal ions with carbohydrates: thin-layer ligand-exchange chromatography and determination of relative stabilities of complexes, *Carbohydr. Res.*, 97, 181, 1981.
2. **Doner, L. W., Fogel, C. L., and Biller, L. M.,** 3-Aminopropyl-bonded-phase silica thin-layer chromatographic plates: their preparation and application to sugar resolution, *Carbohydr. Res.*, 125, 1, 1984.
3. **Doner, L. W. and Biller, L. M.,** High-performance thin-layer chromatographic separation of sugars: preparation and application of aminopropyl bonded-phase silica plates impregnated with monosodium phosphate, *J. Chromatogr.*, 287, 391, 1984.

Data from References 1 to 3 reproduced with permission of authors and Elsevier Science Publishers.

TABLE TLC 5
TLC of Higher Oligosaccharides

Type of oligosaccharide	Malto-saccharides											Sophoro-			Laminari		Gentio-	Cello-	Xylo-	Chito-	(GalA)$_n$[a]
Layer	L1	L1	L1	L1	L1	L2	L3	L4	L4	L4	L4	L2	L3	L4	L3	L4	L4	L2	L2	L5	L6
Solvent	S1	S2	S3	S4	S5+S6	S7	S8	S9	S10	S10	S10	S7	S8	S9	S8	S9	S9	S11	S11	S12	S13
Technique	T1	T2	T3	T4	T5	T6	T6	T6	T6	T7	T7	T6	T6	T6	T6	T6	T6	T8	T8	T8	T9
Reference	1	1	2	2	3	4	4	4	4	4	4	4	4	4	4	4	4	5	5	5	6
	$R_{Glc} \times 100$																		$R_{Xyl} \times 100$	$R_{GNAc} \times 100$	$R_{GalA} \times 100$
DP	C	C	*6	*6	*3	*4	*4	*4	*4	1*5	2*3	*4	*4	*4	*4	*4	*4	*3	*3	*3	*
1	100	100	100	100	100	100	100	100	100	100	100	100	100	100	100	100	100	100	100	100	100
2	92	94	92	95	92	97	96	92	96	97	94	98	97	94	97	97	89	90	89	81	98
3	80	86	82	88	84	95	89	84	91	92	86	97	94	89	94	92	76	83	80	61	95
4	66	78	72	79	76	93	82	76	84	86	78	96	89	84	91	87	69	72	73	43	91
5	54	70	64	67	69	91	76	69	75	80	71	90	84	78	88	82	62	57	67	28	85
6	44	62	56	58	62	90	73	65	68	74	63	85	79	74	84	77	56	44	60	13	78
7	34	54	50	50	56	89	70	61	62	69	58	79	75	70	81	73	50	32	52	—	69
8	26	46	45	41	50	84	66	57	56	63	52	74	72	67	78	70	43	25	44	—	61
9	18	38	40	32	44	79	62	53	50	57	48	69	69	64	76	68	37	—	36	—	54
10	14	30	35	23	38	75	57	49	45	51	44	64	66	62	74	66	30	—	27	—	—
11	—	24	—	—	32	71	51	45	40	46	40	60	63	60	72	63	24	—	—	—	—
12	—	19	—	—	26	67	45	41	35	42	37	56	60	58	71	61	19	—	—	—	—
13	—	—	—	—	21	63	39	37	30	37	34	52	56	56	69	60	14	—	—	—	—
14	—	—	—	—	19	59	33	33	26	33	31	48	52	54	68	58	10	—	—	—	—
15	—	—	—	—	17	55	27	28	23	30	28	45	48	51	66	56	8	—	—	—	—
16	—	—	—	—	—	52	22	24	20	27	24	42	44	48	65	55	6	—	—	—	—
17	—	—	—	—	—	50	17	20	17	24	21	39	40	45	63	54	4	—	—	—	—
18	—	—	—	—	—	47	13	17	14	21	18	36	36	42	62	53	—	—	—	—	—
19	—	—	—	—	—	44	10	14	11	18	15	33	32	40	61	52	—	—	—	—	—
20	—	—	—	—	—	41	8	12	9	16	12	30	29	38	60	51	—	—	—	—	—
21	—	—	—	—	—	38	6	10	8	14	10	27	25	36	—	49	—	—	—	—	—
22	—	—	—	—	—	35	—	8	7	11	9	25	22	34	—	48	—	—	—	—	—
23	—	—	—	—	—	32	—	6	6	8	8	22	20	33	—	47	—	—	—	—	—

TABLE TLC 5 (continued)
TLC of Higher Oligosaccharides

DP	$R_{Glc} \times 100$																		$R_{Xyl} \times 100$	$R_{GNAc} \times 100$	$R_{GalA} \times 100$
	C	C	*6	*6	*3	*4	*4	*4	*4	1*5	2*3	*4	*4	*4	*4	*4	*4	*3	*3	*3	*
24	—	—	—	—	—	29	—	5	5	6	6	20	18	32	—	—	—	—	—	—	—
25	—	—	—	—	—	26	—	4	4	5	5	17	16	30	—	—	—	—	—	—	—
26	—	—	—	—	—	24	—	—	—	—	—	15	14	28	—	—	—	—	—	—	—
27	—	—	—	—	—	—	—	—	—	—	—	13	12	27	—	—	—	—	—	—	—
28	—	—	—	—	—	—	—	—	—	—	—	11	10	26	—	—	—	—	—	—	—
29	—	—	—	—	—	—	—	—	—	—	—	9	9	25	—	—	—	—	—	—	—
30	—	—	—	—	—	—	—	—	—	—	—	8	8	24	—	—	—	—	—	—	—

Note: This table supplements Table TLC 5 in *CRC Handbook of Chromatography: Carbohydrates,* Volume I. HPLC data for oligosaccharides of Reference 4 will be found in Table LC 3 of this volume.

C = continuous development. *n = n-fold development (monodimensional). 1* = two-dimensional development, first direction. 2* = two-dimensional run, second direction, orthogonal to first.

[a] Oligogalacturonic acids.
[b] Mobility relative to 2-acetamido-2-deoxy-D-glucose.

Layer
- L1 = silica gel 60 HPTLC plates (Merck No. 5633).
- L2 = Si 50,000 silica gel HPTLC plates (Merck), conditioned at 110° C for 30 — 60 min and cooled in desiccator before use.
- L3 = silica gel 60 plates with concentration zone (Merck), conditioned as for L2.
- L4 = silica gel 70 plates (Wako, Osaka, Japan), conditioned as for L2.
- L5 = HP-KF silica gel HPTLC plates (Whatman) derivatized with 3-aminopropyltriethoxysilane and then impregnated with NaH_2PO_4 (See Table TLC 4).
- L6 = HP-KF silica gel HPTLC plates, activated at 100°C for 30 min and cooled in vacuum desiccator before use.

Solvent
- S1 = 2-propanol-acetone-water (8:7:5).
- S2 = ethanol-acetone-water (9:6:5).
- S3 = 1-butanol-ethanol-water (2:2:1).
- S4 = acetone-water (3:1).
- S5 = 1-propanol-acetone-water (9:6:5).
- S6 = 1-propanol-acetone-water (5:4:1).
- S7 = 1-butanol-pyridine-water (6:5:4).
- S8 = 1-butanol-ethanol-water (5:5:3).

	S9 =	1-butanol-ethanol-water (5:5:4).
	S10 =	1-butanol-ethanol-water (1:1:1).
	S11 =	1-butanol-pyridine-water (8:5:4).
	S12 =	acetonitrile-water (18:7).
	S13 =	ethanol-25 m*M* aqueous acetic acid (21:29).
Technique	T1 =	continuous development, with end of plate exposed to atmosphere, for 80 min.
	T2 =	continuous development for 85 min.
	T3 =	six successive developments (each for 1 h); plates dried in stream of air for 5 min at 30°C after each development.
	T4 =	six successive developments (each for 30 min); plates dried as for T3.
	T5 =	three successive developments, first and third with S5, second with S6; after each development plate dried with air, then *in vacuo* for 10 min.
	T6 =	four successive developments; after each development plates dried as for T5.
	T7 =	two-dimensional development; five successive developments in first direction, three in second; plate dried between developments as for T5.
	T8 =	three successive developments.
	T9 =	single development at 35° C.

REFERENCES

1. **Nurok, D. and Zlatkis, A.,** The separation of malto-oligosaccharides by high-performance thin-layer chromatography, *Carbohydr. Res.*, 65, 265, 1978.
2. **Nurok, D. and Zlatkis, A.,** The separation of malto-oligosaccharides by high-performance thin-layer chromatography using multiple developments, *Carbohydr. Res.*, 81, 167, 1980.
3. **Würsch, P. and Roulet, Ph.,** Quantitative estimation of malto-oligosaccharides by high-performance thin-layer chromatography, *J. Chromatogr.*, 244, 177, 1982.
4. **Koizumi, K., Utamura, T., and Okada, Y.,** Analyses of homogeneous D-gluco-oligosaccharides and -polysaccharides (degree of polymerization up to about 35) by high-performance liquid chromatography and thin-layer chromatography, *J. Chromatogr.*, 321, 145, 1985.
5. **Doner, L. W.,** High-performance thin-layer chromatography of starch, cellulose, xylan and chitin hydrolyzates, *Methods Enzymol.*, 160, 176, 1988.
6. **Doner, L. W., Irwin, P. L., and Kurantz, M. J.,** High-performance thin-layer chromatographic resolution of oligogalacturonic acids, *Carbohydr. Res.*, 172, 292, 1988.

TABLE TLC 6
TLC of Cyclodextrins and Branched Cyclodextrins on TLC, HPTLC, and Bonded-Phase Silica Gel Plates

Layer	L1	L2	L3	L4	L4
Solvent	S1	S1	S2	S3	S4 + S3
Technique	T1	T2	T3	T4	T5
Reference	1	1	1	1	1
	$R_F \times 100$				
Compound	***5**	***4**	***5**	*****	***2**
Cyclomaltohexoaose[a]	58	59	74	54	65
6-*O*-α-D-glucosyl	45	47	64	34	41
6-*O*-α-maltosyl-	39	40	56	27	31
6-*O*-α-maltotriosyl-	34	34	49	21	24
Cyclomaltoheptaose[a]	52	54	68	66	79
6-*O*-α-D-glucosyl-	42	43	58	52	58
6-*O*-α-maltosyl-	37	38	49	45	49
6-*O*-α-maltotriosyl-	33	33	41	38	42
6,6'''-di-*O*-α-D-glucosyl-	34	33	48	34	40
Cyclomaltooctaose[a]	49	50	62	59	73
6-*O*-α-D-glucosyl-	39	39	51	44	52
6-*O*-maltosyl-	34	33	42	36	42
6-*O*-α-maltotriosyl-	30	29	35	30	35

Note: *n = n-fold development (monodimensional). HPLC data for these compounds will be found in Table LC 4, LC 13, LC 47, and LC 52.

[a] Cyclomaltohexaose, - heptaose, and -octaose known, respectively, as α-, β-, and γ-cyclodextrin.

Layer
- L1 = silica gel 60 plates (Merck; cut into 20 × 5 cm sections).
- L2 = silica gel 60 F_{254} HPTLC plates (Merck; cut into 10 × 5 cm sections).
- L3 = NH_2 F_{254} HPTLC plates (Merck; amine-bonded silica, 10 × 5 cm).
- L4 = Si 50,000 F_{254} HPTLC plates (Merck; pore size 50,000 Å, 10 × 5 cm).

Solvent
- S1 = 1-propanol-ethyl acetate-water-25% aqueous ammonia (3:1:2:1).
- S2 = acetonitrile-water (3:2).
- S3 = 1-butanol-pyridine-water (6:4:3).
- S4 = 1-butanol-pyridine-water (6:3:2).

Technique
- T1 = five successive developments (each for 6 h); plate dried for 30 — 60 min at 40 — 50° C after each development.
- T2 = four successive developments (each for 2.5 h); plate dried as in T1.
- T3 = five successive developments (each for 30 min); plate dried as in T1.
- T4 = plate pre-washed with solvent S3, dried for 1 h at 100° C then *in vacuo* overnight; after single development for 1.5 h plate dried with air, then *in vacuo* for 2 h.
- T5 = plate preconditioned as in T4; developed with S4 then with S3 (each for 1.5 h); dried as in T4 after each development.

REFERENCE

1. **Koizumi, K., Utamura, T., Kuroyanagi, T., Hizukuri, S., and Abe, J.-I.,** Analyses of branched cyclodextrins by high-performance liquid and thin-layer chromatography, *J. Chromatogr.*, 360, 397, 1986.

TABLE TLC 7
TLC of Glycolipids and of Oligosaccharides from Enzymatic Digestion of Glycosaminoglycuronans

Layer	L1	L2	L2	L2	L3	L4	L4
Solvent	S1	S2	S3	S4	S5	S6	S7
Technique	T1	T2	T3	T3	T4	T5	T6
Reference	1	2	3	3	4	5	6
Compound	**$R_F \times 100$**						
Gangliosides							
G_{M4}	89	85	—	—	—	—	—
G_{M3}	83	75; 79	77	71	—	—	—
G_{M2}	—	58	—	—	—	—	—
G_{M1}	73	46	65	49	30	—	—
G_{D3}	62	36; 38	54	65	—	—	—
G_{D2}	—	17	39	48	—	—	—
G_{D1a}	49	29	54	55	23	—	—
G_{D1b}	36	12	36	29	14	—	—
G_{T1a}	—	19	45	48	—	—	—
G_{T1b}	23	6	26	36	—	—	—
G_{Q1b}	15	2	17	25	—	—	—
Neutral glycolipids							
Monohexosyl ceramide	—	—	—	—	89; 93	—	—
Lactosyl ceramide	—	—	—	—	81	—	—
Globoside (Gb4)	—	—	—	—	61; 64	—	—
AGM1 (Gg4)	—	—	—	—	53	—	—
Sulfatide	—	—	—	—	33	—	—
Oligosaccharides from hyaluronic acid digests							
Glca β1-3GlcNAc (HA-2)	—	—	—	—	—	59	—
Dimer (HA-4)	—	—	—	—	—	48	—
Trimer (HA-6)	—	—	—	—	—	37	—
Tetramer (HA-8)	—	—	—	—	—	26	—
GlcA β1-3GlcNAc β1-4GlcA (HA-3)	—	—	—	—	—	39	—
GlcA-(GlcNAc-GlcA)$_2$ (HA-5)	—	—	—	—	—	29	—
GlcA-(GlcNAc-GlcA)$_3$ (HA-7)	—	—	—	—	—	20	—
GlcNAc β1-4GlcA (HA-2′)	—	—	—	—	—	52	—
Dimer (HA-4′)	—	—	—	—	—	41	—
Trimer (HA-6′)	—	—	—	—	—	31	—
GlcNAcβ1-4GlcAβ1-3GlcNAc (HA-3′)		—	—	—	—	59	—
GlcNAc(GlcA-GlcNAc)$_2$ (HA-5′)	—	—	—	—	—	49	—
Unsaturated disaccharides[a,b]							
ΔDi — 0S	—	—	—	—	—	—	72
ΔDi — 4S	—	—	—	—	—	—	56
ΔDi — 6S	—	—	—	—	—	—	61
ΔDi — UA2S	—	—	—	—	—	—	66
ΔDi — diS$_B$	—	—	—	—	—	—	47
ΔDi — diS$_D$	—	—	—	—	—	—	52
ΔDi — diS$_E$	—	—	—	—	—	—	41
ΔDi — triS	—	—	—	—	—	—	30
ΔDi — HA	—	—	—	—	—	—	75

Note: See Tables LC 36 and LC 39 for HPLC of these compounds.

[a] From digestion of chrondroitin sulfates and hyaluronic acid by chrondroitinase ABC.

[b] R_F values are those of dansylhydrazone derivatives.

TABLE TLC 7 (continued)
TLC of Glycolipids and of Oligosaccharides from Enzymatic Digestion of Glycosaminoglycuronans

Layer	L1 =	0.25 mm silica gel G60 precoated on glass plates (5 × 20 cm; Merck).
	L2 =	silica gel 60 HPTLC plates (0.2 mm layer; Merck), activated by heating at 100° C for 5 — 10 min, then cooled in desiccator before use.
	L3 =	biphasic plates (Analtech, Newark, DE), with adjacent layers of amino-silica gel and unmodified silica gel, each 5 cm long and 0.25 mm thick.
	L4 =	0.2-mm silica gel 60 layer precoated on plastic sheets (10 × 5 cm; Merck).
Solvent	S1 =	tetrahydrofuran-0.05 *M* aqueous KCl (5:1).
	S2 =	chloroform-methanol-water (11:9:2) containing 0.02% $CaCl_2 \cdot 2H_2O$.
	S3 =	acetonitrile-2-propanol-50 m*M* aqueous KCL (10:67:23).
	S4 =	acetonitrile-2-propanol-2.5 *M* aqueous ammonia (2:13:5).
	S5 =	chloroform-methanol-water (65:35:8).
	S6 =	2-propanol-water (33:17) containing 0.05 *M* NaCl.
	S7 =	1-propanol-2-propanol-1-butanol-water (6:9:1:4) containing 0.04 *M* NaCl and 0.01 *M* ammonia.
Technique	T1 =	samples applied to plate as bands, each spread over 1.5 cm, 1.8 cm from lower edge, and plate developed until solvent front within 2 cm of upper edge; excess solvent removed by stream of warm air in fume cupboard, then plate dried in oven at 120° C for 3 min.
	T2 =	samples applied as bands, each 7 mm, 2 cm from lower edge, and plate developed for 3.5 h in well-saturated tank lined with Whatman No. 1 paper; plate dried in stream of warm air for 10 min.
	T3 =	samples applied in 4 mm bands, 1.5 cm from lower edge and plate developed 8 — 10 cm.
	T4 =	plate washed by continuous development in chloroform-methanol (1:1) before use; samples dissolved in this solvent and applied as 1-μl aliquots to amino layer, 0.5 cm from bottom, and plate developed to ca.8 cm.
	T5 =	normal ascending development.
	T6 =	20-μl aliquots of 0.75% trichloroacetic acid in ethanol and of 1% dansylhydrazine in ethanol added to 10-μl aliquots of aqueous solution of unsaturated disaccharides from enzymatic digest, and mixture incubated at 40° C for 150 min, then cooled to room temperature; 5-μl aliquots applied to plate; after development spots detected by UV lamp.

REFERENCES

1. **Randell, J. A. J. and Pennock, C. A.,** Brain gangliosides: an improved simple method for their extraction and identification, *J. Chromatogr.*, 195, 257, 1980.
2. **Ando, S., Chang, N.-C., and Yu, R. K.,** High-performance thin-layer chromatography and densitometric determination of brain ganglioside compositions of several species, *Anal. Biochem.*, 89, 437, 1978.
3. **Ando, S., Waki, H., and Kon, K.,** New solvent system for high-performance thin-layer chromatography and high-performance liquid chromatography of gangliosides, *J. Chromatogr.*, 405, 125, 1987.
4. **Alvarez, J. G. and Touchstone, J. C.,** Separation of acidic and neutral glycolipids by diphasic thin-layer chromatography, *J. Chromatogr.*, 436, 515, 1988.
5. **Shimada, E. and Matsumura, G.,** Comparison of relationships between the chemical structures and mobilities of hyaluronate oligosaccharides in thin-layer and high-performance liquid chromatography, *J. Chromotogr.*, 328, 73, 1985.
6. **Shinomiya, K., Hoshi, Y., and Imanari, T.,** Sensitive detection of unsaturated disaccharides from chondroitin sulfates by thin-layer chromatography as their dansylydrazine derivatives, *J. Chromatogr.*, 462, 471, 1989.

Section I.V

ELECTROPHORETIC DATA

This section should be regarded as a supplement to Section I.V in *CRC Handbook of Chromatography: Carbohydrates,* Volume I because there have been few advances in the electrophoretic techniques applied to carbohydrates since the compilation of the earlier volume. In paper electrophoresis sodium metavanadate has been added to the list of complexing electrolytes capable of resolving low-molecular-weight carbohydrates, especially alditols, and electrophoresis on glass-fiber strips seems likely to be superseded by the use of silanized silica gel plates, for which some preliminary results are presented here. Changes in the techniques adopted in electrophoresis of glycosaminoglycuronans on both cellulose acetate membranes and agarose gels have improved resolution to the extent that heterogeneity of heparin and heparan sulfate has become detectable by the resolution of various components. Discontinuous polyacrylamide gel electrophoresis in the presence of denaturing agents, notably sodium dodecyl sulfate, which has assumed great importance in analysis of glycoproteins, is reviewed at the end of this section; as this technique was not included in Volume I, the review in this case covers literature from 1970 onward.

TABLE EL 1
Electrophoresis of Low-Molecular-Weight Carbohydrates on Paper and Silanized Silica Gel

Support	P1	P2	P3
Electrolyte	E1	E2	E3
Technique	T1	T2	T3
Reference	1	2	3
Compound	M_{Glc}[a]	M_{Gal}[b]	M_{Glc}
Alditols			
Glycerol	<1	—	—
Erythritol	1—2	—	—
D-Threitol	1—2	—	—
D-Arabinitol	3.2	—	—
1-deoxy-	2.6	—	—
D-Lyxitol	3.2	—	—
1-deoxy-	1.1	—	—
Ribitol	3.1	—	—
Xylitol	6.0	—	—
1-deoxy-	4.8	—	—
2-Deoxy-D-*erythro*-penitol	<1	—	—
2-Deoxy-D-*arabino*-hexitol	1.2	—	—
2-Deoxy-D-*lyxo*-hexitol	5.4	—	—
2-Deoxy-L-*xylo*-hexitol	11.5	—	—
3-Deoxy-D-*arabino*-hexitol	1.1	—	—
3-Deoxy-D-*ribo*-hexitol	1.1	—	—
3-Deoxy-L-*xylo*-hexitol	3.3	—	—
Allitol	7.0	—	—
D-Altritol	9.8	—	—
1-deoxy-	6.7	—	—
1,6-dideoxy-	3.4	—	—
Galactitol	4.8	—	—
1-deoxy-L-	3.3	—	—
1,6-dideoxy-	3.4	—	—
D-Glucitol	10.0	—	—
L-Iditol	15.0	—	—
D-Mannitol	4.0	—	—
1,6-dideoxy-	2.0	—	—
1,2-di-*O*-methyl-	3.4	—	—
2,5-di-*O*-methyl-	1.1	—	—
D-Talitol, 1-deoxy-	3.7	—	—
Reduced oligosaccharides			
D-Glucitol, 2-*O*-α-D-glucopyranosyl-	<1	—	—
2-*O*-β-D-glucopyranosyl-	<1	—	—
3-*O*-α-D-glucopyranosyl-	<1	—	—
3-*O*-β-D-glucopyranosyl-	3.3	—	—
L-Glucitol, 1-*O*-α-D-glucopyranosyl-	7.6	—	—
1-*O*-β-D-glucopyranosyl-	7.4	—	—
3-*O*-α-D-glucopyranosyl	15.2	—	—
3-*O*-β-D-glucopyranosyl-	7.7	—	—
Monosaccharides and derivatives			
DL-Glyceraldehyde	—	—	0.78
L-Arabinose	<1	—	—
methyl β-pyranoside	<1	—	—
D-Lyxose	1.0	—	—
D-Ribose	4.0	—	—
D-Xylose	1.0	—	1.00
2-Deoxy-D-*erythro*-pentose	<1	—	—
methyl β-pyranoside	<1	—	—

TABLE EL 1 (continued)
Electrophoresis of Low-Molecular-Weight Carbohydrates on Paper and Silanized Silica Gel

Compound	M_{Glc}[a]	M_{Gal}[b]	M_{Glc}
2-Deoxy-D-*arabino*-hexose	1.1	—	—
2-Deoxy-D-*lyxo*-hexose	1.6	—	—
L-Fucose	<1	—	—
D-Galactose	<1	1.00	—
(*R*)-3,4-*O*-(1-carboxyethylidene)-	—	1.21	—
(*S*)-3,4-*O*-(1-carboxyethylidene)-	—	1.24	—
(*R*)-4,6-*O*-(1-carboxyethylidene)-	—	0.88	—
(*S*)-4,6-*O*-(1-carboxyethylidene)-	—	0.84	—
D-Glucose	1.0	—	1.00
1,6-anhydro-β-	—	—	0
6-deoxy-	1.0	—	—
2-*O*-methyl-	1.0	—	—
3-*O*-methyl-	<1	—	—
6-*O*-methyl-	<1	—	—
D-Gulose	1.7	—	—
D-Mannose	<1	—	—
D-Fructose	4.5	—	0.90
1-*O*-methyl-	1.9	—	—
D-Tagatose	3.9	—	—
D-Sorbose	2—8	—	—
D-Glucuronic acid	—	—	1.20
Oligosaccharides			
Kojibiose	29	—	—
Sophorose	1.3	—	—
Nigerose	29	—	—
Laminaribiose	1.0	—	—
Maltose	2.6	—	0.30
Cellobiose	<1	—	0.30
Isomaltose	1.0	—	—
Gentiobiose	1.0	—	—
D-Galactose, 2-*O*-β-D-galactopyranosyl-	—	0.54	—
3-*O*-β-D-galactopyranosyl-	—	0.73	—
4-*O*-β-D-galactopyranosyl-	—	0.49	—
6-*O*-β-D-galactopyranosyl-	—	0.86	—
D-Fructose, 3-*O*-α-D-glucopyranosyl-	2.6	—	—
4-*O*-α-D-glucopyranosyl-	3.1	—	—
5-*O*-α-D-glucopyranosyl-	4.5	—	—
6-*O*-α-D-glucopyranosyl-	3.5	—	—
Maltotriose	—	—	0.35

Note: This table supplements Tables EL 1 — 4 in *CRC Handbook of Chromatography: Carbohydrates*, Volume I.

[a] Mobility relative to D-glucose.
[b] Mobility relative to D-galactose.

Support P1 = Whatman No. 3MM paper.
P2 = Whatman No. 1 paper.
P3 = silanized silica gel (15% w/v) suspended in dichloromethane containing 1-octanol (4%), stirred for 30 min, then solvent evaporated *in vacuo*; dry powder mixed with 2% solution of polyvinylpyrrolidone in E3 (10 ml) + 35 ml pure E3; coated on plates (15 cm × 10 cm, 0.3 mm layer); plates kept in humid atmosphere (over buffer solution E3) for 12 h before use.

TABLE EL 1 (continued)
Electrophoresis of Low-Molecular-Weight Carbohydrates on Paper and Silanized Silica Gel

Electrolyte	E1	=	1.5% aqueous sodium metavanadate, pH 8.7.
	E2	=	50 m*M* aqueous sodium tetraborate, pH adjusted to 10.0 with aqueous NaOH.
	E3	=	0.3 *M* aqueous boric acid, containing Titriplex® III (ethylenediaminetetracetic acid disodium salt, 20 m*M*); pH adjusted to 10.0 with 2 *M* NaOH.
Technique	T1	=	mobilities determined at 45 V/cm; 5-hydroxy-methyl-2 furaldehyde added to correct electro-osmosis.
	T2	=	mobilities determined at 55 V/cm; (1.6 — 2 mA/cm); with CCl_4 as coolant; uniformity of development across paper checked with tracking dye, *p*-bromophenol blue (R_{Gal} 0.95).
	T3	=	plate, on cooling block of electrophoresis cell, connected to electrolyte vessels through strips of glass-fiber paper, soaked with electrolyte solution; plate equilibrated for 30 min under applied voltage (200 V) before application of samples, in 1-cm grooves scratched into layer; electrophoresis at 13 — 26 V/cm for 1 — 2 h.

REFERENCES

1. **Searle, F. and Weigel, H.,** Interaction between polyhydroxy compounds and vanadate ions: electrophoresis and composition of complexes, *Carbohydr. Res.*, 85, 51, 1980.
2. **Fontana, J. D., Duarte, J. H., Iacomini, M., and Gorin, P. A. J.,** Synthesis and chromatographic properties of isomeric *O*-β-D-galactopyranosyl-D-galactoses, and of diastereoisomers of 3,4-*O*- and 4,6-*O*-(1-carboxyethylidene)-D-galactose, *Carbohydr. Res.*, 108, 221, 1982.
3. **Bonn, G., Grünwald, M., Scherz, H., and Bobleter, O.,** Thin-layer electrophoretic behavior of oligo- and monsaccharides, uronic acids and lyhydroxy compounds obtained as biomass degradation products, *J. Chromatogr.*, 370, 485, 1986.

TABLE EL 2
Electrophoresis of Glycosaminoglycuronans on Cellulose Acetate

Support	P1	P2	P2, P3	P2, P3	P4
Electrolyte	E1	E2	E3	E4	E5
Technique	T1	T2	T3	T4	T5
Detection	D1	D2	D3	D3	D4
Reference	1	2	3	3	4
Compound	**R_{ChS}[a] × 100**	**R_f[b] × 100**		**R_{ChS} × 100**	
Chondroitin 6-sulfate	108	—	—	108	—
Chondroitin 4-sulfate	100	—	—	100	100
Dermatan sulfate	53	—	—	82	106
Keratan sulfate	133	—	—	—	83
Heparan sulfate[c]	50; 69	—	—	50; 70	96
Heparin (bovine lung)	28	1.5;[d] 28; 41; 54	0;[d] 22;[d] 27	0;[d] 33	119[e]

TABLE EL 2 (continued)
Electrophoresis of Glycosaminoglycuronans on Cellulose Acetate

Compound	R_{ChS}[a] × 100	R_f[b] × 100		R_{ChS} × 100	
Heparin (porcine mucosa)	—	1.5; 17;[d] 28; 41; 54	22;[d] 27	—	
Hyaluronic acid	85	—	—	—	77

Note: This table supplements Table EL 5 in *CRC Handbook of Chromatography: Carbohydrates,* Volume I.

[a] Mobility relative to chondroitin 4-sulfate.
[b] Mobility relative to total distance between cathode and anode.
[c] Formerly heparitin sulfate.
[d] Major component.
[e] Source not stated.

Support	P1 =	Titan® III Zip Zone cellulose acetate sheets, 6.0 × 7.5 cm (Helena Laboratories, Beaumont, TX).
	P2 =	Sepraphore III (Gelman, Ann Arbor, MI and Milan, Italy) or Microphor (Elvi, Milan) cellulose acetate strips, 16 × 2.5 cm.
	P3 =	Schleicher and Schüll (Keene, NH) or Whatman Cellogel® cellulose acetate strips.
	P4 =	cellulose acetate strips (14.5 × 5.5 cm) or sheets (17 × 17 cm) (Kalex, Manhasset, NY).
Electrolyte	E1 =	1.0 *M* aqueous barium acetate, acidified to pH 5.0 with acetic acid, in electrode compartment; plate immersed in 0.1 *M* barium acetate solution, pH 5.0, before electrophoresis steps (see T1 below).
	E2 =	0.1 *M* aqueous barium acetate, pH adjusted to 5.8 with acetic acid.
	E3 =	1.0 *M* aqueous barium acetate in electrode compartment; strip immersed in 0.1 *M* barium acetate before electrophoresis steps.
	E4 =	freshly prepared aqueous solution of 3,3'-diaminodipropylamine (25 m*M*) containing magnesium acetate (30 m*M*) and acetic acid (60 m*M*).
	E5 =	10 m*M* aqueous EDTA, tetrasodium salt, containing 50 m*M* LiCl, pH 8.4.
Technique	T1 =	electrophoresis performed in specially constructed apparatus, by discontinuous method; one end of sheet immersed in water to height of ca. 1.5 cm, blotted, then opposite end immersed in 0.1 *M* barium acetate, leaving narrow band (2—4 mm) between water and buffer zones; samples (1—10 μl) applied in center of water zone; after electrophoresis for 2—3 min (26 V/cm) sheet blotted, soaked in 0.1*M* barium acetate for 1—2 min, blotted again, and subjected to electrophoresis for 9—12 min; sheet blotted, immersed for 2 min in 0.1 *M* barium acetate containing 15% ethanol, blotted, and electrophoresis continued for 15 min; blotted sheet immersed in 0.1 *M* barium acetate containing 50% ethanol, blotted, and submitted to further electrophoresis for 50 min.
	T2 =	strips immersed for 30 min in 0.1 *M* barium acetate, then blotted before application of heparins as sodium salts (1—3 μl of 0.2% solutions in 0.1 *M* barium acetate); electrophoresis at 4° C for 16 h (voltage giving 0.4 mA/cm).
	T3 =	strips equilibrated with 0.1 *M* barium acetate before loading with samples; after electrophoresis at 1 mA/cm for 30 min strips immersed for 2 min in 0.1 *M* barium acetate containing 30% ethanol, blotted, then submitted to further electrophoresis for 30 min.
	T4 =	strips equilibrated with E4 buffer for at least 3 min before application of samples (0.2—0.4 μl, 2—5 mg/ml in water) in 4-mm bands; electrophoresis for 1 h at 1 mA/cm (ca. 15 V/cm).
	T5 =	strips stored at 4° C in 30% methanol solution; before use strips blotted, immersed in E5 for 10—15 min, to 6 cm from top; after blotting strips loaded with samples (1 μl) in 0.75-cm bands, 3 cm from cathodal end, not previously equilibrated in electrolyte solution; electrophoresis carried out at 4° C for 45 min at 15 V/cm.

TABLE EL 2 (continued)
Electrophoresis of Glycosaminoglycuronans on Cellulose Acetate

Detection D1 = stained with 0.1% aqueous alcian blue, destained in 1% acetic acid, rinsed under running water.
D2 = stained with 1% alcian blue in 0.05 *M* sodium acetate-95% ethanol (1:1), destained in 5% acetic acid.
D3 = stained with alcian blue (concentration not given), destained in 5% acetic acid.
D4 = stained for 1 min in 0.1% aqueous alcian blue at 25° C, destained by two successive immersions in 2% acetic acid (5 min each).

REFERENCES

1. **Cappelletti, R., Del Rosso, M., and Chiarugi, V. P.,** A new electrophoretic method for the complete separation of all known animal glycosaminoglycans in a monodimensional run, *Anal. Biochem.*, 99, 311, 1979.
2. **Oreste, P. and Torri, G.,** Fingerprinting of heparins by low-amperage electrophoresis in barium acetate, *J. Chromatogr.*, 195, 398, 1980.
3. **Johnson, E.A.,** Simple electrophoresis of glycosaminoglycuronans and the distinction of lumg heparins from mucosal heparins, *J. Chromatogr.*, 233, 365, 1982.
4. **Schuchman, E. H. and Desnick, R. J.,** A new continuous, monodimensional electrophoretic system for the separation and quantitation of individual glycosaminoglycans, *Anal. Biochem.*, 117, 419, 1981.

TABLE EL 3
Electrophoresis of Glycosaminoglycuronans on Agarose Gel

Gel	G1	G2	G3	G4	G3
Electrolyte	E1	E2	E3 + E4	E1	E3 + E4
Technique	T1	T1	T2	T3	T3
Detection	D1	D1	D1	D2	D2
Reference	1	1	2	2	2
	R_{ChS}[a] × 100				
Compound	*	*	**	1	2**
Chondroitin 6-sulfate	—	—	100	—	—
Chondroitin 4-sulfate	100	100	100	100	100
Dermatan sulfate	—	—	88	100	86
Keratan sulfate	—	—	—	70	63
Heparan sulfate[b]	77; 83; 85; 115	53; 78; 82	70	80	70
Heparin	100; 115	76	37; 53; 70[c]	100	30; 70

Note: This table supplements Table EL 6 in *CRC Handbook of Chromatography: Carbohydrates*, Volume I.
* = single development; ** = twofold,, discontinuous development, monodimensional; 1* = two-dimensional development, first direction; 2* = two-dimensional development, second direction.

[a] Mobility relative to chondroitin 4-sulfate.
[b] Formerly heparitin sulfate.
[c] Proportions of components vary widely among heparins from different sources; see Table EL 2.

Gel G1 = slide (5.0 × 7.5 × 0.1 cm) of 0.9% agarose in E1.
G2 = as G1, 0.9% agarose in E2.
G3 = slide as in G1 and G2; 0.5% agarose in E3.
G4 = as G3, 0.5% agarose in E1.
Electrolyte E1 = 0.06 *M* barbital (5,5-diethylbarbituric acid) buffer, pH 8.6.

TABLE EL 3 (continued)
Electrophoresis of Glycosaminoglycuronans on Agarose Gel

	E2 =	0.05 *M* 1,3-diaminopropane acetate buffer, pH 8.5.
	E3 =	0.04 *M* barium acetate solution.
	E4 =	0.1 *M* 1,3-diaminopropane acetate buffer, pH 9.0.
Technique	T1 =	samples (1—10 μg in 3—5 μl buffer) applied in 3 mm slots 1 cm from edge of gel slide; electrophoresis at 5° C, 20 V/cm continued until dye indicator (cresol red) had migrated 3 cm (with E1) or 4.5 cm (with E2); after electrophoresis gel immersed in 0.1% Cetavlon solution for 3 h, then covered with strip of Whatman 2MM paper wetted with Cetavlon solution and placed under 250W infrared lamp for 2 h, in current of air, prior to staining.
	T2 =	samples applied, as in T1, to gel slabs prepared in E3, and subjected to electrophoresis at 5° C, 10 V/cm (6 mA/cm) in chamber prepared with same electrolyte; gel slide then transferred to another chamber prepared with E4 and maintained at 5° C for 15 min; electrophoresis continued (at 13 V/cm) in same direction (toward anode) until dye indicator (cresol red) had migrated 4.5 cm from origin (ca. 90 min); after electrophoresis gel treated as in T1.
	T3 =	sample applied, as in T1, to gel slab prepared in E1 and electrophoresis carried out as in T1; after run gel strip (5 × 0.2 cm) containing fractionated glycosaminoglycuronans cut off and placed in slot of same width and length situated transversely 2 cm from edge of another gel slide prepared in E3; electrophoresis then repeated as in T2.
Detection	D1 =	stained with 0.1% solution of toluidine blue in acetic acid-ethanol-water (0.1:5:5), destained with solvent.
	D2 =	stained with 0.1% solution of toluidine blue in 1% acetic acid, destained with solvent.

REFERENCES

1. **Dietrich, C. P., McDuffie, N. M., and Sampaio, L. O.,** Identification of acid mucopolysaccharides by agarose gel electrophoresis, *J. Chromatogr.*, 130, 299, 1977.
2. **Bianchini, P., Nader, H. B., Takahashi, H. K., Osima, B., Straus, A. H., and Dietrich, C. P.,** Fractionation and identification of heparin and other acidic mucopolysaccharides by a new discontinuous electrophoretic method, *J. Chromatogr.*, 196, 455, 1980.

POLYACRYLAMIDE GEL ELECTROPHORESIS

Discontinuous polyacrylamide gel electrophoresis in the presence of sodium dodecyl sulfate (SDS), which prevents association through both electrostatic and hydrophobic interaction and minimizes charge differences by complexing with the charged molecules, has long been a method of choice in studies of proteins and has also been applied extensively to glycoproteins. The contact surface of the denatured molecules is proportional to the molecular weight and, therefore, migration velocity is a linear function of the logarithm of the molecular weight. The method used by Laemmli[1] is that most commonly employed for glycoproteins, with a stacking gel (1 cm) composed of 3 to 4% polyacrylamide in 0.125 *M* Tris-HCl, pH 6.8, and a separation gel (9 cm) of 10 to 12.5% polyacrylamide in 0.375 *M* Tris-HCl, pH 8.8, both gels containing 0.1% SDS. The buffer in the electrode compartment contains 0.025 *M* Tris and 0.192 *M* glycine (pH 8.3), with 0.1% SDS. Before electrophoresis, samples (0.2 to 0.3 ml), in 0.0625 *M* Tris-HCl (pH 6.8) containing 2% SDS, 5% 2-mercaptoethanol, 10% glycerol, and 0.001% bromophenol blue as marker dye, are heated at 100° C for 10 min to denature the proteins. Electrophoresis is carried out at 3 mA per gel, until the marker dye has migrated to the end of the gel (ca. 7 h). Before visualization of protein and carbohydrate in the separated bands it is necessary to fix the components in the gels by treatment with 12.5% trichloroacetic acid for 1 h[2] or with 2-propanol-acetic acid-water (5:2:13) overnight,[3] followed by washing with aqueous acetic acid (7.5 to 15%). Rapid removal of SDS by transverse electrophoresis for 40 min in 2-propanol-acetic acid-water (5:2:13), in the presence of a mixed-bed ion-exchange resin, has been recommended.[4] Protein is visualized by staining with 0.1% Coomassie blue in 50% trichloroacetic acid,[1] methanol-acetic acid-water (23:4:23), or 50% methanol,[4] followed by destaining in 7% acetic acid[1] or methanol-acetic acid-water (5:1:14 or 3:1:7).[3,4] For visualization of the carbohydrate moiety, oxidation with periodic acid (0.2 to 1%) for 1 to 2 h, followed by treatment with Schiff's reagent[2-4] or ammoniacal silver sitrate solution[3] is widely used; greater sensitivity (to 0.4 ng of bound carbohydrate) is claimed for the silver-staining technique. In the periodic acid-Schiff method gels are destained by washing in 7% acetic acid,[2,3] while washing in an aqueous solution containing 0.05% citric acid, 0.019% formaldehyde, and 10% methanol, followed by use of a photographic fixative, is used in the silver-staining technique.[3]

These methods have been applied successfully in electrophoresis of a diversity of glycoproteins[2-4] including plant-gum arabinogalactan proteins,[5] and also lipopolysaccharides and capsular polysaccharides of bacteria.[3,6] Recently, however, replacement of SDS by sodium deoxycholate (DOC) was shown [7] to improve resolution of lipopolysaccharides from various strains of *Salmonella* bacteria, this method having the further advantage that boiling the sample prior to electrophoresis was unnecessary, even for proteins. This technique is likely to be widely adopted in the future. Urea has also been used as an alternative denaturing agent,[2] especially in polyacrylamide gel electrophoresis of the blocks of poly-D-mannuronate (M-blocks) and poly-L-guluronate (G-blocks) obtained by partial acid-hydrolysis of alginates;[8] in this case the presence of Ca^{2+} ions in both gel and buffer is essential for optimal resolution.

REFERENCES

1. **Laemmli, U. K.,** Cleavage of structual proteins during the assembly of the head of bacteriophage T4, *Nature (London)*, 227, 680, 1970.
2. **Kapitany, R. A. and Zebrowski, E. J.,** A high resolution PAS stain for polyacrylamide gel electrophoresis, *Anal. Biochem.*, 56, 361, 1973.
3. **Dubray, G. and Bezard, G.,** A highly sensitive periodic acid-silver stain for 1,2-diol groups of glycoproteins and polysaccharides in polyacrylamide gels, *Anal. Biochem.*, 119, 325, 1982.
4. **Phillips, H. M.,** Method for rapid removal of sodium dodecyl sulfate from polyacrylamide gels, *Anal. Biochem.*, 117, 398, 1981.
5. **Gammon, D. W., Stephen, A. M., and Churms, S. C.,** The glycoproteins of *Acacia erioloba* exudates, *Carbohydr. Res.*, 158, 157, 1986.
6. **Pelkonen, S., Hayrinen, J., and Finne, J.,** Polyacrylamide gel electrophoresis of the capsular polysaccharides of *E-coli* K1 and other bacteria, *J. Bacteriol.*, 170, 2646, 1988.
7. **Komuro, T. and Galanos, C.,** Analysis of *Salmonella* lipopolysaccharides by sodium deoxycholate - polyacrylamide gel electrophoresis, *J. Chromatogr.*, 450, 381, 1988.
8. **Doubet, R. S. and Quatrano, R. S.,** Urea-polyacrylamide gel electrophoresis of alginic acid, *J. Chromatogr.*, 264, 479, 1983.

Section II

Detection Techniques

Section II.I

DETECTION AND IDENTIFICATION OF CARBOHYDRATES IN GAS CHROMATOGRAPHY: USE OF MASS SPECTROMETRY

The use of mass spectrometry coupled to gas chromatography is now such an essential technique in the methodology of carbohydrate analysis that it is considered imperative to list in this volume the mass *m/z* charge (values) of the most important ions in the mass spectra (EI and, where possible, also CI) of the derivatives that have been used in GC-MS of sugars and methylated sugars. This section, therefore, is supplementary to Section I.I, which contains GC retention data for these derivatives.

The alditol acetates remain the derivatives most often used in GC-MS of sugars and methylated sugars. The work of the Lindberg group[1] and also of Klok et al.[2] has produced an extensive database for the coupling of both EI[1] and CI[2] mass spectrometry to gas chromatography of peracetylated alditols and partially methylated alditol acetates derived from neutral sugars. This is supplemented by data on GC-MS of corresponding derivatives from amino sugars, extracted from a review by Wong et al.[3]

In studies of the carbohydrate chains of glycoproteins it is necessary to distinguish between 2-acetamido-2-deoxy-D-galactose occurring as chain residues and that involved in linkages with serine and threonine residues in the polypeptide chain, which is converted to the alditol on detachment of the carbohydrate by alkaline borohydride treatment. This can be achieved by use of sodium borodeuteride to reduce the sugars released after acid hydrolysis of the detached oligosaccharide chains, which labels the alditols derived from these sugars, but not those already present as alditols. The diagnostic ions used to make this distinction[4] are included in the table that follows, as are those listed by Waeghe et al.[5] for deuterium-labeled partially methylated alditol acetates derived from neutral hexoses and hexuronic acids in various complex polysaccharides. The latter are converted to 6,6-dideuteriohexosyl residues by carboxylate reduction of the methyl-esterified residues with sodium borodeuteride or lithium aluminum deuteride. The sensitivity of methylation analysis is greatly increased by the use of selected ion monitoring, the intensities of certain of these deuterium-labeled diagnostic ions being used to monitor the emergence from the column of each individual component. Complete methylation analyses of oligosaccharides and polysaccharides using samples of only 1 and 5 μg, respectively, have been achieved by capillary GC with multiple selected ion monitoring.[5]

Several of the other derivatives used in GC analysis of carbohydrates, details of which are given in Section I.I, have also been examined by mass spectrometry with a view to their application in GC-MS. These include peracetylated aldononitriles[6] and aldononitrile acetates derived from partially methylated sugars;[7,8] oxime derivatives, acetylated,[9] trifluoroacetylated,[10-12] or trimethylsilylated;[13,14] trimethylsilyl ethers of methyl glycosides[15] and of partially methylated methyl glycosides;[16] and some partially methylated methyl glycosides not further derivatized.[17] Certain methylated sugars have been examined by GC-MS of the derived trimethylsilylated diethyl dithioacetals.[18] The new deoxy(methoxyamino)alditol derivatives also have highly characteristic mass spectra after trimethylsilylation[19] or permethylation,[20] which makes possible GC-MS analysis with sensitive detection by selected ion monitoring. Diagnostic ions for all of these derivatives are included in Table IIA.

TABLE IIA
Some Important Diagnostic Ions in Mass Spectra of Volatile Derivatives Used in GLC-MS of Sugars and Methylated Sugars

Derivative	Diagnostic ions (m/z[a]) EI-MS	Diagnostic ions (m/z[a]) CI-MS	Ref.
Alditol acetates			
Tetritol-Ac_4	73, 86, *103*, *115*, 128, *145*, 127		1
Pentitol-Ac_5	73, *85*, *103*, *115*, *127*, *145*, 175, *187*, 217, 289	303, 363	1, 2
Hexitol-Ac_6	73, *85*, *103*, *115*, *127*, *128*, *139*, *145*, *157*, 170, *187*, 217, 259, 289, 315	375, 435	
6-Deoxyhexitol-Ac_5	73, 86, *99*, *103*, *115*, *128*, 145, 157, *170*, 187, 201, 217, 231	317, 377	
2-Acetamido-2-deoxy-hexitol-Ac_5	60, *84*, *102*, 114, 126, 139, *144*, 151, 156, 168, 216, 318, 360, 390		3, 4
From $NaBD_4$ reduction	*85*, *103*, *145*		4
Partially methylated alditol acetates			
2(4)-Me-pentitol-Ac_4	*85*, 99, 103, 115, *117*, *127*, 139, 145, 187, 217	275, 303, 335	1, 2
3-Me-pentitol-Ac_4	*87*, *129*, *189*	275, 303, 335	
5-Me-pentitol-Ac_4	*85*, *87*, *103*, *115*, 127, 128, *129*, 145, 187	275, 303, 335	
2,3(3,4)-Me_2-pentitol-Ac_3	71, *87*, 99, 101, *117*, *129*, 189	247, 275, 307	
2,4-Me_2-pentitol-Ac_3	85, *117*, 127, 159, 173, 201, 233	247, 275, 307	
2,5-Me_2-pentitol-Ac_3	*87*, 101, *117*, *129*, 161, 189, 233		—[b]
3,5-Me_2-pentitol-Ac_3	*87*, 101, *129*, 161, 189	247, 275, 307	1, 2
2,3,4-Me_3-pentitol-Ac_2	87, *101*, *117*, 161	219, 247, 279	
2,3,5-Me_3-pentitol-Ac_2	71, *87*, *101*, *117*, *129*, 161	219, 247, 279	
2-Me-hexitol-Ac_5	87, *97*, *117*, *139*, 259, 333	347, 375, 407	
3(4)-Me-hexitol-Ac_5	*85*, *87*, 99, *127*, *129*, 159, *189*, 201, 261	347, 375, 407	
6-Me-hexitol-Ac_5	*87*, 97, 103, *115*, *129*, *139*, *157*, 184	347, 375, 407	
2,3-Me_2-hexitol-Ac_4	*85*, 87, *99*, *101*, *117*, *127*, 159, 201, 261	319, 347, 379	
2,4-Me_2-hexitol-Ac_4	*87*, *117*, *129*, 159, 189, 201, 233, 305	319, 347, 379	
2,6-Me_2-hexitol-Ac_4	*87*, *117*, *129*, 143, 159, 185, 203, 231, 305	319, 347, 379	
3,4-Me_2-hexitol-Ac_4	*87*, 99, *129*, *189*, 233	319, 347, 379	
3,6-Me_2-hexitol-Ac_4	*87*, 99, *113*, *129*, 189, 233	319, 347, 379	
4,6-Me_2-hexitol-Ac_4	85, 87, 99, 101, 127, *129*, *161*, 201, 261	319, 347, 379	
2,3,4-Me_3-hexitol-Ac_3	*87*, *99*, *101*, *117*, *129*, 173, 189, 233	291, 319, 351	
2,3,6-Me_3-hexitol-Ac_3	*87*, *99*, *101*, *113*, *117*, 129, 131, 161, 173, 233	291, 319, 351	

2,4,6-Me_3-hexitol-Ac_3	87, 99, *101*, *117*, *129*, 161, 201, 233	291, 319, 351	
3,4,6-Me_3-hexitol-Ac_3	*87*, 99, 101, *129*, 145, *161*, 189	291, 319, 351	
2,3,4,6-Me_4-hexitol-Ac_2	71, 87, *101*, *117*, *129*, *145*, *161*, 205	263, 291, 323	
2,3,5,6-Me_4-hexitol-Ac_2	*89*, *101*, *117*, 129, 161, 205	263, 291, 323	
2-Me-6-deoxyhexitol-Ac_4	57, *87*, *99*, *117*, 129, 141, 159, 173, 201	289, 317, 349	
3-Me-6-deoxyhexitol-Ac_4	*87*, *101*, 117, *129*, *143*, 159, 189, 203	289, 317, 349	
4-Me-6-deoxyhexitol-Ac_4	85, 87, *89*, 99, 127, *131*, 159, 201, 261	289, 317, 349	
2,3-Me_2-6-deoxyhexitol-Ac_3	*101*, *117*, 129, *143*, 161, 203	261, 289, 321	
2,4-Me_2-6-deoxyhexitol-Ac_3	*89*, *101*, *117*, 127, *131*, 173, 201, 233	261, 289, 321	
3,4-Me_2-6-deoxyhexitol-Ac_3	*87*, *89*, 99, 115, *129*, *131*, 189	261, 289, 321	
2,3,4-Me_3-6-deoxy-hexitol-Ac_2	59, 71, *89*, *101*, *115*, *117*, *131*, 161, 175	233, 261, 293	
3-*O*-Me-2-*N*-Me-hexaminitol-Ac_4	74, 87, 98, 116, 124, 142, 158, 202, 261		3
4-*O*-Me-2-*N*-Me-hexaminitol-Ac_4	74, 87, 98, 116, 129, 158, 170, 189		
6-*O*-Me-2-*N*-Me-hexaminitol-Ac_4	74, 87, 98, 116, 128, 142, 158, 170		
3,4-*O*-Me_2-2-*N*-Me-hexaminitol-Ac_3	74, 87, 98, 116, 129, 142, 145, 158, 161, 189, 202		
3,6-*O*-Me_2-*N*-Me-hexaminitol-Ac_3	74, 87, 98, 116, 124, 129, 142, 158, 170, 173, 202, 233		
4,6-*O*-Me_2-2-*N*-Me-hexaminitol-Ac_3	74, 87, 98, 116, 142, 158, 161, 170, 230, 274		
3,4,6-*O*-Me_3-2-*N*-Me-hexaminitol-Ac_2	74, 87, 98, 116, 129, 142, 158, 161, 202, 205		
Deuterium-labeled partially methylated alditol acetates[c]			
2-Me-hexitol-Ac_5	118, 333		5
3-Me-hexitol-Ac_5	190, 261		
4-Me-hexitol-Ac_5	189, 262		
6-Me-hexitol-Ac_5	129, 185		
2,3-Me_2-hexitol-Ac_4	118, 261		
2,4-Me_2-hexitol-Ac_4	118, 189, 234, 305		
2,6-Me_2-hexitol-Ac_4	118, 305		
3,4-Me_2-hexitol-Ac_4	189, 190		
3,6-Me_2-hexitol-Ac_4	190, 233		
4,6-Me_2-hexitol-Ac_4	161, 262		
2,3,4-Me_3-hexitol-Ac_3	118, 189, 233		
2,3,6-Me_3-hexitol-Ac_3	118, 162, 233		
2,4,6-Me_3-hexitol-Ac_3	118, 161, 234		
3,4,6-Me_3-hexitol-Ac_3	161, 190		
2,3,4,6-Me_4-hexitol-Ac_2	118, 161, 162, 205		
2-Me-6,6-d_2-hexitol-Ac_5	118, 275, 335		
3-Me-6,6-d_2-hexitol-Ac_5	190, 263		
4-Me-6,6-d_2-hexitol-Ac_5	191, 262		
2,3-Me_2-6,6-d_2-hexitol-Ac_4	118, 203, 263		

TABLE IIA (continued)
Some Important Diagnostic Ions in Mass Spectra of Volatile Derivatives Used in GLC-MS of Sugars and Methylated Sugars

Derivative	Diagnostic ions (m/z^a) EI-MS	CI-MS	Ref.
2,4-Me_2-6,6-d_2-hexitol-Ac_4	118, 191, 234, 307		
3,4-Me_2-6,6-d_2-hexitol-Ac_4	190, 191		
2,3,4-Me_3-6,6-d_2-hexitol-Ac_3	118, 191, 235		
Aldononitrile acetates			
Erythrononitrile-Ac_3	73, 99, 103, 141, 145, 170	184, 261	6
2-Deoxy-D-*erythro*-pentononitrile-Ac_3	112, 115, 125, 145, 154, 167, 184	198, 215, 275	
Arabinononitrile-Ac_4	73, 103, 127, 141, 145, 183, 187, 200, 242, 289	256, 333	
Rhamnononitrile-Ac_4	55, 87, 99, 117, 129, 141, 159, 183, 200, 212, 242, 272, 314	270, 347	
2-Deoxy-D-*arabino*-hexononitrile-Ac_4	73, 83, 103, 112, 115, 125, 127, 142, 145, 154, 157, 175, 184, 187, 214, 217, 256, 289	270, 287, 347	
Mannononitrile-Ac_5	73, 103, 115, 127, 141, 145, 157, 170, 175, 183, 187, 200, 212, 217, 242, 272, 289, 314	328, 345, 405	
2-Acetamido-2-deoxyhexononitrile-Ac_4	73, 85, 98, 103, 115, 127, 140, 145, 164, 169, 182, 187, 211, 224, 289	207, 267, 327, 387, 404	
Partially methylated aldononitrile acetates			
2,3-Me_2-hexononitrile-Ac_3	*85*, 87, 88, *99*, 115, *127*		7
2,4-Me_2-hexononitrile-Ac_3	75, *87*, 96, 99, *112*, 127, *129*, *154*, *159*, *186*, *189*	240, 272, 349	7, 8
2,6-Me_2-hexononitrile-Ac_3	75, *87*, 99, 115, 117, 129, *159*, *169*, 184		7
3,4-Me_2-hexononitrile-Ac_3	*87*, 99, 126, *129*, 142, *189*		
3,6-Me_2-hexononitrile-Ac_3	75, 83, 84, 85, *87*, 98, *99*, *113*, 117, *129*, *131*, *142*, 147, 184, 189, 233		
4,6-Me_2-hexononitrile-Ac_3	*71*, 87, 99, *101*, 112, *129*, 154, 161, 214		
2,3,4-Me_3-hexononitrile-Ac_2	71, 73, *87*, *90*, *99*, 101, 113, *129*, 173, *189*		
2,3,6-Me_3-hexononitrile-Ac_2	*87*, *99*, *113*, *129*, 131, 147, 173, 189, 233		
2,4,6-Me_3-hexononitrile-Ac_2	87, 99, *101*, 112, *129*, 161, 186	244, 272, 304	7, 8
3,4,6-Me_3-hexononitrile-Ac_2	71, 74, *87*, *101*, 119, 126, *129*, 142, *161*		7
2,3,4,6-Me_4-hexononitrile-Ac	71, 73, 85, *87*, *88*, 89, 96, *101*, 113, 114, 119, *129*, 130, 131, *145*, 158, *161*, 162, 205	216, 244, 276	7, 8
Peracetylated keto-oximes (PAKO)			
D-*erythro*-2-Pentulose PAKO	81, *99*, *111*, *129*, *141*, 145, 154, *158*, *171*, 183, 184, 213, 231, *273*	*138*, *198*, 256, *258*, 318, 376	9
D-Fructose PAKO	99, *103*, *111*, *115*, *123*, *141*, *153*, *158*, *183*, 184, 201, 243, 345	*150*, *210*, 268, *270*, 328, *330*, 388, *390*	
D-*manno*-2-Heptulose PAKO	*103*, *115*, *123*, *128*, 139, *145*, *153*, *165*, 187, *195*, 196, 200, 213, *315*, 417		

TFA-O-methyloximes			
Pentose		338, 450, *563*, *564*	10
Hexose		449, 450, *464*, 576, *689*, *690*	11
TFA-*O*-butyloximes			
Pentose		380, 492, 605, *606*	10
Hexose		492, 506, 618, 731, *732*	11
2-Pentulose		492, *605*, *606*	12
3-Pentulose		*380*, 436, 492, *605*, *606*	
2-Hexulose		280, 392, 394, 505, *506*, 617, 618, 619, 620, *731*, *732*	
3-Hexulose		280, 392, 394, 505, *506*, 522, 617, 618, 731, *732*	
2,5-Hexodiulose		367, 478, *479*, 591, 592, 593, *704*, *705*	
TMS oximes			
Hexose	103, 205, 218, 307, 320, 409, 422, 524		13
Hexulose	103, 205, 307, 320, 422, 524		
TMS *O*-methyloximes			
Tetrose	103, 160, 205, 262		14
Pentose	103, 160, 205, 262, 307, 364		
Hexose	103, 160, 205, 262, 307, 364, 409, 466		
Heptose	103, 160, 205, 262, 319, 364, 421, 466, 568		
TMS methyl glycosides			
Pentoside	103, *133*, *204*, 205, *217*		15
6-Deoxyhexoside	103, *133*, *204*, *205*, *217*, 319		
Hexopyranoside	103, *133*, *204*, *205*, *217*, 319		
Hexofuranoside	103, 133, *204*, 205, *217*, 319		
2-Acetamidodeoxyhexoside	103, *131*, 133, *147*, *173*, 204, 205, 217, 247		3, 15
TMS partially methylated methyl glycosides			
Me 2-Me-$(TMS)_2$-rhamnoside	75, 133, *146*, 204		16
Me 3-Me-$(TMS)_2$-rhamnoside	75, 133, *146*, *217*		
Me 4-Me-$(TMS)_2$-rhamnoside	75, 133, 146, *204*		

TABLE IIA (continued)
Some Important Diagnostic Ions in Mass Spectra of Volatile Derivatives Used in GLC-MS of Sugars and Methylated Sugars

Derivative	Diagnostic ions (m/z[a]) EI-MS	CI-MS	Ref.
Me 2,3-Me_2-(TMS)-rhamnoside	75, *88*, 133, 146, 159		
Me 2,4-Me_2-(TMS)-rhamnoside	75, 133, *146*, 159		
Me 3,4-Me_2-(TMS)-rhamnoside	75, 133, *146*, *159*		
Me 2-Me-$(TMS)_3$-glucopyranoside	75, 133, *146*, 204		
Me 3-Me-$(TMS)_3$-glucopyranoside	75, 133, *146*, *217*		
Me 4-Me-$(TMS)_3$-glucopyranoside	*73*, 75, 133, 146, *204*		
Me 6-Me-$(TMS)_3$-glucopyranoside	*73*, 133, 146, *204*, 217		
Partially methylated methyl glycosides			
Me pentopyranoside	57, 59, *60*, *61*, 71, *73*, *74*, 86		17
Me 2-Me-pentopyranoside	57, 59, 60, 61, 71, 73, *74*, 85, *87*, 88		
Me 3-Me-pentopyranoside	55, 57, 58, 59, 60, 61, 69, 71, 73, *74*, *75*, 85, 86, 87, 100, 114		
Me 4-Me-pentopyranoside	57, *58*, *59*, *60*, 69, 73, 74, 75, *87*		
Me 2,3-Me_2-pentopyranoside	57, 59, 61, 69, 71, 73, 74, *75*, 85, 87, *88*, 101, 114, 129, 161		
Me 2,4-Me_2-pentopyranoside	57, 58, 59, 61, 71, 73, *74*, 85, 87, 88, *101*		
Me 3,4-Me_2-pentopyranoside	57, 58, 59, 69, 71, 73, *74*, *75*, *87*, 88, 101		
Me 2,3,4-Me_3-pentopyranoside	55, 58, 71, 73, *75*, *88*, 99, *101*, 115, 176		
TMS diethyl dithioacetals, partially methylated sugars			
2,3,4-Me_3-glucose	*190*, *205*, 234, 249		18
2,3,6-Me_3-glucose	*147*, *190*, *248*		
2,4,6-Me_3-glucose	*147*, 190		
3,4,6-Me_3-glucose	*147*, *190*, 237		
TMS deoxy(methoxyamino)alditols			
Pentose, deoxyhexose	103, 205, 307, 366, 454, 468		19
Hexose	*73*, 102, 147, 205, *217*, *276*, 291, 306, 320, *332*, 360, 421, 450, 525, 539, 556	572	
Permethylated deoxy(methylmethoxyamino)alditols			
Pentose	*74*, 236		20
Deoxyhexose	*74*, 250		
Hexose	*74*, *101*, *131*, *145*, 174, *189*, 221, 280		

[a] Ions of higher intensity are italicized; for alditol acetates, m/z 43 is present in all cases (as the base peak).
[b] Author's data.
[c] Reduced with $NaBD_4$; 6,6-d_2-hexitol indicates carboxylate-reduced uronic acid residue, converted to 6,6-dideuteriohexosyl unit.

REFERENCES

1. **Jansson, P. E., Kenne, L., Liedgren, B., Lindberg, B., and Lönngren, J.,** A practical guide to the methylation analysis of carbohydrates, *Chem. Commun. (Univ. Stockholm)*, No. 8, 1976.
2. **Klok, J., Cox, H. C., de Leeuw, J. W., and Schenck, P. A.,** Analysis of synthetic mixtures of alditol acetates by capillary gas chromatography, gas chromatography-electron impact mass spectrometry and gas chromatography-chemical ionization mass spectrometry, *J. Chromatogr.*, 253, 55, 1982.
3. **Wong, C. G., Sung, S.-S. J., and Sweeley, C. C.,** Analysis and structural characterization of amino sugars by gas-liquid chromatography and mass spectrometry, *Methods Carbohydr. Chem.*, 8, 55, 1980.
4. **Weber, P. L. and Carlson, D. M.,** A simultaneous assay system for *N*-acetylgalactosamine and *N*-acetylgalactosaminitol using gas chromatography/mass spectrometry, *Anal. Biochem.*, 121, 140, 1982.
5. **Waeghe, T. J., Darvill, A. G., McNeil, M., and Albersheim, P.,** Determination, by methylation analysis, of the glycosyl-linkage compositions of microgram quantities of complex carbohydrates, *Carbohydr. Res.*, 123, 281, 1983.
6. **Seymour, F. R., Chen, E. C. M., and Bishop, S. H.,** Identification of aldoses by use of their peracetylated aldononitrile derivates: a g.l.c.-m.s. approach, *Carbohydr. Res.*, 73, 19, 1979.
7. **Seymour, F. R., Plattner, R. D., and Slodki, M.E.,** Gas-liquid chromatography-mass spectrometry of methylated and deuteriomethylated per-*O*-acetylaldononitriles from D-mannose, *Carbohydr. Res.*, 44, 181, 1975.
8. **Seymour, F. R., Knapp, R. D., Chen, E. C. M., Bishop, S. H. and Jeanes, A.,** Structural analysis of *Leuconostoc* dextrans containing 3-*O*-α-D-glucosylated α-D-glucosyl residues in both linear-chain and branch-point positions, or only in branch-point positions, by methylation and by ^{13} c-n.m.r. spectroscopy, *Carbohydr. Res.*, 74, 41, 1979.
9. **Seymour, F. R., Chen, E. C. M., and Stouffer, J. E.,** Identification of ketoses by use of their peracetylated oxime derivatives: a g.l.c.-m.s. approach, *Carbohydr. Res.*, 83, 201, 1980.
10. **Schweer, H.,** Gas chromatography-mass spectrometry of aldoses as *O*-methoxime, *O*-2-methyl-2-propoxime and *O*-*n*-butoxime pertrifluoroacetyl derivatives on OV-225 with methylpropane as ionization agent. I. Pentoses, *J. Chromatogr.*, 236, 355, 1982.
11. **Schweer, H.,** Gas chromatography-mass spectrometry of aldoses as *O*-methoxime, *O*-2-methyl-2-propoxime and *O*-*n*-butoxime pertrifluoroacetyl derivatives on OV-225 with methylpropane as ionization agent. II. Hexoses, *J. Chromatogr.*, 236, 361, 1982.
12. **Schweer, H.,** G.L.C.-M.S. of the oxidation products of pentitols and hexitols as *O*-butyloxime trifluoroacetates, *Carbohydr. Res.*, 111, 1, 1982.
13. **Petersson, G.,** Gas-chromatographic analysis of sugars and related hydroxy acids as acyclic oxime and ester trimethylsilyl derivatives, *Carbohydr. Res.*, 33, 47, 1974.
14. **Laine, R. A. and Sweeley, C. C.,** *O*-Methyl oximes of sugars. Analysis as *O*-trimethylsilyl derivatives by gas-liquid chromatography and mass spectrometry, *Carbohydr. Res.*, 27, 199, 1973.
15. **Aluyi, H. A. S. and Drucker, D. B.,** Fingerprinting of carbohydrates of *Streptococcus mutans* by combined gas-liquid chromatography-mass spectrometry, *J. Chromatogr.*, 178, 209, 1979.
16. **Rivière, M., Fournié, J. J. Monsarrat, B., and Puzo, G.,** Identification of partially methylated methyl glycosides by gas chromatography-mass spectrometry of trimethylsilyl derivatives. Application to mycobacterial glycolipid antigen analysis, *J. Chromatogr.*, 445, 87, 1988.
17. **Mihálov, V., Kováčik, V., and Kováč, P.,** Identification of methyl-*O*-methylpentopyranosides by mass spectrometry, *Carbohydr. Res.*, 73, 267, 1979.
18. **Honda, S. Nagata, M., and Kakehi, K.,** Rapid gas chromatographic analysis of partially methylated aldoses as trimethylsilylated diethyl dithioacetals, *J. Chromatogr.*, 209, 299, 1981.
19. **Das Neves, H. J. C., Riscado, A. M. V., and Frank, H.,** Derivatives for the analysis of monosaccharides by capillary g.l.c.: trimethylsilylated deoxy(methoxyamino)alditols, *Carbohydr. Res.*, 152, 1, 1986.
20. **Dav Neves, H. J. C., Riscado, A. M. V., and Frank, H.,** Single derivatives for GC-MS assay of reducing sugars and selective detection by the nitrogen-phosphorus detector (NPD), *J. High Resolut. Chromatogr. Chromatogr. Commun.*, 9, 662, 1986.

Section II.II

DETECTION OF CARBOHYDRATES IN SUPERCRITICAL FLUID CHROMATOGRAPHY: USE OF MASS SPECTROMETRY

As in gas chromatography, the flame ionization detector (FID) is the detector commonly used in supercritical fluid chromatography (SFC) of carbohydrates and other organic compounds.[1,2] Recently, however, some success has been achieved in coupling SFC of carbohydrates to CI-mass spectrometry.[3,4] The microbore capillary column used in SFC is interfaced with the chemical ionization chamber of the mass spectrometer through an integral pressure restrictor, heated to about 280°C at the tip serving as the probe for MS. With ammonia as the reactant gas, selected ion monitoring of the $[M + NH_4]^+$ ions affords a detection method for trimethylsilylated or permethylated D-gluco-oligosaccharides that is far more sensitive than the FID; a detection limit of about 2 pmol has been claimed[4] for SFC-MS by this technique. The upper limit of molecular weight detectable is dependent upon the range of the mass spectrometer itself. With a low-voltage quadrupole instrument having a limit of 3000 D,[3] trimethylsilylated maltodextrins can be detected only to DP 7, but Reinhold et al.[4] have reported detection of the same oligosaccharides to DP 17 when the SFC column was coupled to a double-focusing magnetic sector mass spectrometer operating at 8 kV. SFC-MS has great potential as a sensitive analytical method in such applications as profiling glycoproteins and glycolipids and monitoring degradation of polysaccharides to homologous series of oligosaccharides.

REFERENCES

1. **Lee, M. L. and Markides, K. E.,** Chromatography with supercritical fluids, *Science,* 235, 1342, 1987.
2. **Kuei, J., Her, G.-R., and Reinhold, V. N.,** Supercritical fluid chromatography of glycosphingolipids, *Anal. Biochem.,* 172, 228, 1988.
3. **Pinkston, J. D., Owens, G. D., Millington, D. S., Burkes, L. J., and Delaney, T. E.,** Capillary supercritical fluid chromatography-mass spectrometry using a "high mass" quadrupole and splitless injection, *Anal. Chem.,* 60, 962, 1988.
4. **Reinhold, V. N., Sheeley, D. M., Kuei, J., and Her, G.-R.,** Analysis of high molecular weight samples on a double-focusing magnetic sector instrument by supercritical fluid chromatography/mass spectrometry, *Anal. Chem.,* 60, 2719, 1988.

Section II.III

DETECTION METHODS FOR LIQUID CHROMATOGRAPHY

Since detection in liquid chromatography was reviewed in Volume I, there have been many improvements in existing methods and some powerful new detectors have been introduced. The emphasis has been on automated techniques applicable to HPLC, rather than methods involving analysis of separated fractions. The various detection systems now available, which have been discussed in excellent, comprehensive reviews of HPLC of carbohydrates by Honda[1] and Hicks,[2] may be classified into three main categories:

1. Direct detection methods depending on physical properties of the carbohydrate solutes or chemical changes such as electrochemical oxidation or fragmentation in the mass spectrometer
2. Methods in which precolumn derivatization is used to permit detection by techniques such as fluorimetry or scintillation counting of radioactively labeled compounds
3. Methods involving postcolumn derivatization, with the introduction of chromophores or fluorescence to the molecule

Methods currently used in LC of carbohydrates are reviewed below under these three headings. The sensitivities of the various techniques are compared wherever possible.

Direct Detection Methods

The differential refractometer remains the detector most often used in HPLC of carbohydrates without appreciable UV absorbance. More sensitive refractometers are now available which can detect as little as 15 to 25 ng of monosaccharide,[2] and precise analyses are possible in the range 200 ng to above 20 μg. Such a refractometer is exemplified by the SE-31 model (Showa Denko, Tokyo, Japan) which was used by Koizumi et al.[3] with full-scale deflection at 1×10^{-5} RI units, permitting detection of homologous series of oligosaccharides to DP 30 to 35 in amino-phase HPLC of hydrolysates or acetolysates of polysaccharides (see Table LC 3). For reversed-phase HPLC of sugars on microbore columns eluted with water a subnanoliter RI detector with a laser light source has been described by Bornhop et al.,[4] who claim a detection limit as low as 40 to 50 ng, with an RI change of ca. 2.8×10^{-6} at the peak maximum. Cullen et al.[5] have reported the use of a sensitive interference refractometer in high-performance SEC of dextrans of clinical interest ($\bar{M}_w$ 1,200 to 250,000; see Table LC 55), which permitted chromatography of small samples (50 to 75 μl) of biological fluids containing as little as 6 μg of dextran. Refractive index detectors are affected by changes in solvent composition, temperature, and pressure, especially the more sensitive detectors, which demand the use of low-pulsation solvent-delivery systems and close control of column temperature. A major disadvantage of RI detectors is that they cannot be used with solvent gradients.

UV detectors operating at 250 to 300 nm give sensitive detection of appropriately derivatized carbohydrates (see Section III.III). Simple sugars and alditols can be detected at 188 to 192 nm, but acetonitrile absorbs strongly in this region and, therefore, direct UV detection is seldom useful in normal-phase partition chromatography of sugars.[6] With water as eluent, as in chromatography on Ca^2-form cation-exchange resin (see Table LC 30), detection limits of 2 to 5 nmol have been reported[7] for sugars and alditols monitored by UV absorbance at 190 nm. Direct UV detection is more successful with carbohydrates having high extinction coefficients, such as hexuronic and aldonic acids (210 to 220 nm), ascorbic acids (254 to 268 nm), oligosaccharides containing 2-acetamido-2-deoxyhexose or *N*-acetylneuraminic acid units (190 to 195 nm), and unsaturated oligosaccharides derived from glycosaminoglycuronans (232 nm), for which detection limits, as listed by Hicks,[2] range from 10 to 50 μg for those containing carboxyl groups only to 10 ng or less for sialylated or unsaturated oligosaccharides. The choice of usable solvents is limited, and very high purity is essential.

Polarimetric detection has been used in some cases, e.g., in HPLC of partially methylated sugars[8] on C_{18}-silica packings in water or aqueous methanol (see Table LC 8), for which detection limits of 100 μg are possible (for sugars with $[\alpha]_D$ above 50°) if the specific rotation is monitored at 365 nm. This form of detection has the advantages that enantiomers and anomers can be distinguished and that any optically inactive eluent can be used, and it is therefore surprising that it is not more widely utilized.

The mass detector (not to be confused with mass-spectrometric detection; see Sections II.I, II.III, and next subsection) involves measurement of light scattering from nebulized solute, the solvent being evaporated in a stream of compressed air. It is a universal detector that can be used with solvent gradients, but not eluents containing inorganic salts. The detector has been applied in normal-phase HPLC of mono- and oligosaccharides, for which detection limits of 2 to 3 μg have been reported,[9] and in HPLC of more complex molecules, such as glycolipids[10] (see Table LC 22) and oligosaccharides of high DP.[11] Compatibility with SFC has also been reported.[11]

In addition to its utilization in this evaporative detector, light scattering *per se* has proved valuable in the special case of high-performance SEC of polysaccharides. The use of a low-angle laser light scattering (LALLS) photometer in series with a differential refractometer for detection permits simultaneous determination of $\overline{M}_w$ of the eluted polysaccharide fractions, so that molecular weight distribution can be calculated rapidly[12,13] (see Table LC 57).

In the moving-wire detector the eluate is coated or sprayed on a wire which passes through an evaporator to remove the solvent, followed by an oxidizing furnace in which the organic residue undergoes combustion to form CO_2 and water. The CO_2 is subsequently reduced by hydrogen gas (in the presence of a nickel catalyst) to methane, which is detected by an FID. This is another universal detector which has been applied to carbohydrates, e.g., by Wells and Lester[14] in reversed-phase chromatography of peracetylated maltodextrins (see Table LC 16). The detector generally has microgram sensitivity[15] and can be used in gradient elution, but not with eluents containing water in high proportion, since solvents must be volatile. The moving-wire detector, one of the first developed for HPLC, has seldom been used for several years. Recent adaptation for use with microbore columns at flow rates of 60 to 100 μl/min has proved successful in normal-phase HPLC of mono- and oligosaccharides, with detection limits ranging from 110 ng for glucose to 400 ng for lactose,[16] and may lead to a resurgence of interest in this detector.

For detection of carbohydrates and other organic compounds in HPLC on microbore columns with aqueous eluents, coupling the microcolumn to an inductively coupled plasma atomic emission spectrometer (ICP-AES) has been proposed.[17] The column is interfaced with the ICP nebulizer, and all organic compounds eluted can be detected by monitoring the carbon emission line at 248 nm. The method has been used successfully in SEC of arabinose, glucose, and raffinose in water, at a 4 μl/min flow rate, which gave a detection limit of almost 800 ng for raffinose.

Coupling of Mass Spectrometry to HPLC

The problems inherent in on-line coupling of mass spectrometry to HPLC and the interfaces that have been developed to overcome them have been discussed in detail in a review by Games[18] and briefly by Hicks.[2] Games[18] favors a moving belt system, analogous to the moving wire system described above for coupling of FID and HPLC, in which all solvent is removed by vaporization in vacuum locks before the sample is flash-vaporized into the ionization chamber of the mass spectrometer. This system has the advantages that the mass spectrometer may be operated in EI or CI mode and that eluents containing nonvolatile salts or ion-pairing reagents can be used, since the residues left on the belt are removed by a scrubber through which the belt passes before reaching the HPLC column again. MS detection by this method, in the NH_3-CI mode, has been applied in HPLC of underivatized mono-, di-, and trisaccharides, and some glycosides such as 4-nitrophenyl-α-D-galactopyranoside and salicin.[18] Full spectra can be obtained from injection of 100 ng of sample into the liquid chromatograph, and selected ion monitoring (see Section II.I) allows detection in the high picogram to low nanogram range.

Direct introduction of a portion of the HPLC eluate into the ionization chamber of the mass spectrometer is an approach adopted by many, notably Albersheim and co-workers,[19-21] who have made extensive use of the method in HPLC of the partially methylated, partially ethylated reduced oligosaccharides obtained in their technique for sequencing complex carbohydrates (see Table LC 17). In fractionation of the derivatized oligosaccharide-alditols by reversed-phase HPLC, a small proportion (ca. 3%) of the effluent from the column was diverted into the mass spectrometer and the remainder was collected in fractions for further examination by GLC-MS of the hydrolysates. The eluate to be examined by MS was directed in a fine jet into the chemical ionization chamber, where the solvent (aqueous acetonitrile) was volatilized (at 150°C) to serve as the CI reactant gas. This permitted not only direct MS detection of the oligosaccharide-alditols, but also partial identification from the CI mass spectra and selected-ion chromatograms (using the M + 1 ion where possible). Analysis of samples of ca. 2 mg of the peralkylated oligosaccharide-alditol mixture from a complex polysaccharide (<10 mg) is possible by this method.[19,21] The same technique has been used by Schauer and associates[22] in the HPLC-MS of underivatized sialic acids and the trisaccharide sialyllactose. HPLC was performed in ion-exchange mode with an eluent containing a volatile buffer, 20 m*M* ammonium formate, pH 6, in water-acetonitrile (4:1); 3% of the column effluent was directed into the mass spectrometer ion source, which was maintained at 250°C to volatilize this largely aqueous eluent. The buffer decomposed to formic acid and ammonia, which became the CI reactant gas. This method affords rapid analysis of samples containing 1 to 10 μg of sialic acids, which give simple, highly characteristic CI-MS spectra.

The combination of HPLC with MS by direct liquid introduction is greatly facilitated by the use of microbore columns in HPLC. Flow rates from such columns are in the range 10 to 50 μl/min, so that the entire effluent stream can be introduced into the mass spectrometer ion source, without splitting. Hirter et al.[23] have designed an interface in which the effluent from the microbore column is fed through a fused-silica capillary (25 μm I.D.) with a heated tip into a desolvation chamber and thence into the ion source. The temperatures of probe tip, desolvation chamber, and ion source are controlled independently. To obtain spectra giving molecular weight information for a thermally labile compound such as sucrose, the pressure in the ion source is reduced. Among the samples used to test this interface were sucrose and the phenolic glycoside naringin, eluted from a C_{18}-silica column (25 cm × 1 mm I.D.) with methanol-water (19:1) at 25 μl/min, and the ribonucleoside adenosine, eluted with water. With the probe tip at 160° C and the desolvation chamber and ion source at 270°C, the negative CI spectra showed minimal fragmentation with a base peak due to the $[M - 1]^-$ ion. The same applied to the positive CI spectrum of adenosine, which was dominated by the peak of the $[M + 1]^+$ ion. In actual analyses by this method, selected ion monitoring using these ions would afford highly sensitive detection.

The thermospray interface, in which ionization occurs during evaporation of microdroplets in a sonic jet of vapor formed from the HPLC effluent, has been widely used in recent years. Hsu et al.[24] tested this method in the HPLC-MS of mono- and disaccharides, methyl glycosides, and permethylated mono- to tetrasaccharides eluted from C_{18}-silica or NH_2-silica columns with 0.1 *M* ammonium formate, pH 5.6. The thermospray probe temperature was maintained by electrical heating at 240°C. The use of an ammonium formate buffer as the chromatographic mobile phase provided a source of NH_4^+ ions that formed abundant ammonium molecular ion complexes (AMIC) with the carbohydrate solutes and gave simple characteristic CI mass spectra with the base peak due to the $[M + NH_4]^+$ ion in most cases (although fragmentation was apparent in the spectra given by oligosaccharides). The $[M + NH_4]^+$ ion was most abundant in the spectra of the methyl glycosides, and selected ion monitoring using this ion gave a detection limit as low as 20 pg for these glycosides, indicating that the technique could afford a highly sensitive means of analysis of methanolysates. Sensitivity for underivatized sugars was about ten times lower, but still at subnanogram level.

Where the use of ammonium buffers in the chromatographic eluent is not desirable, postcolumn addition of ammonium acetate prior to passage of the eluate through the thermospray interface can be used to enhance the sensitivity of CI-MS by the production of abundant diagnostic ions. This method has been applied in the HPLC-MS of maltodextrins,[25] simple sugars and alditols of clinical interest,[26] and permethylated maltotriitol.[27] Esteban et al.[26] reported simultaneous analysis of D-glucose, D-galactose, D-fructose, D-mannitol, and D-glucitol in biological fluids, with detection limits of almost 50 pmol for each obtainable by selected ion monitoring of the abundant $[M + NH_4]^+$ ions. Quantitative assay of D-glucose and D-glucitol was performed by the isotope dilution method, using ^{13}C-labeled D-glucose and D-glucitol and monitoring the ions *m/z* 198 and 204, corresponding to the $[M + NH_4]^+$ complexes of unlabeled and labeled compounds, respectively. Recently the direct-liquid-introduction method favored by the Albersheim group has been modified by similar postcolumn addition of ammonium acetate,[27] with the objective of maximizing the abundance of the AMIC giving the diagnostic $[M + NH_4]^+$ ions. With 50% aqueous acetonitrile as the eluent, postcolumn addition of ammonium acetate at a concentration of 0.3 *M* in the same solvent, prior to introduction of the eluate into the ion source, maintained at 200°C, gave maximum relative abundance of the AMIC and high total ion current. This resulted in sensitivity comparable with that given by thermospray HPLC-MS, with which a direct comparison was made using permethylated maltotriitol as a test substance. The CI mass spectra given by permethylated stachyose, peracetylated cellobiose, and underivatized cellobiose and maltobiose were similarly improved by the presence of ammonium acetate in direct-liquid-introduction HPLC-MS. This technique is recommended by Albersheim et al.[27] over the thermospray method because it requires only a small proportion of the column effluent, leaving the remainder for further chemical and biological assays.

Another method of interfacing HPLC and MS that is applicable to nonvolatile compounds such as sugars and nucleosides is the use of a heated nebulizer and vaporizer with atmospheric pressure ionization mass spectrometry.[28] Both nebulizer and vaporizer can be heated to 400° C, but independent control of temperature is possible. The sample and solvent molecules introduced into the ion source of the mass spectrometer in this way are ionized by corona discharge followed by ion-molecule reactions. Clusters are dissociated by collisions with other molecules at enhanced kinetic energy resulting from application of a drift voltage in one region of the ion source. With the temperatures of nebulizer and vaporizer set at 330 and 380°C, respectively, quasi-molecular ions were observed in the CI mass spectra of mono-, di-, and trisaccharides, but were not abundant owing to considerable fragmentation under these conditions. Detection limits with selected ion monitoring of the most abundant (fragment) ions ranged from 50 ng for monosaccharides to 200 ng for maltotriose, in which respect this technique is much inferior to thermospray HPLC-MS or the direct-liquid-introduction method with postcolumn addition of ammonium acetate.

All methods of MS detection for HPLC can be used with gradient elution, but HPLC eluents containing nonvolatile salts or ion-pairing reagents are compatible with the moving belt interface only. With volatile mobile phases, interfacing with MS by thermospray technique or direct-liquid-introduction with added ammonium acetate gives HPLC a sensitivity approaching that possible with GLC-MS.

Electrochemical Detection

The development of high-efficiency electrodes has made possible sensitive detection of readily oxidizable compounds, such as L-ascorbic acid and D-erythorbic acid. These may be determined in the presence of other carbohydrates that are not electroactive by use of an amperometric detector in which the electroactive acids are oxidized on the surface of a glassy carbon electrode.[29] Reducing sugars and aldonolactones which co-elute with L-ascorbic acid upon HPLC on Aminex HPX-87H with 0.1 *M* formic acid as eluent (see Table LC 28) do not interfere when this detector is used, even if present in large excess. The use of the amperometric detector in the determination of L-ascorbic acid in beer by ion-pair reversed-phase chromatography on a C_{18} -silica column eluted with a citrate buffer solution (pH 4.4) containing 1 m*M* *N*-methyldodecylamine as ion-pairing reagent[30] gave a detection limit of $\leq$1 ng.

Other carbohydrates, including reducing sugars, have a large overpotential toward electro-oxidation at the glassy carbon or carbon paste electrodes; therefore, inordinately high detector potentials would be required for oxidation to occur. Santos and Baldwin[31] have developed an electrocatalytic, chemically modified electrode in which the carbon paste electrode is modified by addition of a reversible redox reagent capable of chemically oxidizing carbohydrates. Detection occurs indirectly by measurement of the reoxidation of the modifier following exposure of the electrode to the reducing carbohydrates. The modifier chosen was cobalt(II) phthalocyanine (CoPC), which was added to the graphite powder (CoPC loading 2.0% by weight) before it was mixed with Nujol oil to make the carbon paste used in the electrode. The CoPC-modified electrode was used with an Ag/AgCl reference electrode and a platinum wire auxiliary electrode as a detector in ion chromatography of mono- and disaccharides on HPIC-AS6, eluted with 0.15 *M* NaOH at 36°C; high pH ($\geq$13) was required for efficient functioning of the electrode. Pulsed operation was also necessary, since the electrode became passive after a few cycles unless a negative potential of −0.3 V or lower was applied periodically to restore it; thus, application of the detection potential of +0.39 V was followed by a negative pulse of −0.3 V in each cycle. Under these conditions, detection limits for the sugars were in the range 100 to 500 pmol. Subsequent modification in the preparation of the electrode, by addition of the CoPC to a premixed carbon paste in which silicone oil replaced Nujol, produced a tenfold improvement in sensitivity.[32] Detection limits of 10 pmol were reported for glucose, xylose, arabinose, galactose, and lactose, 25 pmol for maltose, and 50 pmol for ribose, fructose, and sucrose (possibly due to the presence of furanose rings). The detector also was applied successfully in the ion chromatography of maltodextrins to DP 7, with elution by 0.15 *M* NaOH-0.25 *M* sodium acetate.

The use of metallic sensing electrodes instead of carbon greatly diminishes the overpotential of carbohydrates and enables oxidation to occur at moderate potentials in alkaline solutions. Reim and van Effen[33] have used a nickel working electrode, with an Ag/AgCl reference electrode and a stainless steel tube as auxiliary electrode, as a detector in ion chromatography of alditols and mono- and oligosaccharides on HPIC-AS6 eluted with 0.15 *M* NaOH (see Table LC 45). Under these conditions active nickel(III) oxide, NiOOH, is formed *in situ* on the electrode surface at potentials near 0.45 V so that the electrode is effectively composed of nickel(III) oxide, a strong oxidant. In this case, therefore, no negative pulse is required to restore the original electrode, and simple constant-potential amperometric operation is possible. Detection limits were reported as 23 to 26 ppb for alditols, 16 to 80 ppb for monosaccharides, and 120 ppb for lactose. The detector has been applied to ion chromatography of maltodextrins to DP 7, with 0.15 *M* NaOH-0.15 *M* sodium acetate as eluent at 34°C, under which conditions the detection limits for the individual oligomers approach 5 ng.

It is, however, platinum[34-36] and gold[37,38] electrodes that have been used most for amperometric detection of carbohydrates under the alkaline conditions employed in ion chromatography. In both cases a cycle of three pulses is necessary: the first is the detection potential, the second is to oxidize fully any material adsorbed on the surface of the electrode, and the third is to reduce the surface oxide and restore the metallic surface. The triple-pulse potential waveform differs for the two types of electrodes, since the mechanism involved in carbohydrate oxidation is different. For the platinum electrode,[34,35] the surface-catalyzed reaction is anodic and is believed to be dehydrogenation of the adsorbed organic molecules; the applied potential is −0.4 V. The hydrocarbon products of dehydrogenation can be cleaned from the electrode surface oxidatively, with simultaneous formation of oxide on the platinum, if the potential is stepped to a positive value (+0.8 V) approximately corresponding to that for anodic decomposition of the solvent, and the surface oxide is subsequently reduced by application of a negative potential (−1.0 V) approximating that for cathodic breakdown of the solvent. The entire cycle is complete within 600 ms, which is short enough to permit virtually continuous monitoring of the effluent stream in HPLC. Using this electrode with a miniature calomel electrode filled with a saturated solution of KCl as reference electrode, Hughes and Johnson[35,36] detected sugars and alditols eluted at 80 to 85°C from Ca^{2+}-loaded cation-exchange resins with water (postcolumn addition of NaOH adjusted the pH to 13 for efficient functioning of the electrode) at levels as low as 460 ng. Increasing the detector temperature from 35 to 85°C enhanced the sensitivity by factors ranging from about 1.5 for alditols to 2.1 for lactose.[36]

In the pulsed amperometric detector (PAD) described by Rocklin and Pohl[37] in 1983 a gold working electrode is used, with an Ag/AgCl reference electrode and a glassy carbon counterelectrode to allow the potentiostat to maintain the selected applied potential and to prevent damaging current drain on the reference electrode.[39] In this case direct oxidation of the carbohydrates occurs on the surface of the electrode, under alkaline conditions, at an applied potential of +0.2 V. After 60 ms, the applied potential is stepped to +0.6 V (the cleaning potential) and then, after a further 60 ms, to −0.8 V, the negative cleaning potential that reduces surface oxide back to metallic gold. After 240 ms, the cycle starts again. The total duration of only 360 ms allows virtually continuous monitoring of column effluents. Detection limits of 30 ppb for alditols and monosaccharides, 100 ppb for oligosaccharides have been quoted[37] for analysis by ion chromatography using this detector, which has become standard in the Dionex systems for ion chromatography of carbohydrates[38-40] (see Tables LC 45 to 48). The sensitivity of the commercial instrument can be varied over a very wide range,[38] with full-scale deflection at outputs ranging from 1 nA to 100 μA. This facility is invaluable in detection of the higher members of homologous series of oligosaccharides resolved by ion chromatography[40] (see Table LC 46).

A disadvantage of detectors of this type is that they can be used only with alkaline eluents. Recently, however, Haginaka and Nomura[41] have demonstrated that this limitation can be overcome by incorporation of a cation-exchange membrane reactor into the system, so that the pH of the column effluent can be changed, if necessary, to alkaline values before the effluent reaches the detector. This was achieved successfully with the aqueous eluate from a Pb^{2+}-loaded cation-exchange column at 80°C (see Table LC 32) and a (3:6:1) water-acetonitrile-methanol mixture used to elute sugars from a heavily cross-linked resin in the Na^+ -form at 50° C (see Table LC 35). If NaOH of appropriate concentration was delivered to the membrane reactor before the column effluent passed through it, the eluent was made alkaline by both exchange of Na^+ ions from the membrane and permeation of OH^- ions. This gave improved detection limits for sugars and alditols of ca. 10 pmol with the aqueous eluent and 50 pmol with the aqueous/organic eluent.

A different method of applying amperometric detection to the HPLC of sugars is to oxidize the sugars in the eluate and measure the response when the reagent thereby reduced is reoxidized at a glassy carbon electrode. In a sensitive detection method proposed by Watanabe and Inoue,[42] the column effluent is mixed with a reagent solution containing a complex of Cu(II) with bis(phenanthroline), and the resulting Cu(I) complex is reoxidized at the detector electrode. These authors investigated the effect of various conditions on this redox reaction and concluded that the optimal peak response is obtained when reduction of the Cu(II) complex takes place at pH 11.2 and temperature $\geqslant$95°C, with the flow rates of reagent and column effluent controlled so that the reagent:sugar ratio is between 2 and 3 and the length of the reaction coil such that the reaction mixture is in the reactor for at least 40 s. The applied potential at the working electrode, used with an Ag/AgCl reference electrode, is +40 to +75 mV, very low compared with that generally used in oxidative detection; this minimizes interference due to other organic compounds, most of which have higher oxidation potentials. The reagent solution, prepared by dissolving the Cu(II) complex (1 to 3 m*M*) in 0.1 *M* Na_2HPO_4 (the pH being adjusted to the desired value by addition of 2 *M* NaOH), remains stable for several months. Before the analysis both eluent and reagent solutions should be freed of dissolved oxygen by bubbling with nitrogen followed by degassing, and during the experiment they should be kept under nitrogen. The detection method was applied successfully to the HPLC of mono- and disaccharides eluted with water from a cation-exchange column or with 0.05 *M* borate buffer (containing 0.02 *M* Na_2HPO_4) at pH 8.8 from an anion-exchange column. The detection limit for glucose was 0.2 ng (ca. 1 pmol) under these conditions. Watanabe[43] has subsequently reported that the method is also compatible with HPLC using organic solvents, after applying it to normal-phase partition chromatography of sugars in acetonitrile-water (7:3); in this case, the optimal pH was 10.9 and the detection limit for glucose was 5 pmol. Sensitive responses were obtained with all reducing sugars, including aminodeoxyhexoses.

Another detection method involving oxidation of sugars in solution followed by electrochemical determination of a reagent simultaneously reduced is the enzyme reactor proposed by Marko-Varga[44] for selective detection of glucose and certain other sugars (such as lactose in penicillin fermentation baths). This method is applicable only to HPLC with aqueous eluents and was used by Marko-Varga in conjunction with a cation-exchange resin in the H^+-form (Aminex HPX-87H) eluted at 40°C with 5 m*M* H_2SO_4 (see Table LC 27) and resins in the Pb^{2+}- and Ca^{2+}-forms (HPX-87P and -87C) eluted with water at 65°C (see Tables LC 30 and LC 32). The effluent from the columns is mixed with 0.5 *M* phosphate buffer, pH 6.3, containing 10 m*M* nicotinamide adenine dinucleotide (NAD^+), coenzyme of glucose dehydrogenase, and passed through a packed-bed reactor in which that enzyme is immobilized on porous glass. Oxidation of the carbohydrates produces an equivalent amount of reduced coenzyme (NADH), which is detected at an electrode consisting of a polished piece of spectrographic graphite coated (by dipping) with meldola blue ion (MB^+; 7-dimethylamino-1,2-benzophenoxazinium ion); this reoxidizes NADH to NAD^+ through formation of a charge-transfer complex. The meldola blue (MBH) is rapidly reoxidized electrochemically on application of a potential of about −50 mV (working electrode vs. small calomel electrode). This detection system worked efficiently with all the HPLC systems investigated, but where the Pb^{2+}-form resin was used it was necessary to add MgEDTA to the buffer solution containing NAD^+ to avoid deactivation of the enzyme by Pb^{2+} ions leached from the column. The method is highly specific for D-glucose (detection limit 2 ng) and 2-deoxy-D-*arabino*-hexose (detection limit 1.6 ng), with smaller responses for D-xylose (4 ng), D-mannose, 2-amino-2-deoxy-D-glucose (both ca. 10 ng), and the disaccharides cellobiose and lactose (77 and 130 ng, respectively). A very low response is given by D-ribose and none by D-galactose, D-arabinose, or D-fructose. The method is, therefore, of limited application, except where specific detection of the responding sugars is required.

Conductivity detectors are useful in the HPLC of charged molecules. For example, such a detector has been used in ion chromatography of sugar phosphates eluted with 2.4 m*M* $NaHCO_3$-1.92 m*M* Na_2CO_3, with a postcolumn micromembrane suppressor, continually regenerated with 0.0125 *M* H_2SO_4, replacing the Na^+ ions by H^+ and so removing the eluent anions by conversion to CO_2 and water.[45]

Tanaka and Fritz[46] have reported analyses of mixtures containing sugars together with alcohols and carboxylic acids by chromatography on a cation-exchange resin in the H^+-form eluted with 1 m*M* H_2SO_4, using a conductivity detector that gave positive peaks for the acids and negative peaks for the neutral compounds. The negative peaks were ascribed to suppression of the conductivity of the eluent by these solutes, an effect that increased with molecular weight so that disaccharides gave a greater negative deflection than did monosaccharides. However, sensitivity is relatively low (detection limit for glucose is ca. 54 nmol), and the method is suitable only for rough, qualitative scanning (e.g., of beverages). Okada and Kuwamoto[47] used the formation by some sugars and alditols of ionized borate complexes with boric acid, the dissociation constants of the complexes being higher than that of boric acid (pK_a 9.2), as the basis for a method for HPLC separation with conductometric detection. Due to this effect the carbohydrates, eluted with boric acid at 35° C from a cation-exchange resin in the H^+-form, gave positive peaks with a conductivity detector. In this case, sensitivity depended upon both the configuration of the hydroxyl groups in the carbohydrate molecule, which governs the ease of formation of the borate complexes, and the concentration of boric acid. With 0.05 *M* boric acid, detection limits ranged from 1 ppm or less for D-glucitol, D-mannitol, D-fructose, and D-ribose, through 6 and 10 ppm for D-xylose and D-galactose, respectively, to 20 ppm for D-glucose and 30 ppm for glycerol. Lactose was the only disaccharide detectable (detection limit 100 ppm). Organic acids interfere, but at low boric acid concentrations are eluted earlier than the borate complexes; therefore, this method can be used in simultaneous analysis of carbohydrates and carboxylic acids (e.g., in wines).

Recently, Johnson and co-workers[48] used both a conductivity detector and the PAD in conjunction with ion chromatography of glucose and sucrose on the Dionex HPIC-AS6 column eluted with 2 m*M* $Ba(OH)_2$. The conductivity detector gave negative peaks for the two sugars, analogous to those reported by Tanaka and Fritz[46] for elution from a cation-exchange column with dilute acid. The detection limit for glucose was estimated at 90 ng (0.5 nmol), whereas with the pulsed amperometric detector it was 4.5 ng (25 pmol). However, the range of linearity for the PAD (0.5 to 50 ppm) overlapped that for the conductivity detector (10 to 10^4 ppm); therefore, the use of the two detectors in series, with a combined linear range of four decades, could prove useful in carbohydrate separations in which concentrations differ over a wide range.

Precolumn Derivatization Methods

Radiochemical Detection

A method widely used to enhance sensitivity of detection, especially in the chromatography of complex oligosaccharides, is reduction with $NaB[^3H]_4$, which is quantitative and is a simple reaction, proceeding in aqueous media at ambient temperature. Compounds thus labeled can be detected by scintillation counting. Several examples of the use of this technique are cited in Section I.III: Baenziger and co-workers[49-51] have used it extensively in analyses of homologous series of D-gluco-oligosaccharides and of glycoprotein-derived oligosaccharides by SEC (see Tables LC 51 and LC 53) and by HPLC in ion-exchange (Table LC 40) or normal-phase partition (Table LC 3) modes. The method also has been applied in the high-performance ion-exchange chromatography of oligosaccharides from glycosaminoglycuronans[52] (Table LC 39). This technique is particularly useful in chromatographic analysis of degradation products of glycoproteins[53,54] (see Tables LC 53 and LC 35), which are often available only in very small amounts; nanomolar samples can be analyzed in this way.

Oligosaccharides metabolically labeled with ^3H or ^{14}C, used in studies of proteoglycans and glycoproteins, have been analyzed by HPLC of hydrolysates, monitored by scintillation counting[55] (see Tables LC 32, LC 35, LC 43, and LC 53). The method is especially valuable in studies of photosynthesis by ^{14}C incorporation, and excellent quantitative analyses of ^{14}C-labeled sugars and nucleotides from plant materials have been achieved by HPLC monitored with a CaF_2(Eu) scintillator.[56]

Photometric and Fluorometric Detection

Sensitivity of detection in HPLC can be increased by the introduction of chromophoric or fluorescent groups into carbohydrate molecules prior to HPLC on silica (see Tables LC 19 to 21) or C_{18}-silica (Tables LC 14 to 16). This enhances the sensitivity of UV detection or makes possible highly sensitive fluorometric detection. Of the chromophoric derivatives, benzoates and 4-nitrobenzoates (described in Volume I) remain the most widely used, owing to the relative simplicity of the derivatization procedure (see Section III). Nanomolar sensitivity has been reported[57] in analyses of the sugar constituents of glycoproteins by reversed-phase HPLC of the benzoylated methanolyzates, with even higher sensitivity (at picomolar levels) possible by use of a microbore column.[58] Detection of submicrogram quantities also has been achieved by HPLC (on silica columns) of per(dimethylphenylsilyl) ethers of sugars, alditols, and methyl glycosides (detection limit ca. 250 ng with UV detection at 260 nm)[59,60] and of 2,4-dinitrophenylhydrazones of neutral sugars (detection limit ca. 50 pmol at 352 nm).[61] Aldoses react with *p*-anisidine (c.f. the well-known detection method used in PC) to form *N*-(*p*-methoxyphenyl)glycosylamines, which can be resolved on C_{18}-silica with UV detection.[62] This method has the advantage of simplicity, but sensitivity is relatively low (detection limit ca. 1 μg hexose at 254 nm).

Derivatives that absorb strongly in the visible range are produced by reductive amination of sugars, including hexuronic acids and amino- and acetamidodeoxyhexoses, with the azo dye 4'-*N*,*N*-dimethylamino-4-aminoazobenzene (DAAB). The DAAB-glycamines can be separated by HPLC on silica with detection at 436 nm, under which conditions detection limits of 1 to 5 pmol have been reported.[63] Derivatization with 4'-*N*,*N*-dimethylamino-4-azobenzene-1-sulfonylhydrazine (dabsyl hydrazine) also produces strongly chromophoric compounds, which can be resolved by reversed-phase HPLC.[64,65] The absorbance spectra show maxima at 460 to 470 nm, shifting to 495 nm in acidic medium, and a shoulder at 422 to 425 nm that is independent of pH.[65] Muramoto et al.[64] reported analyses of the neutral sugar compositions of glycoprotein samples as small as 5 μg by HPLC of the dabsylhydrazones on a short column of C_{18}-silica eluted with 1:3-acetone-0.08 *M* acetic acid (see Table LC 14). With this acidic eluent the wavelength used for detection was 485 nm, and linear calibration plots were obtained in the range 10 to 200 pmol for each sugar. Similar sensitivity was obtained by gradient elution with aqueous acetonitrile;[65] detection at 425 nm gave linear calibration plots over the range of 10 to 120 pmol.

Chief among the methods used to produce fluorescent derivatives from reducing sugars has been reaction with 5-dimethylaminoaphthalene-1-sulfonylhydrazine (dansyl hydrazine). The dansylhydrazones are strongly fluorescent and can be analyzed by reversed-phase HPLC[66,67] (see Table LC 14) or HPLC on silica columns[68] (see Table LC 19) with fluorometric detection (excitation wavelength 350 to 360 nm, emission 500 to 550 nm). Detection limits of 5 to 15 pmol have been reported[67] for mono- and disaccharides eluted with aqueous acetonitrile from C_{18}-silica columns and 3 to 20 pmol (linear calibration up to 1 nmol) for HPLC on silica columns eluted with chloroform-ethanol mixtures.[68] Reductive amination with 2-aminopyridine has been applied to amino- and acetamidodeoxyhexoses and *N*-acetylneuraminic acid, as well as neutral sugars, by Takemoto et al.,[69] who have resolved mixtures of the derivatized sugars by HPLC on C_{18}-silica (see Table LC 14) with fluorometric detection (excitation wavelength 320 nm, emission 400 nm) in the concentration range 10 pmol to 10 nmol. Glycoconjugates have been analyzed by this method using samples of only 100 to 200 pmol (glycoproteins) or 1 to 2 μg (gangliosides).[69] Coles et al.[70] also have applied the method in the normal-phase partition chromatography of the D-mannose oligomers isolated from the urine of mannosidosis patients, but it was less successful in this case. Reductive amination with 7-amino-1-naphthol followed by fluorometric detection (excitation wavelength 240 nm, emission 320 nm) gave higher sensitivity due to more complete derivatization and a higher fluorescent response.

For analysis of the neutral sugar constituents of glycoproteins Shinomiya et al.[71] have developed a method involving reversed-phase HPLC of fluorescent derivatives obtained by reductive amination of the sugars followed by reaction of the products with 7-fluoro-4-nitrobenz-2-oxa-1,3-diazole (NBD-F), with which fluorometric detection in the concentration range 40 pmol to 50 nmol is possible. Amino sugars can be labeled directly with NBD-F,[71] and another specific method for these sugars is afforded by derivatization with *o*-phthalaldehyde in alkaline solution in the presence of a strong reducing agent, 2-mercaptoethanol, to yield fluorescent 1-alkylthio-2-alkyl-substituted isoindoles (excitation wavelength 340 nm, emission 455 nm), which after reversed-phase HPLC[72] can be detected in the concentration range 0.1 to 1000 ppm. A highly specific and sensitive analytical method for *N*-acetyl- and *N*-glycolyneuraminic acids in biological fluids involving HPLC on a C_{18}-silica cartridge following derivatization with 1,2-diamino-4,5-dimethoxybenzene (DDB), a fluorogenic reagent for α-keto acids, has been developed by Hara et al.;[73] with fluorometric detection (excitation wavelength 369 nm, emission 453 nm), the detection limit for both neuraminic acids was 40 fmol (12 pg) and linear calibration was obtained up to at least 18 nmol (5.6 μg) per 5 μl. The method can readily distinguish changes in neuraminic acid concentration in cancer and other pathological conditions and is, thus, of importance in clinical investigations, as well as in studies of sialylated glycoproteins and glycolipids.

Postcolumn Derivatization Methods

General Methods

Postcolumn derivatization has the advantage that the carbohydrate solutes pass through the column as such and, therefore, there is no restriction on the mode of chromatography employed, in contrast to the precolumn derivatization methods that give the solute molecules hydrophobic character before chromatography so that polar phases cannot be used. A disadvantage, however, is that postcolumn derivatization demands more sophisticated instrumentation for automation of the derivatization process. For example, the use of strong acids in the automated anthrone-, orcinol-, cysteine-, or carbazole-sulfuric acid assays described previously (Volume I) necessitates acid-resistant equipment; nevertheless, these classical methods, which can detect both aldoses and ketoses with nanomolar sensitivity[1] and are applicable to oligosaccharides because the component monosaccharides are released by acid hydrolysis during the reaction, remain useful in some chromatographic systems with aqueous eluents.

Several methods not requiring chemically aggressive reagents have been adapted for automated analysis of carbohydrates separated by HPLC. Among the most commonly used are those depending upon redox reactions, such as the tetrazolium blue method (see Volume I) that has been applied to series of oligosaccharides up to DP 20 to 30, eluted with acetonitrile-water gradients from NH_2-silica columns[74,75] (see Table LC 3) and to SEC of dextrans,[74] as well as to HPLC of sugars in aqueous acetonitrile on NH_2-silica[75] (see Table LC 1) or a macroporous anion-exchange resin in the phosphate form[74,76] (see Table LC 37). Detection limits for sugars range from 10 to 20 ng for monosaccharides to ca. 100 ng for maltotriose,[75] with linear calibration to 20 μg for monosaccharides in 70% acetonitrile and to 3.5 nmol for monosaccharides (10 nmol for lactose) in 83% acetonitrile.[76] With this detection method in an adapted amino acid analyzer, the neutral sugar constituents in hydrolysates from 50 to 300-μg samples of glycoproteins have been analyzed by partition chromatography on a cation-exchange resin in the trimethylammonium form using 89% ethanolas solvent[77] (see Table LC 35).

The copper-bicinchoninate method, described in Volume I, is well suited to use with HPLC columns eluted with water or aqueous solutions: Ugalde et al.[78] have successfully used this for HPLC of sugars on a Ca^{2+}-form cation-exchange resin eluted with water at 75°C or a C_{18}-silica cartridge eluted with water at ambient temperature. The surfactant Brij 35 was added to the reagent to preclude deposition of calcium carbonate in the reaction coil due to slow release of Ca^{2+} from the resin. Color yield from the lavender Cu(I)-bicinchoninate complex produced on reduction of Cu(II) by the sugars was optimal when the reagent solution was mixed with the column effluent in the volume ratio 0:3:1 at 110°C with a reaction time of 1.8 min (6 m × 0.5 mm I.D. reaction coil with an eluent flow rate of 0.5 ml/min). Nonreducing sugars such as sucrose, raffinose, and stachyose were analyzed successfully by addition of a short column packed with H^+-form cation-exchange resin to this chromatographic system; passage through this column at 110°C resulted in hydrolysis of these oligosaccharides to fructose, together with glucose, melibiose, and manninotriose, respectively, all of which responded to the reagent. Detection limits varied from 0.5 ng for fructose and 1 ng for glucose and sucrose to 5 ng for stachyose, with linear calibration to 2.5 μg for glucose and to 1.5 μg for fructose.

Grimble et al.[79] have developed a method for postcolumn detection of sugars by reaction with the cuprammonium complex ion, which gives a product absorbing in the UV region with a broad maximum at 280 to 290 nm. A stock reagent was prepared by adding ammonia solution (30% w/v; 500 ml) to a solution of cupric sulfate (25 g in 500 ml), and the working solution by mixing this solution (100 ml) with the ammonia solution (250 ml) and distilled water (650 ml). The sugars were separated on a column (11 × 0.46 cm I.D.) of silica, modified by inclusion of 1,4-diaminobutane (0.03%) in the eluent (67% acetonitrile). The eluent flow rate was 0.8 ml/min and that of the reagent 1 ml/min. The peaks were detected with highest sensitivity at 290 nm, with a detection limit of ca. 2.5 nmol for the monosaccharides and for sucrose. The method also has been applied to lactose and malto-oligosaccharides of DP 2 to 8. For HPLC on Cu(II)-loaded silica (see Table LC 25), Leonard et al.[80] adapted this detection method by including the reagent in the mobile phase, which was 75% acetonitrile containing 0.02 m*M* Cu^{2+} and 1.5 *M* NH_3. With a standard (254-nm) UV photometer, detection limits were 8 and 12 nmol for fructose and glucose, respectively, with linear calibration to 1 μmol. The cuprammonium method has the advantage over the tetrazolium blue and copper-bicinchoninate methods in that no heated reaction coil is required, but background noise (ascribed to photosensitivity of cuprammonium complexes[80]) presents a problem. In an interesting variant of the method, Cowie et al.[81] have used a metallic copper electrode as a potentiometric detector for reducing carbohydrates after HPLC on a Ca^{2+}-form cation-exchange column eluted with water at 90°C, with postcolumn addition of a solution containing 1 m*M* Cu^{2+} and 35 m*M* NH_3. The copper electrode responds in a Nernstian manner to the concentration of Cu^{2+} or Cu^+ ions at the surface so that reduction of Cu^{2+} to Cu^+ by a sugar produces changes in potential. This affords an indirect detection method that has been applied to aldoses, ketoses, and the disaccharides maltose and lactose. With the effect of background noise decreased by use of potentiometric detection, higher sensitivity is achieved; detection limits range from 0.6 nmol for fructose to 5 nmol for the disaccharides.

Reactions with hydrazides of benzoic acid derivatives in alkaline media are also applicable to the detection of reducing sugars and can be used in analyses of nonreducing sugars if, as in the application of the copper-bicinchoninate method described above,[78] the sugars eluted from the chromatographic column are passed through a catalytic column of H^+-form cation-exchange resin at elevated temperature to hydrolyze non-reducing oligosaccharides to reducing sugars that respond to the reagent. Evaluation of various strong-acid cation exchangers for this purpose led Vrátný et al.[82] to the conclusion that silica-based ion exchangers were totally unsuitable owing to the band broadening resulting from the high adsorptive capacity of the silica matrix for sugars. Among the cation-exchange resins tested (all polystyrenesulfonate type), one having 4% cross-linking was more efficient than those of higher cross-linking, but the lower stability to pressure of the softer resin could be a disadvantage.

After establishing the feasibility of simultaneous detection of reducing and nonreducing sugars with such a catalytic column inserted between the chromatographic column (Ca^{2+}-form resin eluted with water at 80°C) and a detection system in which 4-hydroxybenzoic acid hydrazide in NaOH was mixed with the column effluent at 100° C and the absorbance measured at 410 nm, which proved effective but relatively insensitive (detection limit 0.5 μg of glucose),[83] Vrátný et al.[84] have developed a detection method using 4-aminobenzoic acid hydrazide (ABH) as postcolumn derivatization reagent. The working solution of the reagent is prepared daily by mixing the stock solution (5% 4-aminobenzoylhydrazine in 0.5 *M* HCl) with 2.4 *M* NaOH (1:2); before use, the solution should be clarified by ultrafiltration and sonication. The reagent is mixed 1:2 with the column effluent (optimal flow rate 30 ml/h) at 95°C for a reaction time of 40 s; the detection wavelength is again 410 nm. When this detection system is used in conjunction with a Ca^{2+}-form resin column, eluted with water at 85°C, and a short column (6 × 0.4 cm) of 4% cross-linked H^+-form resin at the same temperature as hydrolytic catalysis reactor, detection limits are ca. 20 ng for D-glucose and D-fructose, 15 ng for sucrose, and 30 ng for raffinose. The contributions of detector and catalytic column to band broadening are low, so that sharp peaks are obtained under these conditions. Comparison of ABH with other reagents used in the same system[84] has shown this reagent to have higher sensitivity and stability than tetrazolium blue, copper-bicinchoninate, and 2-cyanoacetamide (see below). The ABH reagent also has been applied successfully in HPLC on a Pb^{2+}-form cation-exchange resin eluted with water at 85°C, and on a C_{18}-silica column eluted with water at 25° C, in the analyses of sugars and maltodextrins in foods.[85] Apart from its much higher sensitivity, this postcolumn derivatization system has an advantage over refractive index detection in that interference from other components of food, such as amino acids and alditols, is eliminated owing to the specificity of the reagent for reducing carbohydrates.

A general reagent for carbohydrates having vicinal diol systems in the molecule is periodate. Nordin[86] has proposed a technique for monitoring the elution of carbohydrate in aqueous HPLC by continuous measurement of the periodate consumption of the effluent, based on the decrease in UV absorbance at 223 to 260 nm that accompanies reduction of periodate to iodate. This has long been used in a spectrophotometric method for determination of periodate uptake in, for example, Smith-degradation studies of polysaccharides. When a periodate solution is mixed with the effluent from a column used to separate carbohydrates, the emergence of each solute is sensed by a UV photometer as an absorbance decrease. Sensitivity may be enhanced by the use of an alkaline periodate solution at elevated temperature, since some oligosaccharides (sucrose, for example) undergo overoxidation under these conditions. However, aldoses and alditols can be monitored with sufficient sensitivity at the lower temperatures and pH (ca. 5) normally used in periodate oxidation. Thus, Nordin has recommended operation at 40 to 60°C with a reagent consisting of 2 m*M* $NaIO_4$ or KIO_4 in 0.3 *M* acetic acid, which, when mixed in a 1:1 ratio with the effluent from a borate anion-exchange column, gives a solution of pH 5. Where necessary, 1 m*M* $NaIO_4$ or KIO_4 in 0.5 *M* borate buffer (pH 8.6) is used, at 90 to 100°C. Absorbance is measured at 223 nm (as in the classical method) where detection is carried out at pH 5, but with the alkaline reagent the standard 250 to 260 nm of many simple photometers can be used since the gain in sensitivity at the lower wavelength is marginal. The method is applicable to reducing and nonreducing sugars, alditols, and cyclitols, with detection limits in the range 1 to 10 nmol depending upon the conditions used. It was tested by Nordin using borate anion-exchange chromatography, but should be applicable to any HPLC system with an aqueous eluent.

As discussed in connection with precolumn derivatization, the use of fluorogenic reagents in derivatization can result in detection at picomolar levels. There has, therefore, been considerable interest in the development of postcolumn derivatization methods using such reagents. Addition of ethylenediamine (7.5 m*M*) to the mobile phase (0.7 *M* borate buffer, pH 8.6) used in borate anion-exchange chromatography of carbohydrates has been suggested by Mopper et al.[87] as a sensitive detection method; fluorescent products resulting from isomerization of the sugars (Lobry de Bruyn-van Ekenstein reaction) and the reactions of the intermediates with excess amine are formed at elevated temperatures (110 to 120°C for pentoses, 120 to 130°C for hexoses including hexuronic acids, and 140 to 145°C for heptoses and oligosaccharides). Reducing oligosaccharides are readily analyzed by this method as they are cleaved to monosaccharides by alkaline hydrolysis; nonreducing oligosaccharides also give a slight response on heating at 145° C. Fluorometric detection (excitation wavelength 360 nm, emission 455 nm) gives detection limits in the range of 100 to 400 pmol for most sugars (below 50 pmol for some reducing disaccharides, e.g., lactose, derivatized at 145°C) and linear calibration to 100 nmol (500 nmol for glucose). Alditols give a very weak response and can, therefore, be detected only at much higher concentrations. 2-Deoxyhexoses have lower response factors than the other sugars, probably due to an inhibiting effect of the methyl group on isomerization. The ethylenediamine reagent is not compatible under these conditions with HPLC systems using organic eluents and, since metal cations (especially those of heavy metals) cause quenching of the fluorescent response, it would also be of little value in HPLC on cation-exchange resins in Ca^{2+}-, Ag^{+}-, or Pb^{2+}-forms. The necessity for heating to temperatures above 100° C is a further disadvantage, as it is essential to maintain high back-pressure in the reaction coil to prevent boiling.

Honda et al.[88] have recently reported improved sensitivity and greater versatility in use of ethylenediamine as a reagent for postcolumn derivatization of sugars, achieved by substitution of electrochemical detection for fluorometry. The products of the reaction of sugars with ethylenediamine under alkaline conditions are oxidizable at a glassy carbon electrode at an applied potential of 350 mV (vs. Ag/AgCl reference electrode). With water (containing 0.01% EDTA to sequester any metallic ions) as eluent and 0.1 *M* ethylenediamine sulfate in 0.7 *M* borate buffer, pH 9.0, as reagent, detection limits for aldoses as low as 1 pmol (linearity to 10 nmol) have been reported. The disadvantage of heating (to 150°C) remains, but this detection system, which has been shown to be applicable to amino- and acetamidodeoxyhexoses as well as neutral sugars and uronic acids, has been used in HPLC on Ca^{2+}-form resins as well as borate anion-exchange chromatography and even (although with much lower sensitivity) in HPLC on an amino-silica column in 75% acetonitrile.[88]

Kato and Kinoshita[89] have described the use of ethanolamine in boric acid as a fluorogenic reagent for carbohydrates after elution from an anion-exchange column with borate buffer, pH 8.7, or from NH_2-silica columns with 75% acetonitrile. The reagent, an aqueous solution containing 2% (w/v) of both ethanolamine and boric acid, was mixed with the column effluent in the ratio 1:3 at 150°C, and the resulting products were detected fluorometrically (excitation wavelength 357 nm, emission 436 nm). When used with the borate eluent, linear calibrations for monosaccharides and maltose were observed over the somewhat limited concentration range of 0.3 to 5 nmol; there was no response with alditols, methyl glycosides, 2-deoxy sugars, or nonreducing oligosaccharides. Maltodextrins to DP 5 could be detected with this reagent on elution from the NH_2-silica columns, but the response decreased sharply with increasing molecular weight (decreasing reducing power). Recently, Villaneuva et al.[90]

extended the application of this method to non-reducing oligosaccharides (raffinose and sucrose) after on-line post-column hydrolysis with *p*-toluenesulfonic acid, which also increased the response of maltose by 85%.

Similar detection reagents investigated by Kato and co-workers[91,92] and described in reports published in Japanese journals include 2-aminopropionitrile fumarate (6 to 8% in 50 m*M* borate, pH 8.7), which is particularly sensitive in detecting pentoses (detection limit for D-xylose 140 pmol, linear to 5.3 nmol), and taurine (8% in 50 m*M* borate, pH 8.7), which was used in both HPLC and TLC. If periodate is added to the taurine reagent (0.64 *M* taurine, 2 m*M* $NaIO_4$), which is heated at 140°C with the eluate in borate anion-exchange chromatography, sensitive detection of alditols as well as reducing sugars is achieved (detection limit for D-glucitol is 130 pmol).[93] Detection of glycosides and non-reducing oligosaccharides requires a higher concentration of $NaIO_4$ (0.1 *M*), under which conditions detection limits of 60 pmol for α-methyl-D-glucopyranoside and 30 pmol for sucrose, raffinose, and stachyose have been reported.[93]

The fluorogenic reagent that has been used most widely is 2-cyanoacetamide, which reacts with aldoses in an alkaline borate buffer (pH 8) at 100°C to give fluorescent derivatives (excitation wavelength 331 nm, emission 383 nm) and permit fluorometric detection down to 0.1 nmol.[94] As the reagent is unstable if stored in borate buffer, an aqueous solution (10%) is mixed (1:1) with 0.5 to 0.6 *M* borate buffer (pH 9) before being mixed with the column effluent in borate anion-exchange chromatography of aldoses[95] (see Table LC 44), giving a pH of 8, under which conditions fluorescent response is maximal. Linear calibration is observed over the concentration range 0.5 nmol to 1 μmol for all the common aldohexoses, 6-deoxyhexoses, and aldopentoses (except D-ribose, which gives a slightly higher range, 5 nmol to 5 μmol). The derivatives formed, which are believed to be cyano-pyridones and/or -pyrrolidones,[95] can also be detected by their UV absorbance at 276 nm.[96] In this case, the aqueous solution of 2-cyanoacetamide is mixed with 0.6 *M* borate buffer of pH 10.5 before mixing with the effluent from the borate anion-exchange column (see Table LC 44) to give a pH of 9, at which UV absorbance is maximal. Under these conditions, the detection limit is 1 nmol for all aldoses, and linearity is observed to 500 nmol. There is less variation in relative molar response among the different aldoses with UV monitoring than there is with fluorometric monitoring. Ketoses give very low responses in both cases, and nonreducing carbohydrates do not react with 2-cyanoacetamide.

More recently, Schlabach and Robinson[97] have extended the applicability of this detection method to the HPLC of sugars on NH_2-silica, eluted with 75% acetonitrile, by use of a reagent in which the 2-cyanoacetamide (5%) was dissolved in 0.1 *M* potassium borate (pH 10.4). Sonication was required to obtain a clear solution, which was prepared fresh daily. The use of a lower borate concentration and a higher reaction temperature (135°C) prevented precipitation of borate salts when the reagent was mixed with the aqueous acetonitrile eluent, and the raised temperature also had the effect of increasing the fluorescent response of the sugars from that at 105°C, 5-fold for D-ribose and 12-, 16-, and 20-fold for D-glucose, D-fructose, and maltose, respectively. Detection limits were ca. 0.25 nmol for all sugars tested. Doubling the reaction time by use of a longer or wider reaction coil can improve detection limits by a factor of 2 since the reaction is relatively slow, but substantial band broadening, with concomitant loss of resolution, occurs with reaction times in excess of 100 s; in practice, a 10 m × 0.23 mm coil (residence time 28 s at a flow rate of 0.8 ml/min) was optimal at 135°C. Here again a back-pressure restrictor was necessary to prevent outgassing. With a reaction temperature of 120°C, linear calibration was obtained over the range 1 to 20 μg of reducing sugar for HPLC on NH_2-silica and 0.1 to 10 μg for HPLC on Ca^{2+}-form resin, eluted with water at 75°C.

Honda et al. have successfully applied 2-cyanoacetamide to the monitoring not only of neutral sugars, but also of uronic acids,[98] aminodeoxyhexoses,[99] and the sugar constituents of glycoproteins, including acetamidodeoxyhexoses and neuraminic acids.[100] The uronic acids D-glucuronic, D-galacturonic, D-mannuronic, and L-iduronic acid, resolved by borate anion-exchange chromatography on a microparticulate anion-exchange resin eluted with 1.3 *M* borate buffer (pH 8.55) at 65°C, were detected by addition of a 1% aqueous solution of 2-cyanoacetamide to the eluent (1:2), mixing at 100°C, and passage of the effluent through both a UV photometer at 280 nm and a fluorometer.[98] The responses of D-glucuronic and D-mannuronic acid were approximately the same with both monitors, but D-galacturonic acid, which gave the highest response for photometric monitoring, gave the lowest for fluorometric monitoring. Detection limits, therefore, varied among the four acids according to the monitor used. With photometric monitoring, detection limits were 0.24, 0.29, 0.50, and 0.80 nmol for D-galacturonic, D-mannuronic, L-iduronic, and D-glucuronic acids, respectively, and with fluorometric monitoring, 0.05 nmol for D-mannuronic acid and approximately 0.1 nmol for the others. Linearity was observed in the range 5 to 1000 nmol for D-glucuronic acid and L-iduronic acid, 1 to 1000 nmol for D-galacturonic and D-mannuronic acid with photometric monitoring, and 1 to 1000 nmol for all four with fluorometric monitoring. In samples of acidic polysaccharides, such as gum arabic and pectin, that were not hydrolyzed completely, aldobiouronic acids were also resolved and detected by this method. For resolution of the aminodeoxyhexoses, 2-amino-2-deoxy-D-glucose and -galactose, a microparticulate cation-exchange column eluted at 60°C with 0.04 *M* sodium tetraborate adjusted to pH 7.5 with concentrated HCl was used, and the detection reagent (1% aqueous solution) was mixed with the column effluent and then with a 0.6 *M* borate buffer (pH 8.5) to bring the pH to the optimal value for fluorometric detection.[99] In this case, also, both photometric and fluorometric detection were employed, and detection limits were ca. 2 and 3 nmol for 2-amino-2-deoxy-D-glucose and -galactose, respectively, for both monitors, with linear calibration over the range of 10 to 500 nmol.

Simultaneous analysis of the neutral sugars, acetamidodeoxyhexoses, and neuraminic acids in glycoprotein hydrolysates, which is not possible with ion-exchange chromatography, was achieved by Honda and Suzuki[100] by partition chromatography on a microparticulate cation-exchange resin (H^+-form) eluted with 92% acetonitrile at 30°C (see Table LC 35). At this high concentration of acetonitrile, fluorometric monitoring was impossible; the detection method involved photometric monitoring at 280 nm after mixing the column effluent successively with 0.5 *M* borate buffer (pH 8.5) and a 1% aqueous solution of the reagent. Detection limits for L-rhamnose, D-xylose, L-fucose, D-mannose, and D-galactose were 35, 40, 48, 94, and 104 pmol, respectively. Sensitivity was lower for the acetamidodeoxyhexoses, with detection limits of 390 and 415 pmol for 2-acetamido-2-deoxy-D-galactose and -glucose, respectively. For the faster-eluting aldoses, L-rhamnose, D-xylose, and L-fucose, linear calibration was observed over the range of 0.1 to 80 nmol, but greater peak broadening for the slower-eluting aldoses, D-mannose and D-galactose, resulted in a higher range (0.2 to 200 nmol) and that for the acetamidodeoxyhexoses was higher still (4 to 800 nmol). The neuraminic acids were not analyzed as such, but rather as the *N*-acylmannosamines produced by the action of *N*-acetylneuraminate pyruvate lyase on neuraminidase-digested glycoproteins. These were eluted in the same region as D-mannose and D-galactose, and their response to the 2-cyanoacetamide reagent was of the same magnitude.

As in the case of ethylenediamine (see above), the products of the reaction of 2-cyanoacetamide with reducing sugars are readily oxidizable at a glassy carbon electrode, so electrochemical detection is a third option when this reagent is used. Honda et al.[101] have described a method in which the column effluent is mixed with the reagent solution (1.5% aqueous solution of 2-cyanoacetamide) and supporting electrolyte (0.2 *M* borate buffer, pH 9.5) in the ratio 2:1:1. The reaction takes place in a 10 m × 0.5 mm I.D. reaction coil at 100°C, and the products are detected at the working electrode at an applied potential of 400 mV (vs. Ag/AgCl reference electrode). Under these conditions, the detection limit for glucose is about 20 pmol, and linearity is observed over the range 50 pmol to 2 nmol. There is, however, wider variation in relative molar response among the different reducing sugars than is found with fluorometric and photometric detection, and the variation is also much greater than that reported in electrochemical detection using ethylenediamine as reagent.[88] Ketoses, 2-deoxyhexoses, and *N*-acetylneuraminic acid give very low responses, but the method is applicable to all aldoses, including reducing oligosaccharides, hexuronic acids, and amino- and acetamidodeoxyhexoses. Honda et al.[101] applied it to the detection of 500-pmol amounts of maltodextrins (to DP 3) and of isomaltodextrins (to DP 6) from hydrolysis of 10 μg of dextran, separated by SEC in water, 500 pmol amounts of 2-amino-2-deoxy-D-glucose and -galactose separated by the method described above,[99] and 400 pmol amounts of monosaccharides eluted from a cation-exchange resin (H^+-form) with 90% acetonitrile. The presence of acetonitrile caused fluctuation in background current and, therefore, the applied potential was slightly decreased to 300 mV, under which conditions the sugars were monitored effectively by the electrochemical method. This method which is evidently applicable to all modes of HPLC as is photometric detection with 2-cyanoacetamide.

In a photoreduction fluorescence detector developed by Gandelman et al.,[102] a photochemical reagent added to the eluent absorbs UV radiation in a postcolumn photochemical reactor and in the excited state reacts with the analytes separated by the column, producing a fluorescent product that is then detected fluorometrically. The method can be applied to any hydrogen-atom-donating substrate, such as alcohols (including alditols) and aldehydes (including reducing sugars), as well as amines and ethers. With anthraquinone-2,6-disulfonate as the photochemical reagent added (as a 6.7 m*M* solution in 72% acetonitrile) to the eluent (80% acetonitrile) in the ratio 1:3.33, the method has been applied to the HPLC of various sugars commonly found in food (such as D-glucose, D-fructose, maltose, and lactose) on coupled NH_2- and C_{18}-silica columns. The reagent was added between the former and the latter; there was photochemical reduction of the reagent to fluorescent dihydroxyanthracene-2,6-disulfonate, but sensitivity was relatively low (detection limit 2 to 3 nmol). The method is more sensitive for cardiac glycosides, which can be detected at the picomolar level. Higher sensitivity was achieved by replacing the 360-nm fluorescent lamp used in the photochemical reaction by an intense 254-nm irradiation source and using 2-*tert*-butylanthraquinone as the reagent.[103] This did not affect the amino column as did the earlier reagent and therefore could be added directly (7.6 m*M*) to the eluent (87% acetonitrile) used in HPLC of the sugars on this column. Under these conditions, the detection limits for the sugars were below 0.5 nmol. Simultaneous detection of the cardiac glycoside digoxin and lactose, an inert ingredient in the pharmaceutical tablet, by this method exemplified a possible specific application.

Specific Methods

For Alditols

The analytical method involving periodate oxidation of alditols to formaldehyde, followed by Hantzsch condensation with pentane-2,4-dione and ammonia to a chromophoric lutidine derivative, which was described in Volume I, has been adapted by Honda et al.[104] for use as an on-line detection method in HPLC. The alditols are separated on a microparticulate anion-exchange column by stepwise elution at 65°C with borate buffers (0.3 to 0.5 *M*, pH 7 to 10.5; see Table LC 44), and the eluate from the column (flow rate 1 ml/min) is mixed at room temperature with 0.05 *M* $NaIO_4$ (flow rate 0.5 ml/min) and then at 100° C with a 15% solution of ammonium acetate containing 2% pentane-2,4-dione, with 0.2 *M* soldium thiosulfate to reduce excess periodate (flow rate of second reagent is also 0.5 ml/min). The product absorbs strongly at 412 nm, and there is also intense fluorescence (excitation and emission maxima at 410 and 503 nm, respectively); hence, both photometric and fluorometric detection are possible. The absorbance and the fluorescence intensities vary widely among the different alditols. The method is most sensitive for arabinitol, for which the detection limits for photometric and fluorometric monitoring are ca. 2 and 0.5 nmol, respectively. The 6-deoxyhexitols produce less formaldehyde than the hexitols and, therefore, their responses are much lower. Linear calibration is observed for all alditols in the range of 20 to 500 nmol for photometric detection and 20 to 200 nmol for fluorometric detection. Under the conditions used, interference from aldoses is slight, as molar responses are ca. 5% of that of D-glucitol.

For Aminodeoxyhexoses

Detection with 2-cyanoacetamide (see preceding subsection) is not sufficiently sensitive for analysis of aminodeoxyhexoses at the levels present in many biological and clinical samples. Therefore, Honda et al.[105] have developed a more sensitive method involving Hantzsch condensation with pentane-2,4-dione and formaldehyde. The aminodeoxyhexoses were separated as before[99] on a cation-exchange resin eluted with 0.04 *M* sodium tetraborate adjusted to pH 7.5, and to the eluate (flow rate 0.3 ml/min) was added first a 6% solution of pentane-2,4-dione in 0.10 *M* acetate buffer, pH 4.8 (flow rate 0.5 ml/min), and then a 9% solution of formaldehyde in the same buffer at the same flow rate. The final mixture reacted in a 10 m × 0.5 mm I.D. coil immersed in a glycerol bath at 95°C and was cooled by passage through a 1-m tube of the same diameter before passing into a fluorometer to be monitored at 417 nm (excitation) and 476 nm (emission). Under these conditions, the detection limits for 2-amino-2-deoxy-D-glucose and -galactose were 140 and 230 pmol, respectively, and linear calibration was observed over the range 0.3 to 70 nmol for both. Amino acids interfere in this detection but they are eluted much earlier than the aminodeoxyhexoses in the chromatographic method used, so the carbohydrate peaks are clearly distinguishable. Therefore, the method is applicable to the analysis of the component aminodeoxyhexoses of glycoconjugates.

The Elson-Morgan reaction, described in Volume I, has been automated on the AutoAnalyzer principle by White et al.[106] The pentane-2,4-dione (3.5%) was dissolved in 0.3 *M* sodium orthophosphate, pH 9.6, rather than the sodium carbonate solution used in the manual method, which produces CO_2 on addition of the Ehrlich reagent (4% 4-dimethylaminobenzaldehyde in ethanol containing 15% HCl). The sample solution (flow rate 0.16 ml/min) was mixed with the pentane-2,4-dione solution (flow rate 0.10 ml/min) and heated at 95°C (14-min residence time in heating coil) and then, after cooling and debubbling, with the Ehrlich reagent (flow rate 0.34 ml/min) with heating at 67°C for 7.5 min, before spectrophotometric detection at 530 nm. Linear calibration was observed over the range 10 to 100 μg/ml for 2-amino-2-deoxy-D-glucose and -galactose and 20 to 100 μg/ml for 2-amino-2-deoxy-D-mannose, which showed a lower response. A very low response was obtained with muramic acid. This procedure is far less sensitive than the 2-cyanoacetamide and pentane-2,4-dione-formaldehyde methods and would require further modification to permit its use in HPLC systems.

A modified indole-HCl method,[107] which has been used in manual assay of chitooligosaccharides after SEC fractionation, is more sensitive than the Elson-Morgan reaction. In this procedure, the sample (400 μl, containing 5 to 75 nmol of aminodeoxyhexose) is acidified with concentrated HCl (200 μl), and the mixture is heated at 100°C for 15 min and then cooled in ice water before deamination with 5% sodium nitrite (200 μl) followed, after 10 min at room temperature, by 12.5% ammonium sulfamate (200 μl). The mixture is left at room temperature for 30 min, with occasional shaking; then ethanolic 1% indole (100 μl) is added and the mixture is heated at 100°C for 5 min. After cooling, ethanol (1 ml) is added and the absorbance at 492 nm is measured. The response of 2-amino-2-deoxy-D-galactose is slightly less than that of 2-amino-2-deoxy-D-glucose, and that of 2-amino-2-deoxy-D-mannose, which is converted to D-glucose (not an anhydro sugar) on deamination, is almost negligible. There is slight interference from *N*-acetylneuraminic acid, D-galacturonic acid, and 2-deoxyribose in this analytical method. Automation would require special equipment because of the chemically agressive reagents involved.

For Neuraminic Acids

The periodate-thiobarbituric acid detection method for neuraminic acids has been automated by Krantz and Lee[108] using the AutoAnalyzer system connected to an anion-exchange column. Samples containing bound neuraminic acids were hydrolyzed in 0.05 *M* H_2SO_4 for 1 h at 80°C, and an aliquot of the hydrolysate (containing 1 to 15 nmol of neuraminic acid) was applied directly to the column and eluted with 10% acetic acid at a flow rate of 0.6 ml/min. *N*-Acetyl- and *N*-glycolylneuraminic acid were not separated in this system, but the color yield of the latter in the analytical method was lower. The column effluent was mixed first with 0.08 *M* $NaIO_4$ in 1.8 *M* H_2SO_4 (flow rate 0.05 ml/min) at 65°C, then with sodium arsenite (10% in 0.05 *M* H_2SO_4, flow rate 0.16 ml/min) to destroy excess periodate before mixing with a 1% aqueous solution of thiobarbituric acid adjusted to pH 9.0 with NaOH (the flow rate of thiobarbituric acid reagent was 0.6 ml/min). After the mixture had been heated at 95°C, 1-butanol containing 10% concentrated HCl (flow rate 2.03 ml/min) was added and, after mixing and debubbling, the reaction mixture passed through a phase separator and the organic phase passed to a colorimeter for monitoring at 550 nm. The method allows determination of *N*-acetylneuraminic acid in the range of 1 to 12 nmol and is insensitive to acid, salt, or protein, so acid or enzymic hydrolysates can be analyzed directly without purification.

The periodate-resorcinol reaction, with spectrophotometric detection at 630 nm, has been used to monitor the elution of sialylated oligosaccharides in anion-exchange chromatography[109] (see Table LC 40). This analysis was not automated, but was a manual assay of fractions collected on the basis of UV detection at 210 nm.

For 6-Deoxyhexoses

The thioglycolic acid-sulfuric acid method, which is similar to the L-cysteine-H_2SO_4 method (see Volume I), is specific for 6-deoxyhexoses if the absorbance at 400 nm (maximum for 6-deoxyhexoses) is compared with that at 430 nm (maximum for hexoses). This method has been used to monitor the elution of fucosylated oligosaccharides in anion-exchange chromatography[109] (see Table LC 40) by manual analysis of fractions.

For Hexuronic Acids

Automation of the carbazole-sulfuric acid method (see Volume I) has been improved,[110] mainly by introduction of the carbazole and the borate-sulfuric acid separately into the system instead of using a reagent solution in which the carbazole had previously been dissolved in the borate-H_2SO_4 solution, which is not stable and deteriorates in only 4 to 6 h. The carbazole is dissolved (1 g/l) in 95% ethanol and stored at 4°C. The borate-sulfuric acid reagent is prepared by dissolving potassium tetraborate (19 g) in distilled water (200 ml) and adding concentrated H_2SO_4 (2300 ml) carefully in the cold. In the automated method developed by Jeansonne et al.[110] for continuous-flow monitoring of hexuronic acid in the effluent from a gel column, eluted with the dissociative solvent 6 *M* urea, in the SEC of glycosaminoglycuronans, the carbazole and borate-sulfuric acid solutions are mixed just before the combined solution is mixed with the effluent from the column, and the reaction mixture is heated at 95° C and passed under a sensitizing lamp prior to colorimetric detection at 530 nm. The presence of urea has no effect on the response, which is linear over the range 5 to 30 μg (25 to 150 nmol) for D-glucuronic acid.

For Sugar Phosphates

The addition of a europium complex, tris(2,2,6,6-tetramethyl-3,5-heptanedionato)europium(III) (Eu[DPM]$_3$), to the mobile phase in the chromatography of sugar phosphates[111] results in the formation of an adduct that has strong UV absorbance at 240 nm. This has been used to detect D-glucose- and D-fructose-6-phosphate as well as D-fructose-1,6- and -2,6-diphosphate in anion-exchange chromatography with a 0.03 *M* formate buffer, pH 3.3, as eluent[111] and in ion-pair reversed-phase chromatography, in which tetrabutylammonium hydroxide was added to this buffer.[112] Detection limits were ca. 200 ng (<1 nmol), and linearity was observed over the range 1 to 100 μg.

For Cyclodextrins

Frijlink et al.[113] have developed a rapid, sensitive method for analysis of β- and γ-cyclodextrin in biological fluids for pharmacokinetic studies. Samples from which protein has been removed by precipitation with trichloroacetic acid are neutralized with NaOH and then injected into a μ-Bondapak® Phenyl column and eluted with water containing 10% methanol. The detection system depends upon postcolumn complexation of phenolphthalein by the cyclodextrins and negative colorimetric detection at 546 nm. The stock solution is a 6 m*M* solution of phenolphthalein in 96% ethanol, and the working solution is prepared by addition of stock solution (10 ml) to 8 m*M* aqueous sodium carbonate solution (990 ml), the pH being adjusted to 10.5 with *M* NaOH. This solution (at 60°C) is added in a 1:1 ratio to the effluent from the column, and the formation of inclusion complexes with β- and γ-cyclodextrins gives rise to negative peaks at 546 nm. The high stability constants of the complexes make this detection system very sensitive and impervious to interference from competing ligands. Complete mixing of the phenolphthalein solution with the column effluent is essential for a stable baseline. This is achieved by use of capillary tubing (1.5 m × 1 mm I.D.) in the mixing coil. Residence time in the coil is low and its effect upon peak broadening negligible. With a sample volume of 1 ml, of which 250 μl is injected after treatment with trichloroacetic acid (250 μl), linear calibration is observed over the concentration range of 2.5 to 50 μg/ml for β-cyclodextrin (5 to 50 μg/ml for γ-cyclodextrin). With a sample volume of 100 μl treated with 100 μl of water and 50 μl of trichloroacetic acid, of which 50 μl is injected, linearity is observed over the range of 5 to 40 μg/100 μl for β-cyclodextrin.

For Maltodextrins

Recently, improvement in the sensitivity of detection of maltodextrins separated by ion chromatography has been achieved by use of a postcolumn enzyme reactor that converts the eluted oligomers to D-glucose prior to detection by the pulsed amperometric detector, which is more sensitive to D-glucose than to the oligomers having lower reducing power. Larew and Johnson[114] have reported incorporation of such a reactor in the chromatographic system using a Dionex AS6A column eluted with a linear gradient of sodium acetate in 0.1 *M* NaOH (see Table LC 46) with PAD. The enzyme reactor consisted of a short column (2.5 cm × 2.1 mm I.D.) packed with silica on which glucoamylase was immobilized. It was necessary to include, in addition to the gradient elution system used for the AS6A column, two additional solvent delivery modules, with mixers, as the enzyme activity is maximal at pH 4.5, while the PAD is most sensitive at pH >12. Thus, the eluate from the chromatographic column was mixed with 2 *M* sodium acetate buffer, pH 4.5, prior to passage through the reactor (at 50°C), and the eluate from the reactor was mixed with 0.35 *M* NaOH to raise the pH to the level required for sensitive detection by the PAD. Under these conditions, conversion of the oligomers to D-glucose varied from 87% (for maltose) to 96% (DP ⩾4), so the detection limits for the oligomers approached that for D-glucose. This was, however, ca. 30 times greater than is usually obtained with PAD, owing to the band broadening produced by the additional solvent delivery and mixing systems. Nevertheless, sensitivity of detection of the higher oligomers was enhanced significantly by the presence of the enzyme reactor. Linear calibration was observed in the concentration range 0 to 150 μ*M* for all oligomers in the sample tested (DP up to 10).

SUMMARY

The specificity and sensitivity of the main detection methods described in Section II.III are summarized in Table IIB.

TABLE IIB
Specificity and Sensitivity of Liquid Chromatography Detection Methods for Carbohydrates

Method	Specificity	Detection Limit	Linearity	Ref.
Direct methods				
Differential refractometer	Universal detector	15—25 ng	0.2—20 μg	2
UV photometer	Chromophoric compounds (underivatized)[a]	≤10 ng		2
Polarimetry	Optically active compounds	100 μg		8
Mass (evaporative) detector	Universal detector	2—3μg		9
Moving wire (FID) with microbore column	Universal detector	100—400 ng		16
ICP-AES	Universal detector	800 ng		17
MS (moving belt)	Universal detector	100 ng [≤1 ng (SIM)[b]]		18
MS (thermospray)	Universal detector	20—100 pg[b]		24
Amperometric (direct)	Electroactive compounds	≤1 ng		30
PAD (Au electrode)	Compounds oxidizable at Au electrode at pH ≥12	30—100 ppb		37
		10—50 pmol		41
Cu(II)bis(phenanthroline), with amperometric detection	Reducing sugars	0.2 ng (ca. 1 pmol) for D-glucose		42
Conductivity (direct)	Ionic molecules, e.g., sugar phosphates	20—100 pmol		45
Conductivity (indirect; by suppression of electrolyte)	Nonionic molecules, e.g., glucose, sucrose	50—100 nmol[c]		46
		0.5 nmol[d]		48
Precolumn derivatization				
Benzoylation[a]	All carbohydrates	1—10 nmol		57
		<1 nmol[e]		58
Dimethylphenylsilylation[a]	All carbohydrates	1—2 nmol		59, 60
Reductive amination with DAAB	Reducing sugars	<5 pmol		63
Formation of 2,4-dinitrophenylhydrazones[a]	Reducing sugars	50 pmol	50 pmol—3.3 nmol	61
Formation of dabsylhydrazones	Reducing sugars	10 pmol	10—200 pmol	64
Formation of dansylhydrazones[f]	Reducing sugars	3—5 pmol	50 pmol—1 nmol	67, 68
Reductive amination with 2-aminopyridine[f]	Reducing sugars	10 pmol	10 pmol—10 nmol	69
Labeling with NBD-F[f]	Amino sugars and glycamines from neutral aldoses	40 pmol	40 pmol—50 nmol	71
Labeling with *o*-phthaldehyde[f]	Amino sugars	1 ppb	100—1000 ppm	72
Labeling with DDB[f]	Neuraminic acids	40 fmol	40 fmol—18 nmol	73

TABLE IIB (continued)
Specificity and Sensitivity of Liquid Chromatography Detection Methods for Carbohydrates

Method	Specificity	Detection Limit	Linearity	Ref.
Postcolumn derivatization				
Tetrazolium blue	Reducing sugars	10 ng	10 ng—20 μg	75
Cu(II)-bicinchoninate	Reducing sugars[g]	0.5—5ng	5 ng—2 μg	78
Cuprammonium[a]	Reducing sugars[h]	2.5 nmol	2.5—60 nmol	79
Cuprammonium with potentiometric detection	Reducing sugars	0.6—5 nmol		81
4-Aminobenzoic acid hydrazide	Reducing sugars[g]	15—30 ng		84
Periodate	All carbohydrates with vicinal diol system	1—10 nmol		86
Ethylenediamine[f]	All sugars[h]	50—400 pmol	100 pmol—100 nmol	87
Ethylenediamine with amperometric detection	All sugars	1 pmol	5 pmol—10 nmol	88
Ethanolamine[f]	Reducing sugars[g]	0.3 nmol	0.3—5 nmol	89, 90
2-Cyanoacetamide, with fluorometric detection	Neutral aldoses	0.1—1 nmol	0.5 nmol—1 μmol	94, 95
		0.25 nmol	0.5—55 nmol	97
			5—110 nmol[i]	97
	Hexuronic acids	50—120 pmol	1—1000 nmol	98
	Aminodeoxyhexoses	2—3 nmol	10—500 nmol	99
2-Cyanoacetamide, with photometric detection	Neutral aldoses	1 nmol	5—500 nmol	96
		35—104 pmol[i]	0.1—200 nmol[i]	100
	Hexuronic acids	0.2—0.8 nmol	1—1000 nmol	98
	Aminodeoxyhexoses	2—3 nmol	10—500 nmol	99
	Acetamidodeoxyhexoses	390—415 pmol[i]	4—800 nmol[i]	100
2-Cyanoacetamide, with amperometric detection	Neutral aldoses	5—20 pmol	50 pmol—2 nmol	101
	Hexuronic acids	6—12 pmol	in all cases	
	Aminodeoxyhexoses	10 pmol		
	Acetamidodeoxyhexoses	10—12 pmol		
Periodate oxidation, pentane-2,4-dione + NH_3	Alditols			104
Fluorometric detection		0.5—1.2 nmol	20—200 nmol	
Photometric detection		2—4.4 nmol	20—500 nmol	
Pentane-2,4-dione-HCHO[f]	Aminodeoxyhexoses	140—230 pmol	0.3—70 nmol	105
Indole-HCl	Aminodeoxyhexoses	5 nmol	5—75 nmol	107
Periodate-thiobarbituric acid	Neuraminic acids	1 nmol	1—12 nmol	108

Carbazole-H_2SO_4-borate	Hexuronic acids	25 nmol	25—150 nmol	110
Europium complex, $Eu(DPM)_3$[a]	Sugar phosphates	200 ng	1—100 μg	112
Complexation with phenolphthalein, negative colorimetric detection	β-Cyclodextrin	1 μg	2.5—5.0 μg	113
	γ-Cyclodextrin	5 μg	5—50 μg	

[a] Sensitivity of UV photometry is much enhanced by derivatization; see below.
[b] Selected ion monitoring.
[c] Eluent 1 m*M* H_2SO_4.
[d] Eluent 2 m*M* $Ba(OH)_2$.
[e] With microbore HPLC column.
[f] Fluorometric detection.
[g] Method can be applied to nonreducing sugars if H^+-form cation-exchange resin column is included in detection system, to hydrolyze nonreducing oligosaccharides to reducing sugars.
[h] Nonreducing sugars give response due to alkaline hydrolysis, which releases reducing sugars.
[i] With aqueous acetonitrile eluent.

REFERENCES

1. **Honda, S.**, High-performance liquid chromatography of mono- and oligosaccharides, *Anal. Biochem.*, 140, 1, 1984.
2. **Hicks, K. B.**, High-performance liquid chromatography of carbohydrates, *Adv. Carbohydr. Chem. Biochem.*, 46, 17, 1988.
3. **Koizumi, K., Utamura, T., and Okada, Y.**, Analyses of homogeneous D-gluco-oligosaccharides and polysaccharides (degree of polymerization up to about 35) by high-performance liquid chromatography and thin-layer chromatography, *J. Chromatogr.*, 321, 145, 1985.
4. **Bornhop, D. J., Nolan, T. G., and Dovichi, N. J.**, Subnanoliter laser-based refractive index detector for 0.25 mm I.D. microbore liquid chromatography. Reversed-phase separation of nanogram amounts of sugars, *J. Chromatogr.*, 384, 181, 1987.
5. **Cullen, M. P., Turner, C., and Haycock, G. B.**, Gel permeation chromatography with interference refractometry for the rapid assay of polydisperse dextrans in biological fluids, *J. Chromatogr.*, 337, 29, 1985.
6. **Binder, H.**, Separation of monosaccharides by high-performance liquid chromatography: comparison of ultraviolet and refractive index detection, *J. Chromatogr.*, 189, 414, 1980.
7. **Owens, J. A. and Robinson, J. S.**, Isolation and quantitation of carbohydrates in sheep plasma by high-performance liquid chromatography, *J. Chromatogr.*, 338, 303, 1985.
8. **Heyraud, A. and Salemis, P.**, Liquid chromatography in the methylation analysis of carbohydrates, and the use of combined refractometric-polarimetric detection, *Carbohydr. Res.*, 107, 123, 1982.
9. **Macrae, R. and Dick, J.**, Analysis of carbohydrates using the mass detector, *J. Chromatogr.*, 210, 138, 1981.
10. **Christie, W. W. and Morrison, W. R.**, Separation of complex lipids of cereals by high-performance liquid chromatography with mass detection, *J. Chromatogr.*, 436, 510, 1988.
11. **Lafosse, M., Dreux, M., and Morin-Allory, L.**, Application fields of a new evaporative light-scattering detector for high-performance liquid chromatography and supercritical fluid chromatography, *J. Chromatogr.*, 404, 95, 1987.
12. **Hizukuri, S. and Takagi, T.**, Estimation of the distribution of molecular weight for amylose by the low-angle laser-light-scattering technique combined with high-performance gel chromatography, *Carbohydr. Res.*, 134, 1, 1984.
13. **Vijayendran, B. R. and Bone, T.**, Absolute molecular weight and molecular weight distribution of guar by size-exclusion chromatography and low-angle laser light scattering, *Carbohydr. Polym.*, 4, 299, 1984.
14. **Wells, G. B. and Lester, R. L.**, Rapid separation of acetylated oligosaccharides by reverse-phase high-pressure liquid chromatography, *Anal. Biochem.*, 97, 184, 1979.
15. **Scott, R. P. W.**, Column chromatography, in *Chromatography*, Heftmann, E., Ed., Journal of Chromatography Library Series, Vol. 22A, Elsevier, Amsterdam, 1983, 137.
16. **Veening, H., Tock, P. P. H., Kraak, J. C., and Poppe, H.**, Microbore high-performance liquid chromatography with a moving-wire flame ionization detector, *J. Chromatogr.*, 352, 345, 1986.
17. **Jinno, K., Nakanishi, S., and Nagoshi, T.**, Microcolumn gel permeation chromatography with inductively coupled plasma emission spectrometric detection, *Anal. Chem.*, 56, 1977, 1984.
18. **Games, D. E.**, Combined high-performance liquid chromatography-mass spectrometry, *Biomed. Mass Spectrom.*, 8, 454, 1981.
19. **Åman, P., McNeil, M., Franzén, L.-E., Darvill, A. G., and Albersheim, P.**, Structural elucidation, using HPLC-MS and GLC-MS, of the acidic polysaccharide secreted by *Rhizobium meliloti* strain 1021, *Carbohydr. Res.*, 95, 263, 1981.
20. **Åman, P., Franzén, L.-E., Darvill, J. E., McNeil, M., Darvill, A. G., and Albersheim, P.**, The structure of the acidic polysaccharide secreted by *Rhizobium phaseoli* strain 127 K38, *Carbohydr. Res.*, 103, 77, 1982.
21. **McNeil, M., Darvill, A. G., Åman, P., Franzén, L.-E., and Albersheim, P.**, Structural analysis of complex carbohydrates using high-performance liquid chromatography, gas chromatography and mass spectrometry, *Methods Enzymol.*, 83, 3, 1982.
22. **Shukla, A. K., Schauer, R., Schade, U., Moll, H., and Rietschel, E.Th.**, Structural analysis of underivatized sialic acids by combined high-performance liquid chromatography-mass spectrometry, *J. Chromatogr.*, 337, 231, 1985.
23. **Hirter, P., Walther, H. J., and Dätwyler, P.**, Microcolumn liquid chromatography-mass spectrometry using a capillary interface, *J. Chromatogr.*, 323, 89, 1985.
24. **Hsu, F. F., Edmonds, C. G., and McCloskey, J. A.**, Combined liquid chromatography-mass spectrometry for microscale structural studies of carbohydrates, *Anal. Lett.*, 19, 1259, 1986.
25. **Rajakylä, E.**, Use of reversed-phase chromatography in carbohydrate analysis, *J. Chromatogr.*, 353, 1, 1986.

26. **Esteban, N. V., Liberato, D. J., Sidbury, J. B., and Yergey, A. L.,** Stable isotope dilution thermospray liquid chromatography/mass spectrometry method for determination of sugars and sugar alcohols in humans, *Anal. Chem.*, 59, 1674, 1987.
27. **Lau, J. M., McNeil, M., Darvill, A. G., and Albersheim, P.,** Enhancement of protonated molecular ions and ammonium molecular ion complexes in direct-liquid-introduction-mass spectrometry of carbohydrates, *Carbohydr. Res.*, 173, 101, 1988.
28. **Sakairi, M. and Kambara, H.,** Characteristics of a liquid chromatograph/atmospheric pressure ionization mass spectrometer, *Anal. Chem.*, 60, 774, 1988.
29. **Grün, M. and Loewus, F. A.,** Determination of ascorbic acid in algae by high-performance liquid chromatography on strong cation-exchange resin with electrochemical detection, *Anal. Biochem.*, 130, 191, 1983.
30. **Moll, N. and Joly, J. P.,** Determination of ascorbic acid in beers by high-performance liquid chromatography with electrochemical detection *J. Chromatogr.*, 405, 347, 1987.
31. **Santos, L. M. and Baldwin, R. P.,** Liquid chromatography/electrochemical detection of carbohydrates at a cobalt phthalocyanine containing chemically modified electrode, *Anal. Chem.*, 59, 1766, 1987.
32. **Santos, L. M. and Baldwin, R. P.,** Electrochemistry and chromatographic detection of monosaccharides, disaccharides and related compounds at an electrocatalytic chemically modified electrode, *Anal. Chim. Acta*, 206, 85, 1988.
33. **Reim, R. E. and Van Effen, R. M.,** Determination of carbohydrates by liquid chromatography with oxidation at a nickel (III) oxide electrode, *Anal. Chem.*, 58, 3203, 1986.
34. **Hughes, S. and Johnson, D. C.,** Amperometric detection of simple carbohydrates at platinum electrodes in alkaline solutions by application of a triple-pulse potential waveform, *Anal. Chim. Acta*, 132, 11, 1981.
35. **Hughes, S. and Johnson, D. C.,** High-performance liquid chromatographic separation with triple-pulse amperometric detection of carbohydrates in beverages, *J. Agric. Food Chem.*, 30, 712, 1982.
36. **Hughes, S. and Johnson, D. C.,** Triple-pulse amperometric detection of carbohydrates after chromatographic separation, *Anal. Chim. Acta*, 149, 1, 1983.
37. **Rocklin, R. D. and Pohl, C. A.,** Determination of carbohydrates by anion-exchange chromatography with pulsed amperometric detection, *J. Liq. Chromatogr.*, 6, 1577, 1983.
38. **Edwards, P. and Haak, K. K.,** A pulsed amperometric detector for ion chromatography, *Am. Lab.*, 15(4), 78, 1983.
39. **Edwards, W. T., Pohl, C. A., and Rubin, R.,** Determination of carbohydrates using pulsed amperometric detection combined with anion exchange separations, *Tappi*, 70, 138, 1987.
40. **Olechno, J. D., Carter, S. R., Edwards, W. T., and Gillen, D. G.,** Developments in the chromatographic determination of carbohydrates, *Am. Biotechnol. Lab.*, 5(5), 38, 1987.
41. **Haginaka, J. and Nomura, T.,** Liquid chromatographic determination of carbohydrates with pulsed amperometric detection and a membrane reactor, *J. Chromatogr.*, 447, 268, 1988.
42. **Watanabe, N. and Inoue, M.,** Amperometric detection of reducing carbohydrates in liquid chromatography, *Anal. Chem.*, 55, 1016, 1983.
43. **Watanabe, N.,** Amperometric detection of reducing carbohydrates in high-performance liquid chromatography using an amino-bonded column and acetonitrile-water as the eluent, *J. Chromatogr.*, 330, 333, 1985.
44. **Marko-Varga, G.,** High-performance liquid chromatographic separation of some mono- and disaccharides with detection by a post-column enzyme reactor and a chemically modified electrode, *J. Chromatogr.*, 408, 157, 1987.
45. **Smith, R. E., Howell, S., Yourtee, D., Premkumkar, N., Pond, T., Sun, G. Y., and MacQuarrie, R. A.,** Ion chromatographic determination of sugar phosphates in physiological samples, *J. Chromatogr.*, 439, 83, 1988.
46. **Tanaka, K. and Fritz, J. S.,** Ion-exclusion chromatography of nonionic substances with conductivity detection, *J. Chromatogr.*, 409, 271, 1987.
47. **Okada, T. and Kuwamoto, T.,** High-performance liquid chromatographic determination of electrically neutral carbohydrates with conductivity detection, *Anal. Chem.*, 58, 1375, 1986.
48. **Welsh, L. E., Mead, D. A., Jr., and Johnson, D. C.,** A comparison of pulsed amperometric detection and conductivity detection for carbohydrates, *Anal. Chim. Acta*, 204, 323, 1988.
49. **Natowicz, M. and Baenziger, J. U.,** A rapid method for chromatographic analysis of oligosaccharides on Bio-Gel P-4, *Anal. Biochem.*, 105, 159, 1980.
50. **Baenziger, J. U. and Natowicz, M.,** Rapid separation of anionic oligosaccharide species by high-performance liquid chromatography, *Anal. Biochem.*, 112, 357, 1981.
51. **Mellis, S. J. and Baenziger, J. U.,** Separation of neutral oligosaccharides by high-performance liquid chromatography, *Anal. Biochem.*, 114, 276, 1981.
52. **Delaney, S. R., Leger, M., and Conrad, H. E.,** Quantitation of the sulfated disaccharides of heparin by high-performance liquid chromatography, *Anal. Biochem.*, 106, 253, 1980.
53. **Yamashita, K., Mizouchi, T., and Kobata, A.,** Analysis of oligosaccharides by gel filtration, *Methods Enzymol.*, 83, 105, 1982.

54. **Takeuchi, M., Takasaki, H., Inoue, N., and Kobata, A.,** Sensitive method for carbohydrate composition analysis of glycoproteins by high-performance liquid chromatography, *J. Chromatogr.*, 400, 207, 1987.
55. **Lohmander, L. S.,** Analysis by high-performance liquid chromatography of radioactively labeled carbohydrate components of proteoglycans, *Anal. Biochem.*, 154, 75, 1986.
56. **Nakamura, Y. and Koizumi, Y.,** Radioactivity detection system with a CaF_2 (Eu) scintillator for high-performance liquid chromatography, *J. Chromatogr.*, 333, 83, 1985.
57. **Jentoft, N.,** Analysis of sugars in glycoproteins by high-pressure liquid chromatography, *Anal. Biochem.*, 148, 424, 1985.
58. **Gisch, D. J. and Pearson, J. D.,** Determination of monosaccharides in glycoproteins by reversed-phase high-performance liquid chromatography on 2.1 mm narrow-bore columns, *J. Chromatogr.*, 443, 299, 1988.
59. **White, C. A., Vass, S. W., Kennedy, J. F., and Large, D. G.,** High-pressure liquid chromatography of dimethylphenylsilyl derivatives of some monosaccharides, *Carbohydr. Res.*, 119, 241, 1983.
60. **White, C. A., Vass, S. W., Kennedy, J. F., and Large, D. G.,** Analysis of phenyldimethylsilyl derivatives of monosaccharides and their role in high-performance liquid chromatography of carbohydrates, *J. Chromatogr.*, 264, 99, 1983.
61. **Karamanos, N. K., Tsegenidis, T., and Antonopoulos, C. A.,** Analysis of neutral sugars as dinitrophenylhydrazones by high-performance liquid chromatography, *J. Chromatogr.*, 405, 221, 1987.
62. **Batley, M., Redmond, J. W., and Tseng, A.,** Sensitive analysis of aldose sugars by reversed-phase high-performance liquid chromatography, *J. Chromatogr.*, 253, 124, 1982.
63. **Rosenfelder, G., Mörgelin, M., Chang, J.-Y., Schönenberger, C.-A., Braun, D. G., and Towbin, H.,** Chromogenic labeling of monosaccharides using 4′-*N*,*N*-dimethylamino-4-aminoazobenzene, *Anal. Biochem.*, 147, 156, 1985.
64. **Muramoto, K., Goto, R., and Kamiya, H.,** Analysis of reducing sugars as their chromophoric hydrazones by high-performance liquid chromatography, *Anal. Biochem.*, 162, 435, 1987.
65. **Lin, J.-K. and Wu, S.-S.,** Synthesis of dabsylhydrazine and its use in the chromatographic determination of monosaccharides by thin-layer and high-performance liquid chromatography, *Anal. Chem.*, 59, 1320, 1987.
66. **Alpenfels, W. F.,** A rapid and sensitive method for the determination of monosaccharides as their dansylhydrazones by high-performance liquid chromatography, *Anal. Biochem.*, 114, 153, 1981.
67. **Mopper, K. and Johnson, L.,** Reversed-phase liquid chromatographic analysis of Dns-sugars. Optimization of derivatization and chromatographic procedures and applications to natural samples, *J. Chromatogr.*, 256, 27, 1983.
68. **Takeda, M., Maeda, M., and Tsuji, A.,** Fluoresence high-performance liquid chromatography of reducing sugars using Dns-hydrazine as a pre-labelling reagent, *J. Chromatogr.*, 244, 347, 1982.
69. **Takemoto, H., Hase, S., and Ikenaka, T.,** Microquantitative analysis of neutral and amino sugars as fluorescent pyridylamino derivatives by high-performance liquid chromatography, *Anal. Biochem.*, 145, 245, 1985.
70. **Coles, E., Reinhold, V. N., and Carr, S. A.,** Fluorescent labeling of carbohydrates and analysis by liquid chromatography. Comparison of derivatives using mannosidosis oligosaccharides, *Carbohydr. Res.*, 139, 1, 1985.
71. **Shinomiya, K., Toyoda, H., Akahoshi, A., Ochiai, H., and Imanari, T.,** Fluorometric analysis of neutral sugars as their glycamines by high-performance liquid chromatography and thin-layer chromatography, *J. Chromatogr.*, 387, 481, 1987.
72. **Dominguez, L. M. and Dunn, R. S.,** Analysis of OPA-derivatized amino sugars in tobacco by high-performance liquid chromatography with fluorometric detection, *J. Chromatogr. Sci.*, 25, 468, 1987.
73. **Hara, S., Yamaguchi, M., Takemori, Y., and Nakamura, M.,** Highly sensitive determination of *N*-acetyl- and *N*-glycolylneuraminic acids in human serum and urine and rat serum by reversed-phase liquid chromatography with fluorescence detection, *J. Chromatogr.*, 377, 111, 1986.
74. **Noël, D., Hanai, T., and D'Amboise, M.,** Systematic liquid chromatographic separation of poly-, oligo-,and monosaccharides, *J. Liq. Chromatogr.*, 2, 1325, 1979.
75. **D'Amboise, M., Noël, D., and Hanai, T.,** Characterization of bonded-amine packing for liquid chromatography and high-sensitivity determination of carbohydrates, *Carbohydr. Res.*, 79, 1, 1980.
76. **D'Amboise, M., Hanai, T., and Noël, D.,** Liquid chromatographic measurement of urinary monosaccharides, *Clin. Chem.*, 26, 1348, 1980.
77. **Van Eijk, H. G., Van Noort, W. L., Dekker, C., and Van der Heul, C.,** A simple method for the separation and quantitation of neutral carbohydrates of glycoproteins in the one nanomole range on an adapted amino acid analyzer, *Clin. Chim. Acta*, 139, 187, 1984.
78. **Ugalde, T. D., Faber, J. P. M., and Jenner, C. F.,** Optimizing copper-bicinchoninate carbohydrate detection for use with water-elution high-performance liquid chromatography: a technique to measure the major mono- and oligosaccharides in small pieces of wheat endosperm, *J. Chromatogr.*, 449, 207, 1988.

79. **Grimble, G. K., Barker, H. M., and Taylor, R. H.,** Chromatographic analysis of sugars in physiological fluids by post-column reaction with cuprammonium: a new and highly sensitive method, *Anal. Biochem.*, 128, 422, 1983.
80. **Leonard, J. L., Guyon, F., and Fabiani, P.,** High-performance liquid chromatography of sugars on copper(II) modified silica gel, *Chromatographia*, 18, 600, 1984.
81. **Cowie, C. E., Haddad, P. R., and Alexander, P. W.,** The determination of reducing carbohydrates using a cation-exchange column and potentiometric detection with a metallic copper electrode, *Chromatographia*, 21, 417, 1986.
82. **Vrátný, P., Frei, R. W., Brinkman, U. A. Th., and Nielen, M. W. F.,** Evaluation of various packings for solid-state catalytic reactors used in the liquid chromatographic detection of non-reducing carbohydrates, *J. Chromatogr.*, 295, 355, 1984.
83. **Vrátný, P., Ouhrabková, J., and Čopíková, J.,** Liquid chromatography of non-reducing oligosaccharides: a new detection principle, *J. Chromatogr.*, 191, 313, 1980.
84. **Vrátný, P., Brinkman, U. A. Th., and Frei, R. W.,** Comparative study of post-column reactions for the detection of saccharides in liquid chromatography, *Anal. Chem.*, 57, 224, 1985.
85. **Femia, R. A. and Weinberger, R.,** Determination of reducing and non-reducing carbohydrates in food products by liquid chromatography with post-column catalytic hydrolysis and derivatization. Comparison with refractive index detection, *J. Chromatogr.*, 402, 127, 1987.
86. **Nordin, P.,** Monitoring of carbohydrates with periodate in effluents from high-pressure liquid chromatography columns, *Anal. Biochem.*, 131, 492, 1983.
87. **Mopper, K., Dawson, R., Liebezeit, G., and Hansen, H.-P.,** Borate complex ion-exchange chromatography with fluorometric detection for determination of saccharides, *Anal. Chem.*, 52, 2018, 1980.
88. **Honda, S., Enami, K., Konishi, T., Suzuki, S., and Kakehi, K.,** Use of ethylenediamine sulphate for post-column derivatization of reducing carbohydrates to electrochemically oxidisable compounds in high-performance liquid chromatography, *J. Chromatogr.*, 361, 321, 1986.
89. **Kato, T. and Kinoshita, T.,** Fluorometric detection and determination of carbohydrates by high-performance liquid chromatography using ethanolamine, *Anal. Biochem.*, 106, 238, 1980.
90. **Villaneuva, V. R., Le Goff, M. Th., Mardon, M., and Moncelon, F.,** High-performance liquid chromatographic method for sugar analysis of crude deproteinized extracts of needles of air-polluted, healthy and damaged *Picea* trees, *J. Chromatogr.*, 393, 115, 1987.
91. **Kato, T. and Kinoshita, T.,** Fluorometric detection and determination of carbohydrates in high-performance liquid chromatography with 2-aminopropionitrile fumarate-borate reagent, *Bunseki Kagaku*, 31, 615, 1982 *(Chem. Abstr.*, 98, 83047t, 1983).
92. **Kato, T., Iinuma, F., and Kinoshita, T.,** Fluorometric detection and determination of carbohydrates in thin-layer chromatography and high-performance liquid chromatography using taurine-borate reagent, *Nippon Kagaku Kaishi*, p. 1603, 1982 *(Chem. Abstr.*, 98, 54303t, 1983).
93. **Kato, T. and Kinoshita, T.,** Fluorometry of saccharides by high-performance liquid chromatography using taurine-periodate reagent, *Bunseki Kagaku*, 35, 869, 1986 *(Chem. Abstr.*, 107, 3495d, 1987).
94. **Honda, S., Matsuda, Y., Takahashi, M., Kakehi, K., and Ganno, S.,** Fluorometric determination of reducing carbohydrates with 2-cyanoacetamide and application to automated analysis of carbohydrates as borate complexes, *Anal. Chem.*, 52, 1079, 1980.
95. **Honda, S., Takahashi, M., Kakehi, K., and Ganno, S.,** Rapid, automated analysis of monosaccharides by high-performance anion-exchange chromatography of borate complexes with fluorometric detection using 2-cyanoacetamide, *Anal. Biochem.*, 113, 130, 1981.
96. **Honda, S., Takahashi, M., Nishimura, Y., Kakehi, K., and Ganno, S.,** Sensitive ultraviolet monitoring of aldoses in automated borate complex anion-exchange chromatography with 2-cyanoacetamide, *Anal. Biochem.*, 118, 162, 1981.
97. **Schlabach, T. D. and Robinson, J.,** Improvements in sensitivity and resolution with the cyanoacetamide reaction for the detection of chromatographically separated reducing sugars, *J. Chromatogr.*, 282, 169, 1983.
98. **Honda, S., Suzuki, S., Takahashi, M., Kakehi, K., and Ganno, S.,** Automated analysis of uronic acids by high-performance liquid chromatography with photometric and fluorometric post-column labeling using 2-cyanoacetamide, *Anal. Biochem.*, 134, 34, 1983.
99. **Honda, S., Konishi, T., Suzuki, S., Takahashi, M., Kakehi, K., and Ganno, S.,** Automated analysis of hexosamines by high-performance liquid chromatography with photometric and fluorometric post-column labelling using 2-cyanoacetamide, *Anal. Biochem.*, 134, 483, 1983.
100. **Honda, S. and Suzuki, S.,** Common conditions for high-performance liquid chromatographic microdetermination of aldoses, hexosamines, and sialic acids in glycoproteins, *Anal. Biochem.*, 142, 167, 1984.
101. **Honda, S., Konishi, T., and Suzuki, S.,** Electrochemical detection of reducing carbohydrates in high-performance liquid chromatography after post-column derivatization with 2-cyanoacetamide, *J. Chromatogr.*, 299, 245, 1984.

102. **Gandelman, M. S., Birks, J. W., Brinkman, U. A. Th., and Frei, R. W.,** Liquid chromatographic detection of cardiac glycosides and saccharides based on the photoreduction of anthraquinone-2,6-disulfonate, *J. Chromatogr.*, 282, 193, 1983.
103. **Gandelman, M. S. and Birks, J. W.,** Liquid chromatographic detection of cardiac glycosides, saccharides and hydrocortisone based on the photoreduction of 2-*tert.*-butylanthraquinone, *Anal. Chim. Acta*, 155, 159, 1983.
104. **Honda, S., Takahashi, M., Shimada, S., Kakehi, K., and Ganno, S.,** Automated analysis of alditols by anion-exchange chromatography with photometric and fluorometric post-column derivatization, *Anal. Biochem.*, 128, 429, 1983.
105. **Honda, S., Konishi, T., Suzuki, S., Kakehi, K., and Ganno, S.,** Sensitive monitoring of hexosamines in high-performance liquid chromatography by fluorometric post-column labelling using the 2,4-pentanedione-formaldehyde system, *J. Chromatogr.*, 281, 340, 1983.
106. **White, C. A., Vass, S. W., and Kennedy, J. F.,** An automated Elson-Morgan assay for 2-amino-2-deoxyhexoses, with increased sensitivity, *Carbohydr. Res.*, 114, 201, 1983.
107. **Ohno, N., Suzuki, I., and Yadomae, T.,** A method for monitoring elution profiles during the chromatography of amino sugar-containing oligo- and polysaccharides, *Carbohydr. Res.*, 137, 239, 1985.
108. **Krantz, M. J. and Lee, Y. C.,** A sensitive autoanalytical method for sialic acids, *Anal. Biochem.*, 63, 464, 1975.
109. **Tsuji, T., Yamamoto, K., Konami, Y., Irimura, T., and Osawa, T.,** Separation of acidic oligosaccharides by liquid chromatography: application to analysis of sugar chains of glycoproteins, *Carbohydr. Res.*, 109, 259, 1982.
110. **Jeansonne, N., Radhakrishnamurthy, B., Dalferes, E. R., Jr., and Berenson, G. S.,** Continuous-flow monitoring of hexuronic acid by carbazole reaction during gel filtration of proteoglycans in urea solutions, *J. Chromatogr.*, 354, 524, 1986.
111. **Henderson, S. K. and Henderson, D. E.,** Enhanced UV detection of sugar phosphates by addition of a metal complex to the HPLC mobile phase, *J. Chromatogr. Sci.*, 23, 222, 1985.
112. **Henderson, S. K. and Henderson, D. E.,** Reversed-phase ion pair HPLC analysis of sugar phosphates, *J. Chromatogr. Sci.*, 24, 198, 1986.
113. **Frijlink, H. W., Visser, J., and Drenth, B. F. H.,** Determination of cyclodextrins in biological fluids by high-performance liquid chromatography with negative colorimetric detection using post-column complexation with phenolphthalein, *J. Chromatogr.*, 415, 325, 1987.
114. **Larew, L. A. and Johnson, D. C.,** Quantitation of chromatographically separated maltooligosaccharides with a single calibration curve using a postcolumn enzyme reactor and pulsed amperometric detection, *Anal. Chem.*, 60, 1867, 1988.

Section II.IV

DETECTION REAGENTS FOR PLANAR CHROMATOGRAPHY

Only a few new or modified detection reagents for the planar chromatography of carbohydrates have been described since the compilation of Volume I. Therefore, the list of detection reagents for paper and/or thin-layer chromatography included in that volume remains current. The list of reagents used in recent years that is presented in Section II.IV is intended to serve as a supplement to Section II.III in Volume I. As in liquid chromatography, the emphasis has been on the development of fluorogenic reagents for the planar chromatography of carbohydrates: in some cases, derivatization before chromatography is advantageous.

To facilitate reference, the reagents are described in alphabetical order here, as in Volume I.

SUPPLEMENTARY LIST OF DETECTION REAGENTS FOR CARBOHYDRATES IN PAPER AND/OR THIN-LAYER CHROMATOGRAPHY

DETECTION REAGENTS

Reagents Used in Pre-Derivatization

1. 4'-*N,N*-Dimethylamino-4-aminoazobenzene (DAAB)[1]

Procedure: The solution in methanol (200 μl) containing sugars (0.5 m*M*), DAAB (4 m*M*), $NaBH_3CN$ (37.5 m*M*), pentaerythritol (37.5 m*M*), and acetic acid (4 *M*) is heated at 80° C for about 10 min, until the color changes from green (acid) to orange-yellow (neutral to slightly basic). After evaporation of methanol, the residue is dissolved in water and the solution loaded on a Sep-Pak® C_{18} cartridge (Waters), pretreated with methanol and water. The cartridge is flushed with excess of water (20 ml) under pressure produced by a syringe; the colored products are bound to C_{18}-silica, while unreacted sugar, salts, and pentaerythritol (which complexes borate) are eluted. Excess DAAB is washed off the cartridge with (1:3) chloroform-hexane (5 ml), the cartridge is flushed with N_2, and DAAB-glycamines are eluted with methanol (2 to 5 ml). If necessary, further purification can be achieved by chromatography on a short column of bead-form silica gel (Iatrobeads[2]; Iatron, Tokyo, Japan) eluted with $CHCl_3$-CH_3OH-H_2O (11:9:2). DAAB elutes in the solvent front, and DAAB-glycamines are easily monitored as colored bands.

Results: After TLC, the plates are sprayed with 0.125 *M* H_2SO_4. This gives bright red spots or bands that can be scanned by densitometry at 520 nm. This method is applicable to all reducing sugars, including hexuronic acids, amino- and acetamidodeoxyhexoses, and disaccharides. Detection limit: 20 pmol.

Precautions: DAAB is carcinogenic. Contact of solutions with skin should be avoided and all procedures performed under a well-ventilated hood.

2. 4′-*N*,*N*-Dimethylamino-4-azobenzene-1-sulfonylhydrazine (Dabsyl Hydrazine)[2]

Preparation: Dabsyl hydrazine is prepared by reaction of dabsyl chloride (100 mg), dissolved in tetrahydrofuran (25 ml), with hydrazine (0.2 ml) at room temperature for 30 min. After concentration, orange-yellow crude products (44 mg) are obtained; these are purified by recrystallization in ethanol to give orange crystals (30 mg), m.p. 163 to 164°C.

Procedure: The solution (1 ml) in ethanol (with 0.1% acetic acid), containing 1 to 20 nmol of each sugar, is mixed with a solution (1 ml) of dabsyl hydrazine (0.1% in ethanol), and the mixture is heated at 60° C for 1 h. After cooling to room temperature, the reaction mixture (samples containing 2 to 10 pmol of sugars) is applied directly to TLC plates.

Results: Sugar dabsylhydrazones appear as bright yellow spots that may be scraped off the plate for recovery of hydrazones by extraction into ethanol. If developed TLC plates are briefly exposed to HCl vapor in a closed chamber, the spots turn bright pink, which enhances the sensitivity of densitometric scanning. The method is applicable mainly to aldoses; ketoses react very slowly with dabsyl hydrazine (this reaction can be promoted by lowering the pH to 2 with 0.1 *M* trichloroacetic acid).

Precautions: As for DAAB.

3. 5-Dimethylaminonaphthalene-1-sulfonyl Chloride (Dansyl Chloride)

Procedure: This reagent has been used in analysis of aminodeoxyhexoses in glycoproteins.[3] The dried hydrolyzed sample is dissolved in 0.2 *M* $NaHCO_3$, and dansyl chloride reagent (60 mg/ml in acetone) is added (1 ml to 2 ml $NaHCO_3$ solution). The dansyl chloride:protein ratio should be 9:1 (w/w) for maximal fluorescence. After mixing, the samples are incubated in the dark for 30 min at 37° C. The reaction is terminated by addition of 88% HCOOH (0.1 ml). Dansyl amino acids interfere and must be separated from the dansyl sugars by thin-layer electrophoresis (2 to 10-μl samples on 20 × 20 cm cellulose plates in 0.4% pyridine-0.8% glacial acetic acid, pH 4.4; electrophoresis at 500 V for 4 h at room temperature). Bands containing dansyl sugars are scraped from the plate, and dansyl sugars are extracted overnight with 95% ethanol and then dried under N_2 for application (in a small volume of ethanol) to a silica gel TLC plate. After TLC (in cyclohexane-ethyl acetate-ethanol [6:4:3], the plate is immediately sprayed with triethanolamine-2-propanol (1:4) to stabilize and intensify fluorescence of dansyl sugars.

Results: Fluorometric scanning (excitation wavelength 365 nm, emission 512 nm) gives linear calibration over the range of 50 to 200 ng (0.3 to 1.1 nmol) of amino sugar.

4. 5-Dimethylaminonaphthalene-1-sulfonylhydrazine (Dansyl Hydrazine)

In the presence of acid, dansyl hydrazine reacts with reducing sugars (less readily with ketoses) to give fluorescent dansylhydrazones.

Preparation: Avigad[4] used the following solutions:

Solution a: 1% (w/v) dansyl hydrazine in ethanol
Solution b: 4% (w/v) trichloroacetate acid in ethanol
Solution c: 1% (v/v) acetic acid in ethanol

Procedure: A sample (100 μl) containing 0.04 to 2 μmol reducing sugar is mixed with 200 μl *a* and 100 μl *b*. The mixture is heated at 80° C for 10 min, then cooled to room temperature. Samples (containing 2 to 100 nmol sugar) are applied to silica gel plates. After TLC, dansylhydrazones appear as intensely fluorescent bright yellow spots. These may be scraped off plates and extracted into solution *c* for determination by fluorometry or scanned directly on plates.

Results: Fluorometry (excitation wavelength 360 nm, emission 510 nm) gives linear calibration over the range 1 to 50 nmol for D-glucose and other neutral aldoses. Ketoses and aminodeoxyhexoses give a poor response, except after prolonged heating with the reagent.

This reagent has been applied to TLC detection of unsaturated disaccharides from chondroitinase digestion of chondroitin sulfates[5] (see Table TLC7 for conditions used in derivatization). Detection limits of ca. 90 pmol were reported.

Reagents used after Chromatography

5. Diphenylboric Acid β-Aminoethyl Ester-*p*-Anisidine Phosphate[6]

Preparation: *p*-Anisidine phosphate reagent is prepared as described for aniline phosphate (see Volume I, reagent 9), the aniline being replaced by *p*-anisidine in the same proportion (w/v). Solution *a*, containing diphenylboric acid β-aminoethyl ester (1% in ethanol), is prepared as a prespray for PC.

Procedure: Spray with *a* and then with *p*-anisidine phosphate. Heat at 100 to 120° C.

Results: The color of spots given by aldopentoses with *p*-anisidine phosphate spray, normally red-brown, changes to purple-red with the prespray. Colors given by aldohexoses (yellow-brown) and ketohexoses (lemon yellow) are unchanged. The sensitivity is increased five- to tenfold.

6. Fluorescamine

Fluorescamine {4-phenylspiro[furan-2-(3H),1′-phthalan]-3,3′-dione}, which reacts with primary amines at pH 9 to give highly fluorescent pyrrolines and has been used successfully in the detection and determination of amino acids in TLC, has been adapted as a specific reagent for aminodeoxy sugars.[7]

Preparation: Fluorescence is stabilized and intensified in the presence of base; triethylamine has proved effective. Sprays used are:

Solution a: 10% (v/v) triethylamine in dichloromethane

Solution b: 0.05% (w/v) fluorescamine in acetone

Procedure: After TLC, silica gel plates are sprayed with *a*, air-dried, sprayed with *b*, air-dried again, and then sprayed again with *a*.

Results: Fluorometric scanning (excitation wavelength 390 nm, emission 475 nm) gives linear calibration in the range 0.1 to 25 nmol. Detection limit: 50 pmol.

7. Lead Tetraacetate-2,7-Dichlorofluorescein

This detection method has proved effective in HPTLC of aldoses, ketoses, alditols, hexuronic and aldonic acids, and lactones[8,9] on silica gel plates (see Table TLC 3). It cannot be used in PC, or TLC on cellulose plates, which would be affected by lead tetraacetate oxidation.

Preparation:

Solution a: saturated solution of lead tetraacete in glacial acetic acid

Solution b: 1% solution of 2,7-dichlorofluorescein in water

5 ml of each solution are mixed, and the volume is brought up to 200 ml with toluene. Lower concentrations of 2,7-dichlorofluorescein (0.2 to 0.8%) can be used to reduce background interferences in fluorometric detection.

Procedure: After TLC, the plate is dried and then dipped in the reagent solution for about 10 sec. A narrow, vertical tank should be used for immersion, and the solution should be renewed for each plate. After immersion, the plate is activated by heating in a stream of hot air or in a drying cabinet at ca. 100° C for 3 min. Reaction of 2,7-dichlorofluorescein with the products of lead tetraacetate oxidation of the vicinal diol systems in the carbohydrate molecules produces fluorescence.

Results: Fluorometric scanning (excitation wavelength 313 nm, emission 366 nm) permits detection at nanogram levels. Detection limit; 10 ng.

8. Malonamide

This reagent does not contain acid or oxidant and can, therefore, be used in PC or TLC of reducing sugars.[10]

Preparation: The reagent is a 1% solution of malonamide in 1 *M* sodium carbonate buffer (pH 9.2).

Procedure: After chromatography, the air-dried plate or paper is sprayed with malonamide reagent, then heated in an oven at 120° C for 5 min (paper) or 20 min (silica gel plate). Excitation and emission maxima for fluorescent products are in the ranges 328 to 382 nm and 383 to 425 nm, respectively, with intense fluorescence, visible in the green-blue region, at wavelengths above 360 nm. Irradiation with a 365-nm mercury lamp is effective in detecting fluorescent spots on papers and plates.

Results: On plates, the detection limits are 0.25 nmol for all reducing sugars, including aldoses, ketoses, hexuronic acids, and aminodeoxyhexoses, except 2-deoxyhexoses (0.5 nmol). Reducing disaccharides are also detected down to 0.25 nmol. On paper, the detection limit is higher (ca. 1 nmol) owing to diffusion of spots. The reagent is less sensitive in detection of alditols (TLC limit 10 nmol), methyl glycosides (100 nmol), and nonreducing disaccharides (200 nmol) and very insensitive for aldonic acids and polysaccharides (1000 nmol) except starch and pullulan (20 nmol).

9. Ninhydrin-Cu(II)

In the TLC of amino compounds, including aminodeoxy and -dideoxyhexoses, on cellulose plates[11] (see Table TLC 1), the chromogenicity of the ninhydrin reagent was increased by use of a high concentration of ninhydrin and by the presence of Cu(II).

Preparation:

Solution a: acetic acid (10 ml) and collidine (2 ml) are added to a solution of ninhydrin (1 g) in absolute ethanol (50 ml)

Solution b: $CuNO_3 \cdot 3H_2O$ (0.5 g) in absolute ethanol (50 ml)

Immediately before use, *a* and *b* are mixed in the ratio 50:3 (v/v).

Procedure: After TLC, dried plates are sprayed with reagent, then heated at 105° C for 2 to 3 min.

Results: Aminodeoxyhexoses give intense yellow-brown spots, and 2-amino-2,6-dideoxyhexoses (fucosamine, rhamnosamine, and quinovosamine) give orange-yellow spots, distinguishable from those given by amino acids (grey, red, or purple).

10. Resorcinol-Hydrochloric Acid-Cu(II)

In the TLC of brain gangliosides on silica gel plates (see Table TLC 7), detection based on the specificity of the resorcinol-HCl reagent for neuraminic acids (as well as neutral ketoses; see Volume I, reagent 44) is more sensitive in the presence of Cu(II).[12]

Preparation:

Solution a: resorcinol (200 mg) in 4 *M* HCl (100 ml)

Solution b: aqueous 0.1 *M* solution of $CuSO_4 \cdot 5H_2O$

At least 4 h before use, *a* and *b* are mixed in the ratio 40:1 (v/v).

Procedure: After TLC, the dried plate is sprayed lightly and covered carefully with a clean plate before being dried at 120° C for about 20 min.

Results: Gangliosides containing neuraminic acids are revealed as blue spots against a white background. Densitometric scanning (red filter, 610 to 750 nm) gives linear calibration over the range 50 to 350 μg of total ganglioside extract.

11. Sulfosalicylic Acid

This reagent is used in a strong acid and, therefore, its application is confined to silica gel TLC plates. It detects all sugars, including aldoses, ketoses, hexuronic acids, aminodeoxyhexoses, sugar phosphates, and oligosaccharides, both nonreducing and reducing.[13]

Preparation: The reagent is 2% (w/v) sulfosalicylic acid in 0.5 *M* H_2SO_4.

Procedure: After TLC, the air-dried plate is sprayed with reagent and then heated at 110° C for 15 min.

Results: All sugars except 6-deoxyhexoses give grey to grey-brown spots; 6-deoxyhexoses give a yellow coloration, which can be utilized for their determination in mixtures with other sugars. Detection limits are 0.2 to 0.7 μg for hexoses and 6-deoxyhexoses, 1 to 2 μg for pentoses, 0.3 μg for D-glucose-6-phosphate, 0.25 μg for D-glucuronic acid (but 1 μg for D-galacturonic acid), 0.3 to 0.4 μg for reducing disaccharides, and 0.7 to 0.8 μg for nonreducing disaccharides. The method is less sensitive for detection of aminodeoxyhexoses (detection limits 3.5 and 7 μg for 2-amino-2-deoxy-D-galactose and -glucose, respectively) and very insensitive for D-glucitol and mannitol (20 to 22 μg); galactitol and inositol are not detected.

12. Sulfuric Acid + *tert.*-Butanol in Reagents for Bonded-Phase TLC

In TLC on aminopropyl-bonded-phase plates (see Table TLC 4), many of the traditional spray reagents for sugars cannot be used owing to interference by the aminopropyl groups.[14] This can be minimized by addition of *tert.*-butanol to the spray reagent. Two reagents have been used successfully for detection of sugars, methyl glycosides, and acetals on these plates.

a. **Preparation:** Reagent is *tert.*-butanol-ethanol-H_2SO_4 (6:4:3).
Procedure: After TLC, dried plates are sprayed with reagent and then heated at 180° C for 5 min.
Results: Carbohydrates are detected as dark brown to black spots. Aminopropyl-bonded plates are not charred under these conditions.[14]

b. **Preparation:** Diphenylamine hydrochloride (4.0 g), aniline (4.0 ml), *tert.*-butanol (120 ml), and ethanol (80 ml) are mixed and, after the mixture has been cooled in an ice bath, sulfuric acid (30 ml) is stirred in gradually over 15 min. Reagent can be stored indefinitely at −6° C.
Procedure: After TLC, dried plates are sprayed with reagent and then heated at 120° C for 5 min.
Results: Although H_2SO_4 has been substituted for H_3PO_4 and *tert.*-butanol for acetone, colors of spots are those observed with traditional diphenylamine-aniline-phosphoric acid reagent (see Volume I, reagent 23, Tables IIA, IIB, and IIE), so that different classes of sugars can be distinguished.[15]

REFERENCES

1. **Rosenfelder, G., Mörgelin, M., Chang, J.-Y., Schönenberger, C.-A., Braun, D. G., and Towbin, H.,** Chromogenic labeling of monosaccharides using 4′-*N*,*N*-dimethylamino-4-aminoazobenzene, *Anal. Biochem.*, 147, 156, 1985.
2. **Lin, J.-K. and Wu, S.-S.,** Synthesis of dabsylhydrazine and its use in chromatographic determination of monosaccharides by thin-layer and high-performance liquid chromatography, *Anal. Chem.*, 59, 1320, 1987.

3. **Farwell, D. C. and Dion, A. S.,** A fluorometric assay for the quantitation of amino sugars, *Anal. Biochem.*, 95, 533, 1979.
4. **Avigad, G.,** Dansyl hydrazine as a fluorimetric reagent for thin-layer chromatographic analysis of reducing sugars, *J. Chromatogr.*, 139, 343, 1977.
5. **Shinomiya, K., Hoshi, Y., and Imanari, T.,** Sensitive detection of unsaturated disaccharides from chondroitin sulphates by thin-layer chromatography as their dansylhydrazine derivatives, *J. Chromatogr.*, 462, 471, 1989.
6. **Niemann, G. J.,** Improved amine spray reagent for detection of sugars, *J. Chromatogr.*, 170, 227, 1979.
7. **Jimenez, M. H. and Weill, C. E.,** A sensitive, fluorimetric analysis of amino sugars, *Carbohydr. Res.*, 101, 133, 1982.
8. **Klaus, R. and Ripphahn, J.,** Quantitative thin-layer chromatographic analysis of sugars, sugar acids and polyalcohols, *J. Chromatogr.*, 244, 99, 1982.
9. **Klaus, R. and Fischer, W.,** Quantitative thin-layer chromatography of sugars, sugar acids and polyalcohols, *Methods Enzymol.*, 160, 159, 1988.
10. **Honda, S., Matsuda, Y., and Kakehi, K.,** Use of malonamide as a general spray reagent for the fluorimetric detection of reducing sugars on filter papers and thin-layer plates, *J. Chromatogr.*, 176, 433, 1979.
11. **Ryan, E. A. and Kropinski, A. M.,** Separation of amino sugars and related compounds by two-dimensional thin-layer chromatography, *J. Chromatogr.*, 195, 127, 1980.
12. **Randell, J. A. J. and Pennock, C. A.,** Brain gangliosides: an improved simple method for their extraction and identification, *J. Chromatogr.*, 195, 257, 1980.
13. **Ray, B., Ghosal, P. K., Thakur, S., and Majumdar, S. G.,** Sulphosalicylic acid as spray reagent for the detection of sugars on thin-layer chromatograms, *J. Chromatogr.*, 315, 401, 1984.
14. **Doner, L. W., Fogel, C. L., and Biller, L. M.,** 3-Aminopropyl-bonded-phase silica thin-layer chromatographic plates: their preparation and application to sugar resolution, *Carbohydr. Res.*, 125, 1, 1984.
15. **Doner, L. W. and Biller, L. M.,** High-performance thin-layer chromatographic separation of sugars: preparation and application of aminopropyl bonded-phase silica plates impregnated with monosodium phosphate, *J. Chromatogr.*, 287, 391, 1984.

Section III

Sample Preparation and Derivatization

SAMPLE PREPARATION AND DERIVATIZATION

In this section, some further aspects of degradative methods covered in Volume I are discussed, with any improvements that have arisen from more recent work. Additional methods are also included, mainly those used in the production of oligosaccharides from the carbohydrate moieties of glyco-proteins and other glycoconjugates. In the part dealing with derivatization for GLC, some of the established methods are similarly updated, and the newer methods are introduced. Derivatization is now widely applied in HPLC also, and details of the methods used to increase sensitivity of detection, as discussed in Section II.III, are given in the last part of this chapter.

The methods presented are listed below, in alphabetical order within each of the three groups.

Degradation of Polysaccharides and Glycoconjugates

1. Acetolysis
2. Acid hydrolysis
3. Alkaline hydrolysis
4. Deamination
5. Enzymatic degradation
6. Hydrazinolysis
7. Methanolysis
8. Methylation analysis

Derivatization for Gas Chromatography

9. Alditol acetates
10. Aldononitrile acetates
11. *N*-Alkylaldonamide acetates
12. Chiral derivatives
13. Deoxy(methoxyamino)alditols
14. Diethyl dithioacetals
15. Oximes
16. Trifluoroacetylation
17. Trimethylsilylation

Derivatization for HPLC

18. Acetylation
19. Benzoylation
20. Dimethylphenylsilylation
21. Glycosylamine derivatives
22. Hydrazone derivatives
23. Peralkylated oligosaccharide-alditols
24. Reductive amination

Section III.I

DEGRADATION OF POLYSACCHARIDES AND GLYCOCONJUGATES

1. ACETOLYSIS

The classical procedure for acetolysis of cellulose was presented in Volume I. Recently, Narui et al.[1] reported that addition of tetramethylurea to the reaction mixture allowed the use of higher temperatures (80 to 100°C) in acetolysis without inducing irregular degradation and caramelization. The reaction time was thus greatly decreased. For example, acetolysis of a dextran of high molecular weight range (5 to 40 $\times$ 10^6) required 96 h to reach completion at room temperature, but in the presence of 5% of tetramethylurea, acetolysis at 100°C was possible and required only 20 min for completion. Similar results were obtained with amylose (mol wt 150,000) in the presence of 10% of tetramethylurea. In both cases the product was rich in oligosaccharides of DP 2 to 6, whereas without tetramethylurea the reaction at 100°C produced only D-glucose with isomaltose (dextran) or maltose and maltotriose (amylose), and the mixture was highly caramelized.

In the procedure recommended by these authors, the polysaccharide (1 g) was dissolved in 20 ml of the acetolysis reagent, which contained glacial acetic acid, acetic anhydride, sulfuric acid, and tetramethylurea in proportions (v/v) 8:9.5:1.5:1.0 or 7:9.5:1.5:2.0 (5 or 10% tetramethylurea, respectively). The reaction mixture was stirred in a stoppered round-bottomed flask for 20 min at 100°C, and samples were removed at intervals for examination by PC and HPLC, which verified completion of the reaction after this time. The acetolyzed mixture was extracted into chloroform and the extract mixed with methanol (15 ml) and stirred for 15 min at room temperature. De-*O*-acetylation was effected with sodium methoxide (0.5 g sodium metal dissolved in 100 ml dry methanol) with stirring for 1 h at room temperature. After evaporation, the residue was dissolved in the minimum volume of distilled water, the base was neutralized with H^+-form cation-exchange resin (subsequently removed by filtration), and the filtrate was evaporated to a syrup and dried for 12 h at 40°C *in vacuo*. Although this procedure was applied only to α-D-glucans by Narui et al.,[1] it is generally applicable to other oligosaccharides under the same conditions.

An advantage of acetolysis as a degradative procedure in structural studies of polysaccharides is that certain modes of glycosidic linkage that are readily cleaved by acid hydrolysis, such as α-D-(1 → 4) linkages, are less susceptible to acetolysis than others, notably α-D-(1 → 6) linkages,[2] which have slower rates of acid hydrolysis. Similarly, rhamnosyl and fucosyl linkages, which are very labile to acid, tend to resist acetolysis. Thus, acetolysis is a useful complement to acid hydrolysis in the production of structurally significant oligosaccharides from polysaccharides. The method also has been applied in degradative studies of glycopeptides.[3]

2. ACID HYDROLYSIS

The problems inherent in quantitative liberation of monosaccharides from polysaccharides by acid hydrolysis, due to wide variation in the susceptibility to acid of the various glycosidic linkages between different sugar residues and the tendency of the free sugars to decompose in acid, which were discussed in Volume I, have been reviewed recently and quantitative data for decomposition of the sugars presented.[4] Pentoses, especially D-xylose, are particularly vulnerable. For example, in the author's laboratory, losses of 20 and 38% by weight have been observed on treatment of D-xylose with 2 *M* trifluoroacetic acid at 100°C for 8 and 18 h, respectively, and the losses of L-arabinose were 10 and 25% under the same conditions. Heating at 100°C in 1 *M* sulfuric acid for 18 h causes losses of at least 20% for all sugars in cell-wall polysaccharides.[5] The problem is exacerbated in analysis of natural products, such as plant gums or bark polysaccharides, in which polyphenols (e.g., tannins) occur together with the carbohydrate, since the sugars are further destroyed by reaction with these components.[6,7] Protein has a similar effect, which can be minimized by hydrolysis of glycoproteins (0.5 *M* HCl, 100°C, 7 h) in the presence of a strong acid cation-exchange resin (H^+-form).[8]

The differences in acid lability of different polysaccharides occurring together cause difficulty in finding the optimal conditions for hydrolysis. This is exemplified by the work of Uzaki and Ishiwatari[9] on the cellulosic and noncellulosic carbohydrates of lake sediments.

TABLE III
Recommended Conditions for Acid Hydrolysis of Various Polysaccharides and Glycoconjugates

Type of polymer	Hydrolysis conditions	Ref.
Cell-wall components	72% H_2SO_4, room temperature, 45 min, then 1 *M* H_2SO_4, 100°C, 2—3 h under argon.	10
	Pure TFA,[a] 37°C, overnight; (i) 0.5 ml + 1 ml H_2O + 0.5 ml *meso*-inositol solution (2 mg/ml; internal standard), 100°C, 30 min (hemicellulose hydrolysis); (ii) 0.5 ml + 0.1 ml H_2O, 100°C, 30 min, then further additions of H_2O (0.1 ml) and heating at 100°C for 30 min until total of 0.8 ml added; *meso*-inositol solution (0.5 ml) added, 100°C for further 2 h (cellulose hydrolysis).[b]	11
Plant gum polysaccharides	2 *M* TFA, 100°C, under N_2 or Ar, for 8 h (neutral sugars), 18 h (uronic acid present).	—[c]
Glycoproteins	*Neutral sugars*: 2 *M* TFA, 100°C, under N_2, 6 h; hydrolysate deionized (mixed resins).	12
	Acetamidodeoxyhexoses: 4 *M* HCl, 100°C, under N_2, 6 h; residue treated with saturated aqueous $NaHCO_3$ + acetic anhydride overnight to re-*N*-acetylate; fivefold amounts of resins used to deionize.	
Gangliosides	1 *M* HCl, 100°C, under N_2, 8 h; D-xylose added as internal standard *after* hydrolysis (D-ribose better if L-fucose present, giving GC peak overlapping with xylose peak). Hydrolysis for 12 h gives better yield of acetamidodeoxyhexoses, at expense of D-galactose,	13
Proteoglycans	*Neutral sugars*: 4 *M* TFA, 100°C, 4 h; hydrolysate deionized (mixed resins); cellobiose added after hydrolysis as internal standard for HPLC analysis.	14
	Aminodeoxyhexoses: 4 *M* HCl, 100°C, 10 h; hydrolysate passed through cation-exchange resin (H^+-form), eluted with 4 *M* HCl.	
	N-Acetylneuraminic acid: 50 m*M* H_2SO_4, 80°C, 1 h.	
	Hexuronic acids: 90% HCOOH, 100°C, 5 h, then 0.25 *M* H_2SO_4, 100°C, 16 h; hydrolysate passed through anion-exchange resin ($HCOO^-$-form), eluted with 25% HCOOH. Yield from glycosaminoglycuronans only 45% of theoretical.	

[a] Trifluoroacetic acid.
[b] Gradual addition of H_2O prevented precipitation of cellulose.
[c] Churms, S. C. and Stephen, A. M.; laboratory data.

These workers recommended an initial hydrolysis of the freeze-dried sediment sample in 1 *M* hydrochloric acid at 100°C for 7 h under nitrogen in a sealed ampoule, under which conditions the cellulosic fraction was precipitated and only 3 to 6% was hydrolyzed to D-glucose. After this hydrolysis, the suspension was centrifuged and the precipitate washed twice with 2% HCl. The supernatant, which contained the products of hydrolysis of the noncellulosic fraction, was then removed for derivatization to alditol acetates for GC analysis. The precipitate was completely dried *in vacuo* and then dispersed ultrasonically in 72% sulfuric acid. Heating at 40 to 45°C for 1 h dissolved the cellulose, after which the acid was immediately diluted to 0.5 *M* with distilled water. The solution was then heated at 100°C for 7 h under nitrogen in a sealed ampoule, which resulted in a recovery of ca. 90% of the initial mass of cellulose as D-glucose. In this way, analyses of both fractions were obtained, but as the noncellulosic fraction contained acid-labile D-xylose and D-ribose, substantial losses must have occurred during the initial heating in 1 *M* HCl.

Typical conditions recommended for hydrolysis of various types of polysaccharide and glycoconjugate are listed in Table III. While optimal recoveries of sugars have been obtained under these conditions, it must be stressed that losses *always* occur, and the extent of these losses should be determined by analysis of standard sugar mixtures heated in acid under the same conditions as the samples. An internal standard, such as *myo*- or *meso*-inositol, should be added to both standard mixture and sample prior to treatment with acid. This procedure permits the calculation of correction factors that can be applied to the analytical results to allow for the error due to this factor.

The resistance of aldobiouronic acid linkages to hydrolysis, which results in slow release and low yields of both hexuronic acids and contiguous (interior) sugar units except under very vigorous conditions (see Table III) that may cause decarboxylation, is a problem that can be overcome by carboxyl reduction of acid residues prior to hydrolysis. Reduction is usually effected by the method of Taylor et al.,[15] in which the carboxyl groups are activated by treatment with a carbodiimide at pH 4.75 and then reduced with sodium borohydride at pH 7.0. If sodium borodeuteride is used, the resulting hexose residue is labeled and can be distinguished in GC-MS of the hydrolysate (see Section II.I, Table IIA).

3. ALKALINE HYDROLYSIS

Hydrolysis under alkaline conditions is used to remove the *O*-linked oligosaccharide side chains in degradative studies of glycoproteins and proteoglycans. For the many glycoconjugates in which this linkage is through 2-acetamido-2-deoxy-D-galactose or galactose to serine or threonine in the polypeptide chain, the method used is alkaline borohydride treatment. The borohydride reduces the GalNAc or Gal residue at the position of cleavage and, thus, prevents further alkaline degradation by "peeling" from the reducing end of the oligosaccharide released.[2,16] The usual procedure, exemplified by the preparation by Dua et al.[17] of oligosaccharide-alditols from mucins, is to dissolve the glycoprotein in a solution containing 1.0 *M* $NaBH_4$ and 0.05 *M* NaOH and incubate at 50°C for 16 h. The reaction is stopped by cooling and neutralizing to pH 7.0 with HCl, and the solution is concentrated to dryness on a rotary evaporator and the residue evaporated several times with methanol to remove borate. Alternatively, the glycoprotein may be treated in a sealed tube with 0.4 *M* $NaBH_4$-0.2 *M* NaOH at 20°C for 48 h in the dark;[18] in this case, decomposition of excess borohydride by careful addition, in an ice bath, of 6 *M* acetic acid has been recommended.

Treatment of a glycoprotein or proteoglycan with saturated (ca. 0.22 *M*) $Ba(OH)_2$ at 100 to 105°C for 6 to 8 h results not only in β-elimination of carbohydrate from serine or threonine residues, but also in cleavage of the peptide linkages. However, *O*-glycosylic linkage to 4-hydroxyproline is stable under these conditions. This has been greatly exploited in studies of plant arabinogalactan proteins in which such linkages of sugars to hydroxyproline play a major role.[19] For example, a series of glycosylated hydroxyproline derivatives have been isolated by SEC after treatment of a sample of a proteinaceous plant gum[20,21] with 0.22 *M* $Ba(OH)_2$, first at room temperature for 2 h to dissolve the gum and then at 100°C for 6 to 8 h, followed by neutralization (30% H_2SO_4), centrifugation to remove the precipitated $BaSO_4$, and freeze-drying of the supernatant solution. Addition of $NaBH_4$ is advisable to preserve any carbohydrate chains cleaved from serine or threonine under these conditions, which can be detected by $NaB[^3H]_4$ labeling.[22]

The commonly encountered *N*-glycosylamide linkage of oligosaccharide chains to asparagine through 2-acetamido-2-deoxy-D-glucose is very stable to alkali and is usually cleaved by hydrazinolysis (see 6). However, Kochetkov and co-workers[23] have recently reported a new, mild method for removal of these *N*-linked oligosaccharides from glycoproteins by treatment with alkaline $LiBH_4$ in aqueous *tert*-butanol. The glycoprotein sample (ca. 8 mg) is dissolved in water (0.5 ml) together with trilithium citrate tetrahydrate (28 mg; 50 m*M*); 0.5 *M* LiOH (0.1 ml) and *tert*.-butanol (1.4 ml) are added and then, after cooling in ice, $LiBH_4$ (90 mg) is added gradually. The mixture is incubated at 45°C for 5 h and then cooled to 10°C and diluted with cold water (10 ml) before gradual addition of cold acetone (2 ml) and acetic acid (1 ml). The solution is concentrated to ca. 5 ml, left overnight, then repeatedly evaporated with methanol-acetic acid (5:2). The residue is fractionated on Sephadex® G-15 in 0.1 *M* acetic acid, and carbohydrate-containing fractions are combined and left overnight to achieve complete hydrolysis of the glycosylamine to GlcNAc. The solution is then concentrated to ca. 2 ml, acetic acid (0.4 ml) is added, and the mixture is applied to a short column of cation-exchange resin (H^+-form), from which the oligosaccharide fraction is eluted with water and glycopeptides with 0.7 *M* aqueous ammonia. The oligosaccharide fraction, after repeated evaporation with toluene, is dissolved in 0.1 *M* NaOH and $NaBH_4$ is added to reduce the end group to 2-amino-2-deoxy-D-glucitol. The reduced oligosaccharides obtained on application of this reaction to certain well-characterized glycoproteins such as fetuin and thyroglobulin have been studied by two-dimensional mapping after HPLC on reversed-phase and amino-silica columns[24] (see Section I.III, notes to Table LC 62). This new method of alkaline hydrolysis, although accompanied by intense reductive cleavage of peptide bonds with formation of amino alcohols, does not bring about *N*-deacetylation of hexosamine, so the GlcNAc residues involved in the *N*-glycosylamide linkage are preserved intact, and oligosaccharides corresponding exactly to the native asparagine-linked carbohydrate chains of glycoproteins can be isolated.

4. DEAMINATION

Nitrous acid deamination of aminodeoxyhexose residues, yielding 2,5-anhydrohexoses or D-glucose, is an important degradative technique in studies of glycoproteins and proteoglycans,[16] but this aspect is beyond the scope of this book. In preparation of samples for chromatography the reaction is used mainly to convert aminodeoxyhexoses, which have long retention times on GC as their acetylated alditols or aldononitriles, to the more rapidly eluting 2,5-anhydrohexose or D-glucose derivatives (see Tables GC 1, GC 2, and GC 4). Various procedures have been recommended by different authors. Anastassiades et al.[25] added a freshly prepared solution of $NaNO_2$ (5.5 *M*) to the acid hydrolysate of a glycoprotein, and the mixture, in a sealed tube, was left at room temperature for 1 h, with occasional vortexing. Cation-exchange resin (H^+-form) was then added to the mixture to convert excess $NaNO_2$ to HNO_2 and, after shaking for 30 min, the mixture was passed through a cation-exchange resin (H^+-form) and then an anion-exchange resin (HCO_3^--form), both in short columns. The eluate was evaporated and the residue derivatized to alditol acetates. Turner and Cherniak[26] dissolved the samples in 2 *M* TFA at 0°C and added an equimolar amount of $NaNO_2$. After 1 h, residual nitrous acid was removed by evaporation under reduced pressure. This procedure, however, did not remove inorganic salts, which necessitated the use of larger quantities of pyridine and hydroxylamine in subsequent derivatization to aldononitrile acetates. In the method recommended by Henry et al.,[27] an aqueous solution (0.1 ml) containing up to 2 mg of amino sugar was added to solid $NaNO_2$ (100 mg) and, with the solution in an ice bath, 9 *M* H_2SO_4 (0.1 ml) was added. The reaction mixture was allowed to warm to room temperature after 5 min, and after a further 25 min it was made alkaline with 15 *M* ammonia solution (0.2 ml) and then reduced and acetylated for GC. The ammonium sulfate did not interfere in the method used for conversion to alditol acetates[10] (see Section III.II, 9). Upon capillary GC of these derivatives (see Table GC 2), the products of deamination (2.5-anhydromannose and -talose) from GlcN and GalN are well resolved from each other and from the alditol acetates of the neutral sugars.

5. ENZYMATIC DEGRADATION

Enzymatic degradation of polysaccharides to oligosaccharides amenable to chromatographic analysis is useful in cases where the polysaccharide is not easily cleaved by acid hydrolysis. This is due to low solubility, as in the case of cellulose, or the presence of a high proportion of acid-resistant hexuronic acid and/or aminodeoxyhexose residues, as in pectins, alginates, and glycosaminoglycuronans. The use of cellulases to prepare cellodextrins is exemplified by the work of Mikeš and co-workers,[28] who used borate anion-exchange chromatography on derivatives of Spheron® (see Table LC 44) to examine the products of degradation of cellulose by cellulases from a liquid cultivation medium of *Trichoderma viride*. Incubation of cellulose and of cellopentaose was performed in 0.05 *M* citrate buffer, pH 4.8, at 40°C for 90 and 5 min, respectively, and samples were removed at intervals for HPLC analysis. Voragen et al.[29] used HPLC on Pb^{2+}-form cation exchange resin (Aminex® HPX-87P) to monitor degradation of cellulose by various endoglucanases and an exoglucanase, of xylan by certain endoglucanases, of branched and linear L-arabinans by α-L-arabinofuranoside arabinohydrolase, and an endoarabinase (1,5-α-L-arabinan arabinohydrolase), respectively, of a (1 → 4)-linked β-D-galactan by an endogalactanase (1,4-β-D-galactan galactohydrolase), and of (1 → 4)-β-D-galactotetraose by the same enzyme and a β-galactosidase (β-D-galactopyranoside galactohydrolase). In all cases the substrates were dissolved or suspended in 0.05 *M* sodium acetate buffer, pH 5.0, and incubated with the enzyme at 30°C for times ranging from 45 min to 24 h. The reactions were terminated by centrifugation of residual substrate and inactivation of the enzymes in the supernatant by immersion of the vessel in boiling water for 5 min.

The oligogalacturonic acids obtained by enzymatic digestion of pectin have been examined a great deal by LC and TLC (for examples see Tables LC 29, LC 38, LC 64, and TLC 5). Doner et al.[30] have described the preparation of these oligomers from α-D-polygalacturonic acid, derived from citrus pectin, by digestion with a purified endo-polygalacturonase isolated from a crude pectinase (from *Aspergillus niger*). The pure enzyme (300 U) was added to a solution of the polygalacturonic acid (21 g) in water (2.0 l), the pH of which had been adjusted to 4.4 with NaOH and to which NaCl (11 g, 0.1 *M*) had been added. Incubation at 30°C with shaking proceeded for 4 h, after which the pH was adjusted to 7.0 with NaOH and barium acetate (14 g) was stirred into the solution, which was then cooled to 10°C. The precipitate formed was removed by vacuum filtration through Celite®, (Johns-Manville) and the oligogalacturonic acid mixture in the filtrate was precipitated with acetone. This precipitate was collected by centrifugation and redissolved in water; the solution was then treated with a strong-acid cation-exchange resin (H^+-form) to convert barium salts of the oligogalacturonic acids to the free acid form. After the resin had been removed by filtration and washed with water, the filtrate and washings were combined and freeze-dried to yield the oligogalacturonic acid mixture (16.6 g, 79% yield). This was fractionated on a large column packed with macroporous anion-exchange resin ($HCOO^-$-form) with a step gradient (0.33 → 0.75 *M*) of sodium formate, pH 4.7, to give gram quantities of pure oligogalacturonic acids, DP 2 to 7. The yield obtained by use of the purified enzyme was nearly double that given by the crude pectinase. The purity of the isolated oligomers was checked by UV spectroscopy (thus confirming the absence of unsaturation, which would have resulted from lyase activity; see below) and by fast-atom bombardment mass spectrometry (FAB-MS).

A homologous series of oligogalacturonic acids including members of DP up to 17 was obtained by Jin and West[31] by incubation of polygalacturonic acid with endo-polygalacturonase (from *Rhizopus stolonifer*) at 30°C for only 20 min. In this case, the polygalacturonic acid (2 g) was dissolved in 10 m*M* sodium acetate, pH 4.7 (400 ml), and 17.6 U of the enzyme were added. The reaction was terminated by heat deactivation of the enzyme, after which the solution pH was adjusted to 7.0. The solution was diluted to a conductivity slightly less than that of the starting buffer (100 m*M* KCL, 10 m*M* imidazole, pH 7.0) and was loaded directly onto a column of DEAE-Sephadex A-25 and fractionated by use of a KCl gradient (100 → 300 m*M* KCl) in the imidazole buffer (see Table LC 64). These authors also used FAB-MS to check the homogeneity and structure of the isolated oligomers.

The action of endo-polygalacturonic acid lyase produces oligogalacturonic acids with 4,5-unsaturated residues at their nonreducing termini. Davis et al.[32] prepared a series of such oligomers (DP 3 to 13) by incubation of sodium polypectate with this enzyme in 5 m*M* Tris-HCl, 1 m*M* $CaCl_2$ buffer, pH 8.5, at 30°C for 2 h. The incubation was performed in five plastic bottles each containing 150 ml of the solution (1 mg/ml) of sodium polypectate in the buffer: the reaction in each was monitored at intervals by determination of the UV absorbance (235 nm) of aliquots. The reaction was terminated (after the calculated increase in absorbance had been reached) by adjusting the pH to between 4.8 and 5.2 with 1 *M* HCl, after which the contents of the five bottles were combined and heated at 65 to 80°C for 45 min to deactivate the enzyme. This digest was fractionated by anion-exchange chromatography on QAE-Sephadex® (see Table LC 64).

Enzymes capable of degrading alginate are almost exclusively lyases. The depolymerization of (1 → 4)-β-D-mannuronan derived from alginate has been followed by ion-pair reversed-phase chromatography[33] (see review in Section I.III) of samples removed during the course of digestion of the polysaccharide, incubated at room temperature in 0.1 *M* sodium phosphate, pH 7.0, with an extracellular enzyme isolated from actively growing tissues of *Sargassum fluitans*. This enzyme acted as an endolytic alginate lyase, specifically cleaving the β-D-mannuronan, but not the α-L-guluronan portions of alginate. The unsaturated oligosaccharides (DP 2 to 5) obtained after digestion for 12 h were later isolated by preparative SEC on Bio-Gel® P-2 eluted with 0.1 *M* NH_4HCO_3.

Enzymes having lyase activity are very important in analysis of glycosaminoglycuronans, which form part of the proteoglycans with vital biological functions. Chondroitinase AC digests chondroitin 4- and 6-sulfate, but not dermatan sulfate, whereas all three are digested by chondroitinase ABC. The unsaturated disaccharides thus produced are analyzed by HPLC (see Tables LC 36 and 39) as a means of quantification of the respective glycosaminoglycuronans. Typically, the proteoglycan (1 to 1000 μg) is digested at 37°C for 16 h in Tris buffer (80 μl, pH 6) with the enzyme (0.05 U in 20 μl), after which the reaction is terminated by addition of 1 *M* trichloroacetic acid (20 μl). The reaction vial is centrifuged, and a sample (10 μl) of the supernatant is taken for HPLC analysis.[34] Hyaluronic acid can be quantified by separation and determination of the unsaturated tetra- and hexasaccharide produced on digestion with *Streptomyces* hyaluronidase, which is selective for this polymer. In the procedure used by Gherezghiher et al.[35] the sample, containing 50 to 100 μg of hyaluronic acid in 0.02 *M* acetate buffer, pH 6.0 (100 μl), was incubated with the enzyme (10 to 30 μl) at 60°C for 2 h, after which absolute ethanol (four volumes) was added and the mixture was briefly vortexed, then left overnight at 4°C. After centrifugation, the supernatant was dried under a stream of nitrogen and the residue redissolved in the mobile phase for HPLC analysis (see Table LC 39). Recently, the HPLC profiles of the products of depolymerization of heparin samples by digestion with heparin lyase[36] have been suggested as a means of characterizing this polydisperse glycosaminoglycuronan. In the recommended procedure, heparin (50 μl, 20 g/ml) is added to 5 m*M* sodium phosphate (pH 7.0) containing 0.2 *M* NaCl (425 μl), the enzyme (25 μl 0.015 U) is then introduced, and the mixture is incubated at 30°C for 8 h. The reaction is terminated by heating at 100°C for 1 min, and aliquots are analyzed by HPLC in ion-exchange mode (Table LC 39). Oligosaccharides of DP 2 to 6, each with a 4,5-unsaturated uronic acid residue at the nonreducing terminus, are resolved; these differ widely in degree of sulfation, and their proportions in the mixture vary with the source of the heparin.

Digestion of glycosaminoglycuronans with hyaluronidases, which are endoglycanhydrolases, is less specific and can be applied to the chondroitin sulfates as well as hyaluronic acid to yield the disaccharide repeating unit and oligomers, tetra- to decasaccharides. These have been isolated by successive SEC, preparative PC, and anion-exchange HPLC.[37] Nebinger et al.[38] have prepared different series of oligosaccharides from hyaluronic acid: oligosaccharides containing even numbers of sugar residues, one series with GlcA at the nonreducing end (from digestion with bovine testicular hyaluronidase) and the other with terminal nonreducing GlcNAc (from the action of leech hyaluronidase), and oligosaccharides with odd numbers of sugar residues, obtained by removal of the nonreducing GlcA end groups with β-glucuronidase (from *Patella barbara*) and the nonreducing GlcNAc residues with β-*N*-acetylglucosaminidase A (from bovine spleen). These oligosaccharides have been separated on a preparative scale by ion-exchange chromatography, which is readily monitored by HPLC on an amino-silica column acting as a weak anion exchanger (see Table LC 39), or by TLC (see Table TLC 7).

Among the methods used by Lohmander[14] to liberate the monosaccharide constituents of proteoglycans (see Table IIIA for acid hydrolysis conditions) was an enzymatic digestion procedure directed at cleavage of linkages between *N*-acetylneuraminic acid and β-D-galactose, the latter linked to β-GlcNAc. The sample was incubated first with *Clostridium perfringens* α-neuraminidase in 50 m*M* sodium citrate buffer (pH 5.0) for 24 h, after which carrier *N*-acetylneuraminic acid was added and the mixture heated at 100°C for 3 min. The product was then dissolved in 50 m*M* ammonium acetate (pH 4.0) and incubated with jack bean β-D-galactosidase for 24 h; the reaction was stopped with carrier galactose and heating as described. Jack bean 2-acetamido-2-deoxy-β-D-glucosidase was added to this digest, and incubation was continued for a further 24 h, after which GlcNAc was added and the mixture heated and then frozen. Before application to the HPLC column (Aminex® HPX-87H), samples were freeze-dried and redissolved in a small volume of 3 m*M* H_2SO_4, the eluent. In this way, oligosaccharides resistant to degradation by these enzymes were clearly distinguished from the liberated GlcNAc, Gal, and *N*-acetylneuraminic acid. In particular, Galβ1-3GalNAc was resistant to the action of jack bean β-D-galactosidase.

Honda and co-workers[12,39] have developed a method for determination of neuraminic acids by HPLC or GC-MS analysis of the products of release of the neuraminic acid by *Clostridium perfringens* neuraminidase, followed by degradation by *Escherichia coli N*-acetylneuraminate pyruvate-lyase. The glycoprotein or proteoglycan sample (ca. 200 μg) is dissolved in 0.06 *M* phosphate buffer (pH 7.0, 800 μl) containing neuraminidase (0.5 U) and *N*-acetylneuraminate pyruvate-lyase (0.3 U), and the mixture is incubated at 37°C for 1 h. After heating at 100°C for 1 min to deactivate the enzymes, the solution is deionized using mixed resins, evaporated, and dissolved in a small volume of the HPLC eluent (92% acetonitrile; see Table LC 35) or derivatized for GC as the trimethylsilylated diethyl dithioacetals[39] (see Table GC 19). The products are *N*-acylmannosamines, which are readily amenable to analysis by either technique.

Recently, sequential digestion with pepsin (at pH 2.0, 37°C, for 2 days) and *N*-oligosaccharide glycopeptidase (in 0.1 *M* citrate-phosphate buffer. pH 5.0, 37°C, 15 h)[40] has afforded a useful alternative to hydrazinolysis (see 6) as a method of releasing *N*-linked oligosaccharide chains from glycoproteins. This technique was used by Tomiya et al.[41] in preparing the oligosaccharide standards used to provide a database for the two-dimensional HPLC mapping technique for identification of oligosaccharides (see Table LC 62). Smaller oligosaccharides were obtained from these by sequential digestion with various exoglycosidases (β-D-galactosidase, α-L-fucosidase, and 2-acetamido-2-deoxy β-D-glucosidase).

6. HYDRAZINOLYSIS

Hydrazinolysis has long been accepted as the standard method for removal of *N*-linked oligosaccharide chains from glycoproteins, since the reaction specifically cleaves the linkage between GlcNAc and asparagine. It should be followed immediately by *N*-acetylation of the glycosylamine formed at the point of cleavage to prevent further degradation of the released oligosaccharides, and for the same reason the oligosaccharides are usually then reduced with $NaBH_4$. Labeling with $NaB[^3H]_4$ is advantageous.

The procedure generally used in hydrazinolysis is that recommended by Takasaki et al.[42] Anhydrous hydrazine is essential. This is prepared by mixing hydrazine hydrate (80%) with toluene and CaO (1:10:10 w/w/w) and, after allowing the mixture to stand overnight at room temperature, refluxing it for 3 h under a condenser fitted with a $CaCl_2$ tube. Distillation under anhydrous conditions follows, and the fraction distilling at 93 to 94°C is collected. Anhydrous hydrazine separates from toluene as the denser layer; this is removed and stored in an air-tight container under anhydrous conditions at 4°C in the dark. In hydrazinolysis, a dry sample of glycoprotein or glycopeptide, containing 0.2 to 1 mg of asparagine-linked sugar chain, is suspended in 0.5 to 1 ml of freshly distilled anhydrous hydrazine and heated at 100°C in a sealed tube under N_2 or Ar, usually for 8 to 12 h. Most *N*-linked sugar chains are released completely as oligosaccharides within 10 h, but the rate of release differs in different samples; sometimes 20 h is recommended.[43] After hydrazinolysis, the reaction mixture is evaporated to dryness under reduced pressure over concentrated H_2SO_4 at room temperature, and the residue is freed from hydrazine by repeated evaporation with toluene. It is then dissolved in saturated $NaHCO_3$ solution, and acetic anhydride is added to *N*-acetylate all free primary amino groups. If the sample is an intact glycoprotein, repeated *N*-acetylation is required because more amino groups are produced, from the polypeptide moiety as well as the sugar moiety. The reaction mixture is passed through a column of strong-acid cation-exchange resin (H^+-form), the column is washed with distilled water (five bed volumes), and eluate and washings are combined and evaporated to dryness. An appropriate amount of internal standard (maltotriose or sialyllactose) should be added to the residue at this stage if a quantitative estimate of the amount of oligosaccharides released is required. The oligosaccharides are separated from degradation products of the polypeptide moiety by preparative PC (solvent system 1-butanol-ethanol-water, 4:1:1) for 2 d. The degradation products are more mobile than the oligosaccharides, which remain at or near the origin. This area (to 5 cm from the origin) is excised, and the oligosaccharides are recovered by elution with water.

For reduction, the oligosaccharide fraction is treated with $NaB[^3H]_4$ (3 μmol = 1 mCi) in 0.05 *M* NaOH (200 μl) at 30°C for 4 h, after which $NaBH_4$ (7 mg) in 0.05 *M* NaOH (70 μl) is added and reduction continued at 30°C for a further 2 h. The reaction is terminated by acidification (1 *M* acetic acid), and the mixture is passed through a small column (2-ml bed volume) of strong-acid cation-exchange resin (H^+-form), which is washed with distilled water (five bed volumes). Eluate and washings are combined and evaporated to dryness under reduced pressure, and the residue is freed from borate by repeated evaporation with methanol. Upon PC (mobile phase: ethyl acetate-pyridine-acetic acid-water, 5:5:1:3) for 2 d, the radioactively labeled reduced oligosaccharides released from glycoproteins and glycopeptides remain near the origin, well separated from the labeled internal standard. The appropriate portions of the paper are excised and eluted with water, and the radioactivity in each eluate is determined by scintillation counting, from which the amount of oligosaccharide relative to that of internal standard can be calculated. The reduced oligosaccharide fraction thus obtained is then ready for chromatographic analysis.

If *O*-linked sugar chains are present, there is some degradation to monosaccharides and their alkaline degradation products under the conditions employed in hydrazinolysis. This has been ascribed to the "peeling" reaction, since many of the intersugar linkages in these mucin-type chains are 1 → 3 and are very sensitive to this reaction, which is probably made possible by the presence of trace amounts of water in sample or hydrazine.[42]

7. METHANOLYSIS

Methanolysis, as described in Volume I, remains a useful procedure for degradation of polysaccharides and glycoproteins, especially those containing sugar residues that are appreciably decomposed under acid hydrolysis conditions, or uronic acid residues, which can be analyzed directly as the methyl glycoside methyl esters in the products of methanolysis. Methanolysates are analyzed by GC after trifluoroacetylation (see Table GC 13) or trimethylsilylation (Table GC 20). The production of multiple peaks corresponding to different anomeric and ring forms of the sugars is a complication, but the GC pattern for each sugar derivative is diagnostic and aids identification. This also applies to methanolysates from methylated polysaccharides, which can be analyzed by GC without further derivatization (see Volume I, Table GC 11) or after acetylation of free hydroxyl groups (this volume, Table GC 10). HPLC on C_{18}-silica packings permits direct analysis of methanolysates (see Table LC 7), but benzoylation is advantageous (Table LC 15).

The previously described methanolysis procedure, using methanolic HCl prepared by bubbling HCl gas into dry methanol, is still widely used. It has proved valuable in analysis of a diversity of polysaccharides, including bacterial lipopolysaccharides, which contain some less common sugars such as dideoxyhexoses, heptoses, and 3-deoxy-D-*manno*-2-octulosonic acid (KDO). These were analyzed by GC as their trifluoroacetylated methyl glycosides,[44] a method that has also been applied to whole cells of bacteria[45] (see Table GC 13). In this case, no attempt was made to neutralize the HCl prior to rotary evaporation and, therefore, some losses of sugars may have occurred. The authors[44] reported degradation of the dideoxyhexoses under the methanolysis conditions used (2 *M* methanolic HCl, 85°C, 18 h), which could be obviated by use of a lower temperature (37°C) for analysis of these sugars. KDO also underwent some degradation at 85°C, but it was possible to correct for this by determination of a molar response factor. This procedure was used in quantitation of the other sugar constituents also and is generally employed.

Glycosaminoglycuronans pose special problems in methanolysis, owing to the presence of both hexuronic acids and acetamidodeoxyhexoses as major constituents. The latter are de-*N*-acetylated under methanolysis conditions, which inhibits depolymerization owing to the high stability of a glycuronidic linkage to aminodeoxyhexose. Hjerpe et al.,[46] who were primarily interested in the uronic acid constituents of chondroitin and dermatan sulfates, overcame the problem by deamination before methanolysis (1 *M* methanolic HCl, 100°C, 30 to 50 h). The methanolysates were purified by passage through mixed resins in methanol prior to HPLC on C_{18}-silica (see Table LC 7) or, after hydrolysis of the methyl esters (1% aqueous ammonia, 100°C, 3 h), on a weak-base anion-exchange column (see Table LC 38). Yields of uronic acids were poor under these conditions. Dierckxsens et al.[47] used a high concentration of methanolic HCl (5 *M*) in methanolysis of glycosaminoglycuronans, but yields of the constituent monosaccharides were still low unless the aminodeoxyhexoses were re-*N*-acetylated after one treatment (at 100°C for 24 h) and methanolysis was repeated (same conditions). In the procedure recommended by these authors the acid solution was neutralized (Ag_2CO_3) after 24 h, and re-*N*-acetylation was carried out by addition of acetic anhydride (5 μl, to 0.1 ml methanolic HCl containing up to 60 μg of glycosaminoglycuronan) and incubation of the mixture for 6 h at room temperature in the dark. The precipitate was thoroughly triturated and then removed by centrifugation. The supernatant and washings (in dry methanol) were pooled and evaporated under a stream of air. The residue was dried overnight over P_2O_5 *in vacuo* before repetition of methanolysis and re-*N*-acetylation. Under these conditions, yields of uronic acid and acetamidodeoxyhexose constituents (determined by capillary GLC of the trimethylsilylated methyl glycosides) were above 75% for all glycosaminoglycuronans tested.

For more tractable polysaccharides, the conditions of methanolysis originally recommended suffice. Thier[48] has suggested that reaction time with 2 *M* methanolic HCl at 100°C could be shortened to 4 h; a longer time was of little advantage since the methanolic HCl loses its effectiveness owing to conversion to methyl chloride and water on prolonged heating. This has not been generally accepted, although Jentoft[49] has reported similar results on methanolysis of glycoproteins in 1 *M* methanolic HCl at either 80°C for 4 h or 65°C for 16 h. This author analyzed the products of methanolysis by reversed-phase HPLC of the derived benzoates (see Table LC 15), which obviated the need for the determination of molar response factors, as in the GC methods, since the UV absorbance of each sugar derivative was strictly a function of the number of benzoyl groups per molecule. The pyrolytic losses occurring at the high temperatures used in GC methods were not a source of error in HPLC at ambient temperature. However, nonquantitative aspects of the methanolysis procedure obviously remain a problem, even with a quantitative analytical method. Jentoft, like Dierckxsens et al.,[47] continued methanolysis (a mild treatment with 0.1 *M* methanolic HCl at 65°C for 30 min) after re-*N*-acetylation (with acetic anhydride/pyridine at room temperature) of the aminodeoxyhexose components of the glycoproteins, which resulted in better yields of these sugars and eliminated *O*-acetyl groups introduced by the *N*-acetylation procedure. The other main source of error is the loss of glycosides on evaporation to remove methanolic HCl, for which reason neutralization of the HCl with Ag_2CO_3 prior to evaporation has become standard practice. Jentoft found that residual silver salts interfered with benzoylation and instead favored the modification introduced by Chaplin,[50] the use of methanolic HCl containing methyl acetate and addition of *tert*-butanol after methanolysis, before evaporation in a stream of dry, oxygen-free N_2 at room temperature or under vacuum. The methyl acetate acts as a scavenger for water, with formation of methanol and acetic acid, and *tert*-butanol forms a low-boiling azeotrope with water[49] and also coevaporates with HCl.[4] Coconcentration of trace amounts of water with HCl during evaporation is believed to be the reason for hydrolysis of methyl glycosides to free sugars, hence these modifications directed at removal of water (arising from the decomposition of methanolic HCl, mentioned above).

The need for addition of methyl acetate is obviated by preparing the methanolic HCl not by addition of HCl gas to methanol, but by reaction of dry methanol with the stoichiometric amount of acetyl chloride required to produce HCl in the desired molarity; methyl acetate is produced simultaneously in this reaction. This method, which is simple and eliminates the need for HCl gas, is now widely adopted. It is, however, still necessary to avoid evaporation of HCl together with water. Addition of *tert.*-butanol remains a favored procedure,[51] while some authors[52] advocate use of an anion-exchange resin (HCO_3^--form) to neutralize the HCl. A weak-base anion-exchange resin in the OH^--form was used by Roberts et al.[53] to neutralize the acid in a modified methanolysis procedure in which HCl was replaced by 72% H_2SO_4 to facilitate methanolysis of insoluble polysaccharides, such as cellulose, which require presolubilization in this acid. The traditional use of Ba^{2+} to precipitate sulfate is not possible in this case as some methyl hydrogensulfate, which forms a soluble salt with Ba^{2+}, is produced during methanolysis. The drying agent, Drierite® ($CaSO_4$), was added to the methanolic H_2SO_4 to remove water. High yields of methanolyzed products were obtained, with only slight destruction of sugars.

8. METHYLATION ANALYSIS

Since the compilation of Volume I, several modifications to the Hakomori methylation procedure have been suggested, and some novel methylating agents have been applied to carbohydrates. Many important aspects of the technique of methylation analysis have been discussed in reviews,[54-56] but further details of the refinements introduced into this valuable technique are presented here.

The first consideration in methylating a polysaccharide or glycoconjugate is its solubility in the solvent to be used, which is dimethyl sulfoxide (DMSO) in the Hakomori procedure and its modifications. Glycolipids are soluble in this medium and can be methylated directly, but glycoproteins as such have low solubility,[54] and it is necessary to isolate the carbohydrate moiety as a glycopeptide after proteolytic digestion or as a reduced oligosaccharide after treatment with alkaline borohydride as described above (see 3). Polysaccharides containing uronic acid residues must be deionized, and for those of very high uronic acid content, such as pectins, the classical Haworth methylation procedure, in which the polysaccharide (preferably pretreated with $NaBH_4$ to protect reducing end groups) is dissolved in aqueous 30% NaOH and methylated with dimethyl sulfate, may be preferable. This technique is also more suitable for methylation on a preparative scale.[56]

Various methods have been suggested for solubilization of intractable polysaccharides, such as cellulose and other cell-wall polysaccharides, in which a high degree of hydrogen bonding makes disruption of aggregates difficult. Sonication may be helpful, but it can cause some cleavage of glycosidic bonds within the molecule and, therefore, should be used with caution. To avoid the degradation that usually accompanies extraction processes, Lomax et al.[57] developed a procedure for direct methylation of intact plant cell walls which depended upon reduction of particle size by dry milling in liquid nitrogen. The average particle diameter was thus reduced to 10 to 40 μm, and such particles have been found to swell and disperse in DMSO to an extent permitting their methylation by the Hakomori method. An internal standard, methyl β-D-allopyranoside, was added so that recovery of polysaccharides as methylated products could be quantified. There was a considerable increase in recovery for the milled samples compared to those methylated without this pretreatment; the improvement was particularly notable for highly lignified samples. The results of methylation analysis of untreated and milled samples were similar with respect to the proportions of the various methylated sugars, indicating that the distribution of linkages in the cell-wall polysaccharides was not affected by the milling process.

Joseleau et al.[58] have demonstrated the efficacy of 4-methylmorpholine *N*-oxide monohydrate (MMNO) as a solvent for entire cell walls, including cellulose, which was recovered after dialysis against water in yields of 85 to 98%. This indicates that the solvent action does not involve degradation from the reducing end by a "peeling" mechanism, which would have caused much greater losses from the nondialyzable fraction. Upon hydrolysis and GC analysis of the solubilized samples in the presence of an internal standard, yields of D-glucose were 100%, which showed that the solubilizing agent had caused no significant modification by oxidation or degradation of the constituent D-glucose residues of cellulose. There was, however, a decrease in DP of ca. 50%. The samples were dissolved in MMNO at 120°C and then diluted with DMSO to give a final concentration of 0.7% with a DMSO:MMNO ratio of ca. 4:1. This proved to be an effective solvent for Hakomori methylation of both cellulose and hemicellulose, but pectin did not dissolve completely (see comment above). Tetramethylurea, an antagonist to hydrogen bonding, also promotes the dissolution of such polysaccharides as cellulose and amylose in DMSO and facilitates methylation, for which a solvent containing DMSO and tetramethylurea in equal proportions (v/v) has been recommended.[59]

A major improvement in the original Hakomori procedure has been substitution of potassium for sodium as the cation associated with the methysulfinylmethanide (dimsyl) carbanion responsible for alkoxide formation in carbohydrates in DMSO solutions. Dimsyl potassium[60] is prepared by reaction of potassium hydride and DMSO at room temperature (cf. preparation of sodium dimsyl, which requires heating to 50°C); this is complete in ca. 15 min (cf. 45 min for preparation of sodium dimsyl). Another advantage is that reaction products are free from contamination by salt, as KI has a very limited solubility in chloroform (into which the methylated products are extracted). The use of dimsyl potassium in Hakomori methylation has now been generally adopted. However, recently some authors[61-63] have preferred dimsyl lithium, prepared by reaction of butyllithium with DMSO at 4°C. Fewer contaminant peaks are obtained on capillary GC of derivatives of the partially methylated sugars obtained from carbohydrates methylated using this reagent,[61,62] presumably because butyllithium is available in higher purity than are sodium or potassium hydrides. Disadvantages include the danger of handling butyllithium, which is pyrophoric, and the slower reaction of the lithium alkoxide with methyl iodide; this stage requires 40 to 60 min[61-63] as opposed to 10 min for potassium alkoxide.[64] Another alternative that has been explored in the search for a cleaner reagent is the use of potassium *tert.*-butoxide[54,65] instead of potassium hydride in the preparation of dimsyl potassium. Preparation of the reagent is simple and rapid, commercial potassium *tert.*-butoxide readily dissolving in DMSO without the necessity for special precautions other than exclusion of moisture. However, although it has been shown to be as effective as dimsyl sodium, prepared from the hydride, in methylation of brain gangliosides,[65] the reagent is an equilibrium mixture of dimsyl carbanion and butoxide that is not suitable for methylation of polysaccharides.[64]

If ester groups, such as acetate and phosphate, are to survive methylation, special techniques using solvents of basicity lower than that of DMSO are required. For this purpose, the alternative proposed by Prehm[66] of using methyl trifluoromethanesulfonate (methyl triflate) in the relatively nonpolar solvent trimethyl phosphate, with 2,6-di-(*tert.*-butyl)pyridine as a proton scavenger, which has been applied successfully to bacterial lipopolysaccharides (insoluble in dipolar aprotic solvents) after conversion to their triethylammonium salts, may be useful.

In another modification, aimed at simplifying the methylation procedure by elimination of the necessity for anhydrous conditions, the organic base has been replaced by NaOH or KOH. Fügedi and Nánási[67] demonstrated the applicability to carbohydrates of the KOH-DMSO-CH_3I system (previously applied to methylation of phenols, amides, acids, and some simple alcohols) using as a model compound benzyl 4,6-*O*-benzylidene-β-D-glucopyranoside. In this simple procedure, the compound (5.7 g) was stirred in DMSO (20 ml) with powdered KOH (6.73 g; 6 *M*) and CH_3I (3.75 ml) was added dropwise with external cooling using water. After 30 min, the product was extracted into dichloromethane, and the organic layer was washed with water, dried, and evaporated. The residue was recrystallized from ethyl acetate-light petroleum to give, in 96% yield, a product having the melting point and specific rotation of benzyl 2,3-di-*O*-methyl-4,6-*O*-benzylidene-β-D-glucopyranoside. The method was also applied to benzylation with benzyl bromide. A subsequent systematic study by Ciucanu and Kerek[68] of methylation of simple sugars (mono- to trisaccharides) demonstrated that the use of NaOH or KOH as base in methylation in DMSO gave 98 to 99% yields of methylated sugars, much higher than those given by potassium *tert.*-butoxide (40 to 50%) and comparable with those given by dimsyl sodium prepared from NaH in the usual way, but with fewer by-products. The bases were 2 *M* in all cases and methylation was performed at 25°C, the time required for the reaction with CH_3I varying from 1 h with potassium *tert.*-butoxide as base through 15 min with dimsyl sodium to 6 to 7 min with NaOH and KOH. After a study of the effects on reaction time of the proportions of NaOH, solvent, CH_3I, and sugar, it was concluded that the optimal proportions were 3 mol of NaOH and 3 to 5 mol of CH_3I per 1 mol of replaceable hydrogen in the sugar, the optimal concentration of base being 2 to 3 *M*. Methylation was slightly faster with NaOH than with KOH as base, possibly because of the lower solubility of KOH in DMSO.

Methylation using NaOH as base has been adapted successfully by Isogai et al.[69] to permit its use in methylation of cellulose and other cell-wall polysaccharides, which are insoluble in DMSO alone, but are solubilized in the presence of sulfur dioxide and diethylamine. Quantitative methylation of cellulose and neutral hemicelluloses has been reported, but there is some decomposition of uronic acid residues (and possibly β-elimination of exterior, contiguous residues) as is found also when dimsyl sodium is used in the unmodified Hakomori method. In the procedure recommended by Isogai et al., SO_2-DMSO solution (ca. 0.3 g/ml) is prepared by bubbling SO_2 (15 g, dried over $CaCl_2$) into DMSO (50 ml); the solution remains stable at room temperature in the dark for 2 months. Dry polysaccharide (1 g) is dispersed in DMSO (88 ml); the suspension is heated at 60°C for 30 min and then cooled to room temperature, and the SO_2-DMSO solution (ca. 4 ml, containing 1.2 g of SO_2) is added, followed by diethylamine (2 ml). Cellulose powder gives a clear solution within 10 min with stirring, but complete dissolution of wood celluloses requires ca. 3 h. Wood holocelluloses and cell-wall polysaccharides do not dissolve completely, but in this solvent system swell sufficiently to permit methylation after stirring at room temperature for ca 1 h. Methylation is performed by adding freshly powdered NaOH (12 g) to a solution or suspension containing 1g of polysaccharide, stirring under N_2 at room temperature for 1 h, and then adding CH_3I (ca. 12 ml) dropwise at room temperature. The mixture is stirred for 1 h, and then the temperature is raised gradually (stirring at 40 and 50°C for 30 min at each temperature) to 60°C, at which temperature it is stirred for 1 h. If solidification occurs during stirring, the minimal amount of DMSO required for resuspension is added. The methylated products are isolated by dialysis and freeze-drying.

In a modification of this procedure, applied to highly substituted D-glucans and D-xylans, Mabusela et al.[70] increased the proportion of CH_3I to polysaccharide to four times that used by Isogai et al.,[69] the CH_3I being added to the reaction mixture with the latter frozen in an ice bath. Stirring was continued for 1 h after the mixture had warmed to room temperature, but the increase in temperature to 60°C was less gradual than that in the original method, taking place continuously for 30 min. The methylated product was isolated by chloroform extraction of the freeze-dried nondialyzable fraction.

Harris et al.[64] used two rapid preliminary methylations with dimsyl potassium to solubilize cellulose and cell-wall polysaccharides in DMSO for full-scale methylation by the same reagents. This gave better yields than did the use of MMNO as a solvent,[58] but some degradation is inevitable with three methylations by this method, and repeated methylation cannot be applied to polysaccharides containing uronic acid, which would be degraded by β-elimination.[55] The procedure of Isogai et al.[69] is the method of choice for methylation of such polysaccharides, since the dry-milling technique of Lomax et al.[57] causes depollymerization, as does the MMNO method.[58] IR spectra of polysaccharides methylated in the presence of tetramethylurea[59] indicate physical or chemical linkage of the tetramethylurea to the methylated polysaccharides.

After methylation, the product is normally extracted into chloroform or dichloromethane, but it is advisable to test the aqueous washings for carbohydrate since undermethylated material may partition between the aqueous and organic phases.[56] IR spectroscopy is often used to assess the extent of methylation, but traces of water interfere and the method is relatively insensitive. Measurement of the incorporation of ^{14}C from $[^{14}C]H_3I$ is a more direct method, recommended by Harris et al.,[64] or the degree of methylation may be assessed by PC or TLC of the hydrolysate from a small sample. The presence of unmethylated sugars or a preponderance of those of low degree of methylation indicates that further methylation is necessary.[2,56] For this purpose, successive treatments with CH_3I and dry Ag_2O (in the presence of $CaSO_4$) under reflux (classical Purdie methylation) should be applied[56] until no further yellow deposit of AgI is seen on the surface of the black Ag_2O.

When the final methylated product is extracted into chloroform or dichloromethane, Harris et al.[64] recommend the addition of 2,2-dimethoxypropane and a trace (20 μl) of acetic acid to the organic phase prior to evaporation, since dimethoxypropane reacts quantitatively with water in the presence of acid to form methanol and acetone and thus acts as a scavenger for residual water. Before analysis, methylated polysaccharides or oligosaccharides should be purified by SEC on lipophilic gels, as described in Volume I, or by rapid HPLC using Sep-Pak® C_{18} cartridges (Waters).[71,72] Contaminants such as residual DMSO are eluted from the cartridges with water, methyl glycosides and methylated oligosaccharides (to DP4) with 50% aqueous methanol, and most methylated polysaccharides with pure methanol (methylated cell-wall polysaccharides require chloroform-methanol, 1:1).[71] Recoveries range from 85 to 92%, except for methylated cellulose (70 to 80%). Samples of up to 5 mg can be purified by each cartridge, but microscale purification (samples of 1 to 50 μg) is also possible, and this technique is an important part of the protocol developed by Waeghe et al.[72] for methylation analysis on a microgram scale. In this case, the cartridge was first flushed with water and then with aqueous acetonitrile (15 to 20%) before elution of methylated oligosaccharide-alditols with pure acetonitrile and of methylated polysaccharide-alditols with this solvent followed by absolute ethanol.

Hydrolysis or methanolysis of the permethylated polysaccharide or oligosaccharide releases methylated monosaccharides or their glycosides at varying rates, some only incompletely under practical conditions. The problems posed by resistance to cleavage of linkages involving uronic acids and aminodeoxyhexoses are the same as those found in the case of the parent polysaccharides. Complete depolymerization of methylated polysaccharides containing uronic acid residues (now present as methyl esters) is possible only after reduction of the carboxylate groups. This is accomplished[72] by treatment of the methylated carbohydrate with sodium borodeuteride in 27:73 (v/v) 95% ethanol-oxolane at room temperature for 18 h in a sealed vial. The vial is heated at 70°C for 1 h and, after cooling to room temperature, excess borodeuteride is converted to borate by acidification of the reaction mixture with acetic acid. The borate is removed by coevaporation with 10% acetic acid in methanol and the sodium acetate by passage of the methylated carbohydrate, dissolved (with sonication) in 1:1 (v/v) ethanol-water, through a short column of cation-exchange resin (H^+-form). The acetic acid produced is removed by evaporation. The 6,6-dideuteriohexosyl residues resulting from carboxylate reduction of the uronic acid methyl ester residues in this manner are easily identified by GC-MS of the reduced acetylated hydrolysate (see Table IIA). The same result is achieved by the alternative procedure[73] of reducing the carboxylate ester groups with lithium aluminium deuteride in dry oxolane for 16 h at 70°C. Unreacted $LiAlD_4$ is decomposed with moist ethyl acetate, sodium tartrate is added to complex the cations, and the suspension is filtered, the solid is washed with chloroform, and the filtrate is combined with washings and evaporated to dryness. The crude product is extracted with chloroform, the suspension filtered, and the filtrate evaporated, yielding a product virtually free of salts. Realkylation of a portion of the carboxylate-reduced methylated polysaccharide or oligosaccharide with methyl iodide or ethyl iodide can aid identification of these residues by GC-MS of the derived alditol acetates.[73]

After carboxylate reduction, the methylated polysaccharide can be hydrolyzed under the conditions applicable to neutral carbohydrates (2 *M* TFA, 100°C, 6 to 8 h) or, for rapid analysis, autoclaved in 2 *M* TFA at 121°C for 1 h.[64,72] Such conditions do not fully depolymerize methylated polysaccharides containing aminodeoxyhexose residues or acetamidodeoxyhexose residues that become *N*-deacetylated during hydrolysis; therefore, acetolysis followed by acid hydrolysis is necessary.[2,54] However, this procedure is not suitable for analysis of aminodeoxyhexitol residues (from alkaline $NaBH_4$ hydrolysis of oligosaccharide chains in glycoproteins), which give by-products on acetolysis, of or neuraminic acid residues, which decompose under acid hydrolysis conditions.[54] Methanolysis is preferable for preparation of samples to be analyzed for these constituents; the conditions recommended by Rauvala et al.[54] are 0.5 *M* methanolic HCl at 80°C for 18 h for the aminodeoxyhexitols, but milder conditions (0.05 *M* methanolic HCl, 80°C, 16 h) for the neuraminic acid residues (this permits simultaneous identification of differently substituted *N*-acetyl- and *N*-glycolylneuraminic acids). Conchie et al.[74] have reported that the acetolysis-hydrolysis procedure gives low molar proportions for the triply substituted β-D-mannose residue in the core of some glycopeptides; for this purpose, formolysis prior to hydrolysis is necessary. Because of these differences in the susceptibility to hydrolysis of the diverse residues present, complete methylation analysis of oligosaccharides derived from glycoproteins demands division of the permethylated sample into several aliquots for depolymerization by the different methods required.

Section III.II

DERIVATIZATION FOR GAS CHROMATOGRAPHY

9. ALDITOL ACETATES

Modifications of the standard methods for reduction and acetylation of sugars or methylated sugars (see Volume I) have been directed at decrease in the reaction time and prevention of losses of more volatile components during derivatization. Selvendran et al.[5] decreased the time required for acetylation of alditols with acetic anhydride/pyridine to 20 min by heating at 120°C, but there was evidence from response factors that rhamnitol was not completely acetylated under these conditions. Torello et al.,[13] in a study aimed at optimizing conditions for analysis of the sugar components of gangliosides by the alditol acetate method, concluded that reduction with $NaBH_4$ in 1 *M* aqueous ammonia (2 mg/ml) for 40 min followed by acetylation with acetic anhydride at 100°C for 30 min, in the presence of the sodium acetate formed on reaction of the excess $NaBH_4$ with acetic acid, gave the best yields of the alditol acetates. After acetylation the samples were dried under a stream of N_2 at 38°C to minimize losses of voltile alditol acetates, and this was repeated after removal of the sodium acetate by partitioning between chloroform and water. This procedure permitted analysis of glycolipid sugars in the 1 to 10 μg range.

Blakeney et al.[10] have modified the standard procedure considerably with a view to faster and more quantitative derivatization. In the procedure recommended by these authors, the hydrolysate is neutralized with 15 *M* aqueous ammonia to a final ammonia concentration of 1.0 *M*, and reduction is effected with a solution prepared by dissolving $NaBH_4$ (2 g) in anhydrous DMSO (100 ml) at 100°C. This solution (1 ml) is added to the neutralized hydrolysate in 1 *M* ammonia (0.1 ml) and reduction allowed to proceed for 90 min at 40°C. Excess $NaBH_4$ is decomposed by addition of 18 *M* acetic acid (0.1 ml), and 1-methylimidazole (0.2 ml) is added, followed by acetic anhydride (2 ml). The components are mixed and, after 10 min at room temperature, water (5 ml) is added to decompose excess acetic anhydride. After cooling, dichloromethane (1 ml) is added and the mixture agitated on a vortex mixer. After the phases have separated, the lower one is removed with a Pasteur pipette and injected directly into the gas chromatograph.

The advantages of the use of DMSO as a solvent for $NaBH_4$ are the stability of the solution and the fact that this solvent does not consume acetic anhydride during acetylation as water does. The high concentration (20 mg/ml) of $NaBH_4$ and elevated temperature (40°C) increase the rate of reduction. 1-Methylimidazole is a highly efficient catalyst for acetylation of hydroxy compounds; hence, very rapid acetylation occurs at room temperature in its presence. Borate does not interfere with acetylation under these conditions; consequently, a major advantage of this procedure is that it does not demand removal of borate prior to acetylation. This obviates the necessity for repeated evaporation with methanol after reaction of excess borohydride with acetic acid, which is a major cause of losses of volatile alditol acetates. Reduction and acetylation can be carried out in the same vessel so that handling losses are avoided. This method has been applied successfully in derivatization not only of neutral sugars,[10] but also of aminodeoxyhexoses and the products of nitrous acid deamination of these sugars.[27]

For derivatization of methylated sugars, the same research group[64] has developed another procedure, also very different from the standard method. In this case, the $NaBH_4$ is dissolved in 2 *M* aqueous ammonia, as the partially methylated sugars are incompletely reduced if DMSO is used as the solvent. The hydrolysate is dried by evaporation of the TFA at 40°C under a stream of N_2, and is treated with a freshly prepared solution (1 ml) of $NaBH_4$ (0.5 *M*) in 2 *M* ammonia, and reduction is allowed to proceed for 1 h at 60°C. Acetone (0.5 ml) is added to stop the reaction, and the mixture is again evaporated at 40°C in a stream of

N_2. The residue is dissolved in 18 *M* acetic acid (0.2 ml), and acetylation is accomplished by adding ethyl acetate (1 ml) and acetic anhydride (3 ml) followed, after mixing, by perchloric acid (70%, 100 μl). The mixture is allowed to stand at room temperature for 5 min and then cooled in an ice bath before addition of water (10 ml) followed by 1-methylimidazole (0.2 ml). The mixture is left for a further 5 min, and then dichloromethane (1 ml), containing *myo*-inositol hexa-acetate (internal standard), is added with mixing. After phase separation, the lower phase is removed with a Pasteur pipette for injection into the gas chromatograph.

The reason for the different acetylation conditions in this case is that some partially methylated alditols, notably 2,4,6-tri-*O*-methyl-D-glucitol, are not acetylated quantitatively in the presence of borate by the Blakeney method[10] (see above), possibly because of particularly strong complexing of borate. For this reason, acid- rather than base-catalyzed acetylation is necessary if removal of borate by evaporation is to be avoided. Perchloric acid has been found to catalyze quantitative acetylation of all partially methylated alditols. A large excess of acetic anhydride is required, which gives rise to "tailing" of the solvent front during GC; for this reason, 1-methylimidazole is added after acetylation to catalyze conversion of excess acetic anhydride to acetic acid. This derivatization method, which is again a "one-pot" procedure, has been applied successfully to a number of oligo- and polysaccharides of known structure.

Waeghe et al.[72] avoided evaporation at raised temperatures and under reduced pressure by drying all samples at room temperature under a stream of filtered air. After hydrolysis of the permethylated, carboxylate-reduced polysaccharide, the TFA was removed by evaporation with methanol in this manner prior to reduction, for which a solution of sodium borodeuteride in 95% ethanol containing ammonia (1 *M*) was used. An aliquot (50 μl) of this solution (10 μg $NaBD_4$ per microliter) was added to the vial containing the partially methylated sugars and, after mixing, the vial was sealed and left at room temperature for at least 2 h. Acetic acid (5 μl) was then added and the resulting borate removed by evaporation four times, as described (see 8), with 9:1 (v/v) methanol-acetic acid (50 μl in each case) and then with methanol (75 μl). Acetic anhydride (50 μl) was added to the residue; the vial was then sealed and, after mixing of the contents, heated at 121°C for 3 h. After cooling to room temperature, the contents of the vial were mixed with toluene (50 μl) and evaporated just to dryness; this was repeated once. The partially methylated alditol acetates were extracted into dichloromethane and injected into the gas chromatograph without further purification. Derivatization by this method permitted methylation analysis of 100 μg or less of a complex carbohydrate. The use of sodium borodeuteride in reduction facilitated identification of the products by GC-MS (see Table IIA).

10. ALDONONITRILE ACETATES

In a modification of the standard procedure for derivatization of sugars to aldononitrile acetates (see Volume I). Chen and McGinnis[75] replaced pyridine by 1-methylimidazole as solvent and catalyst for both oximation with hydroxylamine hydrochloride and the subsequent simultaneous dehydration and acetylation with acetic anhydride. To the sugars, dissolved in 1-methylimidazole (0.5 ml), was added 0.5 ml of a stock solution containing hydroxylamine hydrochloride (5 g) and D-glucitol (1 g; internal standard) in the same solvent (100 ml). The mixture was heated at 50°C for 5 min and then cooled to room temperature before addition of acetic anhydride (0.3 ml). As in the case of the Blakeney method for production of alditol acetates[10] (see 9), acetylation took place rapidly at room temperature in the presence of 1-methylimidazole. After 5 min chloroform (1 ml) was added, and the solution was washed with water and then dried (anhydrous Na_2SO_4). The aldononitrile acetates, the identity of which was confirmed by MS, were obtained in yields of 90 to 100% by this method.

Subsequently, McGinnis[76] showed that catalysis of the reaction by 1-methylimidazole was possible in the presence of water and was not affected by residual mineral acid in hydrolysates. In this case, the sugars were dissolved in water, and the stock solution of hydroxylamine hydrochloide in 1-methylimidazole was of lower concentration (20 mg/ml) for derivatization of neutral sugars, but for derivatization of 2-amino-2-deoxy-D-glucose the concentration was as before. The hydroxylamine hydrochloride solution (0.4 ml) was added to the sugar solution (0.2 ml) and the mixture heated at 80°C for either 10 min (neutral sugars) or 8 min (aminodeoxyhexose), then cooled to room temperature. Acetic anhydride (1 ml) was added, and after 5 min the products were extracted into chloroform as before. Methyl α-D-glucopyranoside and xylitol were added as internal standards to the hydroxylamine hydrochloride stock solutions used to derivatize neutral sugars and aminodeoxyhexose, respectively. Rate studies showed that oximation was complete within 6 min under these conditions; further heating did not affect the yield of D-glucose oxime, but that of 2-amino-2-deoxy-D-glucose oxime decreased, indicating some degradation of the product. Thus, the specified time of heating with hydroxylamine hydrochloride is crucial when aminodeoxyhexose is present. The same applies to the time of reaction with acetic anhydride. The aldononitrile acetate derived from D-glucose was formed within 1 min, but that from 2-amino-2-deoxy-D-glucose formed more slowly, the yield reaching a maximum at 4 to 5 min. Longer reaction times resulted again in degradation. The ratio of 1-methylimidazole to water affected the rate of oximation only if it was below 1:1, and the presence of water did not affect dehydration and acetylation provided that an excess of acetic anhydride was present to compensate for some loss from hydrolysis. Under the same conditions, with both 1-methylimidazole and acetic anhydride in excess, the presence of mineral acid did not interfere with the reaction. Analysis of neutral sugars in acid hydrolysates by this method gave similar results, regardless of whether or not the hydrolysate was neutralized before derivatization. Thus, acid hydrolysates of polysaccharides can be analyzed directly without neutralization or removal of water.

A disadvantage of derivatization of sugars to aldononitrile acetates for GLC analysis has become evident from the work of Furneaux,[77] who isolated by-products present with the aldononitrile acetates formed when D-glucose and D-galactose were derivatized by the standard method. These by-products were identified as *N*-hydroxy-D-glycosylamine hexa-acetates, mainly in the β-furanose form with some β-pyranose also present; their proportions in the total product mixture were 22 and 19% for D-glucose and D-galactose, respectively. There was evidence from TLC and LC that such by-products also were formed when the modification of using 1-methylimidazole as solvent and catalyst was employed. Although D-mannose, D-xylose, D-arabinose, and L-rhamnose have been found to give the aldononitrile exclusively when submitted to this reaction, the uncertainty of quantitative analysis of D-glucose and D-galactose by the aldononitrile acetate method, due to the formation of the by-products, is a serious flaw in the technique. However, in GC of partially methylated D-glucononitrile penta-acetates derived from permethylated dextran, only a trace (3 to 5%) of by-product was detected by Slodki et al.,[78] who contend that formation of glycofuranosylamine should be blocked in methylated sugars obtained from sugar residues linked other than 1 → 4 in the parent polysaccharide.

11. *N*-ALKYLALDONAMIDE ACETATES

In a procedure developed by Lehrfeld[79-81] for GC analysis of aldonic acids, the acids are converted to aldonolactones by passage through a short column of cation-exchange resin (H^+-form), elution with water, and evaporation of the eluate to dryness *in vacuo* at 45°C, followed by further heating at 85°C *in vacuo* for 2 h. The residue is dissolved in pyridine (0.5 to 1 ml), and an equal volume of 1-propylamine is added. The mixture is heated in a sealed vial at 50 to 55°C for 30 min and then allowed to cool below 45°C before the seal is removed and the solution evaporated by heating at 55°C under a stream of dry N_2. The dry residue is dissolved in pyridine (0.5 ml), acetic anhydride (0.5 ml) is added, and the mixture is heated at 90 to 95°C for 40 to 60 min with occasional shaking to ensure complete solution. The mixture is then evaporated to dryness at 50°C under a stream of dry N_2, and the residue is dissolved in dry chloroform, dichloromethane, or acetone for injection into the gas chromatograph (see Table GC 6).

This method can be utilized in simultaneous analysis of aldonic acids and neutral aldoses, the latter derivatized as alditol acetates.[80] The mixture is dissolved in 0.1 *M* Na_2CO_3 solution (0.6 ml), and the solution is maintained at 30°C for 1 h and then treated with $NaBH_4$ (4% solution, 0.5 ml) for 90 min at room temperature. Excess $NaBH_4$ is decomposed by addition of acetic acid; Na^+ ions are removed by passage through a short column of cation-exchange resin (H^+-form) and borate by evaporation of the eluate with methanol. The residue is then heated at 85°C *in vacuo* to convert the aldonic acids to aldonolactones, and the procedure described above is followed. The *N*-propylaldonamide acetates are well separated from alditol acetates by GC.

In a modification to permit simultaneous analysis of alduronic acids and aldoses,[82] the mixture is dissolved in water (1 ml) and 0.5 *M* Na_2CO_3 solution (13 µl/mg acid) is added. The solution is maintained at 30°C for at least 45 min to ensure hydrolysis of alduronolactone to alduronate, which is reduced to aldonate by the subsequent treatment with $NaBH_4$. Application of the procedure described to the resulting mixture of alditols and aldonic acids permits simultaneous GC analysis. It should be noted that alduronic acids are converted to corresponding *inverted* aldonates; for example, D-glucuronic acid becomes L-gulonate and L-guluronic acid is converted to D-gluconate (see Table GC 6). Recently, a similar analytical method, in which the aldonolactones derived from the alduronic acids are treated with 1-hexylamine, converting them to *N*-hexylaldonamides (see Table GC 6), has been introduced.[83]

12. CHIRAL DERIVATIVES

Since the successful use of chiral glycosides (see Volume I) as derivatives for GC separation of sugar enantiomers, several other methods of derivatization permitting resolution of enantiomers have been described (see Table GC 23).

Little[84] separated enantiomeric pairs by derivatization to acyclic diastereosomeric dithioacetals. The reagent used was (+)-1-phenylethanethiol, synthesized by reaction of sodium (−)-*O*-methyl dithiocarbonate with racemic 1-phenylethyl bromide and subsequent separation by fractional recrystallization of the diastereoisomers formed. The optically active thiol is released by solvolysis in almost 100% purity. In the derivatization procedure, the dried sugar sample (10 to 100 µg) is dissolved in TFA (5 µl), and (+)-1-phenylethanethiol (20 µl) is added. The reaction is allowed to proceed at room temperature for 30 min with occasional mixing; then cold pyridine (50 µl) is added to stop it. The bis [(+)-1-phenylethyl] dithioacetals formed are either trimethylsilylated or acetylated for GC. For trimethylsilylation, *N,O*-bis(trimethylsilyl)trifluoroacetamide (BSTFA) and trimethylchlorosilane (TMCS), 50 and 25 µl, respectively, are added to the reaction vial and the mixture is heated at 65°C for 2 h. For acetylation, the excess of thiol and pyridine are removed by leaving the open vial overnight in a vacuum desiccator containing concentrated H_2SO_4, KOH pellets, and activated charcoal impregnated with $CuSO_4$. To the pyridinium salt remaining is added a freshly prepared solution (100 µl) of 4-dimethylaminopyridine in acetic anhydride (2.5 mg/ml). The mixture is heated at 65°C for 2 h, cooled, and evaporated under a stream of dry N_2, and the residue is dissolved in dichloromethane for GC.

Oshima et al.[85] resolved enantiomers as diastereoisomeric methylbenzylaminoalditols prepared by reductive amination of sugars in the presence of L-(−)-α-methylbenzylamine. The sugar (1 mg), dissolved in water (50 µl), is mixed with a solution containing the α-methylbenzylamine (MBA) and $NaBH_3CN$ (7 and 0.4 mg, respectively) in ethanol (50 µl), and the mixture is kept at 40°C for 3 h. The solution is then acidified with acetic acid and the mixture evaporated several times with methanol. The oily residue is dried *in vacuo* over P_2O_5 before trimethylsilylation or acetylation for GC. For trimethylsilylation, the residue is dissolved in dry acetonitrile (100 μl) and *N,O*-bis(trimethylsilyl)acetamide (BSA) (25 μl) is added. The mixture is allowed to stand at room temperature in a stoppered tube for 15 min before the supernatant (0.2 to 1 μl) is injected into the gas chromatograph. For acetylation,[86] 1:1 (v/v) acetic anhydride-pyridine (100 μl) is added to the residue, and the mixture is heated in a sealed tube at 100°C for 1 h. Water is added and the mixture evaporated; the oily residue is then partitioned between water and chloroform and the chloroform extract concentrated for GC. The acetates, prepared on a larger scale, have also been resolved by HPLC (see Section III.III, 24).

Other chiral derivatives giving a single peak for each enantiomer are the trimethylsilyl ethers of the methyl 2-(polyhydroxyalkyl)-thiazolidine-4(*R*)-carboxylates obtained by reaction of aldoses with L-cysteine methyl ester. In the derivatization procedure described by Hara et al.,[87] pyridine solutions (100 μl each) of the sugar (40 m*M*) and the hydrochloride of L-cysteine methyl ester (60 m*M*) are mixed and warmed at 60°C for 1 h. The trimethylsilylating reagent hexamethyldisilazane-trimethylchlorosilane (HMDS-TMCS; 150 μl) is added and the mixture is kept at 60°C for a further 30 min. The precipitate formed is removed by centrifugation and the supernatant injected into the gas chromatograph.

Schweer has reported GC separation of enantiomers as the trifluoroacetates of chiral oximes. The (−)-menthyloximes[88] are prepared by reaction of the sugar (0.5 mg) with an aqueous solution (100 μl) containing *O*-(−)-menthylhydroxylamine hydrochloride (4 mg) and sodium acetate (3 mg), and the (−)-bornyloximes[89] are prepared by similar reaction with *O*-(−)-bornylhydroxylamine hydrochloride. The mixture is heated for 1 h at 60°C for the *O*-menthyloximes and 80°C for the *O*-bornyloximes. Water is then removed by evaporation at 60°C in a stream of dry air, and the residue is evaporated with methanol (100 μl) in the same way. The last traces of water are removed as an azeotrope by similar evaporation with benzene (100 μl). This produces a crystalline solid, which is then trifluoroacetylated by reaction with trifluroracetic anhydride (30 μl) and ethyl acetate (15 μl) in a sealed vial at room temperature. After 2 h,the derivatives are ready for injection. Thus derivatized, each enantiomer gives two peaks corresponding to geometrical isomers (see Table GC 23).

13. DEOXY(METHOXYAMINO)ALDITOLS

A novel derivatization method proposed by Das Neves et al.[90] involves reduction of sugar *O*-methyloximes with $NaBH_3CN$ to the corresponding deoxy(methoxyamino)alditols, which are then trimethylsilylated or acetylated. These derivatives have highly characteristic mass spectra (see Table IIA), and the sensitive nitrogen-phosphorus selective detector also may be used for GC of the deoxy(methoxyamino)alditols (see Table GC 8); hence, their use as volatile derivatives makes possible particularly sensitive analyses. The derivatization procedure as described by these authors is as follows.

An aliquot (200 μl) of an aqueous solution of monosaccharides (each 1 to 2 mg/ml) containing D-mannitol (2 mg/ml) as internal standard is transferred to a vial and concentrated by evaporation at room temperature in a stream of N_2. The residue is dried over P_2O_5 *in vacuo*. To this is added 100 μl of a solution of *O*-methylhydroxylamine hydrochloride (250 mg) in dry pyridine (10 ml). The mixture is heated at 80°C for 1 h and then concentrated in a stream of N_2. A solution of $NaBH_3CN$ in methanol (200 μl) is added to the residue and the pH is adjusted to 3 by addition of 1 *M* methanolic HCl (50 μl). The solution is sonicated for 1 h with occasional addition of methanolic HCl to keep it slightly acidic and then evaporated in a stream of N_2. The residue is treated with methanol (150 μl) at 80°C for 30 min, the solvent is evaporated under N_2, and the residue containing the deoxy(methoxyamino)alditols is dried over P_2O_5 *in vacuo*. For trimethylsilylation, the residue is treated with HMDS-TMCS-pyridine (8:3:4, v/v) for 5 min at room temperature and, after centrifugation, aliquots of the supernatant are injected into the gas chromatograph. For acetylation, the residue is treated with pyridine (100 μl) and acetic anhydride (100 μl) at 100°C for 20 min, and aliquots of the resulting solution are used for GC.

Better resolution is obtained if the deoxy(methoxyamino)alditols are permethylated[91] (see Table GC 8). The mass spectra of these derivatives have intense diagnostic ions (see Table IIA), which aid detection by selected-ion monitoring. The dry residue containing the deoxy(methyoxamino)alditols, prepared as described above, is sonicated for 15 min at room temperature with a solution (100 μl) of dimsyl potassium in DMSO, and then CH_3I (40 μl) is added and sonication continued for 20 min. The methylated products are extracted into chloroform (three times, with 100 μl of $CHCl_3$ each time), and the chloroform is evaporated under a stream of N_2. The residue is dissolved in dichloromethane for GC.

This method of derivatization has been extended to the GC of reducing disaccharides, which have been analyzed as the permethylated deoxy(methylmethoxyamino)alditol glycosides[92] (see Table GC 8). The method of preparation is identical to that just described for the analogous derivatives of the monosaccharides.

14. DIETHYL DITHIOACETALS

Derivatization to diethyl dithioacetals by mercaptalation with ethanethiol in the presence of acid followed by trimethylsilylation is another method of obtaining a single peak for each aldose, which can be applied to GC analysis of neutral aldoses and hexuronic acids[93] as well as the products of deamination of aminodeoxyhexoses[94] (see Table GC 19). Partially methylated sugars can also be derivatized in this way and give simple, characteristic mass spectra which aid identification in GC-MS[95] (see Table IIA). In the derivatization procedure described by Honda et al.,[93] a solution (100 μl) containing sugar (10 pmol to 10 μmol) and *myo*-inositol (1 μmol) as internal standard is evaporated to dryness over NaOH in a vacuum dessiccator. A mixture (20 μl) of ethanethiol and TFA (2:1) is added, and the mixture is kept at 25°C for 10 min in a sealed vial. Pyridine (50 μl) is then added, followed by HMDS (100 μl) and TMCS (50 μl), and the mixture is incubated at 50°C for 30 min with occasional shaking. After centrifugation, the supernatant (1 to 10 μl aliquots) is injected into the gas chromatograph. For the products of deamination of aminodeoxyhexoses[94] and for partially methylated sugars[95] the derivatization is identical, but the recommended internal standard is 3-*O*-methyl-D-glucose. This applies also to GC analysis by this method of the *N*-acyl-mannosamines obtained on *N*-acetylneuraminate lyase digestion of neuraminic acids,[39] which has been discussed already (see Section III.I.5).

15. OXIMES

Conversion of sugars to oximes, followed by trimethylsilylation, acetylation, or trifluoroacetylation in order to increase volatility and permit use of these derivatives in GC analysis, has been much used in recent years. The chromatograms are relatively simple (two peaks per sugar) and the mass spectra distinctive enough to permit their use in GC-MS identification (see Table IIA). The preparation of trimethylsilylated oximes and *O*-methyloximes (see Table GC 22) has been described in Volume I. Peracetylated oximes (see Table GC 7) are useful derivatives for GC-MS of ketoses (see Table IIA), analogous to the aldononitrile acetates produced by aldoses. In the derivatization procedure described by Seymour et al.,[96] the ketose (5 to 10 mg) is mixed with pyridine (0.2 ml) and hydroxylamine hydrochloride (mass equal to or in excess of that of the sugar), and the mixture is heated at 70°C with stirring for 20 min. The vial is cooled briefly; then acetic anhydride (0.1 ml) is added, and the heating and stirring are continued for 20 min. The resulting solution can be injected directly into the gas chromatograph, but better results are obtained if the peracetylated ketose oximes (PAKO) are extracted into chloroform by partitioning between chloroform and water.

It is peracetylated *O*-methyloximes, however, that have been used most in GC (see Table GC 7), especially for derivatization of amino- and acetamidodeoxyhexoses. The *O*-methyloxime acetates derived from these sugars can be separated by GC from the aldononitrile acetates[97] or *O*-methyloxime acetates[98] from neutral sugars, a method that has proved useful in analysis of the carbohydrate components of glycoproteins. Mawhinney et al.[97] have recommended an *O*-methyloximation reagent prepared by dissolving *O*-methylhydroxylamine hydrochloride (300 mg) in a mixture of dry methanol (1.0 ml) and pyridine (1.78 ml) followed by the addition of 1-dimethylamino-2-propanol (0.22 ml). Neeser and Schweizer[98] have described a procedure in which a freeze-dried mixture of sugars (1 to 9 mg total) containing as internal standard *myo*-inositol, xylitol, or 3-*O*-methyl-D-glucose is dissolved in this reagent (0.4 ml), and the solution is heated at 70 to 80°C in a sealed vial for 20 min. After cooling to room temperature and drying in a stream of N_2, the resulting *O*-methyloximes are acetylated with 1:3 (v/v) pyridine-acetic anhydride (1 ml) at 70 to 80°C for 30 min. The solution is concentrated under reduced pressure, and the residue is dissolved in dichloromethane (1 ml). After washing with 1 *M* HCl (1 ml) and distilled water (three times, 1 ml each time) and filtration through anhydrous Na_2SO_4, an aliquot (0.1 ml) of the resulting solution is diluted to 1.0 ml with a 9:1 (v/v) mixture of heptane and dichloromethane, and aliquots of this solution are injected for GC. If only small quantities of the sugars are available, the *O*-methyloximation reagent (0.2 ml), containing only 120 mg of *O*-methylhydroxylamine hydrochloride, is added to the freeze-dried mixture of sugars (0.1 to 1 mg total) during 30 min, and only 0.5 ml of the acetylating reagent is used. After workup the dichloromethane solutions of the derivatives are injected directly without dilution.

Guerrant and Moss[99] analyzed mixtures of neutral and aminodeoxy sugars, including muramic acid and KDO, which are important constituents of bacterial cell-wall polysaccharides, by this method (see Table GC 7). They used a reagent different from that described above, in that 4-(dimethylamino)pyridine (DMAP) was substituted for 1-dimethylamino-2-propanol. Their reagent contained *O*-methylhydroxylamine hydrochloride and DMAP, each at a concentration of 66 mg/ml, in pyridine-methanol (2:1, v/v). The reagent (0.3 ml) was added to a dry sample containing ca. 0.15 mg of each carbohydrate. The sample was sonicated for 1 min and then heated at 70 to 80°C for 25 min in a sealed vial. After cooling to room temperature, acetic anhydride (1 ml) was added, and the mixture was sonicated for 1 min and reheated for 15 min. The mixture was cooled, and then 1,2-dichloroethane (2 ml) was added and excess derivatization reagents were removed by two extractions with 1 *M* HCl (1 ml) followed by three with water (1 ml). The extractions were performed as rapidly as possible to minimize hydrolysis. After the final extraction, the vial was centrifuged to separate entrapped water, which was then removed. The dichloroethane extract was evaporated at 50°C in a stream of dry N_2, and the residue was dissolved in ethyl acetate-hexane (1:1 v/v) for injection.

Recently, Biondi et al.[100] introduced a novel derivatization method in which sugars were reacted with *O*-(2,3,4,5,6-pentafluorobenzyl)hydroxylamine (PFBOA) to form oximes that were then acetylated for GC (see Table GC 7). The sugars (5 μmol each), with *meso*-inositol (1 μmol) as internal standard, were heated with a solution of PFBOA (20 mg) in pyridine (1 ml) at 80°C for 20 min, and aliquots (0.2 ml) were withdrawn and acetylated with acetic anhydride (0.4 ml), also at 80°C for 20 min. Each sample was then evaporated to dryness, and the residue was dissolved in dichloromethane (3 ml). After washing with 1 *M* HCl (2 ml) and water (2.2 ml), the organic phase was filtered over anhydrous Na_2SO_4 and dried under a stream of N_2. The residue was dissolved in ethyl acetate for GC. This method has been applied successfully in analysis of the neutral sugar components of glycoproteins.

Trifluoroacetylated *O*-alkyloximes have been used a great deal by Schweer as derivatives for GC (see Table GC 15), and useful CI mass spectral data are available (see Table IIA). The special case of chiral oximes has been discussed above (see Section III.II.12). Schweer has reported GC and MS data for trifluoroacetylated *O*-methyl-, *O*-2-methyl-2-propyl-, and *O*-*n*-butyloximes of all aldopentoses[101] and aldohexoses;[102] derivatization involved reaction of each sugar (1 mg) with *O*-methylhydroxylamine hydrochloride (3 mg), *O*-2-methyl-2-propylhyroxylamine hydrochloride (8 mg), or *O*-*n*-butylhydroxylamine hydrochloride (5 mg) in the presence of sodium acetate (6 mg) in water (100 μl). The mixture is heated at 60°C for 1 h, and trifluroacetylation and workup are as described for the chiral oximes (see Section III.II.12).

16. TRIFLUOROACETYLATION

For trifluoroacetylation of free sugars, *N*-methylbis(trifluoroacetamide) (MBTFA) is now generally preferred to triflouroacetic anhydride (TFAA). In a modification of the procedure described in Volume I in which MBTFA and pyridine were added to the sugar mixture simultaneously, Englmaier[103] recommends dissolution in pyridine prior to addition of MBTFA since sugars are less soluble in pyridine in the presence of this reagent. Pyridine (20 μl) containing the internal standard (phenyl β-D-glucopyranoside was used by Englmaier) is added to the residue obtained on evaporation of an aliquot (100 μl) of a solution of carbohydrate (2 mg/ml) in 40% ethanol, and the carbohydrate is dissolved in the pyridine by heating at 75°C for 20 min in a sealed vial. MBTFA (40 μl) is then added, and the mixture heated at 75°C for 10 min. After cooling the solution is left for 15 min, and aliquots are injected into the gas chromatograph (see Table GC 14).

For alditols, trifluoroacetylation is effected by treatment with TFAA in ethyl acetate (1:1, v/v) at room temperature for 30 min, and aliquots of the reaction mixture are used directly in GC[104,105] (see Table GC 12). The use of a trifluoroacetylating mixture containing TFAA and ethyl acetate in the ratio 2:1 (v/v) for derivatizing oximes[88,89,101,102] has been described above (see Section III.II.12). For trifluoroacetylation of methanolysates, Bryn and Jantzen[44] used 1:1 (v/v) TFAA in acetonitrile; in this case, the methanolysate was mixed with the reagent and the mixture heated to boiling for ca. 2 min. After cooling to room temperature, the reaction mixture was allowed to stand for 10 min and then diluted with acetonitrile to 10% TFAA before injection into the gas chromatograph (see Table GC 13). A similar procedure was used by Brondz and Olsen[45] in derivatization of whole-cell methanolysates.

N-Trifluoroacetylation of 2-acetamido-2-deoxyhexose residues in oligosaccharides derived from glycoproteins or glycolipids increases volatility, permitting GC analysis of such oligosaccharides as permethylated oligosaccharide-alditols (see Table GC 11) to higher DP. Oligosaccharides containing up to seven sugar residues have been examined by GC-MS using this method.[106] The presence of the *N*-trifluoroacetyl group stabilizes ions resulting from primary cleavage of the hexosaminidic linkage during EI mass fragmentation; therefore, sugars containing such linkages often are identified more easily by MS when the hexosamine is thus derivatized. N-Trifluoroacetylation is accomplished by treatment with a mixture of TFA and TFAA under carefully controlled conditions so that cleavage by trifluoroacetolysis does not occur. With a TFA-TFAA ratio of 1:100 (v/v), most hexosamine residues remain intact. Rapid trifluoroacetylation of hydroxyl groups occurs at room temperature, and inductive effects of the highly electronegative trifluoroacetyl groups protect glycosidic bonds from acid-catalyzed solvolysis during subsequent heating at 100°C for 48 h. Under these conditions 2-acetamido-2-deoxyhexoses are transamidated to 2-deoxy-2-trifluoroacetamidohexoses. After heating, the mixture is concentrated to dryness by rotary evaporation, and *O*-trifluoroacetyl groups are removed by evaporation with methanol and then with pyridine. The resulting *N*-trifluoroacetylated oligosaccharides are reduced with $NaBH_4$ or $NaBD_4$, if not already reduced (from alkaline $NaBH_4$ treatment of the parent glycoconjugate), and then the hydroxyl groups are methylated prior to GC analysis.

17. TRIMETHYLSILYLATION

The various reagents used in trimethylsilylation of sugars and derivatives were discussed in Volume I. The original method of Sweeley, using HMDS and TMCS, and the alternative reagents *N,O*-bis(trimethylsilyl)acetamide (BSA), *N,O*-bis(trimethylsilyl) (BSTFA), and *N*-(trimethylsilyl)imidazole (TSIM) all continue to find application. The choice depends upon the aspect emphasized in the particular study. For example, Martínez-Castro and co-workers[107-109] used TSIM, 100 μl of the reagent being added to the freeze-dried residue obtained after the monosaccharides had been left in aqueous solution for 48 h at room temperature to attain anomeric equilibrium; the mixture was heated at 65°C for 30 min.[109] Since the various forms of monosaccharides in tautomeric equilibrium were being examined by GC of the TMS derivatives in this study (see Table GC 21), TSIM, which can be used in the presence of water and does not require pyridine (which would alter the tautomeric equilibrium), was clearly the reagent of choice in this case. Nikolov and Reilly,[110] in examining the GC behavior of the anomeric forms of disaccharides as their TMS derivatives, dissolved each disaccharide in pyridine (1 ml) containing 0.2 *M* 2-hydroxypyridine as a catalyst and left the solution at 40°C for 15 h to attain anomeric equilibrium. The equilibrated sugars were derivatized with HMDS in the presence of TFA (9:1), as described in Volume I. For direct derivatization of sugars in biological fluids, Laker[111] found that HMDS-TMCS was unsatisfactory, as an insoluble precipitate was formed (presumably ammonium chloride) and traces of sulfosalicylic acid, used to precipitate proteins, caused breakdown of the products. BSA or BSTFA, used with TMCS in the presence of pyridine, formed a clear solution, and derivatives from plasma remained stable for several days in airtight containers.

For GC of hydroxy mono- and dicarboxylic acids (aldonic and aldaric acids), products of oxidative alkaline degradation of polysaccharides which are important in alkaline processing of wood, Sjöström and associates[112,113] converted the acids to their ammonium salts prior to trimethylsilylation, in order to eliminate the production of multiple peaks on GC due to coexisting acids and lactones. Under these conditions, the reagent of choice was BSTFA. The presence of TMCS was necessary to avoid partial lactonization of the acids upon derivatization. In the method used by these authors, the ammonium salts of the acids, prepared by passage of the products of alkaline degradation through a weak-acid cation-exchange resin in the NH_4^+-form, were treated with pyridine (0.5 ml) and the silylating reagent (0.25 ml) consisting of BSTFA containing 5% of TMCS, and the mixture was shaken at room temperature for at least 30 min prior to GC analysis (see Table GC 18).

The BSTFA-TMCS silylating reagent also has proved effective in analysis of polyols (including some very uncommon deoxyalditols in urine by GC analysis as the TMS ethers (see Table GC 16). Niwa et al.[114,115] isolated the neutral fraction from the samples by passage through cation- and anion-exchange resins and concentration by evaporation with methanol under a stream of N_2. The dried concentrate was trimethylsilylated with BSTFA-TMCS (9:1) at 60°C for 20 min, and aliquots of the resulting mixture were injected directly into the gas chromatograph.

Section III.III

DERIVATIZATION FOR HPLC

18. ACETYLATION

Acetylation of sugars and glycosides (see Volume I, Tables LC 8 and LC 9, and this volume, Tables LC 15 and LC 21) has seldom been used as a means of derivatization for HPLC since it does not increase sensitivity of detection as does the introduction of chromophoric or fluorescent groups. The greatest success of the method has been in HPLC of the homologous series of malto-oligosaccharides on C_{18}-silica. Thus derivatized, the oligosaccharides can be resolved to DP 30 to 35 by gradient elution at 65°C with aqueous acetonitrile[116,117] (see Table LC 16). The dry carbohydrate sample (1 to 25 mg) is acetylated by treatment with acetic anhydride:pyridine (1:1) at 100°C for 90 min, after which the reagents are removed by evaporation in a stream of N_2 and the residue is dried by repeated evaporation with toluene. The acetylated products are dissolved in acetonitrile for injection.

19. BENZOYLATION

The application of benzoylation and 4-nitrobenzoylation as a means of introducing chromophores into carbohydrate molecules and thus increasing the sensitivity of UV detection in HPLC was discussed in Volume I. The behavior of a number of sugars, methyl glycosides, and alditols upon silica gel HPLC as benzoates (see Tables LC 19 and LC 21) has been investigated by White et al.,[118] using a modification of the benzoylation method described in Volume I in that derivatization with benzoyl chloride in pyridine took place at 4°C. The reagent (one equivalent per OH group plus one equivalent excess) was added to a solution of the sample (1 to 100 mg) in pyridine (1 ml) in three portions during 30 min with periodic shaking, while the temperature was kept at 4°C. The mixture was left overnight at this temperature and then at room temperature for 2 h. Methanol was not used to decompose residual benzoyl chloride, since the resulting methyl benzoate is difficult to remove. Instead, distilled water (2 mol/mol of excess benzoyl chloride) at 0°C was added to the mixture, which was chilled in an ice bath. After a further 2 h at room temperature, dichloromethane (1 ml) was added, and the extract was cooled to 4°C and washed with 1 *M* H_2SO_4 (2 × 2 ml) and saturated aqueous $NaHCO_3$ (2 × 2 ml). The organic layer was dried (Na_2SO_4) and the solvent removed by evaporation. The residue, dissolved in chloroform or ethyl acetate-hexane (1:5), was used for HPLC, which gave multiple peaks for sugars and glycosides. For sugars, the ratio of the various forms is affected by the reaction conditions. The data for neutral monosaccharides (Table LC 19) were obtained after equilibration by heating in pyridine solution at 60°C for 1 h before derivatization. However, it should be noted that the mixture of isomeric benzoates obtained under these conditions will not correspond to the equilibrium mixture if isomerization is more rapid than benzoylation or the various isomers are not benzoylated at the same rate.

The problem of multiple peaks can be obviated by reduction of the sugars to alditols prior to benzoylation. In an HPLC method described by Oshima and Kumanotani[119] that is applicable to both neutral aldoses and amino sugars (aminodeoxyhexoses are distinguishable from their *N*-acetylated analogues) the derived alditol benzoates are separated on a microparticulate silica column by gradient elution with the ternary system *n*-hexane-dioxane-dichloromethane (see Table LC 19). The sugar mixture is reduced with $NaBH_4$ (at 40°C for 2 h) and, after acidification and removal of borate in the usual way, the residue is dissolved in dry pyridine (2 ml). In this case, benzoylation with benzoic anhydride (0.2 ml) catalyzed by 4-dimethylaminopyridine (ca. 100 mg) is preferable to the method using benzoyl chloride, which gives two peaks (due to some *N*-benzoylation) for each of the amino sugars. Benzoylation is carried out at 50°C with stirring for 2 h in a stoppered flask or sealed vial.

Methanol (1 ml) is then added, and the mixture is vacuum distilled at 60°C. The residue is partitioned between chloroform and water and the organic layer evaporated. Residual methyl benzoate may be removed by SEC on a lipophilic gel in chloroform.

Jentoft[49] also recommended benzoylation with benzoic anhydride for benzoylation of methyl glycosides in methanolysates (see Table LC 15). The benzoylation procedure used by this author was overnight incubation at room temperature in a freshly prepared solution (0.1 ml) of benzoic anhydride (10%, w/v) and dimethylaminopyridine (5%) in pyridine. The reaction was stopped by addition of water (0.9 ml), followed by incubation at room temperature for 30 min to ensure hydrolysis of benzoic anhydride. Reaction by-products were separated from the benzoylated sugar derivatives by application of the samples to small reversed-phase columns inserted into a vacuum manifold. The by-products were removed by washing with water, and the sugar derivatives were then eluted with acetonitrile, with the vacuum maintained at ca. 100 mmHg throughout the process. The eluates were evaporated in a stream of N_2 or in a vacuum drier connected to a vacuum pump (not a water aspirator) and redissolved in acetonitrile for HPLC.

For benzoylation of oligosaccharides, Daniel et al.[120] used the reaction with benzoyl chloride in pyridine (at 37°C for 16 h) for neutral oligosaccharides, such as the malto-oligosaccharides, which were resolved to DP 13 by reversed-phase HPLC after $NaBH_4$ reduction followed by perbenzoylation (see Table LC 16). The benzoic anhydride-dimethylaminopyridine method, however, was the method of choice for oligosaccharides containing acetamidodeoxyhexose residues, which (as mentioned above) do not undergo *N*-benzoylation under these conditions. Benzoylation was carried out at 37°C for 4 h; excess reagents and by-products were removed by use of a Sep-Pak® C_{18} cartridge (Waters) eluted with water and acetonitrile as just described.

A similar procedure with a pyridine solution containing 20% (w/v) of benzoic anhydride and 5% of dimethylaminopyridine (at 37°C for 4 h) was applied by Gross and McCluer[121] to benzoylation of neutral glycophingolipids for HPLC on silica (see Table LC 22). In this case, the pyridine was evaporated under a stream of N_2, and the benzoylated products were extracted into *n*-hexane. After washing four times with 8% methanol saturated with Na_2CO_3 and twice with 80% methanol, the *n*-hexose was evaporated under N_2 and the benzoylated samples dissolved in CCl_4 for injection. Benzoylation with benzoyl chloride proved effective in derivatization of monosialogangliosides.[122] Dried samples containing 3 to 20 nmol of ganglioside were heated at 60°C for 1 h with a freshly prepared solution of benzoyl chloride (10% v/v) in pyridine, after which the reaction mixture was dried under a stream of N_2, dissolved in benzene and again evaporated to dryness. The products were purified by application in benzene solution to a short column of silica. The column was washed with benzene to remove reagents and by-products, and the perbenzoylated gangliosides were then recovered by elution with 10% (v/v) methanol in benzene. The solvent was evaporated under N_2, and the purified products dissolved in CCl_4 for HPLC (see Table LC 22).

20. DIMETHYLPHENYLSILYLATION

White et al.[123,124] used dimethylphenylsilylation as a means of introducing a chromophore into carbohydrate molecules (sugars, alditols, and methyl glycosides) for sensitive UV detection in HPLC on silica (see Tables LC 19 and LC 21). The derivatization procedure used by these authors is as follows.

The sample (1 to 10 mg) in dimethylformamide (150 μl) is heated at 100°C for 1 h with a solution (200 μl) of imidazole in dimethylformamide (0.33 g/ml). This mixture is then cooled in ice, and chlorodimethylphenylsilane (70 μl) is added. The reaction is allowed to proceed for 6 h at ambient temperature (or 1 h at 100°C); then the product is extracted into *n*-hexane (2 × 200 μl) and concentrated under N_2. The derivatives may be injected immediately (in *n*-hexane solution) or stored at −20°C.

21. GLYCOSYLAMINE DERIVATIVES

Batley et al.[125] have developed a simple derivatization procedure using *p*-anisidine (widely used as a spray reagent in planar chromatography of carbohydrates; see Volume I) to introduce the chromophore required for UV detection in reversed-phase HPLC of sugars (see Table LC 14). The derivatives formed are *N*-(*p*-methoxyphenyl)glycosylamines. In the procedure described by these authors, the reagent is a 25% (w/v) solution of *p*-anisidine in methanol. Aliquots (20 μl) of this solution are placed in a series of small tubes or vials, and the sugar solutions (50 μl, containing 10 to 500 μg total aldoses) are added. The contents of the tubes are mixed carefully and the tubes centrifuged briefly before being sealed and heated at 60°C for 80 min. After cooling, the tubes are opened, water (200 μl) and ether (1 ml) are added to each, and the mixtures are agitated on a vortex mixer. After brief centrifugation to give clear separation of the layers, the organic phase is removed in each case. An aliquot (100 μl) of the aqueous layer is mixed with an equal volume of 10 m*M* sodium phosphate buffer (pH 8.5) for HPLC. This method is applicable only to aldoses, as ketoses do not give glycosylamines without acid catalysis.

Under acid conditions, 1,2-diamino-4,5-dimethoxybenzene (DDB), a fluorogenic reagent for α-keto acids, reacts specifically with *N*-acetyl- and *N*-glycolylneuraminic acids to form highly fluorescent derivatives. This forms the basis of a very sensitive method of analysis of these neuraminic acids in serum and urine by reversed-phase HPLC after pre-column derivatization[126] (see Tables LC 14 and IIB). This is important in clinical chemistry, since the concentration of *N*-acetylneuraminic acid is significantly increased in, for example, liver cancer with metastases. The derivatization procedure described by Hara et al.[126] is as follows.

The reagent is a 7 m*M* solution of DDB monohydrochloride in water, containing β-mercaptoethanol (0.7 *M*) and sodium hydrosulfite (4.6 m*M*); it should be stored in the dark and used within 1 d of preparation. The β-mercaptoethanol stabilizes the DDB during the reaction, while sodium hydrosulfite stabilizes the fluorescent products. The derivatization reaction is effective only in the presence of acid; sulfuric acid was chosen since this was used in hydrolysis to release the neuraminic acids from glycoproteins in the biological fluids. The optimal concentration of acid, 12 to 20 m*M*, can be achieved by hydrolysis of the sample in 25 m*M* H_2SO_4 (at 80°C for 1 h) followed, after cooling, by addition of an equal volume (200 μl) of the DDB solution, thus diluting the acid to 12.5 m*M*. The mixture is heated at 60°C for 2.5 h in the dark to develop the fluorescence. The reaction is stopped by cooling the mixture in an ice bath, and aliquots (10 μl) of the resulting solution are injected for HPLC. Each neuraminic acid gives two peaks (one major). Neutral sugars, hexuronic acids, aminodeoxyhexoses, and dehydroascorbic acid do not interfere under the conditions described.

22. HYDRAZONE DERIVATIVES

Conversion of reducing sugars to hydrazone derivatives of various types is an effective method of introducing chromophoric or fluorescent groups. Hydrazone derivatives used in HPLC have included 2,4-dinitrophenylhydrazones, dansylhydrazones, and, more recently, dansylhydrazones (see Table IIB).

Earlier work of Honda and Kakehi,[127] in which aldehydes produced on periodate oxidation of glycosides were analyzed by HPLC of their 2,4-dinitrophenylhydrazones, has been extended by Karamanos et al.[128] to analysis of the neutral sugar components of glycoproteins (see Table LC 19). The reagent, a 1.5% (w/v) solution of 2, 4-dinitrophenylhydrazine (DNP) in 1,2-dimethoxyethane, can be stored at 4°C for 3 d. In the derivatization procedure described by Karamanos et al.,[128] an aliquot (10 μl) of an aqueous solution containing 1.5 to 100 nmol of reducing sugar is heated with aliquots (100 μl) of a 2% (v/v) solution of TFA in methanol and the DNP solution. The mixture is heated at 65°C for 90 min and then cooled to room temperature, treated with acetone (200 μl), and evaporated to dryness. The residue is dissolved in 1,2-dimethoxyethane (40 μl), and water (210 to 260 μl) is added to the clear solution. The resulting suspension is left at 4°C for 10 min and then centrifuged to remove the selectively precipitated DNP-hydrazone of acetone and, thus, the excess reagent. Aliquots (up to 15 μl) of the supernatant are injected onto a silica column for HPLC analysis. The derivatization procedure is specific for neutral sugars, and there is no interference from aminodeoxyhexoses, hexuronic acids, alditols, or amino acids. The procedure can, therefore, be applied directly to unfractionated hydrolysates of glycoproteins.

The use of dansyl hydrazine in prederivatization for the TLC of reducing sugars has been described (see Section II.IV). Alpenfels[129] prepared dansylhydrazones for analysis by reversed-phase HPLC (see Table LC 14) in the same way, a Sep-Pak® C_{18} cartridge being used to purify the derivatives before chromatography. Excess reagents were removed by washing the cartridge with water and 10% (v/v) acetonitrile in water, and the purified dansylhydrazones were eluted with a 40% solution of acetonitrile in water. After filtration, the eluate was injected onto the column.

Takeda et al.[130] have reported that the fluorescence response (peak height) of dansylhydrazones on HPLC on silica columns (see Table LC 19) decreases almost linearly with increasing concentration of the trichloroacetic acid used to catalyze their formation and increases with increasing concentration of dansyl hydrazine. However, the concentration of the latter was kept at the originally recommended 1%, as the solubility is limited and unknown subpeaks appear in the chromatogram at higher concentrations. The concentration of the acid was decreased to 0.5% (w/v) in the procedure recommended by these authors. Peak heights were found to reach a maximum at a reaction temperature of 50°C and, for most sugars, to reach a plateau at a reaction time of 45 min. However, for D-glucose the peak height continued to increase with time, leveling off only after 90 min. Thus, a reaction temperature of 50°C and reaction time of 90 min are used in the derivatization procedure of Takeda et al.[130] The sample (50 μl, containing 10 to 100 nmol of reducing sugars) is treated with a 0.5% solution of trichloroacetic acid in ethanol and a 1% solution of dansyl hydrazine in ethanol (100 μl of each) under these conditions. The solvent is evaporated under a stream of N_2 and the residue containing the derivatized sugars redissolved in ethanol for injection.

Mopper and Johnson,[131] after optimization studies of various parameters affecting the fluorescent responses of dansylhydrazones in reversed-phase HPLC (see Table LC 14), adopted yet another derivatization procedure. Since it was found that at least a 20-fold excess of reagent over sugar was required for quantitative analysis, a more concentrated solution of dansyl hydrazine was used, the solvent being acetonitrile, in which the reagent is more soluble than it is in ethanol. The effect on fluorescent response of variation in the proportion of water in the reaction mixture was found to be much less with acetonitrile as solvent than with ethanol or methanol; this was another reason for the choice of acetonitrile. The derivatization procedure recommended by these authors is as follows.

To the aqueous sample solution (100 μl) are added a 10% (w/v) solution of trichloroacetic acid in water (10 μl) and a 5% (w/v) solution of dansyl hydrazine in acetonitrile (50 μl). The mixture is heated in a sealed tube at 65°C for 20 min, and then the reaction is stopped by immersion of the tube in an ice bath. The mixture is diluted 1:1 with water, and aliquots (10 to 300 μl) are injected within 2 h. For removal of excess reagents, an on-line procedure using a precolumn in place of the injection loop has been found preferable to the off-line technique of using Sep-Pak cartridges as recommended by Alpenfels[129] (see above). The sample is loaded on the precolumn (conditioned with acetonitrile and then water), and this column is switched into the mobile-phase stream for 30 to 90 s, after which it is switched out and the analysis column brought on-line. The precolumn is washed with acetonitrile to remove excess reagent and then conditioned with water before a new sample is loaded. This technique avoids large dilutions of the sample, and the sugar derivatives are cleanly separated from excess reagent, recoveries of 100% being reported. By use of this procedure, the sensitivity possible with HPLC of sugars derivatized as dansylhydrazones is maximized (see Table IIB).

Reaction of reducing sugars with dansyl hydrazine to form strongly chromophoric dansylhydrazones has been described as a method of prederivatization for TLC (see Section II.IV). The identical procedure[132] is used in precolumn derivatization for reversed-phase HPLC (see Table LC 14). Samples of the reaction mixture obtained as described are passed through a Millipore filter (0.45 μm), and aliquots (1 to 10 μl) of the filtrates are injected for HPLC. Muramoto et al.[133] have described a similar procedure for derivatization of reducing sugars with dansyl hydrazine, prepared by reaction of a solution of dabsyl chloride in chloroform (8.3 mg/ml) with a solution of hydrazine hydrate (0.62 ml) in methanol (60 ml) at room temperature. The dansyl chloride solution is added dropwise with stirring, and the reaction mixture is stirred for 30 min and then concentrated by rotary evaporation. The resulting precipitate is collected by filtration, washed with chloroform, and dried to yield dansyl hydrazine. A stock 0.1% (w/v) solution in ethanol is stored at 4°C, and a portion is mixed with 10% aqueous trichloroacetic acid (39:1) just before use. In the derivatization procedure used by these authors, dansyl hydrazine reagent (40 μl) is added to a 10 m*M* aqueous solution of the sugar (5 μl) and the mixture heated at 50°C for 2 h in a sealed tube and then centrifuged. The centrifugate is used for HPLC; samples should be stored at 4°C prior to injection. These reaction conditions were adopted after optimization studies of the effect of each variable on peak height in the HPLC of the dansylhydrazones.

23. PERALKYLATED OLIGOSACCHARIDE-ALDITOLS

Reversed-phase HPLC fractionation of a complex mixture of reduced, partially ethylated, partially methylated oligosaccharides is an integral part of the methodology developed by Albersheim and co-workers[134,135] for the sequencing of glycosyl residues in complex carbohydrates. The poly- or oligosaccharide is permethylated, and any methyl ester groups arising from uronic acid residues present are reduced by treatment with $NaBD_4$ or $LiAlD_4$ (see Section III.I.8). After partial acid hydrolysis under conditions which vary according to the nature of the glycosidic linkages present, the resulting partially methylated oligosaccharides are reduced with $NaBD_4$ (thus labeling the oligosaccharides with deuterium at the reducing end, which facilitates subsequent GC-MS analysis; see Table IIA). All free hydroxyl groups, at C-1 and C-5 (or C-4 for furanosyl residues) of the alditols present at what was the reducing end of each oligosaccharide and at the positions of attachment of residues removed by the partial acid hydrolysis, are then ethylated. This is accomplished[134] by treatment of the sample, dissolved in dry DMSO, with dimsyl potassium (3 *M*) at 60°C for 1 h, with sonication. Ethyl iodide is then added and the mixture stirred for 1 h at room temperature, after which further ethyl iodide is added and the solution stirred overnight. The products are purified by chromatography on Sephadex® LH-20 in 1:1 (v/v) chloroform-methanol. The resulting partially ethylated, partially methylated oligosaccharide-alditols are fractionated by HPLC (see Table LC 17) prior to complete hydrolysis and GC-MS analysis as the derived alditol acetates (see Table IIA).

24. REDUCTIVE AMINATION

Reductive amination, in which an aldose is treated with sodium cyanoborohydride in the presence of an amine, which couples to the resulting alditol to form a 1-amino-1-deoxyalditol (glycamine) derivative, has proved a valuable method of derivatization for HPLC. Chiral, chromophoric, or fluorescent groups may be introduced in this way.

The use of reductive amination with L-(−)-α-methylbenzylamine to produce diastereoisomeric derivatives from enantiomeric pairs, which could be resolved on GC as their

TMS or acetyl derivatives,[85] has been described (see Section III.II.12).These diastereoisomers have also been resolved by HPLC as the acetates,[86] chromatography on silica (see Table LC 20) giving higher separation factors than reversed-phase HPLC. Although the observed order of elution of enantiomers in GC could not be correlated with structural features,[88] in HPLC there was, in general, a correlation with the configuration at C-2, the enantiomer having the *S* configuration at that position eluting first. For HPLC, preparation of the 1-(*N*-acetyl-α-methylbenzylamino)-1-deoxyalditol acetates is on a larger scale than that of the method already described for GC derivatization (see Section III.II.12). The procedure recommended by Oshima et al.[86] is as follows.

A solution of L-(−)-α-methylbenzylamine (10 mg, 83 μmol) and $NaBH_3CN$ (3 mg, 32 μmol) in methanol (0.2 ml) is added to an aqueous solution (0.2 ml) of the sugar (10 mg). The mixture is allowed to stand overnight at room temperature and then acidified to pH 3 to 4, by addition of glacial acetic acid, and evaporated with methanol. The resultant oily material is acetylated with 1:1 (v/v) acetic anhydride-pyridine (1 ml) in a sealed tube at 100°C for 1 h, after which water (1 ml) is added and the mixture is evaporated. The derivatized product is recovered from the residue by partitioning between chloroform and water, and the crude product, extracted with chloroform, is purified by chromatography, consecutively by SEC with oxolane as the eluent and on a semipreparative reversed-phase column with 7:3 (v/v) acetonitrile-water as the eluent.

Chromogenic labeling of sugars by reductive amination with 4′-*N,N*-dimethylamino-4-aminoazobenzene (DAAB)[136] has already been described as a method of prederivatization for TLC (see Section II.IV), and its use to increase sensitivity of detection in HPLC has also been discussed (see Section II.III and Table IIB). The method can be applied to all reducing sugars, uronic acids, and amino sugars, as well as neutral aldoses, which is an advantage over many of the others. Neutral oligosaccharides and those containing *N*-acetylneuraminic acid also can be derivatized in this way, since the reaction conditions are sufficiently mild to prevent cleavage of glycosidic linkages. For HPLC of the DAAB derivatives on silica columns, eluted with chloroform-methanol (13:5) containing 4% of a 0.1 *M* borate buffer (pH 3.5) or (for the aminodeoxy sugars, which give broad, tailing peaks under these conditions) a 0.1 *M* acetate buffer at the same pH, the derivatization method is the same as that described for TLC derivatization, but scaled up. The DAAB:sugar ratio must be at least 8:1 for quantitative conversion. Before HPLC, it is essential that the DAAB glycamines be separated from excess reagents and by-products by passage of the reaction mixture through a Sep-Pak® C_{18} cartridge. Flushing the cartridge with water removes all but the DAAB and the derivatives; the excess DAAB is largely eluted with 3:1 (v/v) *n*-hexane-chloroform and the glycamines with methanol. Complete removal of unreacted DAAB is achieved by chromatography on a 20-cm column packed with spherical silica beads, with chloroform-methanol-water (11:9:2) as eluent, under which conditions the DAAB is eluted in the solvent front.

Simultaneous determination of neutral and amino sugars, with sensitivity made possible by fluorometric detection (see Table IIB), is achieved by reversed-phase HPLC (see Table LC 14) of the pyridylamino derivatives obtained by reductive amination with 2-aminopyridine.[137] The method has proved particularly useful for analysis of very small samples (100 to 200 pmol) of glycoconjugates, which are hydrolyzed and the aminodeoxy sugars re-*N*-acetylated before derivatization. The reagent used by Takemoto et al.[137] is a pH 6.2 solution containing 2-aminopyridine (500 mg) and concentrated HCl (0.4 ml) in water (11 ml); this solution remains stable for at least a year if stored at −20°C. The reagent (5 μl for analysis at picomolar level) is added to the dried sample and the mixture heated at 100°C for 15 min in a sealed tube. The tube is then opened, the reducing reagent, a freshly prepared aqueous solution of $NaBH_3CN$ (20 mg/ml, 2μl) is added, the tube is resealed, and heating is continued at 90°C for 8 h. Before HPLC, excess reagents and by-products are removed by SEC of an aqueous solution of the reaction mixture on a support of low porosity (such as TSK G2000 PW) eluted with 0.02 *M* ammonium acetate, pH 7.5. The fractions containing the pyridylamino derivatives are combined and concentrated to dryness, and the residue is dissolved in water (250 μl). Aliquots (5 μl) of this solution are injected for HPLC.

This method has been applied to derivatization not only of monosaccharides, but also of a variety of glycoprotein-derived oligosaccharides,[40,41,138] for two-dimensional HPLC mapping (see Table LC 62). Hase et al.,[138] in derivatizing the products of hydrazinolysis and re-*N*-acetylation of a sample (1.5 mg) of material from mouse brain, used 40 μl of the 2-aminopyridine reagent and, after the initial heating, 4 μl of the $NaBH_3CN$ solution, after which heating was continued for 15 h. A TSK® HW-40F column (34 × 1.5 cm I.D.), equilibrated and eluted with 0.01 *M* ammonium acetate buffer, pH 6.0, was used in the purification step.

A feature of the derivatization procedure just described is that the possible loss of sugars due to reduction before amination is avoided by condensation of carbohydrate and amine before addition of the reducing agent. This is also recommended by Coles et al.,[139] who have used pyridylamination in precolumn derivatization for HPLC of the mannose-containing oligosaccharides isolated from the urine of mannosidosis patients. These authors have also described a new technique involving reductive amination with 7-amino-1-naphthol, which gave higher yields of derivatized products and produced enhanced fluorescence (almost eightfold) with a concomitant increase in sensitivity of fluorometric detection in HPLC. In this derivatization procedure, the dried oligosaccharides were treated with the reagent solution of 7-amino-1-naphthol (2 mg) in DMSO (50 μl) and the mixture heated at 105°C for 30 min. A portion (50 μl) of the reducing solution, in which $NaBH_3CN$ (5 mg) was dissolved in a mixture of methanol (400 μl) and glacial acetic acid (40 μl), was then added, and heating at 105°C was continued for 2.5 h. After removal of the solvents, the derivatized products were removed from the residue by Folch extraction; the upper layer was removed and evaporated and the lower layer re-extracted to ensure complete recovery of all derivatized components. In this case, the products were analyzed by HPLC on an amino-silica column eluted with aqueous acetonitrile (20 to 70%) containing 0.15 *M* ammonia.

Glycamines as such, prepared by reductive amination of sugars with $NaBH_3CN$ (0.2 *M*) in the presence of 1 *M* ammonium sulfate at 100°C for 90 min, followed by addition of 0.2 *M* HCl to stop the reaction and evaporation to dryness under a stream of N_2, have been analyzed by cation-exchange chromatography with fluorimetric detection by postcolumn addition of *o*-phthalaldehyde.[140] Simultaneous analysis of aminodeoxyhexoses and -hexitols with the glycamines derived from the neutral sugar constituents of glycoproteins is possible by this method (see Table LC 43), with detection limits of the order of 50 pmol of the parent sugar. Similar sensitivity has been achieved by Shinomiya et al.[141] by precolumn labeling of the glycamines with 7-fluoro-4-nitrobenz-2-oxa-1,3-diazole (NBD-F), a fluorescent prelabeling reagent for amino acids. In this case, the derivatives were analyzed by reversed-phase HPLC (see Table LC 14). The glycamines were prepared as just described, but it was essential to remove all traces of ammonia before treatment with NBD-F. This was accomplished by successive evaporation with 1 *M* NaOH, water, and 0.2 *M* HCl. The residue was dissolved in 0.1 *M* borate buffer (pH 8.0), an equal volume of 50 m*M* NBD-F in ethanol was added, and the mixture was heated at 60°C for 30 min. The reaction was stopped by addition of 0.1 *M* HCl and the reaction mixture injected for HPLC analysis. TLC analysis has also been reported. Results of the preliminary investigation of this method[141] suggest the feasibility of its use in identification and determination of the neutral sugar components in microgram amounts of glycoconjugates.

REFERENCES

1. **Narui, T., Takahashi, K., and Shibata, S.,** Acetolysis of α-D-glucans in the presence of 1,1,3,3-tetramethylurea, *Carbohydr. Res.*, 168, 151, 1987.
2. **Aspinall, G. O.,** Polysaccharide metholology, in *MTP International Review of Science, Organic Chemistry, Ser. 2,* Vol. 7, Aspinall, G. O. Ed., Butterworths, London, 1976, 201.
3. **Fournet, B., Dhalluin, J.-M., Strecker, G., Montreuil, J., Bosso, C., and Defaye, J.,** Gas-liquid chromatography and mass spectrometry of oligosaccharides obtained by partial acetolysis of glycans of glycoproteins, *Anal. Biochem.*, 108, 35, 1980.
4. **Biermann, C. J.,** Hydrolysis and other cleavages of glycosidic linkages in polysaccharides, *Adv. Carbohydr. Chem. Biochem.*, 46, 251, 1988.
5. **Selvendran, R. R., March, J. F., and Ring, S. G.,** Determination of aldoses and uronic acid content of vegetable fiber, *Anal. Biochem.*, 96, 282, 1979.
6. **Nielson, M. J., Painter, T. J., and Richards, G. N.,** Flavologlycan: a novel glyconjugate from leaves of mangrove *(Rhizophora stylosa* Griff), *Carbohydr. Res.*, 147, 315, 1986.
7. **Churms, S. C. and Stephen, A. M.,** unpublished data.
8. **Gehrke, C. W., Waalkes, T. P., Borek, E., Swartz, W. F., Cole, T. F., Kuo, K. C., Abeloff, M., Ettinger, D. S., Rosenshein, N., and Young, R.C.,** Quantitative gas-liquid chromatography of neutral sugars in human serum glycoproteins. Fucose, mannose, and galactose as predictors in ovarian and small cell lung carcinoma, *J. Chromatogr.*, 162, 507, 1979.
9. **Uzaki, M. and Ishiwatari, R.,** Determination of cellulose and non-cellulose carbohydrates in recent sediments by gas chromatography, *J. Chromatogr.*, 260, 487, 1983.
10. **Blakeney, A. B., Harris, P. J., Henry, R. J., and Stone, B. A.,** A simple and rapid preparation of alditol acetates for monosaccharide analysis, *Carbohydr. Res.*, 113, 291, 1983.
11. **Morrison, I. M.,** Hydrolysis of plant cell walls with trifluoroacetic acid, *Phytochemistry*, 27, 1097, 1988.
12. **Honda, S. and Suzuki, S.,** Common conditions for high-performance liquid chromatographic microdetermination of aldoses, hexosamines, and sialic acids in glycoproteins, *Anal. Biochem.*, 142, 167, 1984.
13. **Torello, L. A., Yates, A. J., and Thompson, D. K.,** Critical study of the alditol acetate method for quantitating small quantities of hexoses and hexosamines in gangliosides, *J. Chromatogr.*, 202, 195, 1980.
14. **Lohmander, L. S.,** Analysis by high-performance liquid chromatography of radioactively labeled carbohydrate components of proteoglycans, *Anal. Biochem.*, 154, 75, 1986.
15. **Taylor, R. L., Shively, J. E., and Conrad, H. E.,** Stoichiometric reduction of uronic acid carboxyl groups in polysaccharides, in *Methods in Carbohydrate Chemistry*, Vol. 7, Whistler, R. L. and BeMiller, J. N., Eds., Academic Press, New York, 1976, 149.
16. **Aspinall, G. O.,** The selective degradation of carbohydrate polymers, *Pure Appl. Chem.*, 49, 1105, 1977.
17. **Dua, V. K., Dube, V. E., and Bush, C. A.,** The combination of normal-phase and reverse-phase high-pressure liquid chromatography with NMR for the isolation and characterization of oligosaccharide-alditols from ovarian cyst mucins, *Biochim. Biophys. Acta*, 802, 29, 1984.
18. **Tsuji, T., Yamamoto, K., Konami, Y., Irimura, T., and Osawa, T.,** Separation of acidic oligosaccharides by liquid chromatography: application to analysis of sugar chains of glycoproteins, *Carbohydr. Res.*, 109, 259, 1982.
19. **Clarke, A. E., Anderson, R. L., and Stone, B. A.,** Form and function of arabinogalactans and arabinogalactan-proteins, *Phytochemistry*, 18, 521, 1979.
20. **Gammon, D. W. and Stephen, A. M.,** Glycosylated hydroxyproline derivatives from *Acacia erioloba* exudates, *Carbohydr. Res.*, 154, 289, 1986.
21. **Gammon, D. W., Stephen, A. M., and Churms, S. C.,** The glycoproteins of *Acacia erioloba* exudates, *Carbohydr. Res.*, 158, 157, 1986.
22. **Bacic, A., Churms, S. C., Stephen, A. M., Cohen, P. B., and Fincher, G. B.,** Fine structure of the arabinogalactan-protein from *Lolium multiflorum*, *Carbohydr. Res.*, 162, 85, 1987.
23. **Likhosherstov, L. M., Novikova, O. S., Piskarev, V. E., Trushikhina, E. E., Derevitskaya, V. A., and Kochetkov, N. K.,** A method for reductive cleavage of *N*-glycosylamide carbohydrate-peptide bond, *Carbohydr. Res.*, 178, 155, 1988.
24. **Arbatsky, N. P., Martynova, M. D., Zheltova, A. O., Derevitskaya, V. A., and Kochetkov, N. K.,** Studies on structure and heterogeneity of carbohydrate chains of *N*-glycoproteins by use of liquid chromatography. "Oligosaccharide maps" of glycoproteins, *Carbohydr. Res.*, 187, 165, 1989.
25. **Anastassiades, T., Puzic, R., and Puzic, O.,** Modification of the simultaneous determination of alditol acetates of neutral and amino sugars by gas-liquid chromatography. Application to the fractionation of sialoglycoproteins from bone, *J. Chromatogr.*, 225, 309, 1981.
26. **Turner, S. H. and Cherniak, R.,** Total characterization of polysaccharides by gas-liquid chromatography, *Carbohydr. Res.* 95, 137, 1981.
27. **Henry, R. J., Blakeney, A. B., Harris, P. J., and Stone, B. A.,** Detection of neutral and aminosugars from glycoproteins and polsaccharides as their alditol acetates, *J. Chromatogr.*, 256, 419, 1983.

28. **Hostomská-Chytilova, Z., Mileš, O., Vrátný, P., and Smrž, M.,** Chromatography of cellodextrins and enzymatic hydrolysates of cellulose on ion-exchange derivatives of Spheron, *J. Chromatogr.*, 235, 229, 1982.
29. **Voragen, A. G. J., Schols, H. A., Searle-Van Leeuwen, M. F., Beldman, G., and Rombouts, F. M.,** Analysis of oligomeric and monomeric saccharides from enzymatically degraded polysaccharides by high-performance liquid chromatography, *J. Chromatogr.*, 370, 113, 1986.
30. **Doner, L. W., Irwin, P. L., and Kurantz, M. J.,** Preparative chromatography of oligogalacturonic acids, *J. Chromatogr.*, 449, 229, 1988.
31. **Jin, D. F. and West, C. A.,** Characteristics of galacturonic acid oligomers as elicitors of casbene synthetase activity in castor bean seedlings, *Plant Physiol.*, 74, 989, 1984.
32. **Davis, K. R., Darvill, A. G., Albersheim P., and Dell, A.,** Host-pathogen interactions. XXIX. Oligogalacturonides released from sodium polypectate by endopolygalacturonic acid lyase are elicitors of phytoalexins in soybean, *Plant Physiol.*, 80, 568, 1986.
33. **Romeo, T. and Preston, J. F.,** Liquid-chromatographic analysis of the depolymerization of (1 → 4)-β-D-mannuronan by an extracellular alginate lyase from a marine bacterium, *Carbohydr. Res.*, 153, 181, 1986.
34. **Macek, J., Krajíčková, J., and Adam, M.,** Rapid determination of disaccharides from chondroitin and dermatan sulphates by high-performance liquid chromatography, *J. Chromatogr.*, 414, 156, 1987.
35. **Gherezghiher, T., Koss, M. C., Nordquist, R. E., and Wilkinson, C. P.,** Rapid and sensitive method for measurement of hyaluronic acid and isomeric chondroitin sulfates using high-performance liquid chromatography, *J. Chromatogr.*, 413, 9, 1987.
36. **Linhardt, R. J., Rice, K. G., Kim, Y. S., Lohse, D. L., Wang, H. M., and Loganathan, D.,** Mapping and quantification of the major oligosaccharide components of heparin, *Biochem. J.*, 254, 781, 1988.
37. **Delaney, S. R., Conrad, H. E., and Glaser, J. H.,** A high-performance liquid chromatography approach for isolation and sequencing of chondroitin sulfate oligosaccharides, *Anal. Biochem.*, 108, 25, 1980.
38. **Nebinger, P., Koel, M., Franz, A., and Werries, E.,** High-performance liquid chromatographic analysis of even- and odd-numbered hyaluronate oligosaccharides, *J. Chromatogr.*, 265, 19, 1983.
39. **Kakehi, K., Maeda, K., Teramae, M., and Honda, S.,** Analysis of sialic acids by gas chromatography of the mannosamine derivatives released by the action of *N*-acetylneuraminate lyase, *J. Chromatogr.*, 272, 1, 1983.
40. **Tomiya, N., Kurono, M., Ishihara, H., Tejima, S., Endo, S., Arata, Y., and Takahashi, N.,** Structural analysis of *N*-linked oligosaccharides by a combination of glycopeptidase, exoglycosidases, and high-performance liquid chromatography, *Anal. Biochem*, 163, 489, 1987.
41. **Tomiya, N., Awaya, J., Kurono, M., Endo, S., Arata, Y., and Takahashi, N.,** Analyses of *N*-linked oligosaccharides using a two-dimensional mapping technique, *Anal. Biochem.*, 171, 73, 1988.
42. **Takasaki, S., Mizuochi, T., and Kobata, A.,** Hydrazinolysis of asparagine-linked sugar chains to produce free oligosaccharides, *Methods Enzymol.*, 83, 263, 1982.
43. **Van Pelt, J., Damm, J. B. L., Kamerling, J. P., and Vliegenthart, J. F. G.,** Separation of sialyl-oligosaccharides by medium pressure anion-exchange chromatography on Mono Q, *Carbohydr. Res.*, 169, 43, 1987.
44. **Bryn, K. and Jantzen, E.,** Analysis of lipopolysaccharides by methanolysis, trifluoroacetylation, and gas chromatography on a fused-silica capillary column, *J. Chromatogr.*, 240, 405, 1982.
45. **Brondz, I. and Olsen, I.,** Whole cell methanolysis as a rapid method for differentiation between *Actinobacillus actinomycetemcomitans* and *Haemophilus aphrophilus*, *J. Chromatogr.*, 311, 347, 1984.
46. **Hjerpe, A., Antonopoulos, C. A., Classon, B., Engfeldt, B., and Nurminen, M.,** Uronic acid analysis by high-performance liquid chromatography after methanolysis of glycosaminoglycans, *J. Chromatogr.*, 235, 221, 1982.
47. **Dierckxsens, G. C., De Meyer, L., and Tonino, G. J.,** Simultaneous determination of uronic acids, hexosamines, and galactose of glycosaminoglycans by gas-liquid chromatography, *Anal. Biochem.*, 130, 120, 1983.
48. **Thier, H. P.,** Identification and quantitation of natural polysaccharides in foodstuffs, in *Gums and Stabilizers for the Food Industry*, Vol. 2, Phillips, G. O., Wedlock, D. J. and Williams, P. A., Eds., Pergamon Press, Oxford, 1984, 13.
49. **Jentoft, N.,** Analysis of sugars in glycoproteins by high-pressure liquid chromatography, *Anal. Biochem.*, 148, 424, 1985.
50. **Chaplin, M. F.,** A rapid and sensitive method for the analysis of carbohydrate components in glycoproteins using gas-liquid chromatography, *Anal. Biochem.*, 123, 336, 1982.
51. **Ha, Y. W. and Thomas, R. L.,** Simultaneous determination of neutral sugars and uronic acids in hydrocolloids, *J. Food Sci.*, 53, 574, 1988.
52. **Cheetham, N. W. H. and Sirimanne, P.,** Methanolysis studies of carbohydrates, using HPLC, *Carbohydr. Res.*, 112, 1, 1983.
53. **Roberts, E. J., Godshall, M. A., Clarke, M. A., Tsang, W. S. C., and Parrish, F. W.,** Methanolysis of polysaccharides: a new method, *Carbohydr. Res.*, 168, 103, 1987.

54. **Rauvala, H., Finne, J., Krusius, T., Kärkäinen, J., and Järnefelt, J.,** Methylation techniques in the structural analysis of glycoproteins and glycolipids, *Adv. Carbohydr. Chem. Biochem.*, 38, 389, 1981.
55. **Aspinall, G. O.,** Chemical characterization and structure determination of polysaccharides, in *The Polysaccharides*, Vol. 1, Aspinall, G. O., Ed., Academic Press, New York, 1982, 35.
56. **Stephen, A. M., Churms, S. C., and Vogt, D. C.,** Exudate gums, in *Methods in Plant Biochemistry*, Vol. 2, Dey, P. M., Ed., Academic Press, London, 1990, 483.
57. **Lomax, J. A., Gordon, A. H., and Chesson, A.,** Methylation of unfractionated, primary and secondary cell-walls of plants, and the location of alkali-stable substituents, *Carbohydr. Res.*, 122, 11, 1983.
58. **Joseleau, J.-P., Chambat, G., and Chumpitazi-Hermoza, B.,** Solubilization of cellulose and other plant structural polysaccharides in 4-methylmorpholine-*N*-oxide: an improved method for the study of cell-wall constituents, *Carbohydr. Res.*, 90, 339, 1981.
59. **Narui, T., Takahashi, K., Kobayashi, M., and Shibata, S.,** Permethylation of polysaccharides by a modified Hakomori method, *Carbohydr. Res.*, 103, 293, 1982.
60. **Phillips, L. R. and Fraser, B. A.,** Methylation of carbohydrates with dimsyl potassium in dimethyl sulfoxide, *Carbohydr. Res.*, 90, 149, 1981.
61. **Blakeney, A. B. and Stone, B. A.,** Methylation of carbohydrates with lithium methylsulphinyl cation, *Carbohydr. Res.*, 140, 319, 1985.
62. **Paz Parente, J., Cardon, P., Leroy, Y., Montreuil, J., Fournet, B., and Ricart, G.,** A convenient method for methylation of glycoprotein glycans in small amounts by using lithium methylsulfinyl carbanion, *Carbohydr. Res.*, 141, 41, 1985.
63. **Kvernheim, A. L.,** Methylation analysis of polysaccharides with butyllithium in dimethyl sulfoxide, *Acta Chem. Scand. Ser. B*, 41, 150, 1987.
64. **Harris, P. J., Henry, R. J., Blakeney, A. B., and Stone, B. A.,** An improved procedure for the methylation analysis of oligosaccharides and polysaccharides, *Carbohydr. Res.*, 127, 59, 1984.
65. **Finne, J., Krusius, T., and Rauvala, H.,** Use of potassium *tert.*-butoxide in the methylation of carbohydrates, *Carbohydr. Res.*, 80, 336, 1980.
66. **Prehm, P.,** Methylation of carbohydrates by methyl trifluoromethanesulfonate in trimethyl phosphate, *Carbohydr. Res.*, 78, 372, 1980.
67. **Fügedi, P. and Nánási, P.,** A convenient method for the alkylation of carbohydrates, *J. Carbohydr. Nucleosides Nucleotides*, 8, 547, 1981.
68. **Ciucanu, I. and Kerek, F.,** A simple and rapid method for the permethylation of carbohydrates, *Carbohydr. Res.*, 131, 209, 1984.
69. **Isogai, A., Ishizu, A., Nakano, J., Eda, S., and Katō, K.,** A new facile methylation method for cell-wall polysaccharides, *Carbohydr. Res.*, 138, 99, 1985.
70. **Mabusela, W. T., Stephen, A. M., Rodgers, A. L., and Gerneke, D.,** A highly substituted glucan, coating *Helipterum eximium seeds, Carbohydr. Res.*, in press, 1990.
71. **Mort, A. J., Parker, S., and Kuo, M.-S.,** Recovery of methylated saccharides from methylation reaction mixtures using Sep-Pak C_{18} cartridges, *Anal. Biochem.*, 133, 380, 1983.
72. **Waeghe, T. J., Darvill, A. G., McNeil, M., and Albersheim, P.,** Determination, by methylation analysis, of the glycosyl-linkage compositions of microgram quantities of complex carbohydrates, *Carbohydr. Res.*, 123, 281, 1983.
73. **Åman, P., Franzén, L.-E., Darvill, J. E., McNeil, M., Darvill, A.G., and Albersheim, P.,** The structure of the acidic polysaccharide secreted by *Rhizobium phaseoli* strain 127 K38, *Carbohydr. Res.*, 103, 77, 1982.
74. **Conchie, J., Hay, A. J., and Lomax, J. A.,** A comparison of some hydrolytic and gas chromatographic procedures used in methylation analysis of the carbohydrate units of glycopeptides, *Carbohydr. Res.*, 103, 129, 1982.
75. **Chen, C. C. and McGinnis, G. D.,** The use of 1-methylimidazole as a solvent and catalyst for the preparation of aldononitrile acetates of aldoses, *Carbohydr. Res.*, 90, 127, 1981.
76. **McGinnis, G. D.,** Preparation of aldononitrile acetates using *N*-methylimidazole as catalyst and solvent, *Carbohydr. Res.*, 108, 284, 1982.
77. **Furneaux, R. H.,** Byproducts in the preparation of penta-*O*-acetyl-D-hexononitriles from D-glucose, *Carbohydr. Res.*, 113, 241, 1983.
78. **Slodki, M. E., England, R. E., Plattner, R. D., and Dick, W. E., Jr.,** Methylation analysis of NRRL dextrans by capillary gas-liquid chromatography, *Carbohydr. Res.*, 156, 199, 1986.
79. **Lehrfeld, J.,** Gas chromatographic analysis of mixtures containing aldonic acids, alditols, and glucose, *Anal. Chem.*, 56, 1803, 1984.
80. **Lehrfeld, J.,** Simultaneous gas-liquid chromatographic determination of aldonic acids and aldoses, *Anal. Chem.*, 57, 346, 1985.
81. **Lehrfeld, J.,** GLC determination of aldonic acids as acetylated aldonamides, *Carbohydr. Res.*, 135, 179, 1985.

82. **Lehrfeld, J.,** Simultaneous gas-liquid chromatographic determination of aldoses and alduronic acids, *J. Chromatogr.*, 408, 245, 1987.
83. **Walters, J. S. and Hedges, J. I.,** Simultaneous determination of uronic acids and aldoses in plankton, plant tissues, and sediment by capillary gas chromatography of *N*-hexylaldonamide and alditol acetates, *Anal. Chem.*, 60, 988, 1988.
84. **Little, M. R.,** Separation, by g.l.c. of enantiomeric sugars as diastereoisomeric dithioacetals, *Carbohydr. Res.*, 105, 1, 1982.
85. **Oshima, R., Kumanotani, J., and Watanabe, C.,** Gas-liquid chromatographic resolution of sugar enantiomers as diastereoisomeric methylbenzylaminoalditols, *J. Chromatogr.*, 259, 159, 1983.
86. **Oshima, R., Yamauchi, Y., and Kumanotani, J.,** Resolution of the enantiomers of aldoses by liquid chromatography of diastereoisomeric 1-(*N*-acetyl-α-methylbenzylamino)-1-deoxyalditol acetates, *Carbohydr. Res.*, 107, 169, 1982.
87. **Hara, S., Okabe, H., and Mihashi, K.,** Gas-liquid chromatographic separation of aldose enantiomers as trimethylsilyl ethers of methyl 2-(polyhydroxyalkyl)-thiazolidine-4(*R*)-carboxylates, *Chem. Pharm. Bull*, 35, 501, 1987.
88. **Schweer, H.,** Gas chromatographic separation of carbohydrate enantiomers as (−)-menthyloxime pertrifluoroacetates, on silicone OV-225, *J. Chromatogr.*, 243, 149, 1982.
89. **Schweer, H.,** Gas chromatographic separation of enantiomeric sugars as diastereomeric trifluoroacetylated (−)-bornyloximes, *J. Chromatogr.*, 259, 164, 1983.
90. **Das Neves, H. J. C., Riscado, A. M. V., and Frank, H.,** Derivatives for the analysis of monosaccharides by capillary GLC: trimethylsilylated deoxy(methoxyamino)alditols, *Carbohydr. Res.*, 152, 1, 1986.
91. **Das Neves, H. J. C., Riscado, A. M. V., and Frank, H.,** Single derivatives for GC-MS assay of reducing sugars and selective detection by the nitrogen-phosphorus detector (NPD), *J. High Resolut. Chromatogr. Chromatogr. Commun.*, 9, 662, 1986.
92. **Das Neves, H. J. and Riscado, A. M. V.,** Capillary gas chromatography of reducing disaccharides with nitrogen-selective detection and selection-ion monitoring of permethylated deoxy(methylmethoxyamino)alditol glycosides, *J. Chromatogr.*, 367, 135, 1986.
93. **Honda, S., Yamauchi, N., and Kakehi, K.,** Rapid gas chromatographic analysis of aldoses as their diethyl dithioacetal trimethylsilylates, *J. Chromatogr.* 169, 287, 1979.
94. **Honda, S., Kakehi, K., and Okada, K.,** Convenient method for the gas chromatographic analysis of hexosamines in the presence of neutral monosaccharides and uronic acids, *J. Chromatogr.*, 176, 367, 1979.
95. **Honda, S., Nagata, M., and Kakehi, K.,** Rapid gas chromatographic analysis of partially methylated aldoses as trimethylsilylated diethyl dithioacetals, *J. Chromatogr.*, 209, 299, 1981.
96. **Seymour, F. R., Chen, E. C. M., and Stouffer, J. E.,** Identification of ketoses by use of their peracetylated oxime derivatives: a g.l.c.-m.s. approach, *Carbohydr. Res.*, 83, 201, 1980.
97. **Mawhinney, T. P., Feather, M. S., Barbero, G. J., and Martinez, J. R.,** The rapid quantitative determination of neutral sugars (as aldononitrile acetates) and amino sugars (as *O*-methyloxime acetates) in glycoproteins by gas-liquid chromatography, *Anal. Biochem.*, 101, 112, 1980.
98. **Neeser, J.-R. and Schweizer, T. F.,** A quantitative determination by capillary gas-liquid chromatography of neutral and amino sugars (as *O*-methyloxime acetates), and a study of hydrolytic conditions for glycoproteins and polysaccharides in order to increase sugar recoveries, *Anal. Biochem.*, 142, 58, 1984.
99. **Guerrant, G. O. and Moss, C. W.,** Determination of monosaccharides as aldononitrile, *O*-methyloxime, alditol, and cyclitol acetate derivatives by gas chromatography, *Anal. Chem.*, 56, 633, 1984.
100. **Biondi, P. A., Manca, F., Negri, A., Secchi, C., and Montana, M.,** Gas chromatographic analysis of neutral monosaccharides as their *O*-pentafluorobenzyloxime acetates, *J. Chromatogr.*, 411, 275, 1987.
101. **Schweer, H.,** Gas-chromatography-mass spectrometry of aldoses as *O*-methoxime, *O*-2-methyl-2-propoxime and *O*-*n*-butoxime pertrifluoroacetyl derivatives on OV-225 with methylpropane as ionization agent. I. Pentoses, *J. Chromatogr.*, 236, 355, 1982.
102. **Schweer, H.,** Gas-chromatography-mass spectrometry of aldoses as *O*-methoxime, *O*-2-methyl-2-propoxime and *O*-*n*-butoxime pertrifluoroacetyl derivatives on OV-225 with methylpropane as ionization agent. II. Hexoses, *J. Chromatogr.*, 236, 361, 1982.
103. **Englmaier, P.,** Trifluoroacetylation of carbohydrates for GLC, using *N*-methylbis(trifluoroacetamide), *Carbohydr. Res.*, 144, 177, 1985.
104. **Shinohara, T.,** Use of a flame thermionic detector in the determination of glucosamine and galactosamine in glycoconjugates by gas chromatography, *J. Chromatogr.*, 207, 262, 1981.
105. **Haga, H. and Nakajima, T.,** Analysis of aldoses and alditols by capillary gas chromatography as alditol trifluoroacetates, *Chem. Pharm. Bull*, 36, 1562, 1988.
106. **Nilsson, B. and Zopf, D.,** Gas chromatography and mass spectrometry of hexosamine-containing oligosaccharidealditols as their permethylated, *N*-trifluoroacetyl derivatives, *Methods Enzymol.*, 83, 46, 1982.
107. **Páez, M., Martínez-Castro, I., Sanz, J., Olano, A., García-Raso, A., and Saura-Calixto, F.,** Identification of the components of aldoses in a tautomeric equilibrium mixture as their trimethylsilyl ethers by capillary gas chromatography, *Chromatographia*, 23, 43, 1987.

108. **García -Raso, A., Martínez-Castro, I., Páez, M. I., Sanz, J., García-Raso, J., and Saura-Calixto, F.,** Gas chromatographic behaviour of carbohydrate trimethylsilyl ethers. I. Aldopentoses, *J. Chromatogr.*, 398, 9, 1987.
109. **Martínez-Castro, I., Páez, M. I., Sanz, J., and García-Raso, A.,** Gas chromatographic behaviour of carbohydrate trimethylsilyl ethers. II. Aldohexoses, *J. Chromatogr.*, 462, 49, 1989.
110. **Nikolov, Z. L. and Reilly, P. J.,** Isothermal capillary column gas chromatography of trimethylsilyl disaccharides, *J. Chromatogr.*, 254, 157, 1983.
111. **Laker M. F.,** Estimation of disaccharides in plasma and urine by gas-liquid chromatography, *J. Chromatogr.*, 163, 9, 1979.
112. **Hyppänen, T., Sjöström, E., and Vuorinen, T.,** Gas-liquid chromatographic determination of hydroxy carboxylic acids on a fused-silica capillary column, *J. Chromatogr.*, 261, 320, 1983.
113. **Alén, R., Niemelä, K., and Sjöström, E.,** Gas-liquid chromatographic separation of hydroxy monocarboxylic acids and dicarboxylic acids on a fused-silica capillary column, *J. Chromatogr.*, 301, 273, 1984.
114. **Niwa, T., Yamamoto, N., Maeda, K., Yamada, K., Ohki, T., and Mori, M.,** Gas chromatographic-mass spectrometric analysis of polyols in urine and serum of uremic patients. Identification of new deoxyalditols and inositol isomers, *J. Chromatogr.*, 277, 25, 1983.
115. **Niwa, T. and Yamada, K.,** Identification of 6-deoxyallitol and 6-deoxygulitol in human urine. Electron-impact mass spectra of eight isomers of 6-deoxyhexitol, *J. Chromatogr.*, 336, 345, 1984.
116. **Wells, G. B. and Lester, R. L.,** Rapid separation of acetylated oligosaccharides by reverse-phase high-pressure liquid chromatography, *Anal. Biochem.*, 97, 184, 1979.
117. **Wells, G. B., Kontoyiannidou, V., Turco, S. J., and Lester, R. L.,** Resolution of acetylated oligosaccharides by reverse-phase high-pressure liquid chromatography, *Methods Enzymol.*, 83, 132, 1982.
118. **White, C. A., Kennedy, J. F., and Golding, B. T.,** Analysis of derivatives of carbohydrates by high-pressure liquid chromatography, *Carbohydr. Res.*, 76, 1, 1979.
119. **Oshima, R. and Kumanotani, J.,** Determination of neutral, amino and *N*-acetyl amino sugars as alditol benzoates by liquid-solid chromatography, *J. Chromatogr.*, 265, 335, 1983.
120. **Daniel, P. F., De Feudis, D. F., Lott, I. T., and McCluer, R. H.,** Quantitative microanalysis of oligosaccharides by high-performance liquid chromatography, *Carbohydr. Res.*, 97, 161, 1981.
121. **Gross, S. K. and McCluer, R. H.,** High-performance liquid chromatographic analysis of neutral glycosphingolipids as their per-*O*-benzoyl derivatives, *Anal. Biochem.*, 102, 429, 1980.
122. **Bremer, E. G., Gross, S. K., and McCluer, R. H.,** Quantitative analysis of monosialogangliosides by high-performance liquid chromatography of their perbenzoyl derivatives, *J. Lipid Res.*, 20, 1028, 1979.
123. **White, C. A., Vass, S. W., Kennedy, J. F., and Large, D. G.,** High-pressure liquid chromatography of dimethylphenylsilyl derivatives of some monosaccharides, *Carbohydr. Res.*, 119, 241, 1983.
124. **White, C. A., Vass, S. W., Kennedy, J. F., and Large, D. G.,** Analysis of phenyldimethylsilyl derivatives of monosaccharides and their role in high-performance liquid chromatography of carbohydrates, *J. Chromatogr.*, 264, 99, 1983.
125. **Batley, M., Redmond, J. W., and Tseng, A.,** Sensitive analysis of aldose sugars by reversed-phase high-performance liquid chromatography, *J. Chromatogr.*, 253, 124, 1982.
126. **Hara, S., Yamaguchi, M., Takemori, Y., and Nakamura, M.,** Highly sensitive determination of *N*-acetyl- and *N*-glycolylneuraminic acids in human serum and urine and rat serum by reversed-phase liquid chromatography with fluorescence detection, *J. Chromatogr.*, 377, 111, 1986.
127. **Honda, S. and Kakehi, K.,** Periodate oxidation analysis of carbohydrates. VII. High-performance liquid chromatographic determination of conjugated aldehydes in products of periodate oxidation of carbohydrates by dual-wavelength detection of their 2,4-dinitrophenylhydrazones, *J. Chromatogr.*, 152, 405, 1978.
128. **Karamanos, N. K., Tsegenidis, T. and Antonopoulos, C. A.,** Analysis of neutral sugars as dinitrophenylhydrazones by high-performance liquid chromatography, *J. Chromatogr.*, 405, 221, 1987.
129. **Alpenfels, W. F.,** A rapid and sensitive method for the determination of monosaccharides as their dansyl hydrazones by high-performance liquid chromatography, *Anal. Biochem.*, 114, 153, 1981.
130. **Takeda, M., Maeda, M., and Tsuji, A.,** Fluorescence high-performance liquid chromatography of reducing sugars using Dns-hydrazine as a pre-labelling reagent, *J. Chromatogr.*, 244, 347, 1982.
131. **Mopper, K. and Johnson, L.,** Reversed-phase liquid chromatographic analysis of Dns-sugars. Optimization of derivatization and chromatographic procedures and applications to natural samples, *J. Chromatogr.*, 256, 27, 1983.
132. **Lin, J.-K. and Wu, S.-S.** Synthesis of dabsylhydrazine and its use in the chromatographic determination of monosaccharides by thin-layer and high-performance liquid chromatography, *Anal. Chem.*, 59, 1320, 1987.
133. **Muramoto, K., Goto, R., and Kamiya, H.,** Analysis of reducing sugars as their chromophoric hydrazones by high-performance liquid chromatography, *Anal. Biochem.*, 162, 435, 1987.
134. **Valent, B. S., Darvill, A. G., McNeil, M., Robertsen, B. K., and Albersheim, P.,** A general and sensitive chemical method for sequencing the glycosyl residues of complex carbohydrates, *Carbohydr. Res.*, 79, 165, 1980.

135. **McNeil, M., Darvill, A. G., Åman, P., Franzén, L.-E., and Albersheim, P.,** Structural analysis of complex carbohydrates using high-performance liquid chromatography, gas chromatography and mass spectrometry, *Methods Enzymol.*, 83, 3, 1982.
136. **Rosenfelder, G., Mörgelin, M., Chang, J.-Y., Schönenberger, C.-A., Braun, D. G., and Towbin, H.,** Chromogenic labeling of monosaccharides using 4'-N,N-dimethylamino-4-aminoazobenzene, *Anal. Biochem.*, 147, 156, 1985.
137. **Takemoto, H., Hase, S., and Ikenaka, T.,** Microquantitative analysis of neutral and amino sugars as fluorescent pyridylamino derivatives by high-performance liquid chromatography, *Anal. Biochem.*, 145, 245, 1985.
138. **Hase, S., Ikenaka, K., Mikoshiba, K., and Ikenaka, T.,** Analysis of tissue glycoprotein sugar chains by two-dimensional high-performance liquid chromatographic mapping, *J. Chromatogr.*, 434, 51, 1988.
139. **Coles, E., Reinhold, V. N., and Carr, S. A.,** Fluororescent labeling of carbohydrates and analysis by liquid chromatography. Comparison of derivatives using mannosidosis oligosaccharides, *Carbohydr. Res.*, 139, 1, 1985.
140. **Perini, F. and Peters, B. P.,** Fluorometric analysis of amino sugars and derivatized neutral sugars, *Anal. Biochem.*, 123, 357, 1982.
141. **Shinomiya, K., Toyoda, H., Akahoshi, A., Ochiai, H., and Imanari, T.,** Fluorometric analysis of neutral sugars as their glycamines by high-performance liquid chromatography and thin-layer chromatography, *J. Chromatogr.*, 387, 481, 1987.

Section IV

Literature References

LITERATURE REFERENCES

The updated bibliography (see Volume I for books published before 1979) lists some recommended books published during the last 10 years. Titles have been selected in an attempt to cover the range of chromatographic methods now available; thus, this should be regarded as a representative selection of books on chromatography rather than a complete bibliography. The books are of general interest, not devoted to carbohydrate chromatography *per se*, although some cite examples drawn from this field. The format used previously, indicating the forms of chromatography discussed in each book, has been adopted in this bibliography.

The list of review articles and research papers that are considered to be major contributions to the field of carbohydrate chromatography covers the years 1979 to 1989, updating that in Volume I. As before, these journal references are grouped according to the form of chromatography involved and, within each group, are arranged in alphabetical order with respect to the surname of the first author as listed in the paper. Where necessary, the mode of chromatography used is indicated after the reference.

BIBLIOGRAPHY

	Gas	SFC	HPLC	SEC	Affinity	Planar	Electrophoresis
Academic Press, New York, London							
Horváth, C., Ed., *High-Performance Liquid Chromatography: Advances and Perspectives*							
Vol. 1, 1980.			X				
Vol. 2, 1980.			X				
Vol. 3, 1983.			X				
Vol. 4, 1987.			X				
Vol. 5, 1988.			X		X		
Jennings, W., *Gas Chromatography with Glass Capillary Columns,* 2nd ed., 1980.	X						
Jennings, W., *Analytical Gas Chromatography,* 1987.	X						
Lawrence J. F., *Organic Trace Analysis by Liquid Chromatography,* 1981.			X				
Macrae, R., *HPLC in Food Analysis,* 2nd ed., 1988.			X				
Whistler, R. L. and BeMiller, J. N., Eds., *Methods in Carbohydrate Chemistry,* Vol. 8, 1980.	X		X	X			
Chapman and Hall, London, New York							
Hamilton, R. J. and Sewell, P. A., *Introduction to High-Performance Liquid Chromatography,* 2nd ed., 1982.			X	X			

	Gas	SFC	HPLC	SEC	Affinity	Planar	Electrophoresis
CRC Press, Boca Raton, FL							
Biermann, C. J. and McGinnis, G. D., Eds., *Analysis of Carbohydrates by GLC and MS, 1989.*	*X*		*X*				
Dionex Corporation, Sunnyvale, CA							
Weiss, J. (Johnson, E. L., Ed.,), *Handbook of Ion Chromatography,* 1986.			X				
Elsevier Science Publishers, Amsterdam							
Deyl, Z., Ed., *Separation Methods,* New Comprehensive Biochemistry Series, Vol. 8, Neuberger, A. and Van Deenen, L. L. M., Eds., 1984.	X		X	X	X	X	X
Deyl, Z., Ed., *Electrophoresis: A Survey of Techniques and Applications,* Journal of Chromatography Library Series, Vol. 18.							
Part A: *Techniques,* 1979.							X
Part B: *Applications,* 1982.							X
Heftmann, E., Ed., *Chromatography: Fundamentals and Applications of Chromatographic and Electrophoretic Methods,* Journal of Chromatography Library Series, Vol. 22.							
Part A: *Fundamentals and Techniques,* 1983.	X		X	X	X	X	X
Part B: *Applications,* 1983.	X		X	X	X	X	X
Heftmann, E., Ed., *Chromatography: Fundamentals and Applications of Chromatographic and Electrophoretic Methods,* Journal of Chromatography Library Series, new edition, in press.							
Part A: *Fundamentals and Techniques.*	X	X	X	X	X	X	X
Part B: *Applications.*	X	X	X	X	X	X	X
Poole, C. F. and Schuette, S. A., *Contemporary Practice of Chromatography,* 1984.	X	X	X	X		X	
Ellis Horwood, Chichester, U.K.							
Mikeš, O., Ed. (Procházka, Z., translator), *Laboratory Handbook of Chromatographic and Allied Methods,* 1979.				X		X	
Snyder, L. R., Glajch, J. L., and Kirkland, J. J., *Practical HPLC Method Development,* 1989.			X				
Plenum Press, New York							
Dallas, F. A. A., Read, H., Ruane, R. J., and Wilson, I. D., Eds., *Recent Advances in Thin-Layer Chromatography, 1988.*						X	

	Gas	SFC	HPLC	SEC	Affinity	Planar	Electrophoresis
Springer-Verlag, Berlin, Heidelberg, New York							
Engelhardt, H., Ed., *Practice of High-Performance Liquid Chromatography,* 1986.			X				
Hostettmann, K., Hostettmann, M., and Marston, A., *Preparative Chromatography Techniques,* 1986.			X	X		X	
John Wiley & Sons, Chichester, U.K.							
Meyer, V. R., *Practical High-Performance Liquid Chromatography,* 1988.			X				
Smith, R. M., *Gas and Liquid Chromatography in Analytical Chemistry,* 1988.	X	X	X				
Wiley-Interscience, New York							
Scouten, W. H., *Affinity Chromatography,* 1981.					X		

REVIEWS AND SELECTED JOURNAL REFERENCES

REVIEW ARTICLES

Gas Chromatography

1. **Laker, M. F.**, Estimation of neutral sugars and sugar alcohols in biological fluids by gas-liquid chromatography, *J. Chromatogr.*, 184, 457, 1980.

Supercritical Fluid Chromatography

2. **Gere, D. R.**, Supercritical fluid chromatography, *Science*, 222, 253, 1983.
3. **Lee, M. L. and Markides, K. E.**, Chromatography with supercritical fluids, *Science*, 235, 1342, 1987.

Liquid Chromatography

4. **Barth, H. G.**, A practical approach to steric exclusion chromatography of water-soluble polymers, *J. Chromatogr. Sci.*, 18, 409, 1980.
5. **Hicks, K. B.**, High-performance liquid chromatography of carbohydrates, *Adv. Carbohydr. Chem. Biochem.*, 46, 17, 1988.
6. **Honda, S.**, High-performance liquid chromatography of mono- and oligosaccharides, *Anal. Biochem.*, 140, 1, 1984.
7. **Lin, J. K., Jacobson, B. J., Pereira, A. N., and Ladisch, M. R.**, Liquid chromatography of carbohydrate monomers and oligomers, *Methods Enzymol.*, 160, 145, 1988.
8. **Olechno, J. D., Carter, S. R., Edwards, W. T., and Gillen, D. G.**, Developments in the chromatographic determination of carbohydrates, *Am. Biotechnol. Lab.*, 5(5), 38, 1987.
9. **Verhaar, L. A. Th. and Kuster, B. F. M.**, Liquid chromatography of sugars on silica-based stationary phases, *J. Chromatogr.*, 220, 313, 1981.
10. **Verzele, M., Simoens, G., and Van Damme, F.**, A critical review of some liquid chromatography systems for the separation of sugars, *Chromatographia*, 23, 292, 1987.
11. **Wood, R., Cummings, L., and Jupille, T.**, Recent developments in ion-exchange chromatography, *J. Chromatogr. Sci.*, 18, 551, 1980.
12. **Yamashita, K., Mizuochi, T., and Kobata, A.**, Analysis of oligosaccharides by gel filtration, *Methods Enzymol.*, 83, 105, 1982.

Thin-Layer Chromatography

13. **Doner, L. W.**, High-performance thin-layer chromatography of starch, cellulose, xylan, and chitin hydrolyzates, *Methods Enzymol.*, 160, 176, 1988.
14. **Klaus, R. and Fischer, W.**, Quantitative thin-layer chromatography of sugars, sugar acids, and polyalcohols, *Methods Enzymol.*, 160, 159, 1988.

General (All Methods)

15. **Beaty, N. B. and Mello, R. J.**, Extracellular mammalian polysaccharides: glycosaminoglycans and proteoglycans, *J. Chromatogr.*, 418, 187, 1987.
16. **Churms, S. C.**, Recent developments in chromatographic analysis of carbohydrates, *J. Chromatogr.*, 500, 555, 1990.
17. **Kakehi, K. and Honda, S.**, Profiling of carbohydrates, glycoproteins and glycolipids, *J. Chromatogr.*, 379, 27, 1986.
18. **Kodama, C., Kodama, T., and Yosizawa, Z.**, Methods for analysis of urinary glycosaminoglycans, *J. Chromatogr.* 429, 293, 1988.
19. **McNeil, M., Darvill, A. G., Åman, P., Franzén, L.-E., and Albersheim, P.**, Structural analysis of complex carbohydrates using high-performance liquid chromatography, gas chromatography and mass spectrometry, *Methods Enzymol.*, 83, 3, 1982.
20. **Robards, K. and Whitelaw, M.**, Chromatography of monosaccharides and disaccharides, *J. Chromatogr.*, 373, 81, 1986.

IMPORTANT PAPERS ON SPECIFIC ASPECTS

Gas Chromatography

If not mentioned in the title of the paper, the type of derivative used is included in parentheses after the reference.

1. **Bacic, A., Harris, P. J., Hak, E. W., and Clarke, A. E.**, Capillary gas chromatography of partially methylated alditol acetates on a high-polarity bonded-phase vitreous-silica column, *J. Chromatogr.*, 315, 373, 1984.

2. **Blakeney, A. B., Harris, P. J., Henry, R. J., Stone, B. A., and Norris, T.,** Gas chromatography of alditol acetates on a high-polarity bonded-phase vitreous-silica column, *J. Chromatogr.*, 249, 180, 1982.
3. **Blakeney, A. B., Harris, P. J., Henry, R. J., and Stone, B. A.,** A simple and rapid preparation of alditol acetates for monosaccharide analysis, *Carbohydr. Res.*, 113, 291, 1983.
4. **Bryn, K. and Jantzen, E.,** Analysis of lipopolysaccharides by methanolysis, trifluoroacetylation, and gas chromatography on a fused-silica capillary column, *J. Chromatogr.*, 240, 405, 1982.
5. **Chaplin, M. F.,** A rapid and sensitive method for the analysis of carbohydrate components in glycoproteins using gas-liquid chromatography, *Anal. Biochem.*, 123, 336, 1982. (Trimethylsilylated methyl glycosides)
6. **Das Neves, H. J. C., Riscado, A. M. V., and Frank, H.,** Single derivatives for GC-MS assay of reducing sugars and selective detection by the nitrogen-phosphorous detector (NPD), *J. High Resolut. Chromatogr. Chromatogr. Commun.*, 9, 662, 1986. (Permethylated deoxy(methylmethoxyamino)alditols)
7. **Das Neves, H. J. C. and Riscado, A. M. V.,** Capillary gas chromatography of reducing disaccharides with nitrogen-selective detection and selected-ion monitoring of permethylated deoxy(methylmethoxyamino)alditol glycosides, *J. Chromatogr.*, 367, 135, 1986.
8. **Elkin, Yu. N.,** Gas chromatographic separation of the methyl ester methylglycopyranoside series hexose, 6-deoxyhexose and pentose acetates, *J. Chromatogr.*, 180, 163, 1979.
9. **Englmaier, P.,** Trifluoroacetylation of carbohydrates for GLC, using *N*-methylbis(trifluoroacetamide), *Carbohydr. Res.*, 144, 177, 1985.
10. **Furneaux, R. H.,** Byproducts in the preparation of penta-*O*-acetyl-D-hexononitriles from D-galactose and D-glucose, *Carbohydr. Res.*, 113, 241, 1983.
11. **García-Raso, A., Martínez-Castro, I., Páez, M. I., Sanz, J., García-Raso, J., and Saura-Calixto, F.,** Gas chromatographic behaviour of carbohydrate trimethylsilyl ethers. I. Aldopentoses, *J. Chromatogr.*, 398, 9, 1987.
12. **Gerwig, G. J., Kamerling, J. P., and Vliegenthart, J. F. G.,** Determination of the absolute configuration of monosaccharides in complex carbohydrates by capillary g.l.c., *Carbohydr. Res.*, 77, 1, 1979. (Trimethylsilylated(−)-2-butyl glycosides of hexuronic acids and 2-acetamido-2-deoxyhexoses)
13. **Guerrant, G. O. and Moss, C. W.,** Determination of monosaccharides as aldononitrile, *O*-methyloxime, alditol, and cyclitol acetate derivatives by gas chromatography, *Anal. Chem.*, 56, 633, 1984.
14. **Haga, H. and Nakajima, T.,** Analysis of aldoses and alditols by capillary gas chromatography as alditol trifluoroacetates, *Chem. Pharm. Bull*, 36, 1562, 1988.
15. **Hara, S., Okabe, H. and Mihashi, K.,** Gas-liquid chromatographic separation of aldose enantiomers as trimethylsilyl ethers of methyl 2-(polyhydroxyalkyl)thiazolidine-4(*R*)-carboxylates, *Chem. Pharm. Bull*, 35, 501, 1987.
16. **Harris, P. J., Henry, R. J., Blakeney, A. B., and Stone, B. A.,** An improved procedure for the methylation analysis of oligosaccharides and polysaccharides, *Carbohydr. Res.*, 127, 59, 1984. (Partially methylated alditol acetates)
17. **Harris, P. J., Bacic, A., and Clarke, A. E.,** Capillary gas chromatography of partially methylated alditol acetates on a SP-2100 wall-coated open-tubular column, *J. Chromatogr.*, 350, 304, 1985.
18. **Henry, R. J., Blakeney, A. B., Harris, P. J., and Stone, B. A.,** Detection of neutral and aminosugars from glycoproteins and polysaccharides as their alditol acetates, *J. Chromatogr.*, 256, 419, 1983.
19. **Klok, J., Cox, H. C., De Leeuw, J. W., and Schenck, P. A.,** Analysis of synthetic mixtures of partially methylated alditol acetates by capillary gas chromatography, gas chromatography-electron impact mass spectrometry and gas chromatography-chemical ionization mass spectrometry, *J. Chromatogr.*, 253, 55, 1982.
20. **König, W. A., Benecke, I., and Bretting, H.,** Gas chromatographic separation of carbohydrate enantiomers on a new chiral stationary phase, *Angew. Chem. Int. Ed. Engl.*, 20, 693, 1981. (Trifluoroacetates on XE-60-L-valine-(*S*)-α-phenylethylamide)
21. **König, W. A., Benecke, I., and Sievers, S.,** New results in the gas chromatographic separation of enantiomers of hydroxy acids and carbohydrates, *J. Chromatogr.*, 217, 71, 1981. (Same as Reference 20)
22. **König, W. A., Lutz, S., Mischnick-Lübbecke, P., Brassat, B., and Wenz, G.,** Cyclodextrins as chiral stationary phases in capillary gas chromatography. I. Pentylated α-cyclodextrin, *J. Chromatogr.*, 447, 193, 1988. (Trifluoroacetates)
23. **König, W. A., Mischnick-Lübbecke, P., Brassat, B., Lutz, S., and Wenz, G.,** Improved gas chromatographic separation of enantiomeric carbohydrate derivatives using a new chiral stationary phase, *Carbohydr. Res.*, 183, 11, 1988. (Trifluoroacetates on pentylated α-cyclodextrin)
24. **Lehrfeld, J.,** Simultaneous gas-liquid chromatographic determination of aldonic acids and aldoses, *Anal. Chem.*, 57, 346, 1985. (Aldonic acids as *N*-propylaldonamide acetates, aldoses as alditol acetates)
25. **Lehrfeld, J.,** Simultaneous gas-liquid chromatographic determination of aldoses and alduronic acids, *J. Chromatogr.*, 408, 245, 1987. (Alduronic acids converted to aldonolactones, then *N*-propylaldonamide acetates; aldoses to alditol acetates)
26. **Little, M. R.,** Separation, by g.l.c., of enantiomeric sugars as diastereoisomeric dithioacetals, *Carbohydr. Res.*, 105, 1,1982. (Bis[(+)-1-phenylethyl]dithioacetals)

27. **Lomax, J. A. and Conchie, J.,** Separation of methylated alditol acetates by glass capillary gas chromatography and their identification by computer, *J. Chromatogr.*, 236, 385, 1982.
28. **Lomax, J. A., Gordon, A. H., and Chesson, A.,** A multiple-column approach to the methylation of plant cell-walls, *Carbohydr. Res.*, 138, 177, 1985. (Partially methylated alditol acetates on three phases of different polarity)
29. **Martínez-Castro, I., Páez, M. I., Sanz, J., and García-Raso, A.,** Gas chromatographic behaviour of carbohydrate trimethylsilyl ethers. II. Aldohexoses, *J. Chromatogr.*, 462, 49, 1989.
30. **Mawhinney, T. P., Feather, M. S., Barbero, G. J., and Martinez, J. R.,** The rapid, quantitative determination of neutral sugars (as aldononitrile acetates) and amino sugars (as *O*-methyloxime acetates) in glycoproteins by gas-liquid chromatography, *Anal. Biochem.*, 101, 112, 1980.
31. **Mawhinney, T. P.,** Simultaneous determination of *N*-acetylglucosamine, *N*-acetylgalactosamine, *N*-acetylglucosaminitol and *N*-acetylgalactosaminitol by gas-liquid chromatography, *J. Chromatogr.*, 351, 91, 1986. (Acetamidodeoxyhexoses converted to *O*-methyloximes; these oximes and acetamidodeoxyhexitols separated as either per-*O*-acetylated or per-*O*-trimethylsilylated derivatives)
32. **McGinnis, G. D.,** Preparation of aldononitrile acetates using *N*-methylimidazole as catalyst and solvent, *Carbohydr. Res.*, 108, 284, 1982.
33. **Neeser, J.-R. and Schweizer, T. F.,** A quantitative determination by capillary gas-liquid chromatography of neutral and amino sugars (as *O*-methyloxime acetates), and a study of hydrolytic conditions for glycoproteins and polysaccharides in order to increase sugar recoveries, *Anal. Biochem.*, 142, 58, 1984.
34. **Neeser, J.-R.,** G.l.c. of *O*-methyloxime and alditol acetate derivatives of neutral sugars, hexosamines, and sialic acids: "one pot" quantitative determination of the carbohydrate constituents of glycoproteins and a study of the selectivity of alkaline borohydride reductions, *Carbohydr. Res.*, 138, 189, 1985.
35. **Nikolov, Z. L. and Reilly, P. J.,** Isothermal capillary column gas chromatography of trimethylsilyl disaccharides, *J. Chromatogr.*, 254, 157, 1983.
36. **Nilsson, B. and Zopf, D.,** Gas chromatography and mass spectrometry of hexosamine-containing oligosaccharide alditols as their permethylated, *N*-trifluoroacetyl derivatives, *Methods Enzymol.*, 83, 46, 1982.
37. **Niwa, T., Yamamoto, N., Maeda, K., Yamada, K., Ohki, T., and Mori, M.,** Gas chromatographic-mass spectrometric analysis of polyols in urine and serum of uremic patients. Identification of new deoxyalditols and inositol isomers, *J. Chromatogr.*, 277, 25, 1983. (Trimethylsilyl ethers)
38. **Niwa, T., Yamada, K., Ohki, T., Saito, A., and Mori, M.,** Identification of 6-deoxyallitol and 6-deoxygulitol in human urine. Electron-impact mass spectra of eight isomers of 6-deoxyhexitol. *J. Chromatogr.*, 336, 345, 1984. (Same as Reference 37)
39. **Oshima, R., Yoshikawa, A., and Kumanotani, J.,** High-resolution gas chromatographic separation of alditol acetates on fused-silica wall-coated open-tubular columns, *J. Chromatogr.*, 213, 142, 1981.
40. **Oshima, R., Kumanotani, J., and Watanabe, C.,** Fused-silica capillary gas chromatographic separation of alditol acetates of neutral and amino sugars, *J. Chromatogr.*, 250, 90, 1982.
41. **Oshima, R., Kumanotani, J., and Watanabe, C.,** Gas-liquid chromatographic resolution of sugar enantiomers as diastereoisomeric methylbenzylaminoalditols, *J. Chromatogr.*, 259, 159, 1983. (L-α-Methylbenzylaminoalditols separated as peracetylated or trimethylsilylated derivatives)
42. **Páez, M., Martínez-Castro, I., Sanz, J., Olano, A., García-Raso, A., and Saura-Calixto, F.,** Identification of the components of aldoses in a tautomeric equilibrium mixture as their trimethylsilyl ethers by capillary gas chromatography, *Chromatographia*, 23, 43, 1987.
43. **Pelletier, O. and Cadieux, S.,** Glass capillary or fused-silica gas chromatography-mass spectrometry of several monosaccharides and related sugars: improved resolution, *J. Chromatogr.*, 231, 225, 1982. (Trimethylsilylated *O*-methyloximes)
44. **Schweer, H.,** Gas chromatography-mass spectrometry of aldoses as *O*-methoxime, *O*-2-methyl-2-propoxime and *O*-*n*-butoxime pertrifluoroacetyl derivatives on OV-225 with methylpropane as ionization agent. I. Pentoses, *J. Chromatogr.*, 236, 355, 1982; II. Hexoses, *J. Chromatogr.*, 236, 361, 1982.
45. **Schweer, H.,** Gas chromatographic separation of carbohydrate enantiomers as (−)-menthyloxime pertrifluoroacetates on silicone OV-225, *J. Chromatogr.*, 243, 149, 1982.
46. **Schweer, H.,** Gas chromatographic separation of enantiomeric sugars as diastereomeric trifluoroacetylated (−)-bornyloximes, *J. Chromatogr.*, 259, 164, 1983.
47. **Selosse, E. J.-M. and Reilly, P. J.,** Capillary column gas chromatography of trifluoroacetyl trisaccharides, *J. Chromatogr.*, 328, 253, 1985.
48. **Seymour, F. R., Chen, E. C. M., and Bishop, H.,** Identification of aldoses by use of their peracetylated aldononitrile derivatives: a g.l.c.-m.s. approach, *Carbohydr. Res.*, 73, 19, 1979.
49. **Seymour, F. R., Chen, E. C. M., and Stouffer, J. E.,** Identification of ketoses by use of their peracetylated oxime derivatives: a g.l.c.-m.s. approach, *Carbohydr. Res.*, 83, 201, 1980.
50. **Slodki, M. E., England, R. E., Plattner, R. D., and Dick, W. E., Jr.,** Methylation analysis of NRRL dextrans by capillary gas-liquid chromatography, *Carbohydr. Res.*, 156, 199, 1986. (Partially methylated aldononitrile acetates)

51. **Tanner, G. R. and Morrison, I. M.,** Gas chromatography-mass spectrometry of partially methylated glycoses as their aldononitrile peracetates, *J. Chromatogr.*, 299, 252, 1984.
52. **Waeghe, T. J., Darvill, A. G., McNeil, M., and Albersheim P.,** Determination, by methylation analysis, of the glycosyl-linkage compositions of microgram quantities of complex carbohydrates, *Carbohydr. Res.*, 123, 281, 1983. (Deuterium-labeled, partially methylated alditol acetates)

Supercritical Fluid Chromatography

53. **Kuei, J., Her, G.-R., and Reinhold, V. N.,** Supercritical fluid chromatography of glycosphingolipids, *Anal. Biochem.*, 172, 228, 1988.
54. **Reinhold, V. N., Sheeley, D. M,. Kuei, J., and Her, G.-R.,** Analysis of high molecular-weight samples on a double-focusing magnetic sector instrument by supercritical fluid chromatography/mass spectrometry, *Anal. Chem.*, 60, 2719, 1988.

Liquid Chromatography

Affinity Chromatography

55. **Borchert, A., Larsson, P.-O., and Mosbach, K.,** High-performance liquid affinity chromatography on silica-bound concanavalin A, *J. Chromatogr.*, 244, 49, 1982.
56. **Borrebaeck, C. A. K., Soares, J., and Mattiasson, B.,** Fractionation of glycoproteins according to lectin affinity and molecular size using a high-performance liquid chromatography system with sequentially coupled columns, *J. Chromatogr.*, 284, 187, 1984.
57. **El Rassi, Z., Truei, Y., Maa, Y.-F., and Horváth, C.,** High-performance liquid chromatography with concanavalin A immobilized by metal interactions on the stationary phase, *Anal. Biochem.*, 169, 172, 1988.
58. **Honda, S., Suzuki, S., Nitta, T., and Kakehi, K.,** Analytical high-performance affinity chromatography of ovalbumin-derived glycopeptides on columns of concanavalin A- and wheat germ agglutinin-immobilized gels, *J. Chromatogr.*, 438, 73, 1988.
59. **Muller, A. J. and Carr, P. W.,** Chromatographic study of the thermodynamic and kinetic characteristics of silica-bound concanavalin A, *J. Chromatogr.*, 284, 33, 1984.
60. **Muller, A. J. and Carr, P. W.,** Examination of kinetic effects in the high-performance liquid affinity chromatography of glycoproteins by stopped-flow and pulsed-elution methods, *J. Chromatogr.*, 294, 235, 1984.
61. **Muller, A. J. and Carr, P. W.,** Examination of the thermodynamic and kinetic characteristics of microparticulate affinity chromatography supports. Application to concanavalin A, *J. Chromatogr.*, 357, 11, 1986.
62. **Narasimhan, S., Wilson, J. R., Martin, E., and Schachter, H.,** A structural basis for four distinct elution profiles on concanavalin A-Sepharose affinity chromatography of glycopeptides, *Can. J. Biochem.*, 57, 83, 1979.
63. **Ohlson, S., Lundblad, A., and Zopf, D.,** Novel approach to affinity chromatography using "weak" monoclonal antibodies, *Anal. Biochem.*, 169, 204, 1988.
64. **Shibuya, N., Goldstein, I. J., Broekaert, W. F., Nsimba-Lubaki, M., Peeters, B., and Peumans, W. J.,** The elderberry (*Sambucus nigra* L.) bark lectin recognizes the Neu5Ac (α2-6)Gal/GalNAc sequence, *J. Biol. Chem.*, 262, 1596, 1987.
65. **Sturgeon, R. J. and Sturgeon, C. M.,** Affinity chromatography of sialoglycoproteins, utilizing the interaction of serotonin with *N*-acetylneuraminic acid and its derivatives, *Carbohydr. Res.*, 103, 213, 1982.
66. **Sueyoshi, S., Tsuji, T., and Osawa, T.,** Carbohydrate-binding specifications of five lectins that bind to *O*-glycosyl-linked carbohydrate chains. Quantitative analysis by frontal affinity chromatography, *Carbohydr. Res.*, 178, 213, 1988.
67. **Sutherland, I. W.,** Affinity chromatography for separation of pyruvylated and non-pyruvylated polysaccharides, *J. Chromatogr.*, 213, 301, 1981.
68. **Yamashita, K., Totani, K., Ohkura, T., Takasaki, S., Goldstein, I. J., and Kobata, A.,** Carbohydrate binding properties of complex-type oligosaccharides on immobilized *Datura stramonium* lectin, *J. Biol. Chem.* 262, 1602, 1987.

High-Performance Liquid Chromatography: Adsorption and Partition on Silica-Based Packings

If not mentioned in the title of the paper, the column packing and mode of chromatography are included in parentheses after the reference.

69. **Alpenfels, W. F.,** A rapid and sensitive method for the determination of monosaccharides as their dansyl hydrazones by high-performance liquid chromatography, *Anal. Biochem.*, 114, 153, 1981. (C_{18}-silica, reversed-phase; see also Reference 88)

70. **Ando, S., Waki, H., and Kon, K.,** High-performance liquid chromatography of underivatized gangliosides, *J. Chromatogr.*, 408, 285, 1987. (NH_2-silica, normal partition + ion exchange; silica, partition, ion-pair)
71. **Armstrong, D. W. and Jin, H. L.,** Evaluation of the liquid chromatographic separation of monosaccharides, disaccharides, trisaccharides, tetrasaccharides, deoxysaccharides and sugar alditols with stable cyclodextrin bonded phase columns, *J. Chromatogr.*, 462, 219, 1989. (α- and β-cyclodextrin bonded to silica; adsorption)
72. **Baust, J. G., Lee, R. E., Jr. Rojas, R. R., Hendrix, D. C., Friday, D., and James, H.,** Comparative separation of low-molecular-weight carbohydrates and polyols by high-performance liquid chromatography; radially compressed amine-modified silica *versus* ion exchange, *J. Chromatogr.*, 261, 65, 1983. (Silica, modified to amino phase by tetraethylenepentamine in eluent, normal partition; see also Reference 83)
73. **Blanken, W. M., Bergh, M. L. E., Koppen, P. L., and Van den Eijnden, D. H.,** High-pressure liquid chromatography of neutral oligosaccharides: effects of structural parameters, *Anal. Biochem.*, 145, 322, 1985. (NH_2-silica, normal partition)
74. **Bremer, E. G., Gross, S. K., and McCluer, R. H.,** Quantitative analysis of monosialogangliosides by high-performance liquid chromatography of their perbenzoyl derivatives, *J. Lipid Res.*, 20, 1028, 1979. (Silica, adsorption)
75. **Cheetham, N. W. H. and Sirimanne, P.,** Separation of partially methylated sugars by reversed-phase high-performance liquid chromatography. *J. Chromatogr.*, 196, 171, 1980. (See also Reference 84)
76. **Cheetham, N. W. H., Sirimanne, P., and Day, W. R.,** High-performance liquid chromatographic separation of carbohydrate oligomers, *J. Chromatogr.*, 207, 439, 1981. (C_{18}-silica, hydrophobic chromatography)
77. **Cheetham, N. W. H. and Sirimanne, P.,** Methanolysis studies of carbohydrates, using HPLC, *Carbohydr. Res.*, 112, 1, 1983. (C_{18}-silica, reversed-phase)
78. **Cheetham, N. W. H. and Teng, G.,** Some applications of reversed-phase high-performance liquid chromatography to oligosaccharide separations, *J. Chromatogr.*, 336, 161, 1984. (C_{18}-silica, hydrophobic chromatography)
79. **D'Amboise, M., Noël, D., and Hanai, T.,** Characterization of bonded-amine packing for liquid chromatography and high-sensitivity determination of carbohydrates, *Carbohydr. Res.*, 79, 1, 1980. (Normal partition)
80. **Gazzotti, G., Sonnino, S., and Ghidoni, R.,** Normal-phase high-performance liquid chromatographic separation of non-derivatized ganglioside mixtures, *J. Chromatogr.*, 348, 371, 1985.
81. **Gisch, D. J. and Pearson, J. D.,** Determination of monosaccharides in glycoproteins by reversed-phase high-performance liquid chromatography on 2.1 mm narrow-bore columns, *J. Chromatogr.*, 443, 299, 1988. (Benzoylated methanolysates on C_{18}-silica; see also Reference 85)
82. **Hase, S., Ikenaka, K., Mikoshiba, K., and Ikenaka, T.,** Analysis of tissue glycoprotein sugar chains by two-dimensional high-performance liquid chromatographic mapping, *J. Chromatogr.*, 434, 51, 1988. (Pyridylaminated oligosaccharides, NH_2- and C_{18}-silica, normal- and reversed-phase partition)
83. **Hendrix, D. L., Lee, R. E., Jr., Baust, J. G., and James, H.,** Separation of carbohydrates and polyols by a radially compressed high-performance liquid chromatographic silica column modified with tetraethylenepentamine, *J. Chromatogr.*, 210, 45, 1981. (Normal partition)
84. **Heyraud, A. and Salemis, P.,** Liquid chromatography in the methylation analysis of carbohydrates, and the use of combined refractometric-polarimetric detection, *Carbohydr. Res.*, 107, 123, 1982. (C_{18}-silica, reversed-phase)
85. **Jentoft, N.,** Analysis of sugars in glycoproteins by high-pressure liquid chromatography, *Anal. Biochem.*, 148, 424, 1985. (Benzoylated methanolysates, reversed-phase chromatography on C_{18}-silica)
86. **Koizumi, K., Utamura, T., and Okada, Y.,** Analysis of homogeneous D-gluco-oligosaccharides and -polysaccharides (degree of polymerization up to about 35) by high-performance liquid chromatography and thin-layer chromatography, *J. Chromatogr.*, 321, 145, 1985. (NH_2-silica, normal partition)
87. **Lin, J.-K. and Wu, S.-S.,** Synthesis of dabsylhydrazine and its use in the chromatographic determination of monosaccharides by thin-layer and high-performance liquid chromatography, *Anal. Chem.*, 59, 1320, 1987. (Dabsylhydrazone derivatives, reversed-phase chromatography on C_{18}-silica; see also Reference 89)
88. **Mopper, K. and Johnson, L.,** Reversed-phase liquid chromatographic analysis of Dns-sugars. Optimization of derivatization and chromatographic procedures and applications to natural samples, *J. Chromatogr.*, 256, 27, 1983. (Dansylhydrazone derivatives on C_{18}-silica)
89. **Muramoto, K., Goto, R., and Kamiya, H.,** Analysis of reducing sugars as their chromophoric hydrazones by high-performance liquid chromatography, *Anal. Biochem.*, 162, 435, 1987. (Dabsylhydrazone derivatives, reversed-phase chromatography on C_{18}-silica)
90. **Nikolov, Z. L. and Reilly, P. J.,** Retention of carbohydrates on silica and amine-bonded silica stationary phases: application of the hydration model, *J. Chromatogr.*, 325, 287, 1985. (Normal partition)
91. **Nikolov, Z. L. Meagher, M. M., and Reilly, P. J.,** High-performance liquid chromatography of disaccharides on amine-bonded silica columns, *J. Chromatogr.*, 319, 51, 1985. (Normal partition)
92. **Nikolov, Z. L., Meagher, M. M., and Reilly, P. J.,** High-performance liquid chromatography of trisaccharides on amine-bonded silica columns, *J. Chromatogr.*, 321, 393, 1985. (Normal partition)

93. **Oshima, R., Yamauchi, Y., and Kumanotani, J.**, Resolution of the enantiomers of aldoses by liquid chromatography of diastereoisomeric 1-(*N*-acetyl-α-methylbenzylamino)-1-deoxyalditol acetates, *Carbohydr. Res.*, 107, 169, 1982. (Silica, adsorption)
94. **Oshima, R. and Kumanotani, J.**, Determination of neutral, amino and *N*-acetyl amino sugars as alditol benzoates by liquid-solid chromatography, *J. Chromatogr.*, 265, 335, 1983. (Silica, adsorption)
95. **Praznik, W., Beck, R. H. F., and Nitsch, E.**, Determination of fructan oligomers of degree of polymerization 2-30 by high-performance liquid chromatography, *J. Chromatogr.*, 303, 417, 1984. (Silica, modified to amino phase by 1,4-diaminobutane in eluent, normal partition; see also Reference 105)
96. **Saadat, S. and Ballou, C. E.**, Separation of *O*-methylhexitols and *O*-methyl-*myo*-inositols by reverse-phase high-performance liquid chromatography, *Carbohydr Res.*, 119, 348, 1983.
97. **Takeda, M., Maeda, M., and Tsuji, A**, Fluorescence high-performance liquid chromatography of reducing sugars using Dns-hydrazine as a pre-labelling reagent, *J. Chromatogr.*, 244, 347, 1982. (Danzylhydrazone derivatives on silica, adsorption)
98. **Takemoto, H., Hase, S., and Ikenaka, T.**, Microquantitative analysis of neutral and amino sugars as fluorescent pyridylamino derivatives by high-performance liquid chromatography, *Anal. Biochem.*, 145, 245, 1985. (C_{18}-silica, reversed-phase)
99. **Tomiya, N., Kurono, M,. Ishihara, H., Tejima, S., Endo, S., Arata, Y., and Takahashi, N.**, Structural analysis of *N*-linked oligsaccharides by a combination of glycopeptidase, exoglycosidases, and high-performance liquid chromatography, *Anal. Biochem.*, 163, 489, 1987. (Pyridylaminated oligosaccharides, C_{18}-silica, reversed-phase)
100. **Tomiya, N., Awaya, J., Kurono, M., Endo, S., Arata, Y., and Takahashi, N.**, Analysis of *N*-linked oligosaccharides using a two-dimensional mapping technique, *Anal. Biochem.*, 171, 73, 1988. (Pyridylaminated oligosaccharides, C_{18}- and amide-silica, reversed-phase and normal partition)
101. **Valent, B. S., Darvill, A. G., McNeil, M,. Robertsen, B. K., and Albersheim, P.**, A general and sensitive chemical method for sequencing the glycosyl residues of complex carbohydrates, *Carbohydr. Res.*, 79, 165, 1980. (Peralkylated oligosaccharide-alditols, C_{18}-silica, reversed-phase)
102. **Verzele, M. and Van Damme, F.**, Polyol bonded to silica gel as stationary phase for high-performance liquid chromatography, *J. Chromatogr.*, 362, 23, 1986. (Normal partition)
103. **Wells, G. B. and Lester, R. L.**, Rapid separation of acetylated oligosaccharides by reverse-phase high-pressure liquid chromatography, *Anal. Biochem.*, 97, 184 1979.
104. **White, C. A., Kennedy, J. F., and Golding, B. T.**, Analysis of derivatives of carbohydrates by high-pressure liquid chromatography, *Carbohydr. Res.*, 76, 1, 1979. (Benzoates on silica, adsorption)
105. **White, C. A., Corran, P. H., and Kennedy, J. F.**, Analysis of underivatized D-gluco-oligosaccharides (DP 2-20) by high-pressure liquid chromatography, *Carbohydr. Res.*, 87, 165, 1980. (Silica, modified to amino phase by 1,4-diaminobutane in eluent, normal partition)
106. **Yang, M. T., Milligan, L. P., and Mathison, G. W.** Improved sugar separation by high-performance liquid chromatography using porous microparticle carbohydrate columns, *J. Chromatogr.*, 209, 316, 1981. (NH_2-silica, normal partition)

High-Performance Liquid Chromatography: Ion-Moderated Partitioning on Microparticulate Resins

If not mentioned in the title of the paper, the resin counter-ion and main mechanism governing chromatographic separation are included in parentheses after the reference.

107. **Baker, J. O. and Himmel, M. E.**, Separation of sugar anomers by aqueous chromatography on calcium- and lead-form ion-exchange columns. Application to anomeric analysis of enzyme reaction products, *J. Chromatogr.*, 357, 161, 1986. (Ligand exchange)
108. **Bonn, G.** High-performance liquid chromatographic elution behaviour of oligosaccharides, monosaccharides and sugar degradation products on series-connected ion-exchange resin columns using water as the mobile phase, *J. Chromatogr.*, 322, 411, 1985. (Ca^{2+}, Ag^+, and H^+, coupled in various combinations; ligand exchange, size exclusion)
109. **Bonn, G., Pecina, R., Burtscher, E., and Bobleter, O.**, Separation of wood degradation products by high-performance liquid chromatography, *J. Chromatogr.*, 287, 215, 1984. (Ag^+ and H^+, singly; ligand exchange, size exclusion)
110. **Derler, H., Hörmeyer, H. F., and Bonn, G.**, High-performance liquid chromatographic analysis of oligosaccharides. I. Separation on an ion-exchange stationary phase of low cross-linking, *J. Chromatogr.*, 440, 281, 1988. (H^+; size exclusion)
111. **Hicks, K. B., Lim, P. C., and Haas, M. J.**, Analysis of uronic and aldonic acids, their lactones, and related compounds by high-performance liquid chromatography on cation-exchange resins, *J. Chromatogr.*, 319, 159, 1985. (H^+; ion and size exclusion)
112. **Hicks, K. B. and Sondey, S. M.**, Preparative high-performance liquid chromatography of malto-oligosaccharides, *J. Chromatogr.*, 389, 183, 1987. (Ag^+ and H^+, singly; size exclusion)

113. **Hicks, K. B., Sondey, S. M., and Doner, L. W.,** Preparative liquid chromatography of carbohydrates: mono- and disaccharides, uronic acids, and related derivatives, *Carbohydr. Res.*, 168, 33, 1987. (Ca^{2+} and H^+, singly; ligand exchange, ion exclusion)
114. **Hicks, K. B. and Hotchkiss, A. T., Jr.,** High-performance liquid chromatography of plant-derived oligosaccharides on a new cation-exchange resin stationary phase: HPX-22H, *J. Chromatogr.*, 441, 382, 1988. (H^+; size exclusion)
115. **Honda, S. and Suzuki, S.,** Common conditions for high-performance liquid chromatographic microdetermination of aldoses, hexosamines, and sialic acids in glycoproteins, *Anal. Biochem.*, 142, 167, 1984. (H^+, in aqueous acetonitrile; normal partition)
116. **Honda, S., Suzuki, S., and Kakehi, K.,** Improved analysis of aldose anomers by high-performance liquid chromatography on cation-exchange columns, *J. Chromatogr.*, 291, 317, 1984. (Na^+ and Ca^{2+}, in aqueous acetonitrile; normal partition and ligand exchange)
117. **Lohmander, L. S.,** Analysis by high-performance liquid chromatography of radioactively labeled carbohydrate components of proteoglycans, *Anal. Biochem.*, 154, 75, 1986. (Pb^{2+} for neutral sugars, by ligand exchange; Ca^{2+}, in aqueous acetonitrile, for alditols, by ligand exchange and normal partition; H^+ for hexuronic acids, aminodeoxy sugars, and *N*-acetyl-neuraminic acid, by ion exclusion and partition)
118. **Owens, J. A. and Robinson, J. S.,** Isolation and quantitation of carbohydrates in sheep plasma by high-performance liquid chromatography, *J. Chromatogr.*, 338, 303, 1985. (Ca^{2+}; ligand exchange)
119. **Pecina, R., Bonn, G., Burtscher, E., and Bobleter, O.,** High-performance liquid chromatographic elution behaviour of alcohols, aldehydes, ketones, organic acids and carbohydrates on a strong cation-exchange stationary phase, *J. Chromatogr.* 287, 245, 1984. (H^+; ion and size exclusion)
120. **Schmidt, J., John, M., and Wandrey, C.,** Rapid separation of malto-, xylo- and cello-oligosaccharides (DP2-9) on cation-exchange resin using water as eluent, *J. Chromatogr.*, 213, 151, 1981. (Ca^{2+}; size exclusion)
121. **Takeuchi, M., Takasaki, S., Inoue, N., and Kobata, A.,** Sensitive method for carbohydrate composition analysis of glycoproteins by high-performance liquid chromatography, *J. Chromatogr.*, 400, 207, 1987. (Pb^{2+}; ligand exchange of derived alditols)
122. **Van Riel, J. A. M. and Olieman, C.,** High-performance liquid chromatography of sugars on a mixed cation-exchange resin column, *J. Chromatogr.*, 362, 235, 1986. (Ag^+, Pb^{2+}, and Ag^+/Pb^{2+}; ligand exchange, size exclusion)
123. **Voragen, A. G. J., Schols, H. A., Searle-Van Leeuwen, M. F., Beldman, G., and Rombouts, F. M.,** Analysis of oligomeric and monomeric saccharides from enzymatically degraded polysaccharides by high-performance liquid chromatography, *J. Chromatogr.*, 370, 113, 1986. (Pb^{2+}; ligand exchange, size exclusion)

Ion Chromatography

124. **Edwards, W. T., Pohl, C. A., and Rubin, R.,** Determination of carbohydrates using pulsed amperometric detection combined with anion-exchange separations, *Tappi*, 70, 138, 1987.
125. **Hardy, M. R., Townsend, R. R., and Lee, Y. C.,** Monosaccharide analysis of glycoconjugates by anion-exchange chromatography with pulsed amperometric detection, *Anal. Biochem.*, 170, 54, 1988.
126. **Hardy, M. R. and Townsend, R. R.,** Separation of positional isomers of oligosaccharides and glycopeptides by high-performance anion-exchange chromatography with pulsed amperometric detection, *Proc. Natl. Acad. Sci. U.S.A.*, 85, 3289, 1988. (See also Reference 129)
127. **Johnson, D. C.,** Carbohydrate detection gains potential, *Nature (London)*, 321, 451, 1986.
128. **Rocklin, R. D. and Pohl, C. A.,** Determination of carbohydrates by anion-exhange chromatography with pulsed amperometric detection, *J. Liq. Chromatogr.*, 6, 1577, 1983.
129. **Townsend, R. R., Hardy, M. R., Hindsgaul, O., and Lee, Y. C.,** High-performance anion-exchange chromatography of oligosaccharides using pellicular resins and pulsed amperometric detection, *Anal. Biochem.*, 174, 459, 1988.

Ion-Exchange Chromatograhy: High-Performance

The type of packing (silica-based or resin) is given in parentheses after each reference.

130. **Baenziger, J. U. and Natowicz, M.,** Rapid separation of anionic oligosaccharide species by high-performance liquid chromatography, *Anal. Biochem.*, 112, 357, 1981. (Silica-based packing; weakly basic anion-exchanger)
131. **Delaney, S. R., Conrad, H. E., and Glaser, J. H.,** A high-performance liquid chromatography approach for isolation and sequencing of chondroitin sulfate oligosaccharides, *Anal. Biochem.*, 108, 25, 1980. (Silica-based packing; strongly basic anion-exchanger)
132. **Delaney, S. R., Leger, M., and Conrad, H. E.,** Quantitation of the sulfated disaccharides of heparin by high-performance liquid chromatogrpaphy, *Anal. Biochem.*, 106, 253, 1980. (Same as Reference 131)

133. **Gherezghiher, T., Koss, M. C., Nordquist, R. E. and Wilkinson, C. P.,** Rapid and sensitive method for measurement of hyaluronic acid and isomeric chondroitin sulfates using high-performance liquid chromatography, *J. Chromatogr.*, 413, 9, 1987. (NH_2-silica, as weakly basic anion-exchanger for unsaturated oligosaccharides from *Streptomyces* hyaluronidase digestion)
134. **Honda, S., Takahashi, M., Kakehi, K., and Ganno, S.,** Rapid, automated analysis of monosaccharides by high-performance anion-exchange chromatography of borate complexes with fluorimetric detection using 2-cyanoacetamide, *Anal. Biochem.*, 113, 130, 1981. (Resin, strongly basic anion-exchanger)
135. **Honda, S. Takahashi, M., Nishimura, Y., Kakehi, K., and Ganno, S.,** Sensitive ultraviolet monitoring of aldoses in automated borate complex anion-exchange chromatography with 2-cyanoacetamide, *Anal Biochem.*, 118, 162, 1981. (Same as Reference 134)
136. **Honda, S., Konishi, T., Suzuki, S. Takahashi, M., Kakehi, K., and Ganno, S.,** Automated analysis of hexosamines by high-performance liquid chromatography with photometric and fluorimetric postcolumn labelling using 2-cyanoacetamide, *Anal. Biochem.*, 134, 483, 1983. (Resin, strongly acidic cation-exchanger)
137. **Honda, S., Suzuki, S., Takahashi, M., Kaketi, K., and Ganno, S.,** Automated analysis of uronic acids by high-performance liquid chromatography with photometric and fluorimetric postcolumn labelling using 2-cyanoacetamide, *Anal. Biochem.*, 134, 34, 1983. (Same as References 134, 135)
138. **Linhardt, R. J., Rice, K. G., Kim, Y. S., Lohse, D. L., Wang, H. M., and Loganathan, D.,** Mapping and quantification of the major oligosaccharide components of heparin, *Biochem. J.*, 254, 781, 1988. (Silica-based packing, strongly basic, in anion-exchange of unsaturated oligosaccharides from heparin lyase digestion)
139. **Nebinger, P., Koel, M., Franz, A., and Werries, E.,** High-performance liquid chromatographic analysis of even- and odd-numbered hyaluronate oligosaccharides, *J. Chromatogr.*, 265, 19, 1983. (NH_2-silica, as weakly basic anion-exchanger)
140. **Perini, F. and Peters, B. P.,** Fluorometric analysis of amino sugars and derivatized neutral sugars, *Anal. Biochem.*, 123, 357, 1982. (Cation-exchange resin)
141. **Shukla, A. K. and Schauer, R.,** Analysis of *N-O*-acylated neuraminic acids by high-performance liquid anion-exchange chromatography, *J. Chromatogr.*, 244, 81, 1982. (Resin, strongly basic anion-exchanger)
142. **Shukla, A. K., Schauer, R., Unger, F. M., Zähringer, U., Rietschel, E. T., and Brade, H.,** Determination of 3-deoxy-D-*manno*-octulosonic acid (KDO), *N*-acetylneuraminic acid, and their derivatives by ion-exchange liquid chromatography, *Carbohydr. Res.*, 140, 1, 1985. (Same as Reference 141)
143. **Šmid, F., Bradová, V., Mikeš, O., and Sedláčková, J.,** Rapid ion-exchange separation of human brain gangliosides, *J. Chromatogr.*, 377, 69, 1986. (Macroporous glycolmethacrylate resin, weakly basic anion-exchanger)
144. **Tsuji, T., Yamamoto, K., Konami, Y., Irimura, T., and Osawa, T.,** Separation of acidic oligosaccharides by liquid chromatography: application to analysis of sugar chains of glycoproteins, *Carbohydr. Res.*, 109, 259, 1982. (Anion-exchange resin)
145. **Van Pelt, J., Damm, J. B. L., Kamerling, J. P., and Vliegenthart, J. F. G.,** Separation of sialyl-oligosaccharides by medium pressure anion-exchange chromatography on Mono Q, *Carbohydr. Res.*, 169, 43, 1987. (Resin, strongly basic anion-exchanger)
146. **Voragen, A. G. J., Schols, H. A., De Vries, J. A., and Pilnik, W.,** High-performance liquid chromatographic analysis of uronic acids and oligogalacturonic acids, *J. Chromatogr.*, 244, 327, 1982. (Silica-based packings; both strong- and weak-base types used; also ion-pair reversed-phase chromatography on C_{18}-silica)

Ion-Exchange Chromatography: Preparative

The type of packing (DEAE or QAE derivatives of gels) is given in parentheses after each reference.

147. **Davis, K. R., Darvill, A. G., Albersheim, P., and Dell, A.,** Host-pathogen interactions. XXIX. Oligogalacturonides released from sodium polypectate by endopolygalacturonic acid lyase are elicitors of phytoalexins in soybean, *Plant Physiol.*, 80, 568, 1986. (QAE-dextran; strongly basic anion-exhanger)
148. **Jin, D. F. and West, C. A.,** Characteristics of galacturonic acid oligomers as elicitors of casbene synthetase activity in castor bean seedlings, *Plant Physiol.*, 74, 989, 1984. (DEAE-dextran; weakly basic anion-exchanger)
149. **Lee, Y. J., Radhakrishnamurthy, B., Dalferes, E. R., Jr., and Berenson, G. S.,** Fractionation of aorta glycosaminoglycans by high-performance liquid chromatography, *J. Chromatogr.*, 419, 275, 1987. (DEAE-agarose-polyacrylamide)
150. **Nebinger, P.,** Comparison of gel permeation and ion-exchange chromatographic procedures for the separation of hyaluronate oligosaccharides, *J. Chromatogr.*, 320, 351, 1985. (DEAE-cellulose)

Steric-Exclusion Chromatography

The type of packing is given in parentheses after each reference.

151. **Beck, R. H. F. and Praznik, W.,** Molecular characterization of fructans by high-performance gel chromatography, *J. Chromatogr.*, 369, 208, 1986. (Agarose gel, highly cross-linked; cf. Reference 161)
152. **Djordjevic, S. P., Batley, M., and Redmond, J. W.,** Preparative gel chromatography of acidic oligosaccharides using a volatile buffer, *J. Chromatogr.*, 354, 507, 1986. (Polyacrylamide gel)
153. **Dreher, T. W., Hawthrone, D. B. and Grant, B. R.,** Comparison of open-column and high-performance gel permeation chromatography in the separation and molecular-weight estimation of polysaccharides, *J. Chromatogr.*, 174, 443, 1979. (Agarose-polyacrylamide gels and silica-based packings)
154. **Haglund, A. C., Marsden, N. V. B., and Östling, S. G.,** Partitioning of oligoglucans in Sephadex G-15 in relation to their conformational structure, *J. Chromatogr.*, 318, 57, 1985. (Dextran gel)
155. **Hizukuri, S. and Takagi, T.,** Estimation of the distribution of molecular-weight for amylose by the low-angle laser-light-scattering technique combined with high-performance gel chromatography, *Carbohydr. Res.*, 134, 1, 1984. (Semirigid gel: cross-linked hydrophilic vinyl polymer)
156. **John, M., Schmidt, J., Wandrey, C., and Sahm, H.,** Gel chromatography of oligosaccharides up to DP 60, *J. Chromatogr.*, 247, 281, 1982. (Polyacrylamide gels)
157. **Kato, T., Tokuya, T., and Takahashi, A.,** Comparison of poly(ethylene oxide), pullulan and dextran as polymer standards in aqueous gel chromatography, *J. Chromatogr.*, 256, 61, 1983. (Semirigid gels; hydrophilic vinyl polymers of different cross-linking; columns in series).
158. **Kuge, T., Kobayashi, K., Tanahashi, H., Igushi, T., and Kitamura, S.,** Gel permeation of polysaccharides: universal calibration curve, *Agric. Biol. Chem.*, 48, 2375, 1984. (Same as Reference 157)
159. **Luchsinger, W. W., Luchsinger, S. W., and Luchsinger, D. W.,** Gel chromatography of (1 → 3), (1 → 4), and mixed linkage (1 → 3), (1 → 4)-β-D-gluco-oligosaccharides, *Carbohydr. Res.*, 104, 153, 1982. (Polyacrylamide gel)
160. **Natowicz, M. and Baenziger, J. U.,** A rapid method for chromatographic analysis of oligosaccharides on Bio-Gel P-4, *Anal, Biochem.*, 105, 159, 1980. (D-Gluco-oligosaccharides and oligosaccharides from glycoproteins, on polyacrylamide gel)
161. **Praznik, W. and Beck, R. H. F.,** Application of gel permeation chromatographic systems to the determination of the molecular-weight of inulin, *J. Chromatogr.*, 348, 187, 1985. (Polyacrylamide and allyl-dextran gels)
162. **Praznik, W., Burdicek, G., and Beck, R. H. F.,** Molecular weight analysis of starch polysaccharides using cross-linked allyl-dextran gels, *J. Chromatogr.*, 357, 216, 1986. (cf. Reference 163)
163. **Praznik, W., Beck, R. H. F., and Eigner, W. D.,** New high-performance gel permeation chromatographic system for the determination of low molecular-weight amyloses, *J. Chromatogr.*, 387, 467, 1987. (Agarose gel, highly cross-linked)
164. **Vandevelde, M.-C. and Fenyo, J.-C.,** Macromolecular distribution of *Acacia senegal* gum (gum arabic) by size-exclusion chromatography, *Carbohydr. Polym.*, 5, 251, 1985. (Allyl-dextran gels)

Planar Chromatography

The type of plate or paper is given in parentheses after each reference.

165. **Ando, S., Waki, H., and Kon, K.,** New solvent system for high-performance thin-layer chromatography and high-performance liquid chromatography of gangliosides, *J. Chromatogr.*, 405, 125, 1987. (HPTLC silica plates)
166. **Bilisics, L. and Petruš, L.,** Chromatographic separation of alditols and some aldoses on *O*-(carboxymethyl)cellulose paper in the lanthanum, calcium, and barium forms, *Carbohydr. Res.*, 146, 141, 1986. (CM-cellulose paper)
167. **Briggs, J., Chambers, I. R., Finch, P., Slaiding, I. R., and Weigel, H.,** Thin-layer chromatography on cellulose impregnated with tungstate: a rapid method of resolving mixtures of some commonly occurring carbohydrates, *Carbohydr. Res.*, 78, 365, 1980. (Cellulose plates)
168. **Doner, L. W. and Biller, L. M.,** High-performance thin-layer chromatographic separation of sugars: preparation and application of aminopropyl bonded-phase silica plates impregnated with monosodium phosphate, *J. Chromatogr.*, 287, 391, 1984. (Bonded-phase silica plates)
169. **Doner, L. W., Fogel, C. L., and Biller, L. M.,** 3-Aminopropyl-bonded-phase silica thin-layer chromatographic plates: their preparation, and application to sugar resolution, *Carbohydr. Res.*, 125, 1, 1984. (Same as Reference 168)
170. **Ghebregzabher, M., Rufini, S., Sapia, G. M., and Lato, M.,** Improved thin-layer chromatographic method for sugar separations, *J. Chromatogr.*, 180, 1, 1979. (Silica plates)
171. **Klaus, R. and Ripphahn, J.,** Quantitative thin-layer chromatographic analysis of sugars, sugar acids and polyalcohols, *J. Chromatogr.*, 244, 99, 1982. (HPTLC silica plates)
172. **Nurok, D. and Zlatkis, A.,** The separation of malto-oligosaccharides by high-performance thin-layer chromatography using multiple developments, *Carbohydr. Res.*, 81, 167, 1980. (HPTLC silica plates)

173. **Papin, J.-P. and Udiman, M.,** Thin-layer chromatography of principal aldoses, *J. Chromatogr.*, 170, 490, 1979. (Silica plates)
174. **Shimada, E. and Matsumura, G.,** Comparison of relationships between the chemical structures and mobilities of hyaluronate oligosaccharides in thin-layer and high-performance liquid chromatography, *J. Chromatogr.*, 328, 73, 1985. (Silica plates)
175. **Würsch, P. and Roulet, Ph.,** Quantitative estimation of malto-oligosaccharides by high-performance thin-layer chromatography, *J. Chromatogr.*, 244, 177, 1982. (HPTLC silica plates)

Electrophoresis

The type of electrophoretic system is given in parentheses after each reference.

176. **Bianchini, P., Nader, H. B., Takahashi, H. K., Osima, B., Straus, A. H., and Dietrich, C. P.,** Fractionation and identification of heparin and other acidic mucopolysaccharides by a new discontinuous electrophoretic method, *J. Chromatogr.*, 196, 455, 1980. (Agarose gel electrophoresis)
177. **Cappelletti, R., Del Rosso, M., and Chiarugi, V. P.,** A new electrophoretic method for the complete separation of all known animal glycosaminoglycans in a monodimensional run, *Anal. Biochem.*, 99, 311, 1979. (Cellulose acetate)
178. **Dubray, G. and Bezard, G.,** A highly sensitive periodic acid-silver stain for 1,2-diol groups of glycoproteins and polysaccharides in polyacrylamide gels, *Anal. Biochem.*, 119, 325, 1982. (Polyacrylamide gel electrophoresis)
179. **Kumoro, T. and Galanos, C.,** Analysis of *Salmonella* lipopolysaccharides by sodium deoxycholate-polyacrylamide gel electrophoresis, *J. Chromatogr.*, 450, 381, 1988. (Polyacrylamide gel electrophoresis)
180. **Searle, F. and Weigel, H.,** Interaction between polyhydroxy compounds and vanadate ions: electrophoresis and composition of complexes, *Carbohydr. Res.*, 85, 51, 1980 (Paper electrophoresis)

Index

INDEX

A

B

C

K

L

M

N

O

Milton Keynes UK
Ingram Content Group UK Ltd.
UKHW051930141024
449569UK00027B/1435